Lichen Biology

Lichens are symbiotic organisms in which fungi and algae and/or cyanobacteria form an intimate biological union. This diverse group is found in almost all terrestrial habitats from the tropics to polar regions. In this second edition, four completely new chapters cover recent developments in the study of these fascinating organisms, including lichen genetics and sexual reproduction, stress physiology and symbiosis, and the carbon economy and environmental role of lichens. The whole text has been fully updated, with chapters covering anatomical, morphological and developmental aspects; the chemistry of the unique secondary metabolites produced by lichens and the contribution of these substances to medicine and the pharmaceutical industry; patterns of lichen photosynthesis and respiration in relation to different environmental conditions; the role of lichens in nitrogen fixation and mineral cycling; geographical patterns exhibited by these widespread symbionts; and the use of lichens as indicators of air pollution. This is a valuable reference for both students and researchers interested in lichenology.

THOMAS H. NASH III is Professor of Plant Biology in the School of Life Sciences at Arizona State University. He has over 35 years teaching experience in Ecology, Lichenology and Statistics, and has taught in Austria (Fulbright Fellowship) and conducted research in Australia, Germany (junior and senior von Humbolt Foundation fellowships), Mexico and South America, and the USA. He has coauthored 5 previous books and over 200 scientific articles.

Lichen Biology

Second Edition

Edited by

Thomas H. Nash III
Arizona State University, USA

CAMBRIDGE
UNIVERSITY PRESS

CAMBRIDGE
UNIVERSITY PRESS

University Printing House, Cambridge CB2 8BS, United Kingdom

Cambridge University Press is part of the University of Cambridge.

It furthers the University's mission by disseminating knowledge in the pursuit of education, learning and research at the highest international levels of excellence.

www.cambridge.org
Information on this title: www.cambridge.org/9780521692168

© Cambridge University Press 1996, 2008

First published 1996
Second Edition 2008
Reprinted with corrections 2010

A catalogue record for this publication is available from the British Library

ISBN 978-0-521-87162-4 Hardback
ISBN 978-0-521-69216-8 Paperback

Contents

Contributors

Dr. Richard Beckett
School of Biological and Conservation Sciences, University of Kwazulu-Natal, Private Bag X01, Pietermaritzburg, South Africa

Professor Dr. Burkhard Büdel
Department of General Botany, Department of Biology, Erwin-Schrödinger-Str. 13 University of Kaiserslautern, D-67663 Kaiserslautern, Germany

Dr. Lena Dahlman
Department of Ecology and Environmental Science, Umeå University, SE-90187 Umeå, Sweden

Emeritus Professor J. A. Elix
Department of Chemistry, Faculty of Science, Australian National University, Canberra, ACT 0200, Australia

Dr. Dianne Fahselt
Department of Plant Sciences, University of Western Ontario, London, Ontario N6A 5B7, Canada

Professor Dr. Thomas Friedl
Department of Experimental Phycology and Sammlung von Algenkulturen, Albrecht-von-Haller-Institute for Plant Sciences, University of Göttingen, Untere Karspuele 2, 37073 Göttingen, Germany

Dr. David Galloway
16 Farquharson St., Opoho, Dunedin, New Zealand

Dr. T. G. A. Green
Department of Biological Sciences, The University of Waikato, Private Bag 3105, Hamilton 3240, New Zealand

Professor Dr. Rosmarie Honegger
Institute of Plant Biology, University of Zürich, Zollikerstrasse 107, CH-8008 Zürich, Switzerland

Dr. Anna Jonsson
Department of Ecology and Environmental Science, Umeå University, SE-90187 Umeå, Sweden

Dr. Ilse Kranner
Seed Conservation Department, Royal Botanic Gardens, Kew, Wakehurst Place, West Sussex, RH17 6TN, UK

Professor Dr. Otto L. Lange
Julius-von-Sachs-Institute of Biosciences, University of Wuerzburg, Lehrstuhl fuer Botanik II, Julius-von-Sachs-Platz 3, D-97082 Wuerzburg, Germany

Dr. Farida Minibayeva
Institute of Biochemistry and Biophysics, P.O.Box 30, Kazan, 420111, Russia

Dr. Thomas H. Nash III
School of Life Sciences, Arizona State University, Box 874501, Tempe, AZ 85287-4501, USA

Professor Kristin Palmqvist
Department of Ecology and Environmental Science, Umeå University, SE-90187 Umeå, Sweden

Dr. Christoph Scheidegger
Swiss Federal Institute for Forest, Snow and Landscape Research, WSL, Zürcherstr. 111, CH-8903 Birmensdorf, Switzerland

Dr. S. Scherrer
Institute of Plant Biology, University of Zürich, Zollikerstrasse 107, CH-8008 Zürich, Switzerland

Professor Mark R. D. Seaward
Department of Geography and Environmental Science, University of Bradford, Bradford, BD7 1DP, UK

Professor Dr. Elfie Stocker-Wörgötter
Department of Organismic Biology (Plant Physiology), University of Salzburg, Hellbrunner Str. 34, A-5020 Salzburg, Austria

Dr. Anders Tehler
Swedish Museum of Natural History, Box 50007, SE 104 05 Stockholm, Sweden

Dr. Mats Wedin
Swedish Museum of Natural History, Box 50007, SE 104 05 Stockholm, Sweden

Preface to the second edition

Twelve years ago the first edition of *Lichen Biology* was published, and brought a new synthesis to the field of lichenology. In the meantime, rapid advances in many areas, particularly in molecular biology, have expanded our horizons and added depth to our knowledge of areas already under investigation. Consequently, it is appropriate that a second edition has now been consummated.

The original edition had 13 chapters, but this edition has 17 chapters and has added an appendix on lichen culturing, which is becoming prominent in the expanding biotechnology area. New chapters include one on sexual reproduction (Chapter 6), summarizing knowledge not available in 1996. As prominent examples of stress-tolerant organisms, lichens have developed a variety of strategies that allow them to occupy both extremely cold and hot environments; consequently, these investigations were meritorious of a chapter of their own (Chapter 8). In addition, a chapter on growth (Chapter 10), a topic briefly covered in the original photosynthesis chapter, is now expanded to cover much new information and the major advances over the past decade. Although many aspects of the ecology of lichens were covered in the first edition, a number of important areas were omitted. This has been rectified in Chapter 14. Of the remaining chapters, the chapter titles remain the same from the first edition, but all chapters have been revised to a greater or lesser degree. For example, the chapter on the individual (Chapter 13) and air pollution (Chapter 15) bear little resemblance to their original counterparts. Altogether 13 additional people have contributed substantially to this edition.

As with the first edition, this book should be of interest to the specialist, whether amateur or professional lichenologist. Furthermore, the book will provide an essential reference for many other people, such as anyone interested in the phenomenon of symbiosis, ecologists interested in the role of lichens in ecosystems, or a land manager charged with assessing the effects of air pollution on natural systems. We also hope it will stimulate the next generation of students and young scientists to advance our knowledge of these wonderful organisms.

1

Introduction

T. H. NASH III

1.1 The symbiosis

Lichens are by definition symbiotic organisms, usually composed of a fungal partner, the mycobiont (Chapter 3), and one or more photosynthetic partners, the photobiont (Chapter 2), which is most often either a green alga or cyanobacterium. Although the dual nature of most lichens is now widely recognized, it is less commonly known that some lichens are symbioses involving three (tripartite lichens) or more partners. The potential relationships of mycobionts and photobionts may in fact be quite complex (Chapter 4), and a rigorous classification of many types of relationships was developed by Rambold and Triebel (1992). In general, lichens exist as discrete thalli and are implicitly treated as individuals in many studies (but see Chapter 13), even though they may be a symbiotic entity involving three kingdoms! From a genetic and evolutionary perspective, lichens can certainly not be regarded as individuals and this fact has major implications for many areas of investigation, such as developmental and reproductive studies (Chapter 5).

The nature of the lichen symbiosis is widely debated and deserves further investigation. Most general textbooks and many researchers refer to lichens as a classical case of mutualism, where all the partners gain benefits from the association. Alternatively, lichens are regarded as an example of controlled parasitism, because the fungus seems to obtain most of the benefits and the photobiont may grow more slowly in the lichenized state than when free-living (Ahmadjian 1993). In fact, the relationships may be much more complex, especially when additional lichenicolous fungi (Lawrey and Diederich 2003) occur on/in lichens. These are different fungi from the dominant mycobiont, and they may have a parasitic, commensalistic, mutualistic or saprophytic/saprobic

Lichen Biology, ed. Thomas H. Nash III. Published by Cambridge University Press.
© Cambridge University Press.

relationship to the lichen (Rambold and Triebel 1992). Parasitic symbiotic fungi may cause extensive damage, resulting in localized necrotic patches or in complete death of the thallus. On the other hand, commensalistic symbiotic fungi apparently share the photosynthetically derived products from the photobiont with the mycobiont of the existing symbiosis. Such secondary fungi are assumed not to benefit their hosts, and they do not appear to damage them, although they may lead to the formation of gall-like growths, the morphology and physiology of which are, as yet, little understood. A few cases are being discovered where the secondary parasitizing fungus progressively eliminates the primary mycobiont, taking over the photobiont to produce a new thallus of its own; the stability of such a union can be fragile, since in some cases it has been discovered that the original photobiont can be exchanged for another preferred photobiont after the takeover.

Within the realm of what we call lichens, the degree of lichenization varies tremendously from a few photobiont cells that seem to be almost haphazardly associated with a fungus (e.g. some Caliciales) to the more typical well-integrated thallus, in which a distinct photobiont layer is found beneath cortical fungal tissue (Chapters 4 and 5). In most of the latter cases, the lichen bears little morphological similarity to the bionts that form it. Because of the differences in degree of lichenization, no single definition may adequately cover the full range of relationships found within lichens.

The morphology of the lichenized thallus is strongly influenced by the photobiont and its direct contact with the mycobiont (Chapters 4 and 5). In nature there are at least a few cases where the same mycobiont, as ascertained with molecular techniques, is able to form two very different, interconnected thalli with respectively a cyanobacterium and a green alga (Armaleo and Clerc 1990). These different morphotypes are called photosymbiodemes, and their occurrence implies ontogenetic control by the photobiont. In culture, unlichenized mycobionts remain relatively amorphous, but initiate thallus development when they first come in contact with their photobiont (Ahmadjian 1993; Chapter 5). Subsequently the mycobiont may completely envelop the photobionts and, particularly in the case of green algae, penetrate the surface of the photobiont with structures called haustoria. Because haustoria are sometimes associated with dead photobiont cells and because parasitic fungi frequently form haustoria, Ahmadjian (1993) interprets lichenization as an example of controlled parasitism. Although there is limited experimental evidence, these haustoria are assumed to facilitate carbohydrate transfer from the photobiont to the mycobiont. In the future it would be interesting to determine whether haustoria can also facilitate nutrient delivery from the mycobiont to the photobiont.

Certainly there is variation in the degree to which the symbiosis is an obligate one for the partners involved. The green algal *Trebouxia*, which occurs in approximately 20% of all lichens, has rarely been found free-living (Chapter 2). In contrast, other photobiont genera, such as *Gleocapsa*, *Nostoc*, *Scytonema*, and *Trentepohlia*, occur commonly both in lichenized and free-living states. In at least some cases, both free-living and lichenized populations occur in the same habitat, such as free-living *Nostoc* and *Scytonema* in desert soils and their lichenized counterparts respectively in the terricolous lichens *Collema* and *Peltula*. The degree to which the same photobiont species occurs in both free-living and lichenized states (Beck 2002) is not well established, because relatively few lichen algae have been definitively identified to species, and more generally, the systematics at the species level of many cyanobacteria and unicellular green algae are not well resolved (Chapter 2). Nevertheless, it appears that most lichens are highly specific in their choice of photobiont (Beck *et al.* 1998; Rambold *et al.* 1998). In contrast, the systematics of the mycobiont is well known. Because isolated mycobionts grow so slowly, they are unlikely to survive well in the free-living state due to competition with other fungi or consumption by other organisms. Thus, most mycobionts are assumed to have an obligate relationship to lichenization, although the specificity of the mycobiont for a particular photobiont may not be as great as one might assume. In addition to the photosymbiodeme example cited above, more than one species of *Trebouxia* have been isolated from the same thallus (Friedl 1989*b*; Ihda *et al.* 1993).

Overall, the lichen symbiosis is a very successful one as lichens are found in almost all terrestrial habitats from the tropics to polar regions (Chapter 14). Certainly as a result of the symbiosis, both photobiont and mycobiont have expanded into many habitats, where separately they would be rare or non existent. For example, most free-living algae and cyanobacteria occur in aquatic or at least very moist terrestrial habitats, but as part of lichens they occur abundantly in habitats that are frequently dry as well. Not only may the fungus enhance water uptake due to its low water potential (see below), but also it substantially reduces the light intensity to which the photobiont is exposed (Ertl 1951). High light intensity adversely affects the photobiont (Demmig-Adams *et al.* 1990), and hence lichenization is one mechanism by which photobionts may expand into high light environments. Thus, there may well be benefits to lichenization from the perspective of the photobiont. Overall, it may be less important to evaluate lichenization from a strict cost/benefit perspective than to recognize it as a prominent example of a successful symbiosis. Additional studies will doubtlessly help to elucidate further our understanding of the symbiosis.

1.2   Systematics

Lichens are classified as fungi (Chapter 17), and estimates of the number of species vary from 13 500 (Hawksworth and Hill 1984) to approximately 17 000 (Hale 1974). Because many regions of the world have been poorly collected, the higher number may well be more reasonable. By far the largest number of lichens are Ascomycetes and in fact almost half of the described Ascomycetes are lichenized (Chapter 17). In addition, there are a few lichenized Basidiomycetes and Deuteromycetes (= Fungi Imperfecti). The latter group is an artificial class, in which sterile species are placed. If fruiting structures are eventually found, then these lichens may in due course be classified as either Ascomycetes or Basidiomycetes. In addition, in the Actinomycetes, Mastigomycetes and Myxomycetes, there are a few symbiotic associations with some properties similar to lichens, but in general these are excluded from lichen classifications.

Although one might hypothesize that cyanobacteria, green algae, and fungi evolved from lichens, it is generally assumed that lichenization occurred subsequent to development of these organisms. In the fossil record there is limited evidence for the occurrence of lichens, but this may be more due to lack of preservation than their absence from earlier eons. In fact, several quite old fossils have recently been interpreted as being lichens (Chapters 5 and 16). The diversity of lichenized fungi and the fact that some groups contain both lichenized and free-living fungi has led to the inference that lichenization and delichenization have occurred more than once and in fact may have occurred several times (Gargas et al. 1995; Lutzoni et al. 2001). The initial inference is supported by the occurrence of lichens in different classes of fungi, and, within the Ascomycetes, by the fact that lichenization occurs exclusively in only five of the 16 orders, in which lichenization has thus far been found (Hawksworth 1988a). If lichenization has occurred multiple times, then in an evolutionary sense lichens cannot be regarded as one group or, as a phylogeneticist would say, lichens are polyphyletic (Chapter 17).

1.3   Diversity and ecological domain of lichens

Among the terrestrial autotrophs of the world, lichens exhibit intriguing morphological variation in miniature (Chapter 4). In color they exhibit a fantastic array of orange, yellow, red, green, gray, brown, and black (Wirth 1995; Brodo et al. 2001). Lichens vary in size from less than a mm² to long, pendulous forms that hang over 2 m from tree branches (Chapter 4). Almost all lichens are perennials, although a few ephemerals (e.g. Vezdaea) are known. At the other extreme some lichens are estimated to survive well over 1000 years and may be useful in dating rock surfaces (Beschel 1961; Section 10.7). Linear growth varies from imperceptible to many millimeters in a year.

Lichens occur commonly as epiphytes on trees and other plants, and in some ecosystems epiphytic lichen biomass may exceed several hundred kg ha^{-1} (Coxson 1995). In addition, they frequently colonize bare soil, where they are an important component of cryptogamic soil crusts in arid and semi-arid landscapes (Evans and Johansen 1999; Belnap and Lange 2003). Furthermore, lichens occur almost ubiquitously on rocks with the most obvious ones occurring as epiliths, either growing over the surface or embedded within the upper few millimeters. A few lichens even occur endolithically within the upper few millimeters of the rock, such as occurs in Antarctica (Friedmann 1982). In the tropics and subtropics, some rapidly growing lichens even colonize the surface of leaves as epiphylls (Lücking and Bernecker-Lücking 2002). Although most lichens are terrestrial, a few occur in freshwater streams (e.g. *Peltigera hydrothyria*) and others occur in the marine intertidal zone (e.g. *Lichina* spp. and the *Verrucaria maura* group).

Lichens occur in most terrestrial ecosystems of the world, but their biomass contribution varies from insignificant to being a major component of the whole ecosystem (Kershaw 1985; Chapter 14). In many polar and subpolar ecosystems, lichens are the dominant autotrophs (Longton 1988). In addition, lichens are conspicuous components of many alpine, coastal and forest ecosystems, such as the temperate rain forests of the southern hemisphere (Galloway 2007) and taiga of the northern hemisphere (Kershaw 1985). Because most lichens grow relatively slowly, their primary productivity contribution is fairly small in most ecosystems (Chapter 10). On the other hand, the more rapidly growing species may increase their biomass by 20–40% in a year and these species may play an important role in the mineral cycling patterns of their ecosystems (Section 12.10), particularly if cyanolichens are the dominant component (Chapter 11).

1.4   Lichens as poikilohydric organisms

Most flowering plants and conifers have developed the capacity to maintain the water status of their leaves or needles at fairly constant levels and hence are referred to as homiohydric organisms. In contrast, lichens are prominent members of poikilohydric organisms, whose water status varies passively with surrounding environmental conditions (Chapter 9). Other poikilohydric organisms include the bryophytes, some ferns and other primitive vascular plants. All of these organisms become desiccated relatively rapidly and, as a consequence, water availability is of prime importance for their

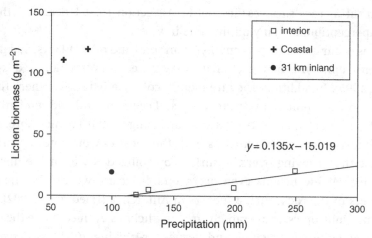

Fig. 1.1 Relationship of biomass of lichen communities within the Sonoran Desert region to mean annual precipitation (Nash and Moser 1982).

survival and in explaining their patterns of occurrence (Chapter 9). One might assume that poikilohydric organisms are highly dependent on precipitation, primarily in the form of rain. Certainly this is true for many lichens, as can be seen for the lichen biomass relationship among interior desert sites (Fig. 1.1, the straight line). On the other hand, lichen biomass near the Pacific Ocean in the western part of the Sonoran Desert vastly exceeds values that would be predicted based on precipitation alone (Fig. 1.1, crosses). This illustrates the ability of lichens to utilize other water sources, such as fog and dew. In addition, lichens have the remarkable ability to extract some moisture from non saturated air under conditions of low temperatures and high humidities. This is essentially the reverse of transpirational water flow occurring through vascular plants and is due to the low osmotic values of lichen thalli. However, under intermediate to high temperatures and intermediate to low humidities, the water potential gradient from the lichen to the atmosphere is reversed and evaporation occurs.

1.5   Practical applications

Many of the secondary products formed by lichens are unpalatable and may serve as defensive compounds against herbivores as well as decomposers (Rundel 1978; Chapter 14). As a consequence, it is not surprising that these secondary products are frequently used by the pharmaceutical industry as antibacterial and antiviral compounds. In addition, lichens have long been used as a source of natural dyes and in the making of perfumes. In both cases the secondary products provide the chemical basis for these applications (Chapter 7).

The differential sensitivity of lichens to air pollution has been recognized for over a century and a half, and the application of lichenological studies to biomonitoring of air pollution is now well developed (Chapter 15). For example, patterns in lichen communities may be correlated with sulfur dioxide levels in the atmosphere (Hawksworth and Rose 1970). In recent years sulfur dioxide levels have been reduced, either by improved controls on emissions or by more efficient dispersion strategies, and, as a consequence, lichens are now reinvading areas from which they had previously disappeared (Rose and Hawksworth 1981; Bates *et al.* 1990). However, the recolonization is incomplete because other factors, such as high nitrate deposition, modify lichen community composition as well. Finally, lichens are efficient accumulators of metals and persistent organic pollutants and are frequently used as surrogate receptors for documenting deposition of these pollutants (Chapter 12).

1.6   Lichens as self-contained miniature ecosystems

The lichen thallus is a relatively stable and well-balanced symbiotic system with both heterotropic and autotrophic components. From this perspective, the lichen can be regarded as a self-contained miniature ecosystem (Farrar 1976c; Seaward 1988), particularly if one considers the parasitic lichenicolous fungi colonizing lichens as this ecosystem's decomposers. The lichen fungus undoubtedly benefits enormously by obtaining its nutrition from the photobiont, but the photobiont's gain from the association is less obvious. Fundamentally, the photobiont gains protection from high light, temperature extremes and to some extent drought, but the premise that the alliance between free-living algae or cyanobacteria and the fungal partner enables them to live together in inhospitable areas where they could not do so independently cannot be fully justified. Pushed to its ultimate limit, this train of thought leads to the fallacy that lichens are the only form of life possible on other planets – a false assumption, because, even supposing that the environment there was capable of supporting life as we know it, then representatives of both symbiotic partners would have to be present in the first instance. However, lichens have recently been put to the test in terms of their ability to cope with extreme conditions of outer space, even Martian conditions, the symbiotic system and germination capacity proving remarkably resistant to UV radiation and vacuum exposure (de Vera *et al.* 2003, 2004).

The lichen symbiosis typically involves a close physiological integration. The usually dominant mycobiont is, of course, a heterotrophic organism that derives its carbon nutrition from the photobiont (Chapter 3). The flux of carbohydrates, as polyols in the case of green algal lichens and glucose in the case of cyanolichens, from the photobiont to the mycobiont is well established (Smith and Douglas 1987). This is a necessary benefit for the mycobiont and is the result of the photobiont's cell walls being more permeable to carbohydrate loss in the lichenized than nonlichenized state (Hill 1976). In addition, the mycobiont gains a nitrogen source in the case of cyanolichens, in which nitrogen fixation occurs in the photobiont (Chapter 11). No comparable flux of nutrients from the mycobiont to the photobiont has been demonstrated. However, the recent demonstation of recycling of nitrogen and phosphorus in a mat lichen is an exciting first step to providing such documentation (Hyvärinen and Crittenden 2000; Ellis *et al.* 2005). Does the fungus in general serve as a reservoir of inorganic nutrients for the photobiont through the haustoria? Certainly other fungi facilitate nutrient uptake in other symbiotic relationships, such as occurs in mycorrhizae and rhizospheric fungi. Another result of close physiological integration is the occurrence of a wide range of secondary products, many of which occur as crystals extracellularly within the lichens (Chapter 7). Most of these are unknown in free-living fungi (or other organisms) and hence their occurrence adds to the uniqueness of the lichen symbiosis.

From an ecological perspective, lichens may be even more complex, as free-living bacteria and non symbiotic fungi may be found associated with an "individual" (Section 13.4) and, as a consequence, some authors regard a lichen as a miniature ecosystem. Further support for accepting the lichen as an ecosystem is provided when one considers the range of other benign or harmful microorganisms associated with one or more of the above bionts; these include fungi and bacteria found both on the surface and within thalli, or in the microenvironment generated beneath thalli or within lichen-weathered substrata (Bjelland and Ekman 2005), and also invertebrates which graze upon them, or seek protection from predators through crypsis or by sheltering beneath thalli; the intimate relationship between the lichen and its substratum in the case of epiphytic, lignicolous and foliicolous species adds to the complexity of the microhabitat generated.

2

Photobionts

T. FRIEDL AND B. BÜDEL

2.1   Major differences in cyanobacteria vs. algae

Nearly 40 genera of algae and cyanobacteria have been reported as photobionts in lichens (Tschermak-Woess 1988; Büdel 1992; Rikkinen 2002). Three genera, *Trebouxia*, *Trentepohlia* and *Nostoc*, are the most frequent photobionts. The genera *Trebouxia* and *Trentepohlia* are of eukaryotic nature and belong to the green algae; the genus *Nostoc* belongs to the oxygenic photosynthetic bacteria (cyanobacteria). Eukaryotic photobionts are also referred to as "phycobionts" while cyanobacterial photobionts are sometimes called "cyanobionts" (Sanders 2004). The vast majority of eukaryotic photobionts belongs to the green algae (phylum Chlorophyta) which share many cytological features and their pigmentation, e.g. the presence of chlorophylls *a* and *b*, with the land plants (Bold and Wynne 1985; van den Hoek *et al.* 1993; Lewis and McCourt 2004). Only two genera of eukaryotic photobionts containing chlorophylls *a* and *c* (phylum Heterkontophyta sensu van den (Hoek *et al.* 1993)) have thus far been reported (*Heterococcus*, Xanthophyceae, and *Petroderma*, Phaeophyceae (Tschermak-Woess 1989; Gärtner 1992; Sanders *et al.* 2005).

Cyanobacteria are prokaryotic and lack chloroplasts, mitochondria and a nucleus, all of which are found in eukaryotic algae. In cyanobacteria, thylakoids lie free in the cytoplasm, often more or less restricted to the periphery. The circular DNA is not associated with histones and is concentrated in areas of the cytoplasm free of thylakoids, which sometimes are called "nucleoplasm."

Metabolite transfer from the autotrophic photobiont to the heterotrophic mycobiont depends on the type of photobiont involved. In lichens with green algal photobionts, the carbohydrates are sugar alcohols; in lichens with cyanobacteria, glucose (Feige and Jensen 1992; Section 10.2.1). The mode of activation

Lichen Biology, ed. Thomas H. Nash III. Published by Cambridge University Press.
© Cambridge University Press.

of CO_2-uptake is another basic feature that varies depending on whether the photobiont is prokaryotic or eukaryotic. In many green algal lichens, positive net photosynthesis is possible after water vapor uptake alone. In contrast, in cyanobacterial lichens no measurable gas exchange occurs because the water content level required to activate photosynthesis is higher and liquid water is needed to obtain such levels (Lange *et al.* 1986).

2.2   Identification, reproduction and taxonomy of photobionts

2.2.1   *Cyanobacteria*

Identification of cyanobacterial photobionts in the intact lichen thallus is often impossible, because the morphology of the photobiont is changed by the influence of the fungal partner. Filamentous forms may be deformed to such a degree that their originally filamentous organization cannot be recognized within the lichen thallus, e.g. in the genus *Dichothrix* (Fig. 2.1). Only the truly branched filamentous cyanobacterial genus *Stigonema* (Fig. 2.2) and the non-branched genus *Nostoc* (Fig. 2.3) can often be identified within the lichen thallus. Furthermore, cyanobacteria do not show all characteristic stages of their life cycles in the lichenized state. Because it is essential to know these stages for positive identification of cyanobacteria, even at the genus level (Komárek and Anagnostidis 1998, 2005), isolation and cultivation of the cyanobacterial photobionts are essential steps for positive identification. The mode of vegetative cell divisions is also important in the delimitation of many unicellular cyanobacteria at the genus level. However, using molecular techniques, identification at least at the generic level is possible directly from the lichen thallus, by using specific PCR amplification primers for cyanobacterial 16S rDNA (e.g. Lohtander *et al.* 2003; O'Brien *et al.* 2005).

Cyanobionts with heterocysts like *Nostoc* (Fig. 2.3) increase heterocyte-frequency up to five times when lichenized compared to the free-living state (Feige and Jensen 1992). Also, cell size of cyanobacterial photobionts may be increased compared with cultured or free-living material, as has been reported for the genera *Gloeocapsa* (Fig. 2.4) and *Chroococcidiopsis* in the lichen genera *Lichinella, Peccania, Psorotichia, Synalissa*, and *Thyrea* (Geitler 1937; Büdel 1982). Increase of cell size can be a result of a very close mycobiont photobiont-contact, as in the deeply penetrating haustoria. This can well be seen in the vegetative trichome cells of *Scytonema* sp. within the lichen *Dictyonema sericeum* (Fig. 2.5). Unicellular cyanobacterial photobiont genera, e.g. *Chroococcidiopsis* and *Myxosarcina*, very rarely show their specific mode of reproduction when lichenized, but frequently show these stages when cultured. For instance, *Myxosarcina* (Fig. 2.6) and *Chroococcidiopsis* (Fig. 2.7) are characterized in culture by multiple fission following one or two binary divisions (Waterbury and Stanier 1978; Büdel and Henssen 1983).

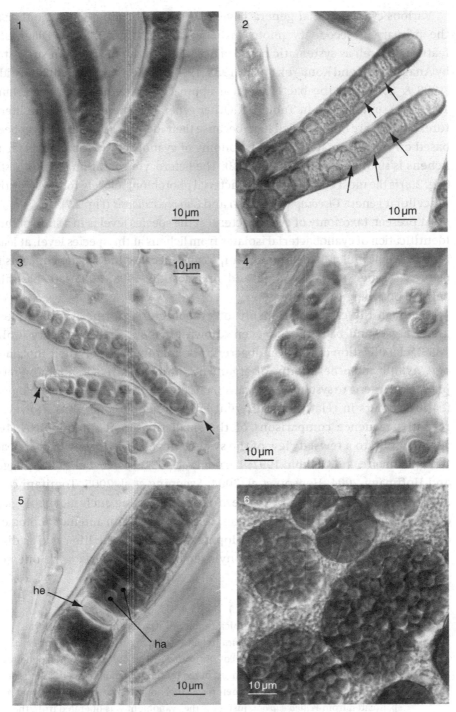

Figs. 2.1–2.6 Light microscopy of cyanobacterial photobionts. Fig. 2.1. *Dichothrix* sp. isolated from *Placynthium nigrum*, showing the characteristic branching mode of the genus. Fig. 2.2. *Stigonema ocellatum*, free-living sample at an early stage of licheni-zation; fungal hyphae indicated by arrows. Fig. 2.3. *Nostoc* sp. from *Peltigera canina*;

Various cyanobacterial genera have been found forming lichens, either as the primary or secondary photosynthetic partner (Figs. 2.1–2.7). Diagnostic features, as well as systematic position according to the new system suggested by Anagnostidis and Komárek (1985, 1988, 1990) and Komárek and Anagnostidis (1998 & 2005) and the bacteriological approach (Castenholz and Waterbury 1989), are summarized in Table 2.1. Other genera are mentioned in the literature, but they are not included here because their identification has not yet been based on cultured material. The taxonomy of cyanobacteria found thus far in lichens is summarized in Büdel (1992). The heterocyte containing genus *Nostoc* (Fig. 2.3) is the most common cyanobacterial photobiont, closely followed by the unicellular genera *Gloeocapsa* (Fig. 2.4) and *Chroococcidiopsis* (Fig. 2.7).

At present, taxonomy of cyanobacteria at the species level is in a state of flux. Identification of cyanobacterial isolates from lichens at the species level, at least for many unicellular taxa, is almost impossible, because the species concepts in use (e.g. Geitler 1932; Komárek and Anagnostidis 1998, 2005) are basically defined on ecological features.

Although the close relationship of cyanobacteria (formerly called blue–green algae) and bacteria has been known for more than a century (Cohn 1853), classification of the cyanobacteria was mainly based on their morphology. As in many other groups of organisms, problems arise in applying morphological criteria to systematics because considerable variation of morphological features occurs in relation to different environmental conditions. However, recently, sequence comparisons of the small ribosomal subunit RNA (16S rRNA) has led to a revised view of the systematics and phylogeny of cyanobacteria (Wilmotte and Golubic 1991; Turner *et al.* 2001; Fewer *et al.* 2002; Gugger and Hoffmann 2004; Henson *et al.* 2004; Svenning *et al.* 2005; Tomitani *et al.* 2006). These data support the modern concept of using such morphological criteria as mode and planes of division, cell differentiation and morphological complexity, and differences in developmental stages in the life cycle as diagnostic features. In the 16S rRNA phylogenies, cyanobacteria with one cell

---

Caption for Figs. 2.1–2.6 (cont.)

primordia of colonies with the typically apically attached primary heterocytes (arrows). Fig. 2.4. *Gloeocapsa sanguinea* in and at the margin of *Synalissa symphorea*. Fig. 2.5. *Scytonema* sp. in the thallus of *Dictyonema sericeum*; vegetative cells penetrated by mycobiont haustoria (ha), arranged in the center of cyanobiont cells along the longitudinal axis of the filament, heterocyte (he) not penetrated. Fig. 2.6. *Myxosarcina* sp. isolated from *Peltula euploca*, just after the cyanobiont was liberated from the thallus; colonies with numerous nanocytes are surrounded by bacteria, typical for early stages of the isolation procedure.

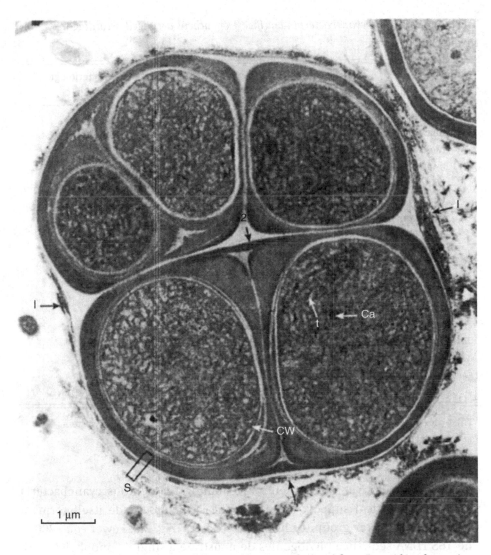

Fig. 2.7 Ultrastructure of *Chroococcidiopsis* sp., isolated from *Psorotichia columnaris*. Colony of daughter cells after two binary divisions (arrows), followed by irregular subsequent multiple fission in the upper part. S, striated sheath, CW, cell wall (outer lipoprotein bilayer, inner murein or petidoglucan layer), t, thylakoids, Ca, carboxysomes.

type and binary division in one plane only are polyphyletic (i.e. do not have a common ancestor). The morphologically more complex groups of genera with cell division in several planes and producing nanocytes ("baeocytes" according to Waterbury and Stanier 1978) form two distinct clades in 16S rDNA sequence phylogenies, i.e. the two *Pleurocapsa* (PLEU)-clades in Fig. 2.8. The unicellular *Chroococcidiopsis* forms a clade of its own and appears as the

Table 2.1. *Genera of cyanobacteria identified from lichens arranged according to taxonomic characters*

Taxonomical character	"Botanical" system	"Bacteriological" system
1. *Unicellular*		
1.1 Binary division only	Chroococcales[a]	Chroococcales
Gloeocapsa, Chroococcus, Cyanosarcina, Entophysalis		
1.2 Binary division + multiple fission	Chroococcales[a]	Pleurocapsales
Nanocytes immotile		
Chroococcidiopsis		
Nanocytes motile		
Myxosarcina, Hyella		
2. *Filamentous with heterocytes*		
2.1 Nonbranched		Nostocales
Nostoc		
2.2 False branching, no tapering trichomes		Nostocales
Scytonema		
2.3 False branching, tapering trichomes		Nostocales
Calothrix, Dichothrix		
2.4 True branching		Stigonematales
Stigonema, Hyphomorpha		

[a] The order Chroococcales is subdivided into seven families in the "botanical" system.
Source: Büdel (1992).

closest living relatives to the heterocyte-foming filamentous cyanobacteria with high statistical support (Fig. 2.8). The *Pleurocapsa*-clade itself forms a well-supported sister group to the Chroococcales (Fig. 2.8; Fewer *et al.* 2002). The 16S rDNA sequence phylogenies demonstrate a single monophyletic origin for the heterocyte-forming filamentous cyanobacteria. Within the latter, the Nostocales as morphologically defined, are divided into three separate groups (Fig. 2.8): one includes the non-branched filamentous taxa (e.g. *Nostoc*), another the filamentous taxa with false branching (e.g. *Scytonema*), and a third comprises (with little statistical support, however) filamentous taxa with false branching and/or tapering trichomes (e.g. *Tolypothrix, Calothrix*). For the morphologically most complex group of cyanobacteria, the former Stigonematales, with true branching, heterocyte-forming trichomes (e.g. *Fischerella, Hapalosiphon,* and *Stigonema*), there is only low statistical support for their monophyletic origin (Fig. 2.8). Another study has conclusively demonstrated multiple origins for this group of cyanobacteria (Gugger and Hoffmann 2004).

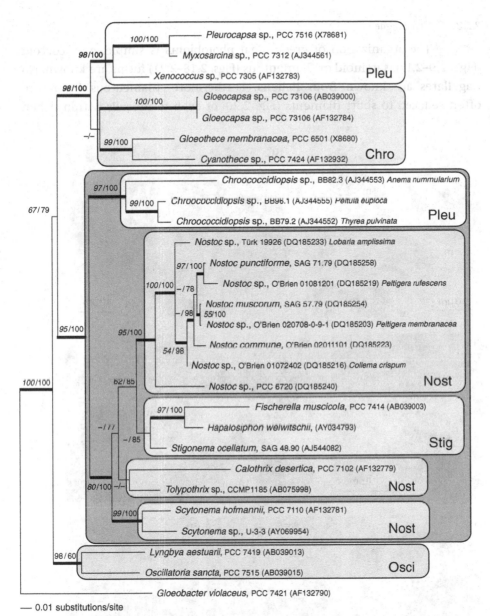

Fig. 2.8 16S rDNA phylogeny of cyanobacterial genera that also occur as cyanobionts (F. Kauff and B. Büdel, unpublished data). *Gloeobacter violaceus* was added as outgroup. Chro, Chroococcales; Nost, Nostocales; Osci, Oscillatoriales; Pleu, Pleurocapsales; Stig, Stigonematales. Bootstrap frequencies (number in italics) and posterior probabilities are indicated above horizontal branches. Phylogram generated with RAxML-HPC-2.1.2 Stamatakis (2006) out of 200 replicates and a GTRMIX model. Bootstrap proportions calculated with 500 replicates. Posterior probabilities estimated with MrBayes3.1.1 Huelsenbeck and Ronquist (2001), generating 20 000 000 generations using a GTR model with gamma distribution and a proportion of invariable sites. GeneBank accession number given in brackets.

2.2.2   Green algae

The organization of green algal photobionts is simple: only coccoid (Figs. 2.9–2.17), sarcinoid or filamentous (Figs. 2.18–2.21) forms are known. No flagellates are known from lichens. Furthermore, filamentous forms are often reduced to short filaments (Fig. 2.20) or even to unicells within lichen

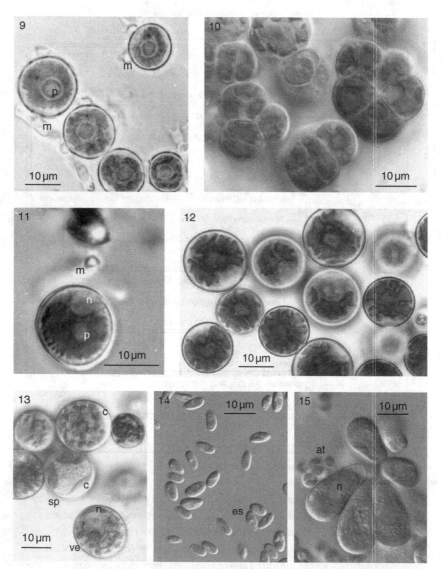

Figs. 2.9–2.15 *Trebouxia* spp. and other coccoid species, lichenized and in culture. Fig. 2.9. *Trebouxia gigantea* within a thallus of *Xanthomaculina hottentotta*. Unicellular stages attached to mycobiont hyphae (m). Note prominent pyrenoid (p) in the center of the algal chloroplast. Fig. 2.10. *Trebouxia gigantea* in culture, isolated from

thalli. In lichens, green algae, except those belonging to the order Trentepohliales (class Ulvophyceae), reproduce only asexually by sporulation (i.e. forming motile zoospores with flagella and/or immotile autospores which lack emergent flagella) or by true cell division (for terminology, see Sluiman *et al.* 1989). If flagellated stages (zoospores) can be formed (e.g. in *Trebouxia*), they are released frequently in culture, but in the lichen thallus they may be observed only occasionally (Slocum *et al.* 1980). Identification of green lichen algae at the genus level is often possible without culturing, e.g. by simple squash preparations of the algal layer. For identification of the species, however, cultures are essential, because important features such as chloroplast morphology or certain stages of the life cycle may be reduced or absent in the lichenized state. As an alternative, sequence analyses of SSU and ITS rDNAs provide a reliable tool for identification independent of culturing, which is often not possible, especially with old dried herbarium specimens of lichens. Most green lichen algae are only facultative photobionts, i.e. they also occur independently as epiphytes, endoliths, or as soil algae.

Many morphologically diverse green algae have been identified from lichens (Tschermak-Woess 1989; Gärtner 1992); the most frequent lichenized green algae and their distribution in different groups of lichenized fungi have been summarized by (Peršoh *et al.* 2004). Examples of green algal photobionts are given below.

Species of *Trebouxia* (Figs. 2.9–2.12, 2.22 and 2.23) and *Asterochloris* (Fig. 2.13) are the most common green photobionts. Vegetative cells, mostly spherical in

---

Caption for Figs. 2.9–2.15 (cont.)

Xanthomaculina hottentotta. Autospore packages are dominant in the culture four weeks after the alga was liberated from the mycobiont. Fig. 2.11. *Trebouxia arboricola* within a thallus of *Omphalora arizonica*, m, mycobiont hypha. Pyrenoid (p) in the center of the chloroplast, nucleus (n) located externally in an invagination of the chloroplast. Fig. 2.12. *Trebouxia gelatinosa*, vegetative cells in culture. Note crinkled chloroplast. Fig. 2.13. *Asterochloris* sp. isolated from *Diploschistes albescens* in culture. Vegetative cells (ve) with deeply incised and crinkled chloroplast, nucleus (n) located externally in an invagination of the chloroplast. Smooth chloroplast appressed to cell wall in early stage of sporangium development (sp). Sporangia with typical cap-like wall thickening (c). Fig. 2.14. Vegetative cells of *Coccomyxa subellipsoidea* isolated from *Botrydina vulgaris* in culture (strain SAG 216–13). Vegetative cells elongated and of irregular shape with a flat chloroplast appressed to the cell wall and without pyrenoids. (es) empty walls of autosporangia. Photograph taken by I. Kostikov. Fig. 2.15. Vegetative cells of *Myrmecia biatorellae* isolated from *Lobaria linita* in culture (strain SAG 8.82). Chloroplast smooth and attached to the cell wall, nucleus (n) with nucleolus located centrally; at, autospore package. Photograph taken by T. Darienko.

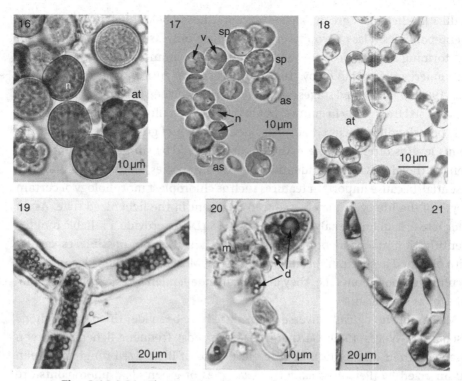

Figs. 2.16–2.21   Fig 2.16. *Dictyochloropsis reticulata* isolated from *Brigantiaea ferruginea* in culture (strain SAG 2150). Reticulate chloroplast attached to the cell wall, nucleus (n) located in the center of the cells. at, autospore package. Fig. 2.17. *Elliptochloris bilobata* in culture, isolated from subaerial habitats (strain SAG 245.80). Flat chloroplast appressed to cell wall, center of the cell with nucleus (n) and several large vacuoles (v). Sporangia at different developmental stages (sp). Note autospores (as) of different shape which is characteristic for the genus, elongated (right) and more spherical (middle). Fig. 2.18. Filamentous stages of *Leptosira obovata* in culture, isolated from freshwater (strain SAG 445–1). at, autosporangium. Fig. 2.19. Branched filament of *Trentepohlia* sp., free-living sample. Cells filled with carotenoid droplets covering the chloroplasts. Note lamellate cell wall (arrow). Fig. 2.20. Short and unbranched filament of a lichenized *Trentepohlia* sp. within a thallus of *Enterographa subpallidella*. Note carotenoid droplets (d) and mycobiont particle (m). Fig. 2.21. Young branched filament of *Dilabifilium arthropyreniae* in culture, isolated from *Arthropyrenia kelpii* (strain SAG 467–2).

Trebouxia and more irregularly shaped in *Asterochloris*, exhibit a central massive or star-shaped chloroplast that is wrinkled in different ways (Figs. 2.10, 2.12, 2.13 and 2.22). The nucleus is located in an invagination of the chloroplast (Figs. 2.11 and 2.12). In *Asterochloris*, the chloroplast is often rather deeply incised (Fig. 2.13), but it becomes smooth and assumes a parietal position shortly before protoplast division. In *Trebouxia*, the chloroplast is more or less massive with several distinct patterns of chloroplast lobes seen in surface view (Ettl and Gärtner 1984), and

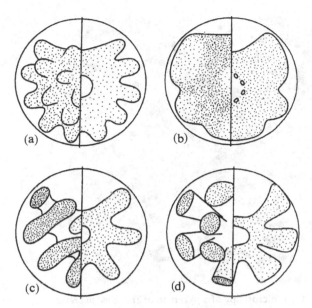

Fig. 2.22   Schematic drawings of some chloroplast types in *Trebouxia*. Left part of cells, surface view; right part of cells, optical section. (a) *T. crenulata*, (b) *T. corticola*, (c) *T. jamesii*, (d) *T. irregularis* (from S. Takeshita, unpublished data).

these are used to identify species of *Trebouxia* (Fig. 2.22; Gärtner 1985). In both genera, the chloroplast contains several pyrenoids with a diverse ultrastructure (Fig. 2.23; Friedl 1989). Within lichen thalli, only non-motile stages with a reduced chloroplast morphology occur (Figs. 2.9 and 2.11). Another important difference between both genera is the pattern of autospore formation that sometimes is even visible within the lichen thallus (Friedl 1993). Packages of large autospores where sometimes several generations of autospores are held together by remnants of old sporangial walls (Fig. 2.10) are formed only in *Trebouxia*, whereas autosporangia with many small autospores are present in both genera. In culture motile stages (zoospores) are also formed for reproduction in both genera. In *Asterochloris*, cap-like one-sided cell wall thickenings are present in zoo- and autosporangia, but these are absent in most species of *Trebouxia*. It is not yet clear whether *Trebouxia* and *Asterochloris* are obligate symbiotic genera of green algae or whether they also occur free-living (Ahmadjian 1988, 1993). The small colonies that have been observed (e.g. Tschermak-Woess 1978) on bark and soil a few times could also have been escaped from damaged lichen thalli. However, there is some indirect evidence that free-living *Trebouxia* cells must be present in order to enable the establishment of a lichen thallus for those lichens that reproduce exclusively sexually by ascospores (Beck *et al.* 1998).

Twenty-five species of *Trebouxia* are recognized by Gärtner (1985), but several of them were subsequently demonstrated to be synonyms by more recent

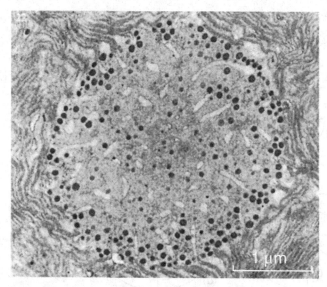

Fig. 2.23   Electron microscopy of a pyrenoid of *Trebouxia impressa*, lichenized phycobiont within a thallus of *Parmelia sulcata*.

morphological and molecular studies, or need to be assigned to *Asterochloris* (see below). On the other hand, recent investigations e.g. Kroken and Taylor 2000; Hauck *et al.* 2007) added additional *Trebouxia* species, but most of them still need to be described formally. Within *Trebouxia* in its present circumscription, a taxonomic problem is evident: the genus is paraphyletic with *Myrmecia* in SSU rDNA analyses (Fig. 2.24) and Internal Transcribed Spacer (ITS) rDNA sequences clearly separate the genus into two groups. ITS sequences from those species that align well with the corresponding sequence of the type species, *T. arboricola*, represent *Trebouxia*. However, ITS rDNA sequences from the other species are almost identical with the corresponding sequence of *Asterochloris phycobiontica*, the type of *Asterochloris*. Therefore, we infer that they should belong to *Asterochloris*, although a formal assignment of the species to that genus is still pending. ITS rDNA sequencing became common practice to identify species of *Trebouxia* directly from lichen thalli without culturing, and thus facilitates studies of photobionts. PCR amplification primers are available that specifically anneal to the algal rDNA, but do not amplify fungal sequences (Kroken and Taylor 2000; Helms *et al.* 2001; Guzow-Krzeminska 2006; Nelsen and Gargas 2006). Identification to species level of *Trebouxia* is possible either by simple comparison with available sequences from cultured reference strains or by phylogenetic analyses. As references, authentic strains should be used which represent the culture material, on which the taxonomic description of a *Trebouxia* species has been based. Phylogenies, inferred from ITS rDNA

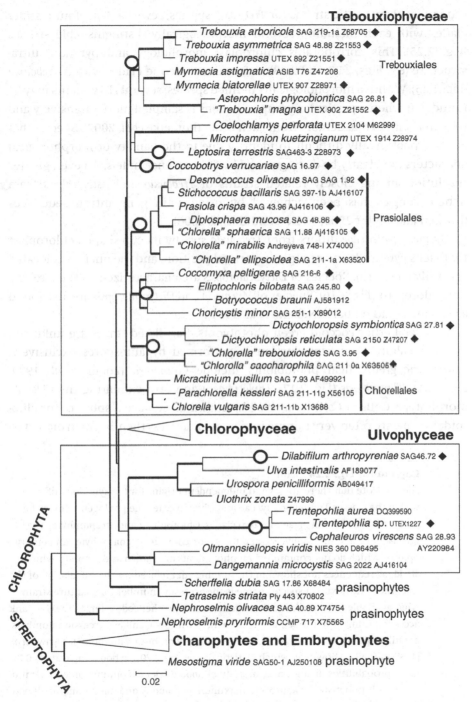

Fig. 2.24 Phylogeny of green-algal photobionts and nonlichenized green algae inferred from 18S rDNA sequence analyses. Species representing lichen photobionts are marked with a filled bar. A circle indicates a lineage of green algae in which lichen symbionts have evolved. Lichen photobionts are closely related to nonlichenized

sequences of authentic strains for *Trebouxia* species, reveal at least four distinct clades with each including several morphologically distinguishable species (Fig. 2.25). This corroborates chloroplast morphology and pyrenoid ultrastructure (e.g. Figs. 2.22 and 2.23) as features to define clades within *Trebouxia* (Fig. 2.25). Within a morphologically defined species, several ITS variants may be found, e.g. when a particular group of lichens is sampled more intensively and they may differ considerably from each other (Romeike *et al.* 2002). Species that are morphologically difficult to identify due to the paucity of morphological characters can clearly be distinguished by ITS sequence indels. An even greater resolution among closely related strains of *Trebouxia* and *Asterochloris* from lichen samples has been achieved by using actin gene intron sequences (Kroken and Taylor 2000; Nelsen and Gargas 2006).

Myrmecia biatorellae differs from *Trebouxia* spp. by its cup-shaped chloroplast that lacks pyrenoids and occupies a parietal position, and the nucleus is located centrally (Fig. 2.15). Species of *Dictyochloropsis* are characterized by their reticulate chloroplast (Fig. 2.16). Species of *Myrmecia* and *Dictyochloropsis* are also found as soil algae and on bark (Nakano *et al.* 1991).

In several other coccoid green photobionts, flagellated stages are unknown even in culture, and reproduction is performed by autospores exclusively. Autosporic green algal photobionts include *Coccomyxa* (Zoller *et al.* 1999; Fig. 2.11), *Elliptochloris* (Fig. 2.17), *Diplosphaera* and taxa exhibiting the *Chlorella* morphotype. Cells of *Coccomyxa* and *Elliptochloris* algae have a spherical or ellipsoidal shape, are often very minute in size (some are less then 10 μm in diameter)

Caption for Fig. 2.24   (cont.)

species. Note that there are several independent origins for the symbiotic lifestyle within the classes Trebouxiophyceae and Ulvophyceae. The phylogeny shows the deep division of green algae into two clades, Chlorophyta and Streptophyta. The green-algal class Chlorophyceae and the group comprising charophytes and embryophytes which do not contain lichen photobionts are shown as diamonds, which stand for sequences not shown in the graphic but are used for the calculation of the tree. The code next to a species name is the accession number for a culture strain when available from a public culture collection. The other code refers to the GenBank accession number of the sequence; sequences without GenBank accession numbers are still unpublished. A distance phylogeny (neighbor-joining method; Hasegawa *et al.* [1985] model) is shown, on which statistical support (> 70% in bootstrap tests, > 0.7 a priori probabilities) using a maximum-likelihood model in conjunction with a minimum evolution distance approach, maximum parsimony and maximum likelihood (Bayesian inference; Huelsenbeck and Ronquist 2001) is indicated by thick internal branches. The phylogeny was rooted by two species of Glaucophyta which have been pruned away from the graphic.

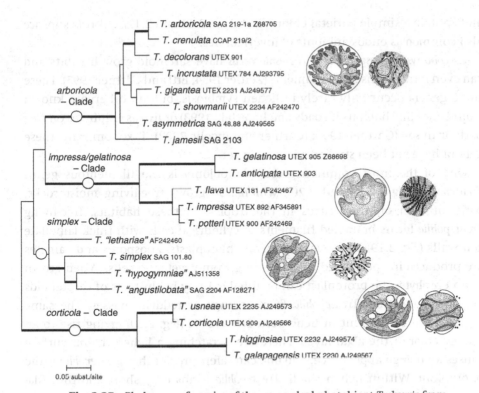

Fig. 2.25 Phylogeny of species of the green-algal photobiont *Trebouxia* from culture inferred from ITS rDNA sequence analyses. Four deeply diverging clades of species (marked by circles on internal branches) are resolved and are named according to a certain species from the clade. Schematic drawings for chloroplast morphology as viewed by light microscopy (left) and pyrenoid ultrastructure (right; Friedl 1989*a*) are shown next to each clade; these features define clades within *Trebouxia*. Only sequences from authentic cultures were used for this phylogeny. Authentic cultures are valuable as references; they represent the culture material on which the taxonomic description of a species was based. Species names in quotation marks indicate that a formal taxonomic description for this species is still pending. The code given next to a species name represents the accession numbers for a culture strain if available from public culture collections. The other code refers to the GenBank accession number of the sequence if available from public sequence data bases. Sequences without GenBank accession numbers are still unpublished. An unrooted maximum likelihood phylogeny is shown on which statistical support (> 70% in bootstrap tests, > 0.7 a priori probabilities) using a maximum-likelihood model in conjunction with a minimum evolution distance approach, maximum parsimony and maximum likelihood (Bayesian inference; Huelsenbeck and Ronquist 2001) is indicated by thick internal branches.

and contain a simple parietal chloroplast (Figs. 2.14 and 2.17). *Chlorella* spp. are also common as endosymbionts of invertebrates.

Coccobotrys, *Diplosphaera*, and *Desmococcus* have sarcinoid growth habits and can even form multiseriate filaments (Zeitler 1954; Ettl and Gärtner 1995). These three genera occur only rarely in lichen symbiosis, but are otherwise known from endolithic habitats (Broady and Ingerfeld 1993) or live as epiphytes on tree bark or in soil (Gärtner 1994; (Gärtner and Ingolic 2003). Taxonomically, these genera have not been studied adequately yet.

One of the most common green photobionts is the filamentous genus *Trentepohlia* (Figs. 2.19 and 2.20), which also grows free-living moist rocks, bark, or even on tree leaves in the tropics. In these habitats, free-living *Trentepohlia* forms branched filaments of cylindrical cells with thick lamellate cell walls (Fig. 2.19) and several parietal chloroplasts. Zoospores and gametes are produced in specialized cells differing from vegetative cells. When grown as an epiphyte, the protoplast can be filled entirely by droplets of carotenoids (Fig. 2.19) and then *Trentepohlia* forms orange or reddish masses. The same pigmentation is present in lichenized *Trentepohlia* (Fig. 2.20) giving rise to an orange color of the algal layer. Therefore, scratching a lichen thallus surface causes an orange appearance, and one can safely predict that *Trentepohlia* is the photobiont. Within lichen thalli, *Trentepohlia* forms only short and thin filaments (Fig. 2.20) or consists of unicellular stages.

Other filamentous green photobionts are species of *Dilabifilium* (Fig. 2.21) and *Leptosira terrestris*, formerly *Pleurastrum terrestre* (Fig 2.18). *Dilabifilium* (including *Pseudopleurococcus*) forms branched filaments of cylindrical cells and has been found in primitively organized crustose aquatic lichens (*Verrucaria* and *Arthropyrenia*; Tschermak-Woess 1976), but is also known as an epiphytic alga in marine habitats and freshwater (Johnson and John 1990; Ihda *et al.* 1993). *Leptosira terrestris* forms short uniseriate filaments of elongated cells with a cup-shaped chloroplast in liquid culture media (Fig 2.18), but is unicellular within lichen thalli. *Leptosira terrestris* is known as a photobiont only from the lichen genera *Vezdea* and *Thrombium* (Tschermak-Woess and Poelt 1976; Tschermak-Woess 1989), but has often been isolated from soil samples (Tupa 1974).

The phylogenetic relationships of green algal photobionts has mainly been assessed by SSU rDNA sequence phylogenies on which modern taxonomic concepts of the green algae are based (Friedl 1997; Lewis and McCourt 2004). The molecular data have mostly corroborated earlier findings from ultrastructure work, i.e. they substantiated the significance of ultrastructural characters found in flagellated stages (zoospores or gametes) of green algae. Because in lichen photobionts such stages are mostly absent, SSU rDNA sequence analyses became essential for tracing the origins of green algal photobionts.

Within the Viridiplantae, the green algae have evolved in two major lineages (Friedl 1997; Lewis and McCourt 2004): the chlorophyte clade comprises the major diversity of green algae representing the Chlorophyta (*sensu* Sluiman 1985) and the charophyte clade that unites a smaller number of green algal taxa (charophytes) with the embryophytes representing the Streptophyta (*sensu* Bremer 1985; Fig. 2.24). Within the Chlorophyta, mainly three lineages that radiate almost simultaneously from each other are currently distinguished, i.e. the classes Trebouxiophyceae, Chlorophyceae and Ulvophyceae (Fig. 2.24). Monophyly for each of these lineages is sometimes difficult to resolve with single gene phylogenies and, for instance, varies greatly in SSU rDNA analyses in proportion to the number and phylogenetic diversity of the taxa included. Basal positions in both the Chlorophyta and Streptophyta are occupied by several lineages representing prasinophytes, i.e. green flagellates with scales that coat their body and flagella (Nakayama *et al.* 1998; Marin and Melkonian 1999). On the basis of SSU rDNA phylogenetic analyses, we infer that there are multiple origins for lichen symbioses within the green algae. Green lichen photobionts so far are only known from the Chlorophyta, and, within that division, only from two phylogenetically rather distant lineages, the classes Trebouxiophyceae and Ulvophyceae. No photobionts have thus far been reported from the Chlorophyceae and the several lineages of prasinophytes. The Trebouxiophyceae appears to contain most of the green algal photobionts, and, within that lineage, the symbiotic lifestyle has arisen independently multiple times. In the SSU rDNA phylogeny shown in Fig. 2.24 at least six independent origins of lichen green algal photobionts are resolved within the Trebouxiophyceae. *Trebouxia* spp; together with *Asterochloris,* form lichenized as well as free-living species of *Myrmecia,* a clade of its own representing the Trebouxiales. *Chlorella* has multiple origins with most members distributed on several independent lineages within the Trebouxiophyceae (Huss *et al.* 1999). The genus *Chlorella* may actually be restricted to *C. vulgaris* and a few species closely related to it; they are within a lineage representing the Chlorellales, from which, however, no lichen photobionts have been reported so far. "*Chlorella*" *sphaerica* forms, together with *Diplosphaera, Desmococcus, Stichococcus,* and *Prasiola*. The latter genus has been recently reported as the photobiont of the antarctic lichen *Mastodia tesselata* (Kovacik and Pereira 2001). Together with two non-lichenized green algae of the *Chlorella*-morphotype, it forms another monophyletic lineage (Prasiolales, Fig. 2.24). *Coccobotrys* has no clear affinities with any known member of the Trebouxiophyceae. "*Chlorella*" *saccharophila* and "*C.*" *trebouxioides* are two closely related lichen photobionts, but are phylogenetically removed from *C. vulgaris* and thus cannot be kept in the same genus. Species of *Dictyochloropsis* form another independent lineage, but appear to be more closely allied with *Chlorella*-like algae in some phylogenies (Fig. 2.24). *Coccomyxa* and *Elliptochloris* are closely related sister-taxa that form another monophyletic lineage of lichenized trebouxiophytes.

Filamentous green algal lichen photobionts are distributed on with different classes of green algae. A *Trentepohlia* strain isolated from lichens is most closely related to the non-symbiotic *Cephaleuros*, which lives epiphytically on leaves in the tropics; both genera represent a lineage of the Ulvophyceae (López-Bautista and Chapman 2003). *Dilabifilum* is closely related to marine *Ulva* species in the SSU rDNA phylogenies, while the filamentous *Leptosira terrestris* represents a lineage with unclear phylogenetic affinities within the Trebouxiophyceae. The SSU rDNA sequence analyses also suggest that green algal photobionts have their closest relatives among free-living terrestrial algae (e.g. epiphytes on tree bark or from soil). For several photobionts, e.g. species of *Coccomyxa* and *Dictyochloropsis*, the closest relatives, i.e. species of the same genus, are known to occur only free-living, and even the same species may exhibit both lifestyles.

2.3   Occurrence within lichens

Most lichen species contain green algae as photobionts. Among the lichenized orders and families of the Lecanoromycetes, *Trebouxia* and *Asterochloris* are by far the most frequent photobionts (Peršoh *et al.* 2004). Although for the order Lecanorales a total of nine different green algal genera has been reported, *Trebouxia* and *Asterochloris* are dominant in the families of this order. Whether *Dictyochloropsis*, *Elliptochloris*, *Myrmecia* or green algae of the *Chlorella* morphotype are the photobionts in addition to *Trebouxia* and *Asterochloris* appears to be distinctive feature of families of the Lecanorales. In a more detailed study on the photobiontal selection, the type of photobiont has been found a valuable marker for taxonomy of the Lecanorales at the level of suborders and families (Peršoh *et al.* 2004). In the Ostropales (e.g. *Graphis*, *Diploschistes*), both *Trebouxia* and *Trentepohlia* are equally frequent. Green algae of the *Chlorella* morphotype are most frequent in the Mycocaliciales, but also occur in some crustose lichens of the Lecanorales (e.g. *Lecidella*, *Micarea*, and *Trapelia*) and in *Stereocaulon*. *Coccomyxa* is common in the orders Baeomycetales (Peršoh *et al.* 2004) and Peltigerales (Peltigeraceae and Nephromataceae *sensu* (Miadlikowska and Lutzoni 2004) as well as in lichenized Basidiomycetes, (e.g. *Omphalina*; Zoller and Lutzoni 2003). *Dictyochloropsis* is the most common green algal photobiont in the families Lobariaceae (Peltigerales) and Stictidaceae (Ostropales). *Myrmecia* is a common photobiont in the Pannariaceae (Peltigerales; Miadlikowska and Lutzoni 2004) and in lichenized genera of the Chaetothyriomycetes, e.g. *Dermatocarpon*, Verrucariales. The order Verrucariales

(Chaetothyriomycetes) is rather diverse with respect to their photobionts. Here, the green algae *Coccobotrys*, *Desmococcus*, and *Dilabifilium* as well as the heterokont alga *Heterococcus* (Xanthophyceae) and *Petroderma* (Phaeophyceae; Sanders *et al.* 2005) have all been found as photobionts in species of *Verrucaria*. *Trentepohlia* is more frequent in lichen genera of the Gyalectales (Lecanoromycetes, e.g. *Coenogonium*) and Arthoniomycetes, e.g. *Opegrapha* and *Roccella*. Members of the Trentepohliales are also common green algal photobionts of epiphyllic lichens in the tropics.

Furthermore, the same lichen species can contain different species of *Trebouxia*, i.e. one mycobiont can form morphologically identical thalli with different algal species. For example, two morphologically distinct species of *Trebouxia* (*T. arboricola* and *T. jamesii*) have been isolated from *Parmelia saxatilis* (Friedl 1989b). Other examples are known from species of *Anzia* (Ihda *et al.* 1993), *Diploschistes* (Friedl and Gärtner 1988) and *Umbilicaria* (Romeike *et al.* 2002). Within a lichen genus or even within a family of lichenized ascomycetes, the diversity of green algal photobionts found is often limited to very few species of *Trebouxia* or *Asterochloris*. However, within a photobiont species, a considerable genetic diversity is present (Dahlkild *et al.* 2001; Piercey-Normore and DePriest 2001; Opanowicz and Grube 2004; Cordeiro *et al.* 2005; Hauck *et al.* 2007).

Only about 10% of all lichen species contain a cyanobacterium as the primary photobiont. The Collemataceae, Heppiaceae, Gloeoheppiaceae, Lichinaceae, Peltulaceae, and Placynthiaceae have cyanobacteria as the only photosynthetic partner. In contrast, lichens of the families Arctomiaceae, Arthopyreniaceae, Bacidiaceae, Brigantiaceae, Cladoniaceae, Coccocarpiaceae, Coccotremataceae, Corticiaceae, Ectolechiaceae, Lobariaceae, Nephromataceae, Pannariaceae, Placynthiaceae, Peltigeraceae, Porpidiaceae, Pyrenulaceae, Sphaerophoraceae, Stereocaulaceae, and Stictidaceae, may have either green algal or cyanobacterial photobionts.

Some lichen genera have a green alga as the primary photobiont and a cyanobacterium as a secondary one. In such cases, the cyanobacterium is located either in external or internal cephalodia (Chapter 5). In addition to heterocyte-producing filamentous cyanobacteria, a number of strains of the genus *Chroococcidiopsis* are also capable of N_2-fixation under microaerobic or anaerobic conditions (Stewart 1980; Boison *et al.* 2004).

2.4   Isolation and maintenance of photobionts

Techniques for the isolation of photobionts have been described several times, and the interested reader is referred to special publications like the

Handbook of Lichenology (Galun 1988) or Ahmadjian (1967) for reference. Here, we only can give some general instructions for the isolation procedures.

2.4.1   Cyanobacteria

From a washed thallus, fragments of the cyanobacterial layer are transferred under sterile conditions to agar plates containing a mineral medium. The agar plates are then kept under low light intensities (10–30 μmol m^{-2} s^{-1} PPFD) and moderate temperatures (15–25 °C). After a few days or weeks, cyanobiont cells start to develop free from the mycobiont, and then can be transferred to fresh agar plates under a dissecting microscope using a micropipette or a needle. For long-term cultivation, strains of isolated cyanobionts are kept on agar slants at low light intensities and temperatures at about 15–17 °C to keep growth at slow rates.

2.4.2   Green algae

Isolation of green algal photobionts is similar to that of cyanobacteria. From washed lichen thalli, a squash preparation of the algal layer is made. From this suspension of algal cells and mycobiont fragments, either single algal cells are isolated under microscopical observation using micropipettes or the whole suspension is transferred on agar plates. The "micropipette method" (Warén 1918–19) usually results in clonal and axenic cultures. The "whole suspension" method is easier, and increases the chances that the algae will grow better in culture, but the resulting algal colonies need further purification. Most green algal photobionts grow easily in culture. They are best maintained on liquid or agarized mineral media (Bischoff and Bold 1963; see web sites of service culture collections below). In comparison to their growth on a mineral medium alone, some green algal photobionts (e.g. *Trebouxia*) grow much faster and in larger quantities after the addition of glucose and proteose peptone to the culture medium, and consequently are considered facultative heterotrophs (Ahmadjian 1967). Most green algal photobiont cultures require low light intensities and moderate temperatures (about 10–30 μmol m^{-2} s^{-1} PPFD and 15 °C (Lorenz *et al.* 2005).

The following culture collections maintain a large variety of lichen photobionts: the "Culture Collection of Algae at the University of Texas at Austin (UTEX)," Austin, Texas, U.S.A. (www.bio.utexas.edu/research/utex), the "American Type Culture Collection (ATCC)," Rockville, Maryland, USA (www.lgcpromochem-atcc.com; cyanobacterial photobionts only), the "Culture Collection of Algae and Protozoa (CCAP)," Oban, Scotland, UK (www.ccap.ac.uk), the "Sammlung für Algenkulturen (SAG)," Göttingen, Germany (www.epsag.uni-goettingen.de), and the "Pasteur Culture Collection (PCC)," Paris, France (www.pasteur.fr/recherche/banques/PCC; cyanobacterial photobionts only).

3

Mycobionts

R. HONEGGER

Lichen-forming fungi (also termed lichen mycobionts) are, like plant pathogens or mycorrhizal fungi, a polyphyletic, taxonomically heterogeneous group of nutritional specialists (Tables 3.1 and 3.2) but otherwise normal representatives of their fungal classes. Long after the discovery of the dual nature of lichens by Schwendener (1867; Honegger 2000) and his proposal to include lichens in fungi, most biologists and even the majority of lichenologists considered lichens as a group of organisms that differ so fundamentally from all others that they had to be treated as a separate group, e.g. as a phylum "Lichenes"; this term is nowadays obsolete. Even in the early twenty-first century, many scientists consider lichens as plants, thus ignoring the fact that species names of lichens refer to the fungal partner, fungi forming a separate kingdom. It is the heterotrophic mycobiont of morphologically advanced lichens that mimics plant-like structures. In this chapter the similarities and differences between lichen-forming and nonlichenized fungi are discussed at the phylogenetic, morphological and cytological levels and also with regard to different nutritional strategies.

3.1 Lichenized versus nonlichenized fungi

3.1.1 Lichenization: a successful nutritional strategy

Fungi, as heterotrophic organisms, have developed various nutritional strategies for acquiring fixed carbon (Table 3.1). Lichenization, i.e. the acquisition of fixed carbon from a population of minute, living algal and/or cyanobacterial cells, is a common and widespread mode of nutrition. One out of five fungal species is lichenized (Table 3.1). Some lichen-forming fungi belong to

Lichen Biology, ed. Thomas H. Nash III. Published by Cambridge University Press.
© Cambridge University Press.

Table 3.1. *Acquisition of fixed carbon by Fungi and fungus-like Protoctista*

Degradation of dead organic matter	
Saprobes	*c*. 45–50%
Symbioses with C-autotrophs or C-heterotrophs:	
Parasitic symbioses (biotrophic or necrotrophic) with	
Cyanobacteria	
Algae	
Plants	
Fungi (lichenized and nonlichenized)	*c*. 20%
Animals	
Humans	
Mutualistic symbioses	
Mycorrhizae (*c*. 8%)	
Lichens (*c*. 21%)	*c*. 30%
Mycetocytes, etc.	

Sources: Lewis (1973); Hawksworth *et al.* (1983).

Table 3.2. *Orders of Ascomycota and Basidiomycota, which include lichenized taxa*

Phylum Subphylum Class Subclass Order	**Nutritional strategies** (predominant form in bold characters)	**Thallus anatomy** in lichenized taxa[9]
Ascomycota[1,2,3]		
Pezizomycotina		
Arthoniomycetes		
Arthoniales	**l**, nl (sap, lp)	**ns**, s
Dothideomycetes		
Patellariales	**nl** (sap, lp), l	**ns**
Eurotiomycetes		
Pyrenulales	**nl** (sap), l	**ns**
Verrucariales	l, nl (sap, lp)	**ns**, s
Mycocaliciales	**nl** (sap), l	
Lecanoromycetes		
Acarosporomycetidae		
Acarosporales	l	**ns**
Ostropomycetidae		
Agyriales	l, nl (sap)	**ns**
Gyalectales	l	**ns**

Table 3.2. (*cont.*)

Ostropales	**nl** (sap, pp, lp), 1	**ns**
Pertusariales	1	**ns**
Trichotheliales	1	**ns**
Lecanoromycetidae		
Lecanorales[8]	1, nl (sap, lp)	**ns, s**
Peltigerales	1	**s, ns**
Teloschistales	1, nl (lp)	**ns, s**
Lichinomycetes		
Lichinales	1	**ns, s**
Basidiomycota[4,5]		
Hymenomycetes		
Agaricales	**nl** (sap, myc, pp, lp), 1	**ns, s**
Polyporales	**nl** (sap, myc, pp, f, lp), 1	**ns, s**
Anamorphic fungi[6]	**nl**, 1	**ns**
Sterile taxa with no known reproductive structures[7]		**ns**

Abbreviations: f, fungicolous; l, lichenized; lp, lichenicolous (lichen parasite); myc, mycorrhiza; nl, nonlichenized; ns, nonstratified; pp, plant pathogens; s, internally stratified; sap, saprotrophic.

[1] *c.* 98% of lichen-forming fungi are Ascomycetes.

[2] *c.* 42% of Ascomycetes are lichenized (*c.* 13 500 spp.), all belonging to subphylum Pezizomycotina.

[3] 15 out of 52 orders of Pezizomycotina include lichenized taxa, 5 of them being exclusively lichenized. *c.* 0.4% of lichen-forming fungi are Basidiomycetes.

[5] Only *c.* 0.3% of Basidiomycetes are lichenized (*c.* 50 spp.).

[6] *c.* 1.5% of lichen-forming fungi (*c.* 200 spp.) belong to the Anamorphic fungi.

[7] *c.* 75 species of lichen-forming fungi are sterile (disperse via thallus fragmentation). With molecular data sets the taxonomic affiliation of some of these taxa will be identified.

[8] Lecanorales are the largest order, comprising *c.* 5500 species, >99% of them lichenized.

Sources: Hawksworth (1988a); Honegger (1992); Hawksworth and Honegger (1994); Kirk *et al.* (2001); Eriksson (2006*b*).

The majority of lichen-forming fungi (>55%) form nonstratified (crustose, microfilamentous, etc.) thalli, *c.* 20% form squamulose or placodioid thalli, and *c.* 25% form morphologically advanced, foliose or fruticose thalli with internal stratification.

orders with uniform nutritional strategies; others belong to orders with diverse strategies (Table 3.2).

A high percentage of lichen-forming fungi are ecologically obligate, but physiologically facultative biotrophs (organisms that obtain nutrients from a living host). In other words they can be cultured in the aposymbiotic ("free-living") state but in nature almost exclusively the symbiotic phenotype is found. Nonlichenized germ tubes or other free hyphae of lichen mycobionts certainly exist in natural ecosystems, but, due to their notoriously slow growth rates, they cannot be recovered with conventional isolation techniques.

Molecular phylogenies elucidate taxonomic relationships among lichenized and nonlichenized fungi. Until recently lichenization was thought to be an ultimate state. Among the most fascinating mycological discoveries of recent years is the finding that lichenization can be transient. Based on thorough analyses of large sets of molecular data from lichenized and nonlichenized ascomycetes using a Bayesian phylogenetic tree sampling methodology, combined with a statistical model of trait evolution, Lutzoni *et al.* (2001) demonstrated that (1) lichens evolved earlier than previously assumed, (2) gains of lichenization were distinctly less frequent during ascomycete evolution than previously assumed, and (3) lichen symbiosis was lost several times. Consequently, numerous taxa of nonlichenized ascomycetes, such as Eurotiomycetidae, derive from lichen-forming ancestors. This particular group comprises economically important taxa such as the genera *Penicillium* and *Aspergillus*, with numerous species used in biotechnology because of their interesting secondary metabolism, a trait shared with many of their ancestors. The "fungal branch of the tree of life" is currently under construction, and many new insights in phylogenetic relationships of lichen-forming and nonlichenized fungi are expected in the near future (Lutzoni *et al.* 2004).

3.1.2   *Fossil records*

Fossil records of lichens are extremely rare. Perhaps palaeontologists have not yet adapted their eyes to recognizing lichen-forming fungi and their photobionts. Two recent discoveries of lichen-like fossils support the view that lichenization might be a very ancient nutritional strategy. In marine phosphorites of the Doushantuo Formation (approx. 600 million years before present [MaBP]) from South China, and in the famous Early Devonian Rhynie chert beds in Scotland (approx. 480 MaBP), colonies of coccoid cyanobacteria or unicellular algae with mucilaginous extracellular sheaths were found, which are invaded by fungal hyphae (Taylor *et al.* 1995*b*, 1997; Yuan *et al.* 2005). This situation resembles the mycobiont–photobiont interface in various genera of Lichinaceae (Henssen 1963, 1986; Büdel 1987; Henssen 1995), all with cyanobacterial photobionts (examples in Honegger 2001), or in *Epigloea* spp. (ascomycetes *incertae sedis*), which are symbiotic with or at least live within the mucilaginous colonies of *Coccomyxa dispar*, a unicellular green alga forming massive gelatinous sheaths (Jaag and Thomas 1934; Döbbeler 1984; David 1987). *Winfrenatia reticulata*, the Early Devonian fossil, was assumed to be formed by a zygomycete (Taylor *et al.* 1995*a*, *b*, 1997). As extant zygomycetes are not forming lichen symbioses, some investigators hesitate to interpret *Winfrenatia* as a lichen. However, lichenization can be lost in the course of time, as shown in ascomycete evolution (Lutzoni *et al.* 2001, 2004). Among the Early Devonian Rhynie chert bed fossils are beautifully preserved arbuscular mycorrhizae (Taylor *et al.* 1995*a*), strikingly similar to the ones which are nowadays symbiotic with more than 70% of higher plants. These arbuscular mycorrhizae represent a fungal symbiosis that had already reached an astonishing level of morphological and physiological complexity, although the early vascular plants (Rhyniales) were just starting to colonize terrestrial ecosystems. Terrestrial soil and rock surfaces had certainly been colonized by cyanobacteria and green algae long before the advent of vascular plants. With high probability groups of fungi have formed manifold interactions with these early photoautotrophic inhabitants of terrestrial ecosystems, ranging from parasitism to mutualism. The Protolichenes hypothesis of Eriksson (2005, 2006*a*) proposes the origin of the subphylum Pezizomycotina among nutritional specialists symbiotic with algae and/or cyanobacteria, so-called Protolichenes, whence extant lichenized and nonlichenized groups evolved. Fossil arbuscular mycorrhizae show that complex fungal interactions with photoautotrophic partners were already differentiated 480 Ma ago.

Presently known fossils of morphologically advanced, foliose or fruticose lichens come from Tertiary (65–1.5 MaBP) deposits, i.e. are comparatively young; older ones certainly exist but have not yet been discovered. An easily recognizable impression of a *Lobaria* thallus (resembling *L. pulmonaria*) was found in early to middle Miocene deposits (24–12 MaBP) of a humid conifer forest from Trinity County, California (MacGinitie 1937; Peterson 2000), a site which might harbour many more lichen fossils (Peterson 2000). Amber, fossilized tree resins from the Old World (Baltic amber; 55–35 MaBP) and New World (Dominican amber; 20–15 MaBP) contain extremely well-preserved organisms. Most investigators focused on vertebrate, invertebrate or higher plant fossils, but also a few well-preserved lichens were detected (Mägdefrau 1957). Recently two species of calicioid lichens were found in Baltic amber (Rikkinen 2003), and beautifully preserved thalli of two *Parmelia* species in Dominican amber (Poinar *et al.* 2000).

3.1.3   Cytological aspects

There is no evidence for any fundamental difference between lichenized and nonlichenized fungi. Cell wall structure and composition of lichen-forming ascomycetes occur within the range of variation observed in all Ascomycota (Honegger and Bartnicki-Garcia 1991). Lichen-forming ascomycetes and basidiomycetes produce and secrete the same type of hydrophobic cell wall surface compounds, the hydrophobins, as nonlichenized fungal taxa (see Chapter 5; Scherrer *et al.* 2000; Trembley *et al.* 2002*a*, *b*).

For quite a while lichen-forming ascomycetes were supposed to differ from nonlichenized taxa by the possession of concentric bodies, semicrystalline cell organelles (Figs. 3.6, 3.7) of approximately 0.3 μm diameter, comprising a proteinaceous, electron-dense shell around a gas-filled center. Concentric bodies are not membrane-bound and usually occur in clusters near the cell periphery. They are neither a peculiarity of lichen-forming ascomycetes nor of the symbiotic way of life, since they were found in a range of plant pathogens and saprobes, some of which occur in climatically extreme habitats in microbial communities of desert varnishes (Honegger 1993, 2001, 2006). Concentric bodies were found in the cytoplasm of all types of vegetative cells, in paraphyses, ascogenous hyphae, asci, and in mature ascospores. Longevity and a considerable desiccation tolerance seem to be the features shared by all ascomycetous cells that harbor concentric bodies; these were hypothesized to be remains of drought stress-induced cytoplasmic cavitation events (see Chapter 4), as regularly experienced by long-living fungal structures that are subjected to continuous wetting and drying cycles (Honegger 1995, 2001, 2006; Honegger *et al.* 1996).

Multiperforate septa, a rather unusual feature among ascomycetes, have been observed even by early light microscopists and later with electron microscopy techniques (Wetmore 1973) in medullary hyphae of some Peltigeraceae and Parmeliaceae (Figs. 3.8, 3.9). Only some, but not all, medullary hyphae of a thallus reveal this structural peculiarity. It remains unknown whether these particular hyphae are functionally different from the rest of the medullary hyphae.

3.1.4   *Symbiotic versus aposymbiotic phenotypes*

Aposymbiotic phenotypes in culture

Lichen-forming fungi differ from nonlichenized taxa by their nutritional strategy and by their manifold adaptations to the cohabitation with a population of minute photobiont cells. In pure culture aposymbiotic lichen mycobionts form thallus-like colonies with no morphological resemblance to the symbiotic phenotype (Figs. 3.4, 3.5). The central, microaerobic part of such aposymbiotic thalli is usually composed of cartilaginous, conglutinate cell masses while filamentous growth, i.e. aerial hyphae, are formed at the periphery (Fig. 3.5; further examples in Ahmadjian, 1973; Honegger and Bartnicki-Garcia, 1991; Stocker-Wörgötter, 1995).

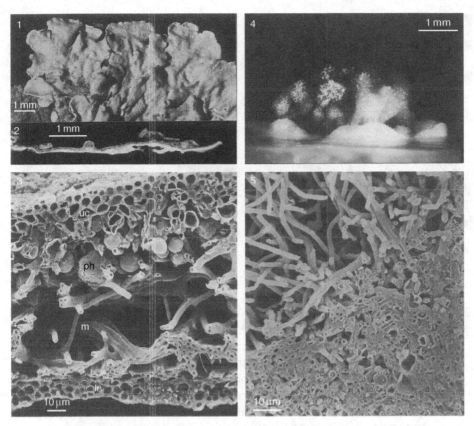

Figs. 3.1–3.5 Symbiotic and aposymbiotic phenotypes of the dorsiventrally organized macrolichen *Xanthoria parietina* (Teloschistales). Fig. 3.1. Laminal view, and Fig. 3.2. vertical cross section of the marginal lobes of the leaf-like thallus. Fig. 3.3. Detail of a vertical cross section: uc, conglutinate upper cortex; ph, photobiont layer harboring the globose cells of the green alga *Trebouxia arboricola*; m, gas-filled medullary layer built up by aerial hyphae; lc, conglutinate lower cortex. Fig. 3.4. Aposymbiotic phenotype on an agar medium. Fig. 3.5. Detail of the peripheral part of a cross section of the thallus-like fungal colony showing filamentous hyphal growth at the periphery and conglutinate zones in the microaerobic central part.

Symbiotic phenotypes in nature

Most biologists consider lichens as being shaped like leaves (foliose; Fig. 3.1) or tiny, erect or pendulous shrubs (fruticose) and having an interesting internal stratification with the photobiont cell population incorporated in a thalline layer (e.g. Figs. 3.3 or 3.14). However, only one out of four lichen-forming fungi has such an impressive morphogenetic capacity. The majority of lichen mycobionts (Table 3.2) overgrow or ensheath photobiont cells on or

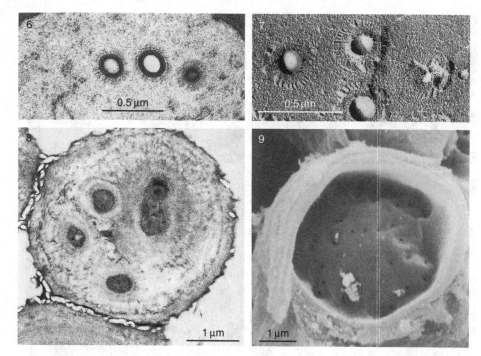

Figs. 3.6–3.9   Figs. 3.6–3.7. Concentric bodies, semicrystalline cytoplasmatic organelles of unknown origin and function in an ultrathin section and a freeze-fracture preparation of *Peltigera canina* (Peltigerales). Figs. 3.8–3.9. Multiperforate septa in medullary hyphae as observed in an ultrathin section of *Parmelia tiliacea* (Lecanorales) and a SEM preparation of *Peltigera canina*.

within the substrate and form microfilamentous (Fig. 3.13), microglobose (Fig. 3.10), or crustose thalli (Figs 3.11, 3.12), some of which are quite inconspicuous. About 20% of the lichen-forming fungi form squamulose or placodioid thalli with an internal stratification, but usually these thalli remain in close contact with the substrate. In only about 25% of lichen species do the mycobionts grow above the substrate and enter the third dimension by differentiating either a foliose or fruticose thallus with internal stratification (Figs. 3.3, 3.14; Table 3.2). These morphologically complex symbiotic phenotypes are formed by a range of functionally and morphologically different fungal cells (see Chapters 4 and 5).

The photobiont cell population is housed, maintained, and controlled within the fungal thallus. It is arranged similarly to the palisade parenchyma (i.e. the photosynthetically most active parts) in vascular plants: either in a plane, as in foliose lichens (Figs. 3.2, 3.3), or at the periphery of either erect (e.g. reindeer

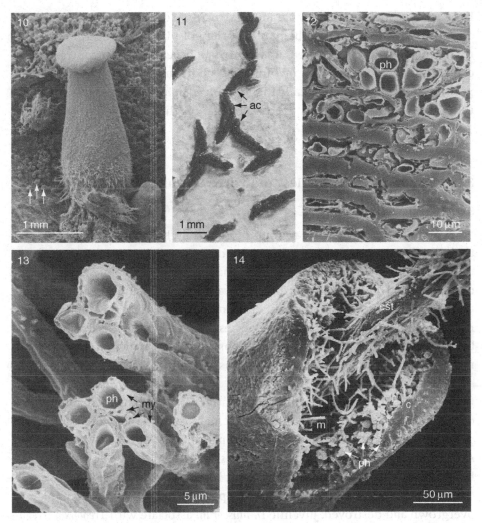

Figs. 3.10–3.14 Structural and taxonomic diversity in lichen-forming fungi. Fig. 3.10. *Omphalina ericetorum* (Agaricales), a lichenized basidiomycete growing on detritus. Arrows point to microglobular lichenized structures (vivid green in fresh samples) on the surface of a decaying leaf. Figs. 3.11–3.12. *Graphis elegans* (Graphidales) produces its crustose, grayish thallus within the smooth bark of *Ilex europaeus*. ac, lirelliform ascomata; ph, coccoid green algal photobiont cells (*Trentepohlia* sp.). Fig. 3.13. Microfilamentous thallus of the tropical *Coenogonium subvirescens* (Gyalectales). The filamentous green-algal photobiont (*Trentepohlia* sp.) (ph) is ensheathed by mycobiont hyphae (my). Fig. 3.14. Cross section of *Usnea rubicunda* (Lecanorales), a radially organized, fruticose lichen with internally stratified thallus. c, conglutinate cortical layer; ph, photobiont cell population (*Trebouxia* sp.); m, gas-filled medullary layer; cst, conglutinate central strand.

lichens) or pendulous, radially organized structures (e.g. beard lichens; Fig. 3.14) analogous to the vegetative body of a variety of plants (e.g. *Ephedra*, rushes, many succulent plants in the Cactaceae, Stapeliaceae, etc.). In contrast to all other mutualistic symbioses of fungi and photoautotrophs, it is the fungal partner of these morphologically highly evolved lichens that secures adequate illumination and facilitates gas exchange of the photobiont (Honegger 1991*a*, 1992).

3.2   Specialized "lifestyles" of lichens and associated fungi

3.2.1   *Parasitic lichens*

Parasitic lichens are a group of obligately lichenized fungi that start their development on or within the thallus of another lichen species. Depending on the degree of colonization, the host thallus is either locally or completely overgrown and destroyed. Accordingly, parasitic lichens may either be confined to host thalli throughout their lifetime or outgrow their host and become independent. In the former case the species are readily recognized as parasites, but in the latter case, extensive observations on all developmental stages may be required before the species is recognized as a parasite.

Parasitic lichen-forming fungi acquire their photobiont either "by theft" from the host lichen or separately. In the latter case the compatible photobiont of the parasite is usually taxonomically not identical with the photoautotrophic partner of the host lichen (Poelt and Doppelbauer 1956; Hawksworth 1988*b*). A complicated mode of photobiont acquisition was observed in the crustose lichen *Diploschistes muscorum*. This parasitic lichen starts its development in the thallus squamules and podetia of *Cladonia* spp. (Fig. 3.18) that may be completely overgrown and destroyed. Juvenile *D. muscorum* associate with *Trebouxia irregularis*, the photobiont of the host lichen. However, large, independent thalli of *D. muscorum* were invariably found to have replaced *Trebouxia irregularis* by *T. showmanii* (Friedl 1987).

3.2.2   *Bryophilous and foliicolous lichens*

A large number of crustose lichens favor decaying bryophytes as a substrate (e.g. the *Buellietum olivaceobruneae* in the Antarctic; Kappen 1985), but a small, taxonomically heterogeneous group (approx. 40 species) is parasitic on live bryophytes that are overgrown or even intracellularly infected (Figs. 3.15, 3.16; Poelt 1985). Compatible photobionts are often found under the cuticle between leaf and stem cells (Fig. 3.16) or, in some combinations, even within leaf cells (Döbbeler and Poelt 1981).

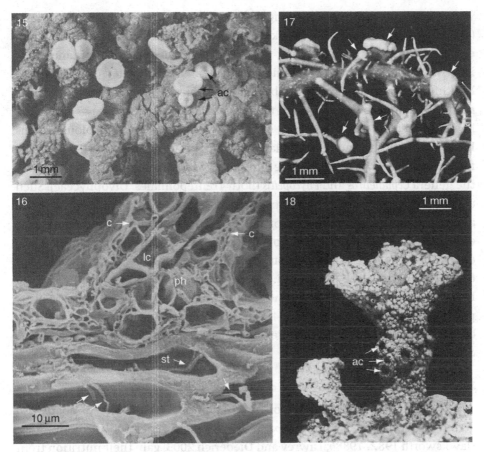

Figs. 3.15–3.18 Examples of multiple symbioses. Figs. 3.15–3.16. *Dimerella lutea*
(Gyalectales), a bryophilous lichen developing its inconspicuous crustose thallus
between the cuticle (c) and leaf cells (lc) of the foliose liverwort *Frullania dilatata*. Leaf
and stem cells of the hepatic are invaded by the lichen mycobiont (arrows). ac,
ascomata (pale orange); ph, coccoid cells of the green-algal photobiont, a *Trentepohlia*
sp.; st, central strand. Fig. 3.17. *Biatoropsis usnearum* (Tremellales), a cecidogenous
lichenicolous fungus induces gall formations (arrows) on the fruticose thallus of
Usnea cornuta. Fig. 3.18. The parasitic lichen *Diploschistes muscorum* starts its development in the thallus squamules and cup-shaped podetia of *Cladonia pyxidata* and
acquires its green-algal photobiont (*Trebouxia irregularis*) by theft from the host lichen
(for details see Friedl, 1987). ac, ascomata of the parasitic lichen.

In tropical and subtropical areas, the diverse, taxonomically heterogeneous
group of foliicolous lichens (*c.* 600 species; Farkas and Sipman 1993, 1997) has
attracted considerable interest in recent years. Their crustose, microfilamentous or squamulose thalli develop, usually quite unspecifically, on perennial
leaves of a very wide range of vascular plants from sea level to montane areas,

with their greatest diversity being found in humid submontane rain forests (Gradstein and Lücking 1997; Lücking *et al.* 2003). Foliicolous lichens are bioindicators of microclimate (Lücking 1997). In one hectare of a submontane forest in Costa Rica *c.* 200 species of foliicolous lichens were detected (Gradstein and Lücking 1997). Because large numbers of economically important crops, such as many spice-producing shrubs and trees, *Camellia* (tea), various *Citrus* spp., *Coffea*, *Hevea* (rubber), *Theobroma* (cacao), etc., are colonized, these foliicolous lichens have attracted the interest of plant pathologists (Hawksworth 1988*c*). Most foliicolous lichen mycobionts are symbiotic with filamentous green algae of the genera *Cephaleuros* and *Phycopeltis* (both Trentepohliales, Ulvophyceae). Such algae occur abundantly in the aposymbiotic state on perennial leaves and are also considered as pests (Hawksworth 1988*c*). These potential photobionts grow either on the cuticle or below. Some of them even occur in the palisade parenchyma of the leaf. In association with lichen mycobionts, the growth rate of these foliicolous algae is reduced. Most epicuticular lichens probably use the leaf merely as a substrate and grow equally well on artificial substrates such as plastic tape or slides (Sipman 1994; Sanders 2001*a*; Sanders and Lücking 2002), but subcuticular ones are quite likely to benefit from nutrients of vascular plant origin.

3.2.3   *Lichenicolous fungi and fungal endophytes of lichen thalli*

A considerable number of fungi (approx. 1250 species in about 280 genera of ascomycetes and approx. 62 species in 10 genera of basidiomycetes; Hawksworth 1982, 1988*b*; Lawrey and Diederich 2003) gain their nutrition from lichens and a formal classification of relationships has been developed by Rambold and Triebel (1992). Beside these approx. 1300 lichenicolous species there are about 260 doubtfully and/or infrequently lichenicolous taxa. Some lichenicolous fungi are necrotrophic, i.e. have a devastating effect on either the lichen mycobiont (mycoparasites; Diederich 1996; de los Ríos and Grube 2000), or on the photobiont (algal parasites; Grube and Hafellner 1990) or on both (e.g. the widespread basidiomycete *Athelia arachnoidea*). Others sporulate abundantly without causing major damage to either the fungal or the photoautotrophic partner of the host thallus. In the literature the former group is referred to as parasites, the latter as "parasymbionts" or "commensalists." However, as very oligotrophic heterotrophs, these non-destructive inhabitants of lichens drain their nutrition from the thallus and thus are best regarded as mild parasites. Because experimental data are missing, it is often very difficult to interpret the biology of these multiple symbioses. With light and electron microscopic techniques it is often impossible to distinguish foreign hyphae within lichen thalli from the mycobiont proper. A few lichenicolous fungi are quite catholic with regard to host preference (i.e. they have been found on different, unrelated taxa). However, the majority of lichenicolous fungi (an estimated 95%) shows a remarkable host specificity (Clauzade and Roux 1976; Hawksworth 1983; Lawrey and Diederich 2003). A group of approximately 40 species of lichenicolous ascomycetes and basidiomycetes is particularly interesting because they induce more or less species-specific gall formations on their host thalli (Fig. 3.17). Some of these cecidogenous (gall-inducing) taxa stimulate, in an unknown manner, the growth of the mycobiont and photobiont (Triebel and Rambold 1988; Hawksworth and Honegger 1994); others parasitize the photobiont, as observed in the heterobasidiomycete *Biatoropsis usnearum* (Grube and de los Ríos 2001).

The majority of lichenicolous fungi is likely to be exclusively lichenicolous, but a considerable number are also known as saprobes or as parasites of non-symbiotic aerophilic algae (Hawksworth 1982). Transitions between modes of nutrition have been recorded. An example is *Chaenothecopsis consociata* (Mycocaliciales), a lichenicolous fungus that invades thalli of *Chaenotheca chrysocephala* (Caliciaceae, Lecanorales; photobiont: *Trebouxia simplex*), but, if available, it associates with *Dictyochloropsis symbiontica* to form its own crustose thallus (Tschermak-Woess 1980). The widespread heterobasidiomycete *Athelia arachnoidea*, necrotrophic on various lichen taxa, free-living algae and bryophytes especially in areas with severe air pollution (Yurchenko and Golubkov 2003), was shown to be the sexual state of *Rhizoctonia carotae*, a postharvest disease of carrots (Adams and Kropp 1996). Other *Athelia* spp. are lichenized (Oberwinkler 2001).

Finally, there is a wide range of very inconspicuous, symptomless inhabitants of lichen thalli that were discovered by culturing experiments only (Petrini *et al.* 1990; Miadlikowska *et al.* 2004*a*, *b*). Among these fungal endophytes of lichen thalli are taxa that occur also as saprobes outside lichens. Others are known as endophytes or pathogens of higher plants. In addition, a range of partially characterized species has so far not been found outside lichen thalli (Petrini *et al.* 1990). The lichen thallus as an ecological niche for fungi not yet known to science is an interesting field for future research.

4

Thallus morphology and anatomy

B. BÜDEL AND C. SCHEIDEGGER

Symbiosis is now widely accepted as a source of evolutionary innovation (Margulis and Fester 1991) that has stimulated an enormous morphological radiation in ascomycetes. Vegetative structures have especially developed to a complexity that is not reached elsewhere in the fungal kingdom (Honegger 1991b). Lichen morphology and anatomy are now understood as being highly adapted to constraints imposed by the environment on the mutualistic symbiosis, where the mycobiont is the exhabitant and the cyanobacterial or green-algal photobiont is the inhabitant (Hawksworth 1988b). A very wide range of different thallus structures have been described and a complete outline of lichen morphology is not the scope of this chapter. However, detailed reviews are given by Henssen and Jahns (1973) and Jahns (1988). Common mycological terms also used in lichenology are not always explained here. Readers are referred to recent mycological textbooks, to Hawksworth *et al.* (1983), or to a glossary of a recent lichen flora. Irrespective of lichen growth form, it must function as a photosynthetically active unit in a manner that allows positive net photosynthesis and subsequently sufficient growth rates. This implies that the photobiont has to be supplied with just the right amount of light, even in the deep shade of rain forests or under fully exposed conditions of deserts. Carbon dioxide (CO_2) diffusion to the photobiont needs to occur readily, even under fully hydrated conditions. Water loss should be adapted to the specific environment: minimized in dry environments, and maximized in very wet environments. Thereby optimal CO_2 gain may be realized.

4.1 Growth forms

The appearance of a lichen thallus is primarily determined by the mycobiont. Only a few cases are known where the photobiont determines the

Lichen Biology, ed. Thomas H. Nash III. Published by Cambridge University Press.
© Cambridge University Press.

habit of the whole thallus, e.g. in the filamentous genera *Coenogonium*, *Ephebe*, *Cystocoleus*, and *Racodium*. However, knowledge of the influence of the photobiont on the lichen morphogenesis is important, because only after the establishment of the symbiosis is the characteristic thallus of a lichen developed.

On the basis of their overall habit, lichens are traditionally divided into three main morphological groups: these are the **crustose**, **foliose** and **fruticose** types. There are a number of additional special types, such as the **gelatinous** lichens that all have cyanobacteria as photobionts (but not all lichens with cyanobacteria are gelatinous lichens!). However, all of them can be integrated within the threefold scheme of the main growth types.

4.1.1   Crustose lichens

Crustose lichens are tightly attached to the substrate with their lower surface and may not be removed from it without destruction. Water loss is restricted primarily to the upper, exposed surface only. When growing on inclined rock surfaces, they profit from surface water flow. These features allow these organisms to tolerate extreme habitats such as bare, exposed rock surfaces. Although the crustose growth type seems to be clearly defined, variation of the basically crustose type is abundant. The following subtypes can be distinguished: powdery, endolithic, endophloeodic, squamulose, peltate, pulvinate, lobate, effigurate, and suffruticose crusts. Their thallus organization may be either **homoiomerous** or **heteromerous**.

In terms of complexity in thallus structure, the **powdery crusts**, as found in the lichen genus *Lepraria*, are the simplest and lack an organized thallus. Fungal hyphae envelop clusters of photobiont cells and do not have a distinct fungal or algal layer. They have a powdery appearance and are also referred to as **leprose**. Even more simple is the construction of thalli of the epiphyllic, epiphytic, and terricolous genus *Vezdaea*, in which the vegetative, photobiont-containing thallus consists of single, globose soredia-resembling granules, which are usually less than 1 mm in diameter. The granules occur either on the surface or underneath the cuticle of bryophytes or other plant material. They are called **goniocysts** (Sérusiaux 1985) and are corticate and often have distinct spines (Fig. 4.4).

The construction of **endolithic** (growing inside the rock) and **endophloeodic** (growing underneath the cuticle of leaves or stems) lichens seems to be more organized. In most cases, an upper cortex is developed, e.g. in *Lecidea* aff. *sarcogynoides* (Wessels and Schoeman 1988). The upper cortex can consist of densely conglutinated hyphae forming a dense layer named "**lithocortex**", as for example in *Acrocordia conoidea*, *Petractis clausa*, *Rinodina immersa*, *Verrucaria baldensis*, and *V. marmorea* (Pinna *et al.* 1998). Other endolithic lichens like *Verrucaria rubrocincta* form a micrite layer with only a few hyphae involved

(Bungartz *et al.* 2004). In *Lecidea* aff. *sarcogynoides*, the photobiont layer as a part of the medulla is located underneath the cortex within the rock. The medullary hyphae may extend up to 2 mm deep into the sandstone matrix. By penetrating the sandstone, *L.* aff. *sarcogynoides* weathers the sandstone at a rate of 9.6 mm per 100 years in the semihumid climate of South Africa (Wessels and Schoeman 1988). In contrast, Antarctic endolithic lichens apparently do not form a strongly stratified thallus (Friedmann 1982). In those lichens, thallus structure is mainly determined by the rock matrix itself (de los Ríos *et al.* 2005).

Active weathering by etching can be found in the marine species *Arthopyrenia halodytes*, living on calcareous substrates in the littoral. *A. halodytes* thalli can often be found in the shell of *Balanus* species (Fig. 4.1) and of molluscs. Crustose lichens (e.g. *Leproplaca chrysodeta*, *Dirina massiliensis*) contribute to the biodeterioration of a wide range of building materials and historical monuments (Nimis *et al.* 1992).

Epilithic and epiphloeodic lichens comprise the vast majority of the crustose growth type and a number of special thallus types are developed. Most crustose lichens are stratified and show some, if not all, of the layers that are described under Section 4.2.2. The margin of the thallus may be clearly delimited or indistinct. In the genus *Rhizocarpon* or *Placynthium*, for example, a **prothallus** is developed. This is a photobiont-free, white or dark brown to black zone, visible between the areolae and at the growing margins of such crustose lichens. The corticolous lichen *Cryptothecia rubrocincta* in the rain forest of French Guiana collects water in its hydrophilic, nonlichenized prothallus and drains it through little channels passing underneath the thallus, thus avoiding periods of supersaturation. This was frequently observed after rainwater exuded continuously from experimentally induced injuries of the surface of the lichenized thallus part (Lakatos 2002; Lakatos *et al.* 2006).

An **areolated thallus** consists of numerous areolae, which are polygonal parts of the thallus containing both symbiotic partners. In the dry state, the areolae are clearly distinguishable from each other, but under wet conditions they swell and the cracks between them close. The areolae might either develop from a primary thallus with a closed surface that splits secondarily from the surface to the bottom into several areolae, or from single thallus primordia developing on a prothallus. A thallus is called **effigurate** when the marginal lobes are prolonged and are radially arranged, as in many species of the genera *Caloplaca*, *Dimelaena*, *Acarospora*, and *Pleopsidium*.

The **squamulose** type of crustose lichens is the most complex. The areolae are enlarged in their upper part and become partially free from the substrate. Often they develop overlapping scale-like squamules. This is the case, for example, in the genera *Catapyrenium*, *Peltula*, *Psora*, and *Toninia*. Flat scales of squamulose thalli, with a more or less central attachment area on the lower surface, are called **peltate** (e.g. *Peltula euploca*, *Anema nummularium*; Fig. 4.2). The peltate type of squamulose crusts is often developed in lichens colonizing soils or rock surfaces in hot, arid regions of the world. *Peltula radicata* (Fig. 4.22), for example, colonizes soils in deserts of the world. It is completely immersed in the soil, exposing its flat upper surface only. Lichens with this type of growth were called "Fensterflechten" (i.e. window) by Vogel (1955), who first observed such lichens in the genera *Eremastrella* and *Toninia*. Extremely inflated squamules in the lichen genus *Mobergia* (Fig. 4.3) are called **bullate**. In other cases, coralloid tufted cushions, designated **suffruticose**, are formed. Within the genera *Caloplaca* and *Lecanora*, a **lobate** thallus is developed by some species. This is the case when a thallus becomes radially striate with marginal, at least partially raised lobes.

4.1.2   Foliose lichens

Foliose lichens are leaf-like, flat and only partially attached to the substrate. Foliose thalli are either homoiomerous (gelatinous lichens) or heteromerous. Typically they have a dorsiventral organization with distinct upper and lower surfaces. Often the thallus is divided into lobes, which show various degrees of branching (Figs. 4.5–4.8). Foliose lichens develop a great range of thallus size and diversity.

Laciniate lichens are the typical foliose lichens. They are lobate and vary considerably in size; they may either be gelatinous-homoiomerous (e.g. *Collema*, *Leptogium*, *Physma*; Fig. 4.8) or, as in most cases, heteromerous. The lobes can be radially arranged (e.g. in *Parmelia* species, Figs. 4.5, 4.6) or overlapping like tiles on a roof (e.g. *Peltigera*, *Hypocoenomyce*). In some genera, thallus lobes can become inflated, having a hollow medullary center (e.g. *Menegazzia*; Fig. 4.7). The lower surface is often covered by rhizinae, cilia or a tomentum, which, to a limited degree, may also serve as attachment structures. Conspicuous lichens, such as the genera *Sticta* and *Pseudocyphellaria* in the understory of tropical and temperate rain forests, *Lobaria* and *Nephroma* in alpine and oceanic forests, or *Peltigera* in arctic tundras, belong to this growth type.

Umbilicate lichens have circular thalli, consisting either of one single, unbranched lobe or multilobate thalli with limited branching patterns. All are attached to the substrate by a central umbilicus from the lower surface. This often can be recognized by a navel-like depression on the upper side. The umbilicus usually consists of tightly packed, parallely arranged and conglutinated hyphae without photobiont cells. Umbilici have apparently evolved several times in such unrelated groups as the Dermatocarpaceae, Parmeliaceae, Physciaceae, and Umbilicariaceae.

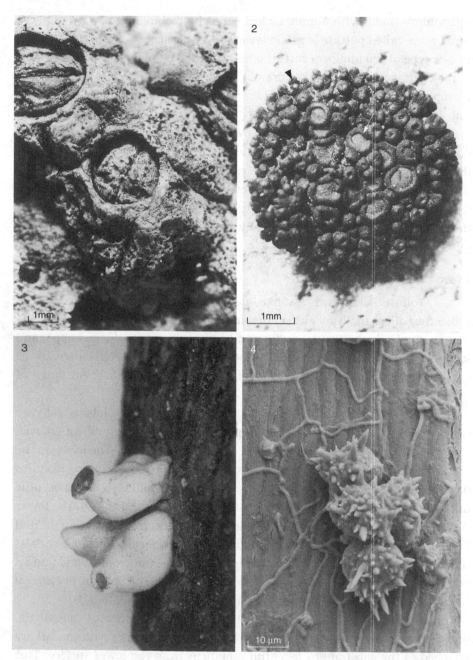

Figs. 4.1–4.4 Growth forms, crustose lichens. Fig. 4.1. *Arthopyrenia halodytes* (Spain, Mediterranean Sea, on *Balanus* sp.), an endolithic lichen within calcareous substrates with perithecia (arrow). Fig. 4.2. *Anema nummularium* (Spain, on conglomerate), a gelatinous lichen with unicellular cyanobionts and lecanorine apothecia (white arrow); the thallus is squamulose-umbilicate and effigurate at the margins (arrowhead). Fig. 4.3. *Mobergia calculiformis* (Baja California, on rock), bullate thallus. Fig. 4.4. *Vezdaea rheocarpa* (Switzerland, over epiphytic mosses), goniocysts (SEM micrograph).

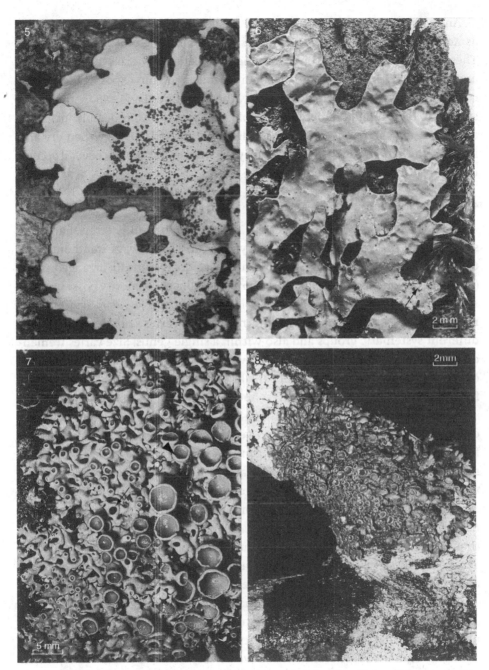

Figs. 4.5–4.8 Growth forms, foliose lichens. Fig. 4.5. *Parmelia pastillifera* (Switzerland, on *Acer pseudoplatanus*), thallus with black, knob-like isidia. Fig. 4.6. *Parmelia sulcata* (Germany, on bark), flat thallus lobes with soralia (arrow). Fig. 4.7. *Menegazzia terebrata* (New Zealand, temperate rain forest, on *Nothofagus menziesii*), thallus with inflated, hollow lobes with pores (arrow) and slightly stalked, lecanorine apothecia (arrowhead). Fig. 4.8. *Physma* sp. (Australia, tropical rain forest, on bark), gelatinous foliose lichen.

An ecologically interesting group of foliose lichens are the vagrant lichens, such as *Xanthomaculina convoluta* and *Chondropsis semivirdis* in deserts and semideserts. These lichens show hygroscopic movement (Büdel and Wessels 1986; Lumbsch and Kothe 1988). In the dry state the thalli are rolled up, thus exposing their lower cortices. When they take up liquid water, the thalli unroll and expose the upper surface to the sunlight. In both lichens photosynthesis is considerably increased after exposing the upper surface to the light (Lange *et al.* 1990b). When dry and inrolled, the lichens can easily be blown by the wind and as soon as dewfall occurs, they unroll and expose the upper surface again.

4.1.3   Fruticose lichens

The thallus lobes of fruticose lichens are hair-like, strap-shaped or shrubby and the lobes may be flat or cylindrical. They always stand out from the surface of the substrate. Some groups have dorsiventrally arranged thalli (e.g. *Sphaerophorus melanocarpus*, *Evernia prunastri*), but the majority possess radial symmetric thalli (e.g. *Sphaerophorus globosus*, *Usnea* species, *Ramalina* species; Figs. 4.11, 4.12). The branching pattern of lobes varies considerably among different systematic groups and also within a single genus. Size varies tremendously, from some *Usnea* species that grow several meters long to minute species only 1 or 2 mm high. Fruticose lichens are found in a wide range of climates, from the desert to the wet rain forest and on various types of substrates.

Genera like *Baeomyces* or *Cladonia* (Fig. 4.10) develop a twofold thallus, which is differentiated into a fruticose **thallus verticalis** and a crustose-squamulose to foliose **thallus horizontalis**. In the genus *Cladonia*, the thallus verticalis originates from primordia of the fruit body and results in apothecia-bearing stalks, which are termed **podetia** (Fig. 4.10). In cases where the thallus verticalis develops from primordia of the thallus horizontalis, e.g. in the genus *Stereocaulon*, it is named **pseudopodetium**. The genus *Cladia* has developed **reticulate** podetia, thus exposing the medulla and parts of the algal layer (Fig. 4.9). As a consequence, uptake and loss of water is rapid. This feature probably reflects the wet soils and moss cushions within which many of the reticulate *Cladia* species grow.

Another peculiar type of anatomy is that of the beard-like-growing genus *Usnea* (Fig. 4.12), which has a strong central strand of periclinally arranged, conglutinated hyphae that provide mechanical strength along the longitudinal axis.

Highly branched fruticose lichens have a high surface to volume ratio, which results in a more rapid drying and wetting pattern compared with lichens having lower surface to volume ratios. Fruticose growth forms can be found preferentially either in very wet, humid climates, e.g. *Usnea xanthophana* and *U. rubicunda* in the temperate rain forest, or in arid climates with regular dew and fog events, e.g. *Teloschistes capensis* in the Namib Desert. While *Usnea*

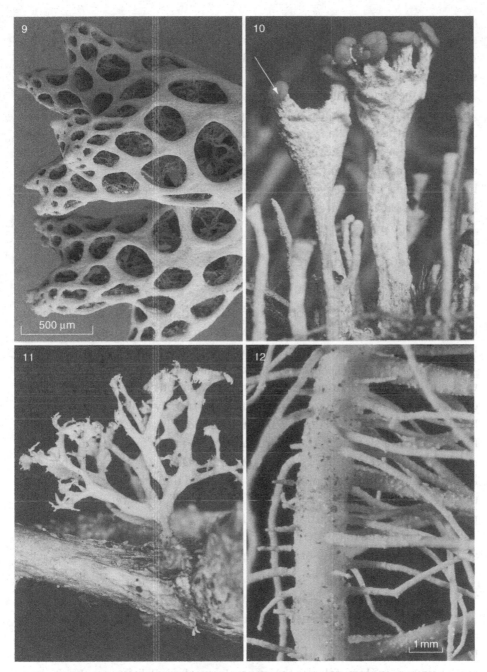

Figs. 4.9–4.12   Growth forms, fruticose lichens. Fig. 4.9. *Cladia retipora* (New Zealand, temperate rain forest, on soil), fenestrate lobes (SEM micrograph). Fig. 4.10. *Cladonia coccifera* (Russia, Lake Baikal, on soil), vertical thallus (= podetia) with red hymenium (arrow). Fig. 4.11. *Ramalina pollinaria* (Switzerland, on *Abies alba*), typical fruticose growth form. Fig. 4.12. *Usnea filipendula* (Germany, *Fagus* forest, on *Fagus sylvatica*), radial symmetric thallus with numerous fibrils.

xanthophana and *U. rubicunda* hardly showed any depression in CO_2-uptake due to slower CO_2-diffusion under water-supersaturated conditions (Lange *et al.* 1993*b*), *Teloschistes capensis* is able to collect sufficient fog and early morning dewfall to sustain positive net photosynthesis for considerable time periods (Lange *et al.* 1990*a*). In the former habitat, the hanging "beard" of the *Usnea* thalli is advantageous in that rapid water loss occurs. Together with the numerous hydrophobic airspaces in the medulla, supersaturation with water is avoided (Lange *et al.* 1993*b*). In the second case, the large tufts of *Teloschistes capensis* act like a comb for dew and fog condensation. In a similar manner, increased water uptake occurs in many other fruticose genera, such as *Ramalina menziesii* (Larson *et al.* 1985; Boucher and Nash 1990*a*) and the genera *Bryoria* and *Usnea*.

4.2   Vegetative structures

4.2.1   *Homoiomerous thallus*

Mycobionts and photobionts are evenly distributed in homoiomerous thalli, as is often found in thin crustose lichens, in gelatinous crustose and foliose lichens, for example of the genera *Caloplaca*, *Pyrenopsis* or *Collema* (Fig. 4.13).

The homoiomerous gelatinous lichens absorb much more water in relation to their dry weight than nongelatinous, heteromerous lichens do. As a consequence, CO_2 gas diffusion to the photobiont is strongly limited or may even be blocked in supersaturated thalli. Carbon dioxide can become the limiting factor for photosynthesis under these circumstances (Lange and Tenhunen 1981). However, Lange *et al.* (1993*b*) showed for *Collema laeve* and other cyanobacterial lichens that CO_2 uptake is only slightly depressed under supersaturated conditions, a fact that may be due to the functioning of the recently described photosynthetic CO_2 concentration mechanism. A steeper concentration gradient from air outside to the CO_2 fixation site within the cyanobacterium leads to an increasing diffusion rate of CO_2 (Badger *et al.* 1993).

4.2.2   *Stratified thallus*

The majority of lichens including many crustose species develop internally stratified thalli. The main subdivisions are into upper cortex, photobiont layer, medulla, and lower cortex. These layers may include various tissue types, and their terminology follows the general mycological literature. In the case where the hyphae are conglutinated to an extent that single hyphae are not usually distinguishable, the main tissue types are **pseudoparenchymatous** and/or **prosoplectenchymatous**. The former case involves more or less isodiametrical cells (Fig. 4.18) resembling true parenchyma of vascular plants. The latter

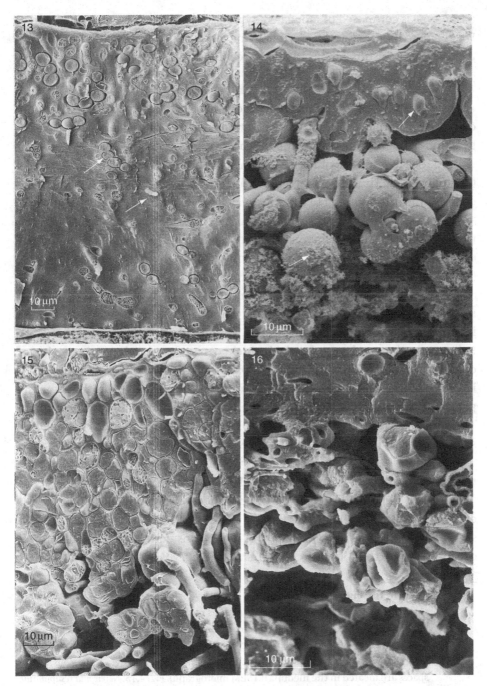

Figs. 4.13–4.16   Thallus anatomy, heteromerous versus homoiomerous thalli, water storage. Fig. 4.13. *Collema nigrescens* (Switzerland, on *Acer pseudoplatanus*), a wet, homoiomerous thallus of the gelatinous type with the filamentous *Nostoc* cyanobiont (upper arrow). Water is located in the jelly of fungal and cyanobacterial walls and

tissue type is composed of elongated hyphal cells that are arranged anticlinally. For the description of dense plectenchyma in ascoma, a terminology introduced by Korf (1973) for nonlichenzed discomycetes is useful (Scheidegger 1993).

Cortex, epicortex, and epinecral layer

Only a few growth forms are known where the photobiont layer immediately reaches the surface of the lichen thallus. Mostly the algal layer is covered by a thin to thick (up to several hundred micrometers) cortical layer (Fig. 4.14). In many dark lichens pigmentation is confined to fungal cell walls of cortical hyphae (Esslinger 1977; Timdal 1991) or the **epinecral** layer. In gelatinous lichens the color is primarily confined to the outer wall layers of the mycobiont (Büdel 1990).

In many foliose or fruticose lichens a cortex is formed by pseudoparenchymatous or a prosoplectenchymatous fungal tissue. Usually living or dead photobiont cells are completely excluded from the cortex, but in the so-called **phenocortex** (Poelt 1989) collapsed photobiont cells are included (e.g. *Lecanora muralis*).

In Parmeliaceae some species have a 0.6 to 1 μm thick **epicortex**, which is a noncellular layer secreted by the cortical hyphae. This epicortex can be pored, as in *Parmelina*, or nonpored, as in *Cetraria*. In a broad range of foliose to crustose lichens an epinecral layer of variable thickness is often developed. It consists of dead, collapsed and often gelatinized hyphae and photobiont cells. Thalli often have a whitish, flour-like surface covering, the so-called **pruina** that consists primarily of superficial deposits, of which calcium oxalate is the most common. The amount of calcium oxalate is probably dependent on ecological parameters, such as calcium content of the substrate and the aridity of the microhabitat.

Functions of the upper cortex and/or its pruina include mechanical protection, modification of energy budgets (Kershaw 1985), antiherbivore defense (Reutimann and Scheidegger 1987), and protection of the photobiont against excessive light (Ertl 1951; Büdel 1987; Jahns 1988; Kappen 1988). Light- and shade-adapted thalli of several species differ considerably in the anatomical organization of their upper cortical strata (e.g. *Peltigera rufescens*; Figs. 4.17,

---

Caption for Figs. 4.13–4.16 (cont.)

sheaths (lower arrow). There are no airspaces in the thallus (LTSEM micrograph). Fig. 4.14. *Anaptychia ciliaris* (Switzerland, on *Acer pseudoplatanus*), a wet, heteromerous thallus with unicellular *Trebouxia* phycobionts in the algal layer (lower arrow). Water is located in the cell walls of the mycobiont mainly (upper arrow). Numerous airspaces are located in the medulla (LTSEM micrograph). Fig. 4.15. *Nephroma resupinatum* (Switzerland, on *Acer pseudoplatanus*), a wet, heteromerous thallus with filamentous *Nostoc* cyanobionts (LTSEM micrograph). Fig. 4.16. *Cetraria islandica* (Sweden, terricolous), a dry, heteromerous thallus with *Trebouxia* photobiont. Note the collapsed phycobiont cells (LTSEM micrograph).

4.18). In a nearby fully sun exposed habitat, the thickness of the cortex is reduced but a thick, epinecral layer with numerous air spaces is formed (Fig. 4.17), giving the thallus a grayish-white surface due to a high percentage of light reflection (Dietz *et al.* 2000). This cortical organization results in decreased transmission of incident light by 40% in the sun-adapted thallus measured at the upper boundary of the algal layer. Additionally, epinecral layers of *Peltula* species contain airspaces that may also function as CO_2 diffusion paths under supersaturated conditions (Büdel and Lange 1994).

Photobiont layer and medulla

In most foliose or fruticose thalli, the medullary layer occupies the major part of the internal thalline volume. Usually it consists of long-celled, loosely interwoven hyphae forming a cottony layer with a very high internal airspace. The upper part of the medulla forms the photobiont layer. In many lichens, the hyphae of the photobiont layer are anticlinally arranged and may sometimes form short or globose cells.

Supporting tissue is often formed within the medullary layer of fruticose lichens and to a secondary degree in other lichens. It consists of thick-walled, conglutinated hyphae. This special tissue may be formed as irregularly arranged hyphal strands (e.g. in maritime, placodioid crustose lichens), as a central cylinder (*Cladina*) or as a central, thread-like elastic strand (*Usnea*).

The hyphal cell walls of the algal and medullary layer are often encrusted with crystalline secondary products. These crystals and/or tessellate outer cell wall layers (Honegger 1991*c*) make the medullary hyphae hydrophobic. Therefore, during wet periods the medullary and algal layer remain air filled and capillary water is probably not present in internal parts of the thalli (Figs. 4.42, 4.43; Brown *et al.* 1987; Scheidegger 1994*a*). Water transport to the photobionts seems to be restricted to mycobiont cell walls. Under water-saturated conditions, photobiont and mycobiont cells are fully turgid (Figs. 4.14, 4.15). But in the air-dry state photobiont cells are collapsed following water loss (Fig. 4.16). Green-algal symbionts have the ability to become turgid (Brown *et al.* 1987; Büdel and Lange 1991; Scheidegger 1994*a*) and achieve positive net photosynthesis at high relative air humidity (>85 %) and low temperatures. However, cyanobacterial photobionts only become turgid in the presence of liquid water (Büdel and Lange 1991; Scheidegger 1994*a*). Although some thin-walled medullary hyphae collapse during water loss, cortical and thick-walled medullary fungal hyphae usually show a different reaction to water loss. These cells cavitate and keep their shape more or less unaltered during the desiccation process (Figs. 4.37–4.39). Cavitation is an explosion-like formation of bubble-like structures in the symplasts of fungal hyphae and spores (Scheidegger *et al.* 1995*a*).

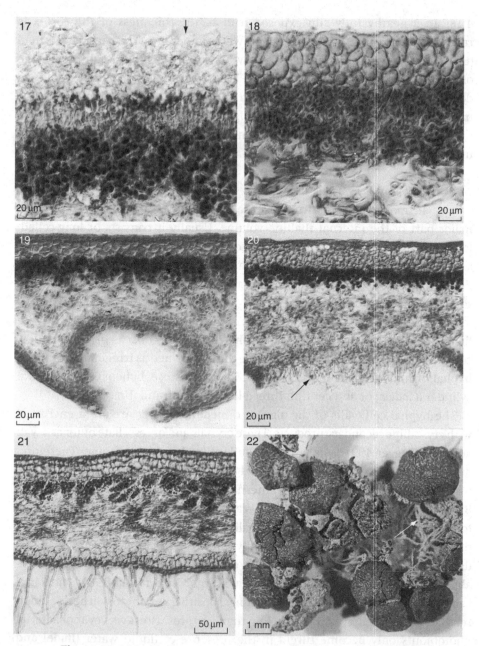

Figs. 4.17–4.22 Thallus anatomy, functional anatomy, vegetative structures. Figs. 4.17–4.18. *Peltigera rufescens* (Germany, xerothermic steppe, on soil, from Dietz *et al.* 2000). Fig. 4.17. Specimen from a sunny, fully exposed habitat, upper cortex with an additional epinecral layer (arrow). Fig. 4.18. Specimen from a permanently shaded habitat; the pseudoparenchymatic cortex is thicker compared with the fully exposed specimen and lacks an epinecral layer. Fig. 4.19. *Sticta latifrons* (New Zealand, temperate rain forest), cross section with cyphellae with the typical corticated anatomy at

Cavitation bubbles take a major part of the volume of the symplast and compensate the volume of the water that has evaporated during desiccation. During rehydration, cavitation bubbles are refilled within seconds. Because cavitation does not change the shape of the cells, desiccation and rehydration processes that change water content of lichen thalli between <20% and >150% of the lichen dry weight do not induce shearing forces between fungal hyphae, for example in a multilayered cortex (Scheidegger *et al.* 1995a). Even in freezing and melting cycles of hydrated lichens the same morphological processes can be observed. In *Umbilicaria aprina* cooling of hydrated lichen thalli below the freezing point leads to extracellular freezing and an accumulation of ice crystals in the intercellular space of the lichen medulla. The effects of this freezing process are identical to dehydration and lead to collapsed photobionts and cavitated fungal hyphae (Schroeter and Scheidegger 1995).

Free, capillary water has only been found thus far within the hollow podetia of *Cladonia* and *Cladina* species, in which an internal, hydrophilic central cylinder is developed.

Lower cortex

In some foliose lichens such as *Peltigera* or *Heterodermia* the medulla directly forms the outer, lower layer of the thallus. However, typical foliose lichens of the Parmeliaceae and many other groups have a well-developed lower cortex. As is the case with the upper cortex, it is either formed by pseudoparenchymatous or a prosoplectenchymatous tissue. But unlike the upper cortex, the lower cortex is often strongly pigmented. Its ability to absorb water directly is well documented. Only low water conductance has been found thus far. However, it may play a major role in retaining extrathalline, capillary water (Jahns 1984).

Attachment organs and appendages

An impressive variety of attachment organs may be developed from the lower cortex and also rarely from the thallus margin or the upper cortex. In

Caption for Figs. 4.17–4.22 (cont.)

the lower surface. Fig. 4.20. *Pseudocyphellaria filix* (New Zealand, temperate rain forest), cross section with pseudocyphellae showing the typical protrusion of medullary hyphae into the opening (arrow). Fig. 4.21. *Pseudocyphellaria dissimilis* (New Zealand, temperate rain forest), cross section through the heteromerous thallus with *Nostoc* photobiont and with a tomentum forming hyphae originating from lower cortical cells. Fig. 4.22. *Peltula radicata* (Saharan Desert, soil), squamules connected with rhizines (arrow) to each other forming one thallus.

foliose lichens attachment is mainly by simple to richly branched rhizines, mostly consisting of strongly conglutinated prosoplectenchymatous hyphae. Umbilicate lichens as well as *Usnea* and similarly structured fruticose lichens are attached to the substrate with a holdfast, from which hyphae may slightly penetrate into the substrate. Deeply penetrating rhizine strands are found in some squamulose, crustose or fruticose lichens growing in rock fissures, over loose sand and sod.

In crustose lichens a prosoplectenchymatous prothallus is often formed around and below the lichenized thallus. It establishes contact with the substrate. From there bundles of hyphae penetrate among soil particles. Members of various growth forms produce a loose web of deeply penetrating hyphae, growing outwards from the noncorticate lower surface of the thalli. Various attachment organs with high amounts of extracellular gelatinous material establish tight contacts to the substrate. In this manner attachment to loose substrates, such as sand, is possible.

Cilia are fibrillar outgrowths from the margins or from the upper surface of the thallus. A velvety **tomentum** consisting of densely arranged short, hair-like hyphae may be formed on the upper or lower cortex. Tomentose surfaces are mainly reported from broad-lobed genera such as *Pseudocyphellaria*, *Lobaria*, and *Sticta* but are also found in a few *Leptogium* and others.

Cyphellae and pseudocyphellae

Upper or lower cortical layers often bear regularly arranged pores or cracks. **Pseudocyphellae**, as found on the upper cortex of *Parmelia sulcata* (Fig. 4.24) or on the lower side of *Pseudocyphellaria* (Figs. 4.20, 4.21), are pores through the cortex with loosely packed medullary hyphae occurring to the interior. **Cyphellae** are bigger and anatomically more complex than pseudo-cyphellae. In the interior portions of the cyphellae, hyphae form conglutinated, globular terminal cells, and this is the main difference from pseudocyphellae. They are only known from the genus *Sticta* (Fig. 4.19).

Pseudocyphellae and young cyphellae are thought to considerably lower gas diffusion resistance of the cortex. Because pseudocyphellae and cyphellae are hydrophobic structures, they may act as pathways for gas diffusion into thalli. However, under supersaturated conditions they can no longer function in this way (Lange *et al.* 1993*b*).

Cephalodia (Photosymbiodemes)

Representatives of the foliose *Peltigerales* with green algae as the primary photobiont, and members of such genera as *Stereocaulon*, *Amygdalaria*, *Chaenotheca*, *Micarea*, and *Placopsis* (Fig. 4.25) usually possess an additional

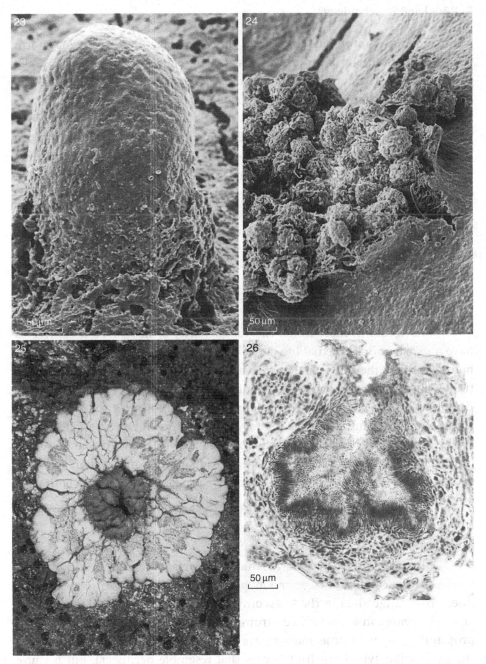

Figs. 4.23–4.26. Vegetative structures and anamorph structures. Fig. 4.23. *Parmelia crinita* (Switzerland, saxicolous), thallus with finger-like isidia (LTSEM micrograph, courtesy of S. Geissbühler). Fig. 4.24. *Parmelia sulcata* (Switzerland, on *Acer pseudoplatanus*), soralium with numerous soredia (LTSEM micrograph). Fig. 4.25. *Placopsis gelida* (Sweden, saxicolous), thallus with a centrally arranged, gall-like external cephalodium, containing the N-fixing cyanobiont. Fig. 4.26. *Heppia lutosa* (Arizona, Sonoran Desert, terricolous), cross section exposing the immersed, flask-shaped pycnidium with pycnospores.

cyanobacterial photobiont. In *Solorina* this secondary photobiont may form a second photobiont layer underneath the green-algal layer, but usually it is restricted to minute to several millimeters wide **cephalodia**. Cephalodial morphology is often characteristic on a species level and ranges from internal verrucae to external warty, globose, squamulose or shrubby structures on the upper or lower thallus surfaces. Cephalodial morphology usually differs completely from the green-algal thallus, and this emphasizes the potential morphogenetical influence of the photobiont on the growth form of the mycobiont–photobiont association (Chapter 5). Because many cyanobacterial photobionts are nitrogen-fixing (Chapter 11), these lichens may considerably benefit from cephalodia, especially in extremely oligotropic habitats.

4.3   Reproductive structures

As occurs in most fungi, the vast majority of lichenized ascomycetes have a sexual and an asexual life cycle. Within lichens, usually only the mycobiont expresses the full sexual and, to a certain degree, also asexual reproduction. The reproduction mode of the photobiont is, however, reduced in the lichenized state. The principal problem with lichenization is the necessity of fungal spores meeting the proper photosynthetic partner for the re-establishment of the symbiosis. In addition to the typical sexual (**teleomorph**) and asexual (**anamorph**) fruiting structures of the individual symbionts, lichenized ascomycetes have evolved a number of vegetative propagules, by which both partners are distributed.

4.3.1   *Generative reproduction: ascoma (pl. ascomata)*

In contrast to the vegetative thallus, ascomata are composed of haploid hyphae and dicaryotic ascogenous hyphae. Two main ascocarp development lines are distinguished in the Euascomycetidae. In the **ascolocular** type, asci arise in cavities in a preformed stroma. True paraphyses and an excipulum proprium (proper or true margin) are lacking. Many lichenized fungi with the ascolocular type have fruit bodies that resemble perithecia, but because of their ascolocular origin, they are called **pseudothecia** (e.g. Pleosporales, Hysteriales).

The second and most common ascocarp development in lichenized ascomycetes is the **ascohymenial** type, which is initiated with the development of an ascogonium, followed by the establishment of dicaryosis. It is still unknown how dicaryosis occurs in lichenized ascomycetes. The ascocarp is composed of ascogenous hyphae and haploid hyphae from the base of the ascogonia-bearing hyphae and the asci are developed from ascogenous hyphae. Together with the sterile paraphyses (**hamathecium**), they form the **hymenium**. Underneath the hymenium a generative layer (i.e. **subhymenium**) is present, which gives rise to the hymenium. Sometimes a **hypothecium**, underlying the subhymenium, can be developed. The hymenium itself may be overlaid by a distinct **epithecium**. The development of a fruit body is a **gymnocarp**, when the hymenium is exposed from the earliest stage on. An initially enclosed fruiting body that opens before forming a fully mature hymenium is typical of **hemi-angiocarp** development. If the hymenium remains closed until the spores are mature, then it is called **angiocarp** development. For more detailed descriptions the reader is referred to basic mycological and lichenological works.

There is considerable variation among ascomata and according to their morphology and anatomy several types of fruit bodies exist. In **apothecia** the hymenium is exposed at maturity and the hamathecium is either lacking or consists of paraphyses, paraphysoids or pseudoparaphyses; whereas **perithecia** remain closed, not exposing the hymenium.

Perithecia

Perithecia open with a small tube-like **ostiolum** and have peri-physes and sometimes paraphyses (hamathecium). They are globose to flask–shaped and are more or less immersed (e.g. *Arthopyrenia, Catapyrenium, Dermatocarpon*; Fig. 4.1). The exciple is carbonized in some genera, as in *Verrucaria*, and the ostiolum may be surrounded by a shield-like, carbonized layer.

Apothecia

Apothecia are cup- or disk-shaped (Figs. 4.2, 4.7, 4.8) and two main morphological and developmental types are distinguished. Apothecia with a margin originating from the thallus (margo thallinus) are **lecanorine** (Fig. 4.31). In other cases where the margin develops from the tissue of the fruit body (*margo proprius*), it is either called **lecideine** (Fig. 4.32) when the margin is carbonized or **biatorine** when noncarbonized. In some genera, both margins are present, and this type is called **zeorin**. Within the *margo proprius*, two layers can be distinguished: the inner part is formed by the **parathecium**, giving rise to the outer layer, the **amphithecium**.

Thallinocarpia

Thallinocarpia are derived from a strongly, if not totally, reduced generative tissue. In this special type, the ascogonia develop from thallus hyphae underneath the lichen surface, between mycobiont and photobiont cells. Subsequently, generative tissue is developed, from which true paraphyses are formed (Henssen *et al.* 1981). Parts of the thallus, including photobiont cells, are dispersed in between the hymenial parts so that the fruiting structure is subdivided into thalline and generative compartments (Fig. 4.33). Thallinocarpia are only known from the family Lichinaceae, e.g. the genus *Lichinella* and *Gonohymenia*.

Pycnoascocarpia

Pycnoascocarpia are a special ontogenetical type of apothecia, originating by transformation of pycnidia into apothecia. They occur in the genera *Ephebe*, *Paulia*, or *Thyrea* (Henssen *et al.* 1981).

Hysterothecia

Hysterothecia are elongated, small fruit bodies with a split-like hymenium (Fig. 4.34). They may be either derived from perithecia or, as in all lichenized ascomycetes, from apothecia. Hysterothecia are found in the genera *Graphis* and *Opegrapha*.

Asci

The ascus structure and function plays an important role in ascomycete systematics. On the basis of electron microscopical and traditional light microscopical investigations, it is evident that characteristic types of asci occur within the lichenized ascomycetes. A basic difference between ascus types seems to be prototunicate asci and unitunicate/bitunicate ones. These latter two terms are derived from the original light microscopy studies, from which walls were thought to be either one- or two-layered. However, electron-microscopical studies have subsequently demonstrated that this terminology is misleading. To reduce confusion we distinguish here between anatomically and functionally single- or two-layered asci.

Prototunicate asci have anatomically and functionally single-layered, thin walls without any special mechanisms for spore dehiscence. Spores are either released by apical splitting or disintegration of the wall (Figs. 4.35, 4.36H, I).

From an anatomical perspective **unitunicate** asci have a two-layered wall, but it behaves functionally as a single layer. Spore release is usually supported by an apical apparatus (part of the inner wall), which shows a high degree of

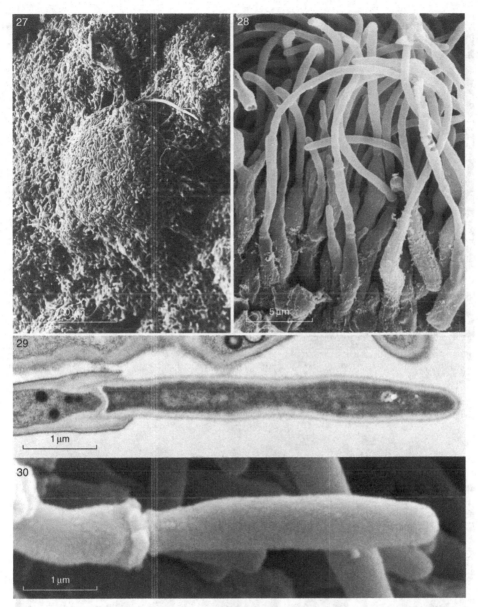

Figs. 4.27–4.30 Conidia. Fig. 4.27. *Micarea adnata* (Switzerland, on *Abies alba*),
sporodochium with conidiospores (LTSEM micrograph). Fig. 4.28. *Amandinea coniops*
(Norway, saxicolous) conidiogenous cells with filiform conidia (SEM micrograph).
Figs. 4.29–4.30. *Parmelia tiliacea*, part of conidiogenous cell with enteroblastic
formation of the primary conidium. A collarette is seen at the neck of the phialide.
TEM (Fig. 4.29) and SEM (Fig. 4.30) micrographs (from Honegger 1984a, with
permission from the author and publisher).

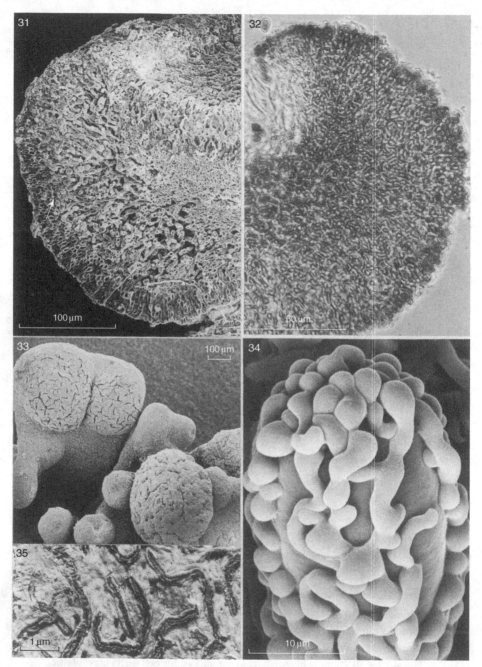

Figs. 4.31–4.35 Types of ascomata. Fig. 4.31. *Pyxine physciaeformis* (Brasil, epiphytic), lecanorine apothecium with algal cells in the excipulum (arrow; SEM micrograph). Fig. 4.32. *Buellia leptocline* (Sweden, saxicolous), section of the excipulum of the lecideine apothecium (from Scheidegger 1993). Fig. 4.33. *Lichinella intermedia* (Baja California, volcanic rock), typically cracked thallinocarp. The hymenium is divided

variability (Fig. 4.36C–G) and seems to be of great value in systematics. Prior to spore release, the outer wall opens and the apical apparatus elongates towards the surface of the hymenium. This type is termed **nonfissitunicate**, because the wall layers do not split (Honegger 1982*b*).

Bitunicate asci have two wall layers that also function like two layers. The outer wall (exoascus) is not expandible and opens apically. The inner wall (endoascus), which is at least partly liberated from the exoascus (Fig. 4.36A, B), is highly expandible and elongates substantially towards the hymenial surface, prior to spore release. This type is termed **fissitunicate** because the inner, expandible wall is liberated from the outer nonexpandible wall (Honegger 1982*b*). Variation of the two basic types (fissitunicate and nonfissitunicate) have undergone evolutionary radiation. The systematics chapter (Chapter 17) in this book demonstrates the use of ascus structure as a phylogenetic marker.

Basidioma (pl. basidiomata)

To date only a few species of the basidiolichenes are known and none forms a specific lichen thallus. All basidiolichens belong to the Holobasidiomycetidae, which is characterized by nondivided basidia. The fruiting structures are typical of the Aphyllophorales or Agaricales. They are either crust-like (resupinate, e.g. *Dictyonema*), cylindrical or club-shaped (clavarioid, e.g. *Multiclavula*) or have the shape of a lamellated mushroom fruit body (agaricoid, *Omphalina*). For further information see Oberwinkler (1984).

4.3.2   *Vegetative reproduction*

Aposymbiotic (apo = non) propagules

Conidiomata occur in many lichenized ascomycetes. **Pycnidia** are the main type of anamorph structures and are pear-shaped or globose receptacles (Fig. 4.26), within which conidia are formed on a special hyphal type, called conidiophores (Figs. 4.28, 4.29, 4.30). Pycnidia typically have a plectenchymatous wall. Several types of pycnidia are distinguished by the arrangement and morphology of the conidiophores (Vobis and Hawksworth 1981; Hawksworth 1988*d*). Conidiogenesis seems to be phialidic in all cases investigated so far

---

Caption for Figs. 4.31–4.35 (cont.)

into numerous small partial hymenia that are interrupted by photobiont-containing, vegetative thallus parts (from Henssen *et al.* 1986). Fig. 4.34. *Vezdaea aestivalis* (Switzerland, over saxicolous bryophytes), ascus with covering paraphysoids. (LTSEM micrograph, from Scheidegger 1994*b*). Fig. 4.35. *Graphis scripta* (Germany, on *Fagus sylvatica*), hysterothecia with a letter-like shape.

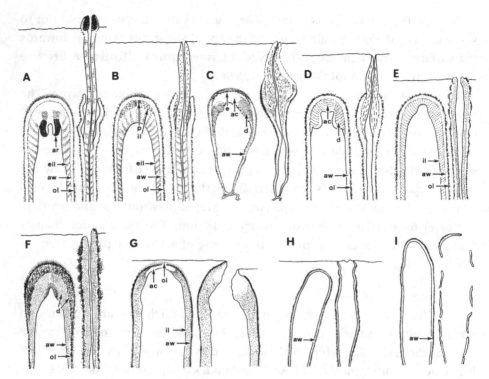

Fig. 4.36   Ascus types before (left) and after (right) dehiscence: A, *Peltigera*-type; B, *Rhizocarpon*-type; C, *Euopsis*-type; D, *Lecanora*-type; E, *Pertusaria*-type; F, *Teloschistes*-type; G, *Anzina*-type; H, *Heppia*-type; I, *Chaenotheca*-type. The horizontal line indicates the surface of the hymenium. *Chaenotheca* has a mazaedium (i.e. the ascus is covered by a thick layer of spores). ac, apical cushion; ar, eversible amyloid ring; aw, ascus wall; d, apical dome; e, electron dense particles; il, inner layer; ol, outer layer; p, preformed porus. They are redrawn from Honegger's (1982*b*) TEM micrographs (A,B,D,E,F); Henssen *et al.* 1987 (C); Scheidegger 1985 (G); Büdel 1987 (H); and Honegger 1985 (I) with permission from the authors.

(Figs. 4.28, 4.29, 4.30). The shape of the conidia, their mode of formation and the construction of the receptacles often have taxonomic value and may be used in characterizing species or even genera.

Campylidia are erect, helmet-shaped conidiomata known from several foliicolous lichens. In the *Badimia pollilensis* aggregate, campylidia develop from primordia identical to those of the apothecia (Sérusiaux 1986).

Only one lichen (*Micarea adnata*) is known where the anamorph is a **sporodochium**. Sporodochia are characterized by palisade-like arranged conidiophores, developing on or reaching the surface of the lichen where they release conidia directly (Fig. 4.27).

Hyphophores are a special anamorph structure known from the myco-biont of many foliicolous lichens. They have a multihyphal stalk and often an apical thickening, from which conidiophores grow downwards (Vězda 1979).

Thallospores are special asexual spores having no conidiophore. They are known from *Umbilicaria* (Hasenhüttl and Poelt 1978; Hestmark 1990), as well as from various crustose species (Poelt and Obermayer 1990a).

Propagation of the aposymbiotic photobionts is seemingly rather rare (Tschermak-Woess 1988). **Flagellate stages** of unicellular green algae have been reported from *Flavoparmelia caperata* (Slocum *et al.* 1980) and from *Anzina carneonivea* (Scheidegger 1985). However, motile stages of the photo-bionts are usually absent or very rarely found in lichens, although they are regularly found in the free-living, cultured state of algae (Tschermak-Woess 1988).

Hormogonia are short chains of cyanobacterial cells that act as diaspores. The cyanobacterial photobiont of *Placynthium nigrum* forms hormogonia after periods of high rainfall, and these may escape the lichen thallus (Geitler 1934).

Symbiotic propagules

Vegetative diaspores provide a means of propagating the whole lichen-ized ascomycetes and basidiomycetes without relichenization. A wide range of ontogenetically and/or structurally different vegetative propagules have been described (Poelt 1993). However, only a few are routinely considered, and of these, isidia and soredia are the most important (Table 4.1).

In most lichens, **fragments** are not normally able to establish and regen-erate a thallus. However, many fruticose lichens are highly adapted to propa-gate by thallus fragments. Beard-like, epiphytic thalli of the genera *Bryoria* and some *Ramalina* are torn and dispersed by strong winds. The fragments with lengths often exceeding 10 cm get entangled with the foliage and branches of their new substrate. Some *Cladonia* species are very brittle in the dry state and are often fragmented after trampling. Dibben (1971) has successfully used such fragments as innoculum for growing *Cladonia* species in phytotron experiments.

Goniocysts are minute granular thalli (Fig. 4.4) with external monolayered paraplectenchyma and numerous photobiont cells. The formation of goniocysts in the genus *Vezdaea* may occur by serial *de novo* lichenization (Scheidegger 1994b) or by ongoing proliferation of a subcuticular thallus (Tschermak-Woess and Poelt 1976) where soredia-like propagules are formed (Vězda 1980; Sérusiaux 1985).

Table 4.1. *Symbiotic and aposymbiotic vegetative propagules*

Structure	Definition
Goniocyst	Granular thalli with external monolayered paraplectenchymatous fungal plectenchyme and numerous photobiont cells
Fragment	Detached terminal or marginal parts of thallus
Phyllidium	Corticate, dorsiventral protuberance of upper cortex
Bulbil	Multilayered paraplectenchymatous globular outgrowth with only a few algal cells
Blastidium	Pseudocorticate budding proliferations of upper or lower phenocortex
Isidium	Corticate protuberance of upper cortex and photobiont layer
Pseudoisidium	Isidia-like structures which lack photobiont cells
Polysidium	Corticate protuberance of thalline outgrowth
Schizidium	Flakes of upper cortex and algal layer
Soredium	Noncorticate clump originating in the medulla and photobiont layer
Consoredium	Aggregations of incompletely separated soredia
Parasoredium	Clump of disintegrating (pheno?)cortex and photobiont layer
Isidioid soredium	Secondarily corticate protuberance produced in soralia-like clusters
Dactylidium	Phenocorticate nonbudding protuberance of upper cortex and photobiont layer
Thlasidium	Isidia-like structure internally producing soredia-like granules which are squeezed out
Hormocyst	Hormogonium with adhering mycobiont
Conidium	Aposymbiotic, nonmotile, asexual fungal spore; formed within or on a conidioma
Thallospore	Aposymbiotic, asexual fungal spore produced on the thallus surface or on attachment organs
Hormogonium	Aposymbiotic, few-celled motile trichomes without heterocysts
Zoospore	Aposymbiotic, motile daughter cells of the phycobiont

Isidia are scattered across the thallus surface, and their height ranges from around 30 μm to more than 1 mm. Isidia are often cylindrical and simple (Figs. 4.5, 4.23) or branched, but warty or coralloid forms are also known. Although they may serve as diaspores in many species, isidia may also play an important role in increasing thallus surface area. Because numerous isidia are usually present on a thallus surface without being detached, they both increase the photosynthetically active surface (Chapter 9) and enhance other interactions with the atmosphere (e.g. trace gas exchange, aerosol deposition; Chapters 11 and 12).

Pseudoisidia differ from isidia by the lack of photobiont cells (Walker 1985). Therefore, they would be better treated as fungal (aposymbiotic) propagules, although their function is not fully understood.

Phyllidia resemble isidia in very early developmental stages but they soon bend over the parent thallus and become dorsiventral. Usually phyllidia are constricted at the base and become easily detached. In contrast, **lobules** are similar morphologically, but do not usually act as reproductive propagules. Phyllidia are stratified in a similar manner to the parent thallus (i.e. with upper and lower cortices, a photobiont layer and a medulla).

Polysidia are clustered isidia, formed on thalline outgrowths. They have only been reported from *Pyxine* (Kalb 1987).

Thlasidia morphologically resemble pseudoisidia in their terminal ends but contain soredia-like patches with photobiont cells in their bases. Ontogenetically they emerge from the thlasidium. They are only reported from the crustose, epiphytic lichen *Gyalideopsis anastomosans* (Vězda 1979; Poelt 1986).

Bulbils are multilayered paraplectenchymatous globular outgrowths of the thallus and have been found in lichenized Basidiomycetes. They contain only a few photobiont cells (Poelt and Obermayer 1990*b*).

Schizidia are formed from upper thalline layers by disintegrating flakes of cortical and algal layers. Schizidia are found in crustose lichens (e.g. the genus *Fulgensia*) and also occur in foliose lichens, such as *Flavoparmelia caperata*, *Hypogymnia* and *Xanthoria* (Poelt 1980, 1994).

Soredia consist of a few photobiont cells enveloped by a loose, spherical mantle of hyphae. Soredia are formed by proliferation of the algal and medullary layers. They are very small and often range from 20 to 50 μm in diameter. Soredia either develop diffusely on the upper surface of the thallus or in delimited areas, called **soralia** (Figs. 4.6, 4.24). Soralia covered by small soredia are **farinose** and those with larger soredia are **granular**. According to their position soralia are classified as laminal, marginal, fissural, cuff-shaped, vaulted, labriform or terminal. Soredial masses are loosened from soralia and scattered by hygroscopical movements of the cortical tissue (Jahns *et al.* 1976). Soralia are often hydrophobic, but in repelling raindrops soredia may be removed as well. In many species of *Lobaria*, *Melanelia* and other genera, soredia may remain attached in the soralium after their formation. Such soredia may later develop an outer layer of more or less agglutinated hyphae until a cortex or a phenocortex is formed. Such structures are soredia in origin and are called **isidioid soredia** (Esslinger 1977).

Consoredia, a term recently introduced by Tønsberg (1992), refer to aggregated soredia that are formed by **incomplete division** of parent soredia.

Parasoredia have a different ontogeny from soredia. They are formed in layers from clumps of disintegrating cortical and photobiont material and are often bigger than soredia (Codogno *et al.* 1989).

Blastidia are **pseudocorticate**, budding proliferations of the upper or the lower cortex, and sometimes they resemble soredia (Poelt 1980, 1994).

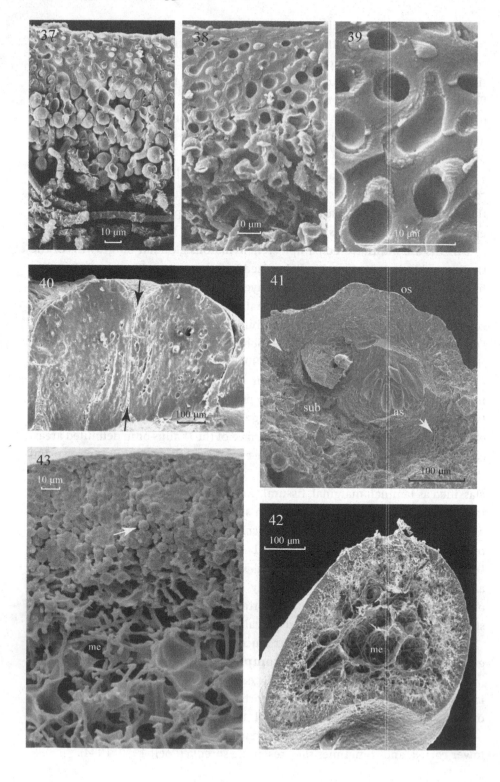

The structures described above and summarized in Table 4.1 are often multi-functional organs, for which structure and function may change during their ontogeny. For instance *Parmotrema crinitum* develops cylindrical isidia (Fig. 4.23) that may serve as diaspores after they become mechanically detached from the thallus. The same structures, however, may develop further on the thallus into multibranched isidia or even into dorsiventral lobules (Ott *et al.* 1993) that may be interpreted as phyllidia. In *Lobaria pulmonaria* laminal and marginal lobules develop from isidioid soredia. In the process their function is changed from that of a dispersal propagule to a regenerative structure that enables the re-establishment of a new thallus where the old one grew.

4.4   Evolution of the lichen thallus morphology and resulting functional aspects

Among all symbiotic associations, the lichen symbiosis is unique as the host (mycobiont) expresses its specific phenotype only in association with an adequate photobiont (see also Chapter 5). Thallus morphology is optimized for water uptake and/or loss within a specific habitat, and the anatomical structure of the thallus must follow principles that allow optimal carbon dioxide gain of the photobiont under a given habitat moisture regime.

4.4.1   *Morphology and anatomy of fossil lichens*

Since there is only a very limited record on fossil lichens, we do not know much about the morphology and anatomy of early lichens and how they evolved. The oldest fossil lichen on record is 600 million years old and comes from

Caption for Figs. 4.37–4.43
Functional morphology, fossil record. Figs. 4.37–4.39. *Lobaria pulmonaria* (Switzerland, corticolous), cross fractures of dry thallus showing cortical and medullary hyphae with cavitations. (LTSEM.) Fig. 4.40. *Spongiophyton minutissimum*, fossil lichen, vertical section through upper half of the thallus of two neighboring lobes (arrows show lobe margins). (SEM, with permission of *International Journal of Plant Science* and W. A. Taylor.) Fig. 4.41. *Verrucaria elaeina* (Germany, saxicolous, submerged), wet thallus of the aquatic lichen, vertical section through perithecium showing asci (as) with spores and the ostiole (os). Paraplectenchymatic thallus structure with airspaces (arrows); sub, substratum. (LTSEM, courtesy of Dr. H. Thues.) Fig. 4.42. *Peltula tortuosa* (South Africa, saxicolous), cross section through an erect thallus lobe (submerged in water for 6 hours). Medulla (me) still filled with airspaces. (LTSEM.) Fig. 4.43. *Peltula tortuosa* (South Africa, saxicolous), cross section through thallus margin and medulla (submerged in water for 6 hours). Cyanobiont layer and medulla remain free of capillary water, cyanobiont cells turgid (arrow). (LTSEM.)

China (Yuan *et al.* 2005). In terms of morphology, the most convincing report came from the Lower Devonian Rhynie Cherts (*c.* 400 million years old) and showed the structure of a gelatinous thallus of the lichen *Winfrenatia reticulata*, formed by the interaction of nonseptate fungal hyphae surrounding clusters of cyanobiont cells resembling unicellular cyanobacteria of the genera *Gloeocapsa* or *Chroococcidiopsis* (Taylor *et al.* 1997). Another Devonian lichen, *Spongiophyton minutissimum*, is reported from the Middle Devonian of Brazil and Ghana, and the Lower Devonian of Bolivia and Canada (Taylor *et al.* 2004). The thallus of this fossil formed by dichotomously to irregularly branching lobes about 1 mm wide between branching points is also of the compact gelatinous type (compare also *Collema*; Fig. 4.13), with no or only minor airspaces included (Fig. 4.40). From this, one could conclude that the first lichen thalli had cyanobionts and evolved in amphibious habitats where they frequently became submerged. The fact that many modern aquatic lichens, e.g. the genus *Verrucaria*, still have a compact and sometimes even gelatinous structure (Fig. 4.41) supports that hypothesis.

There is a big gap between the Devonian lichen records and the next available fossils. A younger record from tuff rock of the middle Miocene (12–24 million years) can be attributed to the Lobariaceae (Peterson 2000). Another rich source for lichen fossils are amber inclusions which date back as far as 55 to 35 million years and less. However, these lichens very much resemble recent species and already have the morphological features of modern lichens (e.g. Rikkinen 2003).

4.5   Outlook

Although the principal aspects of morphology have been known for many years, functional aspects of lichen morphology are still poorly known. Understanding the functional role of morphology is of major importance to ecophysiology (Chapter 9), reproductive biology (Chapter 6), and systematics (Chapter 17).

5

Morphogenesis

R. HONEGGER

This chapter focuses on the development of the lichen thallus and on key factors playing a role in this fascinating process. The term morphogenesis is derived from Greek, in which "morph" means form and "genesis" means origin or creation.

5.1 Acquisition of a compatible photobiont

Lichen-forming fungi express their symbiotic phenotype (produce thalli with species-specific features) only in association with a compatible photobiont. About 85% of lichen mycobionts are symbiotic with green algae, about 10% with cyanobacteria ("blue-green algae") and about 3–4%, the so-called cephalodiate species, simultaneously with both green algae and cyanobacteria (Tschermak-Woess 1988; Peršoh et al. 2004). Lichen photobionts are extracellularly located within lichen thalli. The enigmatic Geosiphon pyriforme, the only representative of Glomeromycota which does not form arbuscular mycorrhizae with plants (Schüssler et al. 2001; Schüssler 2002), is not normally considered a lichen. In this endocyanosis the uptake of cyanobacterial filaments (Nostoc punctiforme) into plasma-membrane-bound vesicles by the fungal protoplast can be readily studied, a feature of considerable interest in cell biology with regard to chloroplast acquisition (Mollenhauer 1992; Schüssler et al. 1995; Gehrig et al. 1996; Mollenhauer et al. 1996).

A wide range of so-called cyanotrophic lichens form loose, associative symbioses with free-living cyanobacteria (Poelt and Mayrhofer 1988), and bacterial films are regularly found on thallus surfaces of lichens (Fig. 5.15). The biology of these associative symbioses has not been experimentally explored. It is highly probable that many of these prokaryotic epibionts are diazotrophic, i.e. capable

Lichen Biology, ed. Thomas H. Nash III. Published by Cambridge University Press.
© Cambridge University Press.

of fixing nitrogen (N$_2$) and providing thalli with fixed nitrogen. Other bacterial epibionts might produce hormones or other biologically active compounds. Plant beneficial bacteria and mycorrhization helper bacteria were shown to play important roles in ecosystems (Davison 1988; Garbaye 1994; Haas and Keel 2003; Aspray *et al.* 2006), but future investigations are needed to establish whether there are also lichen beneficial or lichenization helper bacteria. Only recently was a first molecular analysis on the taxonomic affiliation of lichen-associated bacteria published (Cardinale *et al.* 2006).

The majority of lichen-forming fungi reproduce sexually and thus have to re-establish the symbiotic state at each reproductive cycle. Compatible photobiont cells are not normally dispersed together with ascospores; exceptions are found in a few species of Verrucariales with hymenial photobionts (e.g. *Endocarpon pusillum*). Many tropical lichen-forming fungi associate with green-algal taxa that are widespread and common in the free-living state (e.g. *Cephaleuros*, *Phycopeltis*, *Trentepohlia*; see Chapter 3), but contradictory views are found in the literature concerning the abundance of representatives of the genus *Trebouxia* in the free-living state. This is the most common and widespread genus of lichen photobionts in temperate regions and especially in extreme climates such as arctic, alpine, antarctic or desert ecosystems. For a long time *Trebouxia* spp. were postulated to be rare or even missing outside lichen thalli (Ahmadjian 1988, 2001). Based on the assumption of scarcity of free-living photobionts some common and widespread, sexually reproducing *Xanthoria* species were hypothesized to acquire compatible photobionts "by theft," their ascospore-derived germ tubes entering the symbiotic propagules or thalli of adjacent *Physcia* spp. and associating with their photobiont (Ott 1987a,b; Ott *et al.* 2000b). However, in the phycological literature *T. arboricola*, type species of this genus and photobiont of innumerable lichen-forming ascomycetes, is common and widespread in the free-living state; the same seems to apply for other *Trebouxia* species (Ettl and Gärtner 1984; John *et al.* 2002; Rindi and Guiry 2003). In a series of elegant *in situ* studies on relichenization processes, carried out on microscopy slides that were incubated in lichen communities, Sanders and Lücking (2002) and Sanders (2005) showed that free-living *Trebouxia* cells are abundant and readily available to germ tubes derived from ascospores of lichen mycobionts, *Xanthoria* spp. included. These authors confirmed earlier reports on the abundance of free-living *Trebouxia* spp. (Bubrick *et al.* 1984; Mukhtar *et al.* 1994). Thus there is no need for "cleptobiosis" in *Xanthoria* spp., i.e. theft of photobiont cells from adjacent *Physcia* spp., as postulated by Ott and coworkers (Ott 1987a, b; Ott *et al.* 2000b). Molecular analyses clearly show that *Xanthoria* spp. do not associate with photobiont taxa contained in thalli of *Physcia* spp. of the Physcietum adscendentis (Beck *et al.* 1998; Dahlkild *et al.* 2001; Helms *et al.* 2001; Helms 2003; Nyati 2006).

A large number of lichen-forming fungi produce asexual symbiotic propagules such as soredia, isidia, or blastidia (see Chapter 4), and dispersal of the symbiotic system via thallus fragmentation appears to be more common than has been previously assumed. The role of vertebrates and invertebrates as vectors of symbiotic propagules, thallus fragments or living cells of lichen-forming fungi and their photobionts remains to be explored. Fecal pellets of the almost ubiquitous lichenivorous mites were shown with culturing experiments and molecular techniques to contain viable ascospores and *Trebouxia* cells (Meier *et al.* 2002); they seem to be almost ideal propagules for short and long distance dispersal of the symbiotic system. Intact cells of lichen photobionts were also found in fecal pellets of lichenivorous snails (Fröberg *et al.* 2001).

5.2   Recognition and specificity

About 100 species of cyanobacterial and algal photobionts are known to science. Among the few marine lichen-forming ascomycetes a small number of species associate with Xanthophyceae or Phaeophyceae (Tschermak-Woess 1988; Sanders *et al.* 2004). Lichen photobionts are less intensely studied than their fungal partners. In less than 2% of lichens has the photobiont ever been identified at the species level; fairly often not even the generic affiliation is known. Many algal or cyanobacterial species are compatible photobionts of large numbers of lichen-forming fungal species from different families; others have only rarely been found in lichen symbiosis. The taxonomic range of acceptable photobionts per lichen-forming fungus can be explored in different ways. Most lichen-forming fungi and their photoautotrophic partners can be isolated into sterile culture and grown in the aposymbiotic state (see the Appendix). Coculturing of fungal and algal isolates would be an elegant mode of studying the range of compatible photobionts per fungal species. Unfortunately the symbiotic phenotype is not routinely expressed in axenic cultures under laboratory conditions, as the factors triggering morphogenesis in compatible combinations are very poorly understood. Due to these problems most investigators explore the range of compatible photobiont species per fungal species by examining the algal or cyanobacterial partner contained in lichen thalli that had been collected in the wild, ideally from a very wide area of distribution. Before the advent of molecular techniques this was a difficult and time-consuming task, which included isolation and sterile culturing of the photobiont under defined conditions and comparison with axenically cultured type strains using light microscopy. Thus only a few experts worldwide were able to identify lichen photobionts at the species level. Today cyanobacterial- or algal-specific primers are used to amplify especially informative photobiont of fixing nitrogen (N_2) and providing thalli with fixed nitrogen. Other bacterial epibionts might produce hormones or other biologically active compounds. Plant beneficial bacteria and mycorrhization helper bacteria were shown to play important roles in ecosystems (Davison 1988; Garbaye 1994; Haas and Keel 2003; Aspray *et al.* 2006), but future investigations are needed to establish whether there are also lichen beneficial or lichenization helper bacteria. Only recently was a first molecular analysis on the taxonomic affiliation of lichen-associated bacteria published (Cardinale *et al.* 2006).

The majority of lichen-forming fungi reproduce sexually and thus have to re-establish the symbiotic state at each reproductive cycle. Compatible photobiont cells are not normally dispersed together with ascospores; exceptions are found in a few species of Verrucariales with hymenial photobionts (e.g. *Endocarpon pusillum*). Many tropical lichen-forming fungi associate with green-algal taxa that are widespread and common in the free-living state (e.g. *Cephaleuros, Phycopeltis, Trentepohlia*; see Chapter 3), but contradictory views are found in the literature concerning the abundance of representatives of the genus *Trebouxia* in the free-living state. This is the most common and widespread genus of lichen photobionts in temperate regions and especially in extreme climates such as arctic, alpine, antarctic or desert ecosystems. For a long time *Trebouxia* spp. were postulated to be rare or even missing outside lichen thalli (Ahmadjian 1988, 2001). Based on the assumption of scarcity of free-living photobionts some common and widespread, sexually reproducing *Xanthoria* species were hypothesized to acquire compatible photobionts "by theft," their ascospore-derived germ tubes entering the symbiotic propagules or thalli of adjacent *Physcia* spp. and associating with their photobiont (Ott 1987*a,b*; Ott *et al.* 2000*b*). However, in the phycological literature *T. arboricola*, type species of this genus and photobiont of innumerable lichen-forming ascomycetes, is common and widespread in the free-living state; the same seems to apply for other *Trebouxia* species (Ettl and Gärtner 1984; John *et al.* 2002; Rindi and Guiry 2003). In a series of elegant *in situ* studies on relichenization processes, carried out on microscopy slides that were incubated in lichen communities, Sanders and Lücking (2002) and Sanders (2005) showed that free-living *Trebouxia* cells are abundant and readily available to germ tubes derived from ascospores of lichen mycobionts, *Xanthoria* spp. included. These authors confirmed earlier reports on the abundance of free-living *Trebouxia* spp. (Bubrick *et al.* 1984; Mukhtar *et al.* 1994). Thus there is no need for "cleptobiosis" in *Xanthoria* spp., i.e. theft of photobiont cells from adjacent *Physcia* spp., as postulated by Ott and coworkers (Ott 1987*a, b*; Ott *et al.* 2000*b*). Molecular analyses clearly show that *Xanthoria* spp. do not associate with photobiont taxa contained in thalli of *Physcia* spp. of the Physcietum adscendentis (Beck *et al.* 1998; Dahlkild *et al.* 2001; Helms *et al.* 2001; Helms 2003; Nyati 2006).

A large number of lichen-forming fungi produce asexual symbiotic propagules such as soredia, isidia, or blastidia (see Chapter 4), and dispersal of the
symbiotic system via thallus fragmentation appears to be more common than
has been previously assumed. The role of vertebrates and invertebrates as
vectors of symbiotic propagules, thallus fragments or living cells of lichen-
forming fungi and their photobionts remains to be explored. Fecal pellets of
the almost ubiquitous lichenivorous mites were shown with culturing experiments and molecular techniques to contain viable ascospores and *Trebouxia* cells
(Meier *et al.* 2002); they seem to be almost ideal propagules for short and long
distance dispersal of the symbiotic system. Intact cells of lichen photobionts
were also found in fecal pellets of lichenivorous snails (Fröberg *et al.* 2001).

5.2   Recognition and specificity

About 100 species of cyanobacterial and algal photobionts are known to
science. Among the few marine lichen-forming ascomycetes a small number of
species associate with Xanthophyceae or Phaeophyceae (Tschermak-Woess
1988; Sanders *et al.* 2004). Lichen photobionts are less intensely studied than
their fungal partners. In less than 2% of lichens has the photobiont ever been
identified at the species level; fairly often not even the generic affiliation is
known. Many algal or cyanobacterial species are compatible photobionts of
large numbers of lichen-forming fungal species from different families; others
have only rarely been found in lichen symbiosis. The taxonomic range of
acceptable photobionts per lichen-forming fungus can be explored in different
ways. Most lichen-forming fungi and their photoautotrophic partners can be
isolated into sterile culture and grown in the aposymbiotic state (see the
Appendix). Coculturing of fungal and algal isolates would be an elegant mode
of studying the range of compatible photobionts per fungal species.
Unfortunately the symbiotic phenotype is not routinely expressed in axenic
cultures under laboratory conditions, as the factors triggering morphogenesis in
compatible combinations are very poorly understood. Due to these problems
most investigators explore the range of compatible photobiont species per
fungal species by examining the algal or cyanobacterial partner contained in
lichen thalli that had been collected in the wild, ideally from a very wide area of
distribution. Before the advent of molecular techniques this was a difficult and
time-consuming task, which included isolation and sterile culturing of the
photobiont under defined conditions and comparison with axenically cultured
type strains using light microscopy. Thus only a few experts worldwide were
able to identify lichen photobionts at the species level. Today cyanobacterial- or
algal-specific primers are used to amplify especially informative photobiont

sequences in DNA extracts of entire thalli. Resulting sequences are compared with sequence data of type strains and others contained in gene data banks. With molecular techniques not only the range of compatible photobionts per fungal taxon can be elucidated, but also the phylogeny of the fungal and photo-autotrophic partners of lichen symbiosis (examples in Kroken and Taylor 2000; Helms 2003; Yahr *et al.* 2004, 2006).

The majority of genera of lichen-forming fungi associate with photobionts of one genus (Rambold *et al.* 1998). The situation of cephalodiate species is discussed under Section 5.5. Among the noteworthy exceptions are *Euopsis granatina* (Lichinaceae), whose thalli harbor colonies of the cyanobacterium *Gloeocapsa sanguinea* and of the green alga *Trebouxia aggregata* beside bacterial colonies (Büdel and Henssen 1988), or *Muhria urceolata* (Stereocaulaceae), which comprises areoles with green-algal photobionts overlying a basal hypothallus-like mat containing cyanobacterial filaments (*Microcystis* and *Stigonema* spp.; Jørgensen and Jahns 1987). Both examples differ from cephalodiate taxa by not keeping their cyanobacterial partners in confined gall-like structures (see Section 5.5). A remarkable situation was observed in the genus *Chaenotheca*, whose representatives associate with green-algal photobionts from different genera within the Trebouxiophyceae (*Dictyochloropsis*, *Trebouxia*, *Stichococcus*) and Ulvophyceae (*Trentepohlia*); however, each *Chaenotheca* species associates with representatives of only one algal genus (Tibell 2001; Tibell and Beck 2001).

Some investigators kept claiming that lichen symbiosis is not specific because sterile cultured lichen-forming fungi tend to overgrow whatever they encounter (Ahmadjian 1988). However, the symbiotic phenotype is expressed only in symbiosis with a compatible photobiont (see Table 5.2). Based on the observation that very common and widespread species of aerophilic algal communities are not compatible partners of lichen-forming fungi and on the assumption that *Trebouxia* spp., the most common lichen photobionts, are rare outside lichen thalli the term selectivity was introduced by Galun and Bubrick (1984). Accordingly, specificity refers to the degree of taxonomic relatedness of compatible partners (Smith and Douglas 1987), and selectivity to their availability in natural ecosystems. In the recent literature both terms are confused (e.g. Rambold *et al.* 1998). Yahr *et al.* (2006) define specificity as the range of genetically compatible photobionts; selectivity as the choice of the ecologically optimal genotype among the range of compatible taxa. Data sets from molecular investigations and from field studies, as performed in the last 10 years, significantly improved our knowledge on the availability of photobionts and on the specificity of the lichen symbiosis. However, quite often data sets cannot be compared or combined when investigators focus on different molecular marker sequences. Considerable progress is expected in the coming years.

Table 5.1. *Specificity and selectivity in morphologically complex macrolichens*

Specificity	(levels of specificity as summarized by Smith and Douglas, 1987)		
Very high	Acceptable partners found within a subspecific taxon (strain, variety)		
	No known examples in lichens		
High	One species is acceptable.		
Examples:	*Parmelia sulcata*[1]	Photobiont:	*Trebouxia impressa*
	n = 27; Europe, N. America		
	Pleurosticta acetabulum[1]	Photobiont:	*Trebouxia arboricola*
	n = 19; central Europe		
Moderate	Acceptable partners found with one genus.		
Examples:	*Anzia opuntiella*[2]	Photobionts:	*Trebouxia gelatinosa* (12/39)
	n = 39; Japan		*T. potteri* (16/39) *T. showmanii* (11/39)
	Punctelia subrudecta[1]	Photobionts:	*Trebouxia arboricola* (3/13)
	n = 13; central Europe		*T. gelatinosa* (9/13) *T. jamesii* (1/13)
Low to very low	Acceptable partners found within the same phylum or within different phyla.		
Examples:	Photosymbiodemes and cephalodiate species: lichen mycobionts (c. 3–4% of lichen-forming fungi) which associate simultaneously or consecutively with selected green-algal and cyanobacterial photobionts (see text for further explanations).		
Selectivity	(see Galun and Bubrick, 1984)		
Low	Large numbers of unrelated taxa are acceptable partners.		
	No known examples in the lichen symbiosis		
High	Few, possibly unrelated, often quite rare taxa are acceptable partners.		
Examples:	(a) *Trebouxia* spp. (unicellular green algae) are photobionts of > 50% of lichens, rarely found outside lichen thalli. Their mycobionts are moderately to highly specific and highly selective with regard to photobiont acquisition (see above).		
	(b) Lichen mycobionts which form photosymbiodemes or cephalodia associate with very few, selected green-algal and cyanobacterial photobionts, thus revealing low specificity but high selectivity.		

Most mycobionts of morphologically complex macrolichens are HIGHLY to MODERATELY SPECIFIC and HIGHLY SELECTIVE with regard to photobiont acquisition.

n, Number of thalli investigated.

Sources: 1. Friedl (1989b); 2. Ihda *et al.* (1993).

Presently available data indicate that morphologically advanced taxa of lichen-forming fungi (the ones which form foliose or fruticose thalli with internal stratification) are moderately specific to specific; most green-algal lichens associate with different genotypes of one to a few morphospecies belonging to the same *Trebouxia* clade or subclade *sensu* Helms *et al.* (2001). Examples include various *Letharia* spp., all symbiotic with genotypes (referred to as phylospecies) of *Trebouxia jamesii* (Kroken and Taylor 2000), or *Physcia, Phaeophyscia, Phaeorrhiza, Physconia* and *Hyperphyscia* spp., all symbiotic with genotypes of *T. impressa* in subclade I1 (Dahlkild *et al.* 2001; Helms 2003). Lower specificity was detected in *Umbilicaria antarctica* and *U. kappeni* collected along a transect on the Antarctic Peninsula. These morphologically advanced, umbilicate species associated with *Trebouxia* spp. from two different *Trebouxia* clades in different regions of the Antarctic (Romeike *et al.* 2002). In temperate regions, morphologically advanced cyanobacterial lichens select *Nostoc* spp. from the same monophyletic lineage within the genus; they share some of their photobiont genotypes with representatives of other cyanobacterial symbioses, such as liverworts, hornworts, or angiosperms (Rikkinen *et al.* 2002; O'Brien *et al.* 2005). A lower cyanobiont selectivity was documented in gelatinous and squamulose cyanobacterial lichens and in cephalodiate green-algal lichens in the Antarctic (Wirtz *et al.* 2003).

Some of the morphologically less advanced fungal species, which form crustose, nonstratified thalli, were less specific since they associate with photobiont taxa from different clades within the genus *Trebouxia sensu stricto*. Examples include (1) *Rinodina oxydata*, which is symbiotic with photobionts of the *impressa* clade (subclade I5) and the *arboricola* clade (subclades A6 and A10) (Helms 2003), (2) *Lecanora rupicola*, which associates with photobionts of the *arboricola* clade (*T. decolorans, T. incrustata* and unidentified *Trebouxia* spp.), the *impressa* clade (*T. impressa*), and the *simplex* clade (*T. simplex*) (Blaha *et al.* 2006).

Molecular techniques facilitate the study of photobiont diversity within lichen communities (Beck *et al.* 1998, 2002; Beck 1999; Schaper and Ott 2003); Representatives of the same algal clade, species or even genotype may be contained in thalli of unrelated lichen-forming ascomycetes within the community. An example is the Physcietum adscendentis on eutrophicated bark (Beck *et al.* 1998), in which *Trebouxia arboricola* of the *arboricola* clade *sensu* Helms *et al.* (2001) is the photobiont of ascospore-producing taxa such as *Lecania cyrtella, Lecanora* spp., *Lecidella elaeochroma* (all Lecanoraceae) and *Xanthoria parietina* (Teloschistaceae). Adjacent sorediate, usually sterile *Physcia adscendens* and *Phaeophyscia orbicularis* associate with *T. impressa* of the *impressa* clade. With high probability populations of free-living *T. arboricola* and related taxa from the *arboricola* clade are present on such substrates and available to ascospore-derived germ tubes of lichen-forming fungi (Beck *et al.* 1998; Sanders 2005). Soredia were shown to be wind-dispersed over long distances (Marshall 1996; Tormo *et al.* 2001; Muñoz *et al.* 2004). Molecular tools facilitate the analysis of genetic variation among mycobionts and photobionts within populations of sexually reproducing (re-lichenizing) and vegetatively propagating lichen species, i.e. sorediate, isidiate taxa, or species dispersing mainly by thallus fragmentation in undisturbed and disturbed collecting sites (examples in Piercey-Normore 2004, 2006; Yahr *et al.* 2004, 2006; Ohmura *et al.* 2006).

Why are the majority of lichen-forming ascomycetes symbiotic with repre-sentatives of Trebouxiophyceae, especially Trebouxiales? This class of green algae, also referred to as "lichen algae group" (Lewis and McCourt 2004), com-prises also symbionts of protists (Chlorellales) and invertebrates beside numer-ous nonsymbiotic taxa. *Chlorella* photobionts of freshwater invertebrates and protists differ from nonsymbiotic *Chlorella* spp. by their continuous release of photosynthates. As shown in resynthesis experiments, this key feature has a signaling effect in the recognition process, but is a serious disadvantage in competitive interactions with nonsymbiotic algae outside the C-heterotrophic exhabitant (Smith and Douglas 1987; Reisser 1992). So far it has not been possible to identify such key features by which lichen photobionts differ from other algae and cyanobacteria. In the symbiotic state, the majority of green-algal lichen photobionts produce and release acyclic polyols such as ribitol, erythri-tol, and sorbitol, depending on their taxonomic affiliation (Smith and Douglas 1987), but it remains unclear whether they do so in the aposymbiotic state in natural ecosystems. Lichen-forming fungi also produce acyclic polyols (mainly mannitol; Smith and Douglas 1987), but different ones from their photobionts. Acyclic polyols serve as compatible solutes in drought-stressed lichen-forming and nonlichenized fungi, plants, algae, archae- and eubacteria (Jennings 1995; Kets *et al.* 1996; Magan 1997; Martin *et al.* 1999). It is particularly interesting that both partners of green-algal lichens produce the same group of compounds as osmoregulators.

Establishment of a symbiotic relationship is a multistep process, involving preformed compounds and newly synthesized ones. It necessitates the induc-tion and suppression of gene expression in both partners. Only a few studies have explored relichenization events with molecular tools. SEM techniques and cDNA-AFLP were used to investigate gene expression profiles during early re-lichenization of sterile cultured *Baeomyces rufus* and *Elliptochloris bilobata* (Trembley *et al.* 2002c). After one day of coculturing, the photobiont cells were nonspecifi-cally bound to hyphal surfaces by mucilaginous secretions of the mycobiont. After 12 days coculturing the mycobiont started to enwrap the photobiont, and after 28 days the algal cells were enclosed in soredia-like clusters. The analysis of total mRNA on the twelfth day of resynthesis revealed induction of a few, unidentifiable genes and suppression of many others in both partners (Trembley *et al.* 2002*c*). Among potentially preformed substances mediating an initial, often unspecific binding are mycobiont-derived lectins (sugar-specific, cell agglutinating proteins) and other algal-binding proteins (Bubrick and Galun 1980*a*, *b*; Hersoug 1983; Kardish *et al.* 1991; Legaz *et al.* 2004). One of these unidentified algal-binding proteins was located with immunocytochemical techniques on the wall surface of ascospore-derived germ tubes of *Xanthoria parietina* (Bubrick *et al.* 1981). Later developmental stages of morphogenesis in lichens are known from purely descriptive studies only and remain unexplored at the molecular level.

5.3   Morphogenesis

The majority of lichen-forming fungi produce morphologically simple, crustose thalli with no internal stratification (see Chapter 3; Table 5.2). Macrolichens, including foliose and fruticose species with their internally stratified thallus, are particularly interesting because they are the morphologically most complex vegetative structures in the fungal kingdom. Such internally stratified lichen thalli are the result of an amazing hyphal polymorphism, which includes either filamentous (polar) or globose (apolar) hyphal growth in combination with hydrophilic or hydrophobic (water-repellent) wall surfaces. Main building blocks of internally stratified lichen thalli are (1) tissue-like areas (pseudoparenchyma) formed by fungal cells which are tightly adhering (conglutinate) to each other by means of gelatinous, hydrophilic cell wall material, and (2) loosely interwoven aerial hyphae (plectenchyma) with hydrophobic wall surfaces, some of which are in close contact with the photobiont cells. The gelatinous material (mostly β-glucans) of conglutinate pseudoparenchyma is soft, flexible and translucent when wet but firm, opaque and often quite brittle when dry. Conglutinate pseudoparenchyma in cortical layers and/or in internal strands (Figs. 3.3, 3.14, 5.15, 5.16) provide mechanical stability to the thallus while the system of aerial hyphae with hydrophobic wall surfaces creates gas-filled zones either in the medullary layer in the thalline interior or at the periphery (Figs. 5.14–5.16). As an adaptation to the symbiotic way of life, the thalli of most foliose and fruticose lichens have aerial hyphae in the interior and conglutinate, tissue-like zones at their periphery. The same fungi, when isolated and grown in sterile culture, form conglutinate thallus-like structures with aerial hyphae at their periphery (Fig. 3.3). In the symbiotic phenotype, photobiont cells are optimally positioned with regard to illumination and gas exchange at the periphery of the gas-filled thalline areas right underneath the

Table 5.2. *Lichenization and morphogenesis in morphologically complex macrolichens*

→ germ tubes, free hyphae, rhizomorphs
↓

NON SPECIFIC CONTACT with either
↓
   compatible aposymbiotic or symbiotic[a] photobiont cells, or with
     incompatible algae or cyanobacteria
     ↓ ↓

→ PRETHALLUS stage[b]: an inconspicuous, nonstratified crust
    Incompatible associations will not develop beyond this stage
    unless compatible photobiont cells become available

compatible associations only

> **Unknown stimuli, triggering the phenotypic expression (morphotype, chemotype) of the fungal genotype**

SYMBIOTIC PHENOTYPE: STRATIFIED THALLUS
    stratification: differentiation of **conglutinate** zones (usually as peripheral
      cortex) and **gas-filled** thalline areas (as internal medulla and algal layer)
    polarization of the thalline primordium; differentiation of a growing **apical**
      and a nongrowing **basal** thalline area
    short distance shifting of photobiont cells into and within the algal layer
    induction of secondary metabolism in the fungal partner, possibly as a
      response to the establishment of an appropriate **nutritional basis**.
      Secondary metabolites crystallize either at the surface of aerial hyphae or
      within the gelatinous matrix of the cortical layer
  → *regulation of growth and cell turnover* in both partners

MATURE THALLUS: REPRODUCTION AND DISPERSAL
    MYCOBIONT:
  — **ascospores:** *sexual reproductive stages*
  — **conidia[c], brood grains:** *asexual reproduction*
    SYMBIOTIC SYSTEM *(asexual symbiotic propagules):*
  — **soredia, nondifferentiated isidia**
  — **internally stratified isidia, phyllidia, thallus fragments**

---

[a] Symbiotic photobionts of either prethallus stages, thalli or symbiotic propagules of vicinal
   lichen species.

[b] The prethallus stage of development has also been termed: thalle primaire, soredial stage,
   presquamules, disque primaire, Grundgewebe, basal tissue, preliminary phase.

[c] Macroconidia are seldom produced in macrolichens but have been reported in numerous
   crustose species, some of which belong to the conidial lichen-forming fungi.

Sources: As summarized by Honegger (1993).

conglutinate peripheral cortex. It is interesting to see that these main building blocks are formed by unrelated lichen-forming ascomycetes and basidiomycetes in symbiosis with unrelated green-algal or cyanobacterial photobionts.

Wall surface properties of lichen-forming fungi play key roles in thalline water relations and thus in the functioning of the symbiotic relationship. Hydrophilic areas of hyphal walls passively absorb and retain water, but hydrophobic wall surface layers, as typically found on aerial hyphae of lichenized and nonlichenized fungi, prevent free water from accumulating; thus the medullary and algal layers of lichen thalli stay gas-filled at any level of hydration, and this is a prerequisite for efficient gas exchange (Honegger 1997, 2001, 2006). Already Goebel (1926) realized that medullary hyphae of internally stratified lichen thalli have water-repellent wall surfaces, which he assumed to result from depositions of mycobiont-derived, crystalline secondary metabolites. However, wall surface hydrophobicity is also evident in the thalline interior of lichens with no medullary secondary compounds; examples are *Xanthoria* and *Peltigera* species. In freeze-fracturing preparations, a thin, proteinaceous wall surface layer was observed not only on hyphae, but also on algal wall surfaces in the medullary and algal layers of various taxa, which revealed a peculiarly fine structure: groups of minute rodlets, lying in parallel (Fig. 5.15, lower panel), create a hydrophobic discontinuity (Honegger 1982*b*, 1984*a*; Scherrer *et al.* 2000; Trembley *et al.* 2002*a*). The same wall surface layer, termed rodlet layer, was found on aerial hyphae of numerous nonlichenized fungi. In biochemical and molecular studies the rodlet layers of nonlichenized and lichenized ascomycetes and basidiomycetes were identified as hydrophobins, a class of fungal surfactants with very peculiar properties (Kershaw and Talbot 1998; Wessels 1999; Scherrer *et al.* 2000, 2002; Wösten 2001; Trembley *et al.* 2002*a, b*; Scherrer and Honegger 2003). Hydrophobins are small, secreted fungal proteins (approx. 100 amino acids long), which self-assemble at liquid–air or hydrophilic–hydrophobic interfaces to a thin, amphiphilic film with a distinct rodlet (i.e. small rods) structure on its hydrophobic side (Fig. 5.15, lower panel; Wösten *et al.* 1993; Wessels 1999; Wösten 2001; Linder *et al.* 2005). Once assembled, a hydrophobin film cannot be dissolved with techniques that are usually applied for protein solubilization. Hydrophobins reveal very low sequence homology except eight cysteine residues in a conserved pattern (Fig. 5.15, lower panel); these form intramolecular disulfide bonds (Kershaw *et al.* 2005; Linder *et al.* 2005). Many fungal species form several hydrophobins, which are all differerent from each other. Only one hydrophobin was found in each of different *Xanthoria* spp. (Scherrer *et al.* 2000; Scherrer and Honegger 2003), but three were found in the dikaryotic hyphae of the lichenized basidocarps of *Dictyonema glabratum*; these share between 54% and 66% amino acid identity (Trembley *et al.* 2002*a*).

Hydrophobins play important roles in the establishment of fungal interactions with plants and animals (Whiteford and Spanu 2002; Kershaw *et al.* 2005; Linder *et al.* 2005). In the lichen symbiosis, they seal the apoplastic continuum of both partners with a hydrophobic coat, which prevents free water from accumulating on these hydrophobic wall surfaces and forces the passive fluxes of solutes from the thalline surface to the algal layer and vice versa to flow underneath this hydrophobic lining (Honegger 1985, 1991*a*, *c*, 1997, 2001). As in nonlichenized fungi, hydrophobin gene expression is developmentally regulated in lichen-forming ascomycetes and basidiomycetes. *XPH1* is expressed in the aerial hyphae of the medullary and algal layers, but not in pycnidia or in the hymenial and adjacent subhymenial layers of *Xanthoria parietina* (Scherrer *et al.* 2002). *DGH1* and *DGH2* are expressed in aerial hyphae of the photobiont layer and *DGH3* in hyphae of the boundary layers in the lichenized basidiocarps of *Dictyonema glabratum* (Trembley *et al.* 2002*b*). Due to their low sequence homology even within closely related species, hydrophobins would be excellent molecular markers in phylogenetic analyses (Scherrer and Honegger 2003), but they are often too difficult to characterize. There are no conserved regions for which primers could be designed to amplify hydrophobin gene sequences.

Thin hydrophobic surface layers often overlie very thick hydrophilic areas of the fungal cell wall. For example, the very thick walled medullary hyphae of *Xanthoria* spp. form hydrophilic, conglutinate strands with hydrophobin-derived hydrophobic surface coats, and the very thick walled aerial hyphae of the medullary layer in *Cetraria islandica* ("Icelandic moss") comprise a massive, hydrophilic layer composed mainly of lichenin (a linear $(1\rightarrow3)$ $(1\rightarrow4)$ β-glucan), which overlies the cell wall proper (Honegger and Haisch 2001); both absorb and retain water. Such hydrophilic wall layers are thick in the hydrated state and thin in the desiccated state.

The subsequent developmental steps in the ontogeny of macrolichens are summarized in Table 5.2. Starting with either free hyphae or symbiotic propagules many mycobionts first produce a prethallus, i.e. an inconspicuous crustose structure with no internal differentiation (Fig. 5.4). These prethallus stages may be formed even with ultimately incompatible algal cells (Ahmadjian and Jacobs 1981; Ott 1987*a*, *b*; Schaper and Ott 2003). The most fascinating event in lichen morphogenesis is the least understood: the onset of thallus formation (Figs. 5.5–5.10). How does it become morphologically and physiologically differentiated from the prethallus stage? The factors that trigger the morphotypic and chemotypic expression of the fungal genotype are unknown, but the series of events is impressive (Table 5.2). Even the earliest developmental stages show a distinct polarity. The growing apical and the nongrowing basal poles are

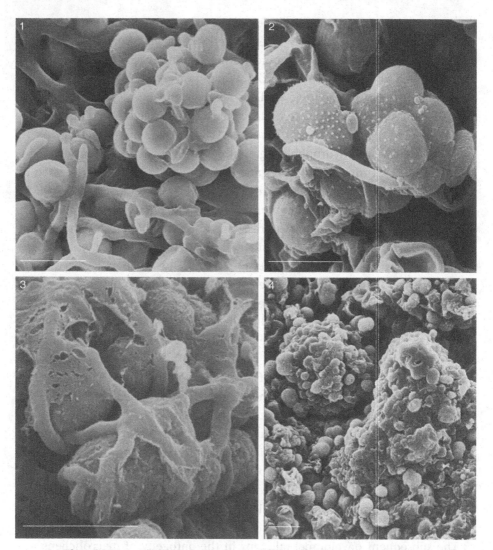

Figs. 5.1–5.4 Thallus ontogeny in the foliose macrolichen *Xanthoria parietina*. Fig. 5.1. Prethallus stage in an axenic, artificial combination of a cultured multispore isolate of the mycobiont and an isolate of the compatible photobiont, *Trebouxia arboricola*. Fig. 5.2. A germ tube of the mycobiont in search of compatible photobiont cells in a natural association of aerophilic green algae on bark. The fungal hypha is seen in close contact with cells of *Desmococcus* sp. (spiny cell wall surface), an incompatible green alga. Fig. 5.3. Mycobiont hyphae in a natural resynthesis, secreting mucilaginous material in initial response to the contact with compatible algal cells. Fig. 5.4. Prethallus stages of development in nature (Note the different magnifications in Figs. 5.3 and 5.4!). Magnification bars represent 10 μm.

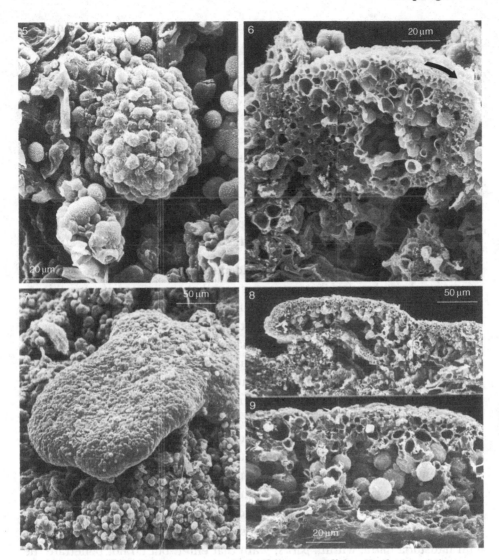

Figs. 5.5-5.9 Thallus ontogeny in the foliose macrolichen *Xanthoria parietina* (*continued*). Figs. 5.5, 5.6. External view and longitudinal cross section of primordial stages of the symbiotic phenotype growing out of the prethallus stage of development. These minute primordia are polarized (growing/nongrowing poles being defined) and recognizable by their bright yellow, cortical anthraquinones. The arrow in Fig. 5.6 points to the direction of growth. Figs. 5.7-5.9. External view and longitudinal cross section of juvenile thallus lobules. An internal stratification (upper and lower cortical layers and algal layer) is obvious. The medullary layer (see Fig. 3.3) will be differentiated in a later developmental stage.

differentiated (Figs. 5.5, 5.6). Such minute primordial stages reveal an internal stratification: the algal cells are kept in the gas-filled internal zone which is surrounded by a peripheral conglutinate cortex (Figs. 5.6, 5.8, 5.9). The medullary layer (Fig. 3.3) is not yet fully differentiated. Species-specific secondary metabolites are already produced; the example shown in Figs. 5.5 and 5.6 are bright yellow owing to the presence of the cortical anthraquinone parietin. Neighboring, potentially genetically heterogeneous prethalli and/or primordia may fuse (Fig. 5.10). The resulting thallus rosette may therefore be genetically heterogeneous.

5.4   Structural and functional aspects of the mycobiont–photobiont interface

Depending on the taxonomic identity of the partners, a variety of appressorial and haustorial structures is found at the immediate mycobiont–photobiont interface (Honegger 1991a, 1992). The structure and composition of the algal cell wall is centrally important to the outcome of the symbiotic relationship. Algal walls with enzymatically nondegradable biopolymers (sporopollenin-like compounds, as found in the genera *Coccomyxa* and *Elliptochloris*) are not normally invaded by fungal haustoria (Brunner and Honegger 1985; Honegger 1991a). Correlations between thallus morphology and type of mycobiont–photobiont relationship are obvious in Lecanorales with photobionts of the genera *Trebouxia* and *Asterochloris* (Tschermak 1941; Plessl 1963; Honegger 1986a, 1992, 2001). Mycobionts of morphologically simple crustose taxa often pierce the algal cell wall with finger-shaped protrusions ("intracellular" haustoria; Fig. 5.11), that are in intimate contact with the invaginating plasma membrane of the photobiont. Mycobionts of morphologically advanced, foliose or fruticose Lecanorales and Teloschistales form very peculiar intraparietal haustoria (*intra*: within; *paries*: wall) that enter, but do not penetrate the cellulosic wall of *Trebouxia* photobionts (Figs. 5.13, 5.15). Intermediate types of interactions occur where the fungal partner forms a short, almost globose protrusion that is enveloped by the locally enlarged algal wall (Fig. 5.12). Such intimate contact sites are established between juvenile cells: growing hyphal tips of the fungal partner meet young algal cells when these are still ensheathed by the degrading mother cell wall (Fig. 5.15). Subsequently, the haustorial complex on the fungal side and the algal cell grow and develop coordinately to reach maturity.

At the first contact of the growing hyphal tip with the wall surface of juvenile photobiont cells, the thin, proteinaceous, water-repellent cell wall surface layer (which was identified as hydrophobin in some species; see above)

Fig. 5.10 Thallus ontogeny in the foliose macrolichen *Xanthoria parietina* (*continued*). Vicinal, juvenile thallus lobules, growing out of different prethalli, fuse to form a rosette-like thallus which is likely to be genetically inhomogenous. Arrows point to further prethallus stages.

of the mycobiont spreads over the algal cell wall surface, thus sealing the apoplastic continuum of both partners with a hydrophobic coat (Honegger 1984*a*, 1986*b*, 1991*a*, 2001). In many taxa, especially in Lecanorales, the hydrophobicity of this mycobiont-derived, water-repellent wall surface is increased by secondary metabolites that crystallize on and within this water-repellent surface layer (Fig. 5.14; Honegger, 1986*b*, 2001, 2006). Water and dissolved nutrients and soluble metabolites of fungal and algal origin are passively translocated within the apoplastic continuum underneath this hydrophobic coat. The regularly occurring, often quite dramatic wetting and drying cycles are the main driving forces in more distant translocation processes (Honegger 1991*a*, 1992).

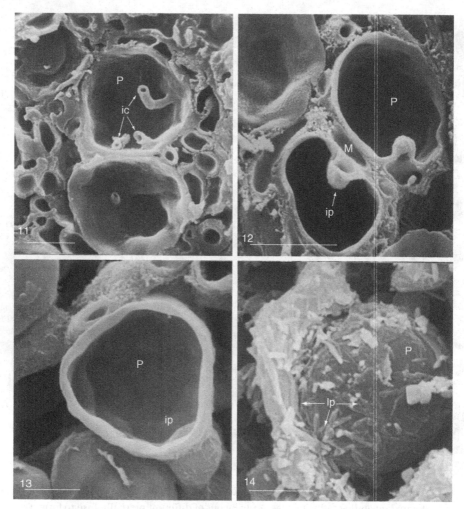

Figs. 5.11–5.14 Mycobiont (M) – photobiont (P) relationships in Lecanorales with *Trebouxia* spp. Fig. 5.11. Finger-like, intracellular fungal haustoria (ic) are typically found in photobiont cells of the crustose *Lecanora conizaeoides*. Fig. 5.12. Almost globose, intraparietal haustorial structures (ip) are produced at the mycobiont–photobiont interface in *Cladonia arbuscula*. The algal cell wall is not pierced, but grows around the haustorial complex. The arrow points to the border between algal and fungal cell walls. Fig. 5.13. Intraparietal haustorium (ip) at the mycobiont–photobiont interface in *Alectoria ochroleuca*. This haustorial type is found in foliose and fruticose lecanoralean macrolichens (except Cladoniaceae). Fig. 5.14. Low temperature scanning electron micrograph of the algal layer in a water-saturated, frozen-hydrated sample of *Parmelia sulcata*. Mycobiont and photobiont cells are coated by a mycobiont-derived, hydrophobic wall surface layer. Secondary metabolites of fungal origin (mainly salazinic acid) crystallize in and on this water-repellent coat, thus increasing its hydrophobicity. lp: crystalline lichen products (mycobiont-derived secondary metabolites). (For further details see Honegger, 1986*a*, *b*, 1991*a*). Magnification bars represent 5 μm.

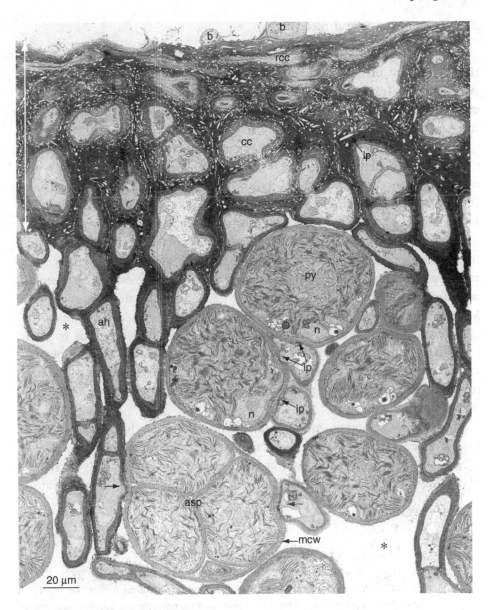

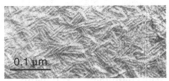

Hydrophobins: small, secreted proteins of
lichen-forming and non-lichenized fungi

$$X_{26-85}\text{-}C\text{-}X_{5-8}\text{-}C\text{-}C\text{-}X_{17-39}\text{-}C\text{-}X_{8-23}\text{-}C\text{-}X_{5-6}\text{-}C\text{-}C\text{-}X_{6-8}\text{-}C\text{-}X_{2-13}$$

Fig. 5.15 (Upper panel) TEM micrograph of a vertical cross section of the conglutinate upper cortex (uc) and algal layer in the growing, marginal zone of the foliose macrolichen *Parmelia tiliacea* (photobiont: *Trebouxia impressa*). ah, aerial hyphae of the thalline interior. Their cell wall surfaces were originally covered with crystalline,

5.5   Genotypes and phenotypes

One of the most fascinating aspects of lichen biology is the impact of the photobiont on the expression of the symbiotic phenotype in the fungal partner. Complex morphotypes are expressed exclusively in compatible associations (Table 5.1). Particularly interesting are the 3–4% of lichen-forming fungi that associate simultaneously (cephalodiate species) or consecutively (photosymbiodemes) with both green-algal and cyanobacterial photobionts. Cephalodiate lichen mycobionts have a green alga as primary photobiont and additionally incorporate nitrogen-fixing cyanobacteria in gall-like structures, the cephalodia. These are located either on the thallus surface (external cephalodia; Fig. 5.16) or in its interior (internal cephalodia). The fungal partner forms a dense, conglutinate cortical layer around cephalodia, thus creating microaerobic conditions in its interior. Cephalodia-bound cyanobacteria often reveal an increased heterocyst frequency and thus an elevated nitrogenase activity compared with the aposymbiotic ("free-living") state (Englund 1977). Most cephalodia are morphologically simple globose, sacculate, lobate or coralloid structures that differ from the rest of the vegetative thallus by their coloration. But even morphologically simple cephalodia illustrate a biologically most fascinating phenomenon: the different response of the mycobiont to a selected cyanobacterial partner compared with the green-algal photobiont (Fig. 5.16).

Mycobionts of photomorph pairs (also termed morphotype pairs, photosymbiodemes, etc.) form either a cephalodiate green-algal morphotype (usually vivid green with dark cephalodia) or a cyanobacterial morphotype (usually grayish to black). Both morphotypes of the same fungal species derive their coloration

---

Caption for Fig. 5.15 (cont.)

secondary metabolites (see Fig. 5.14). asp, autosporangium of the photobiont; the thin-walled, young autospores are ensheathed by the degrading mother cell wall (mcw). Arrows point to adjacent mycobiont hyphae entering the autosporangium in order to establish contact sites with the juvenile algal cells. b, bacterial epibionts; cc, cortical cells; ip, intraparietal haustoria entering, but not piercing, the algal cell wall (see Fig. 5.13). lp, impressions of crystalline lichen products (mainly atranorin) in the conglutinate upper cortex. The crystals were dissolved during dehydration. n, nucleus of the algal cell; py, pyrenoid of the large, lobate chloroplast; rcc, remains of decaying cortical cells. Asterisks mark the gas-filled space in the thalline interior. (Lower panel) Rodlet layer, built up by self-assembled hydrophobin protein on the wall surface of aerial hyphae in the medullary and algal layers of asco- and basidiomycetes. See text for further explanations.

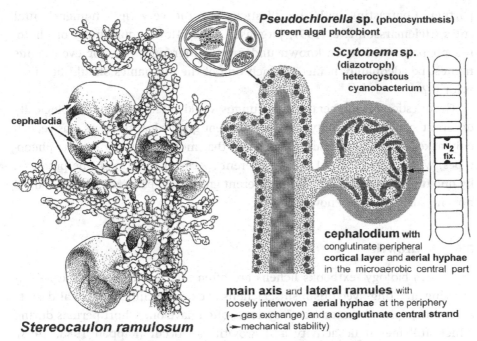

Pseudochlorella sp. (photosynthesis)
green algal photobiont

Scytonema sp.
(diazotroph)
heterocystous
cyanobacterium

N₂
fix.

cephalodium with
conglutinate peripheral
cortical layer and **aerial hyphae**
in the microaerobic central part

cephalodia

main axis and **lateral ramules** with
loosely interwoven **aerial hyphae** at the periphery
(➤gas exchange) and a **conglutinate central strand**
(➤mechanical stability)

Stereocaulon ramulosum

Fig. 5.16. Diagram illustrating the impact of the photobiont on the expression of the symbiotic phenotype in the mycobiont as observed in the cephalodiate lichen *Stereocaulon ramulosum*. Line drawings by Sibylle Erni.

from their photobiont. Photomorph pairs may be morphologically similar (isomorphic), as seen in various *Peltigera* spp. (Brodo and Richardson 1978; Tønsberg and Holtan-Hartwig 1983; Goffinet and Bayer 1997) or dissimilar (heteromorphic) as observed in *Sticta/Dendriscocaulon*, *Pseudocyphellaria/Dendriscocaulon* or *Lobaria/Dendriscocaulon* photomorph pairs (reviewed in James and Henssen 1976). Heteromorphic photomorph pairs comprise a foliose, dorsiventrally organized green-algal (chloromorph) and a shrubby (small fruticose) cyanobacterial phenotype (cyanomorph), both being described under different genera and species names in the lichenological literature. Intermediate and intermixed growth forms were described as "lichen chimerae": green, foliose thalli with large, blackish shrubby outgrowths resembling outsized cephalodia, or shrubby, blackish cyanomorphs with green, dorsiventrally organized "leaflets." Molecular analyses confirmed the taxonomic identity (conspecificity) of *Peltigera* spp. and *Lobaria* spp. involved in photomorph pairs (Goffinet and Bayer 1997; Stenroos *et al.* 2003). Thus photomorph pairs mirror an astonishing phenotypic plasticity, which allows a few lichen-forming ascomycetes to express different phenotypes in association with either a green-algal or a cyanobacterial partner. Contrasting with this plasticity are the very rigid nomenclatural rules (Heiðmarsson *et al.* 1997; Jørgensen 1998). How should fungi of photo-morph pairs, which are known under two different species or even genus names, be adequately named? Which one of the old names should be given priority?

The question of whether the same fungus would differentiate morphologically different symbiotic phenotypes with different green-algal partners was repeat-edly asked. It is theoretically imaginable that morphologically similar pheno-types, which are described under different species names, might be produced by one fungus in association with different green-algal partners. However, until now no such cases are known.

5.6   Growth

In biology textbooks lichens are often referred to as extremely slow-growing and long-lived organisms. Lichens of extreme climates such as deserts and arctic/alpine or antarctic ecosystems often have only short periods during which full metabolic activity and growth can occur (Kappen 1988, 1993). Consequently, only very low cell turnover rates and minimal annual size increases are recorded. However, as shown in a 20-year survey of antarctic lichen communities, not all lichen species from extreme climates are extreme slow-growers (Lewis Smith 1995). Some lichens are assumed to be very long-lived. If true, then their growing thalline areas must be at least minimally active for decades or centuries. The most extreme age estimates are in the range of millennia, as concluded from annual measurements of different size classes of thalli. However, in very large thalli, one can never exclude the possibility that previous fusion of neighboring thalli has occurred. Many lichen-forming fungi do not form a distinct rim due to vegetative incompatibility, when bordering upon other thalli of the same species. Instead they fuse and consequently appear as one entity. In contrast, short-lived lichen species terminate their full devel-opment within months or a few years (Poelt and Vězda 1990).

The majority of lichens of temperate or subtropic to tropic climates have annual radial growth (or annual linear elongation in the case of fruticose species) in the range of millimeters to a few centimeters. The highest growth rates are recorded in moist, coastal-influenced regions, including such species as *Ramalina menziesii*, the "lace lichen" (Figs. 5.24–5.25; Boucher and Nash 1990*a*) and lungworts like *Lobaria oregana* (Rhoades 1977) and *L. pulmonaria* in rain forests of the Pacific Northwest of North America. Under favourable climatic conditions these taxa reveal growth patterns that allow rapid size increase in relatively short periods of time (Table 5.3).

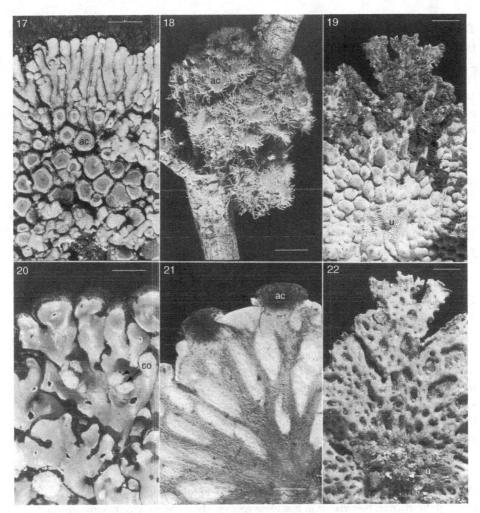

Figs. 5.17–5.22 Growth patterns in macrolichens. Bars represent 2 mm.
Fig. 5.17 *Caloplaca thallincola*, Fig. 5.18 *Teloschistes chrysopthalmus*, and Fig. 5.20
Menegazzia pertusa: thalli with predominantly apical/marginal growth, youngest
portions at the marginal pseudomeristems and ascomata (ac) or soredia (so) being
produced in nongrowing subapical areas. Figs. 5.19, 5.22. *Lasallia pustulata*, upper
and lower surface: irregular, patchy intercalary growth, some of the oldest thalline
areas are located at the fuzzy, isidiate margins. u, central umbilicus (holdfast)
by which the whole thallus is fixed to the substratum. Fig. 5.21. *Peltigera venosa*,
lower surface: predominantly apical/marginal growth, ascomata (ac) are produced
in apical thalline areas which terminate growth once that reproductive stages are
formed.

Table 5.3. *Growth patterns in foliose and fruticose macrolichens with internally stratified thallus (as summarized by Honegger, 1993)*

(1) Predominantly apical/marginal growth

Laminal size increase is accomplished by an apical/marginal pseudomeristem and the adjacent elongation zone; highest cell turnover rates occur in the apical/marginal pseudomeristem. Limited cell turnover and fully differentiated mycobiont and photobiont cells are typically found in the nongrowing subapical area. Increasing numbers of dead photobiont cells occur in senescent basal thalline areas. Central senescent parts of foliose thalli may break off, thus giving the thallus a ring-shaped outline.

Producton of sexual or (symbiotic) asexual propagules (soredia, isidia) either

- in apical/marginal pseudomeristematic zones of lobes or branches which terminate growth once that reproductive stages are formed.

    examples: *Peltigera venosa* and other *Peltigera* spp.
    Cetraria islandica ("Iceland moss")
    Teloschistes chrysophthalmus

- in subsenescent, nongrowing thalline areas;

    examples: *Caloplaca thallincola* and other *Caloplaca* spp.
    Xanthoria parietina
    Menegazzia pertusa

(2) Combined apical/marginal and intercalary growth, best visible in reticulate thalli with increasing mesh size towards the basal/central parts in either erect or pendulous fruticose species and in foliose lichens which adhere to the substratum with only a small central part of the thallus. Laminal size increase due to the activities of an apical/marginal pseudomeristem and intercalary growth processes.

    examples: *Lobaria oregana*, *L. pulmonaria* ("lungwort") and other *Lobaria* spp.
    Ramalina menziesii ("lace lichen")

(3) Regular or irregular ("patchy") intercalary growth in foliose, umbilicate thalli of the Umbilicariaceae; best visible in pustulate species. These thalli typically show irregular intrathalline gradients of photosynthetic activity, often highest at the central umbilicus. Some of the oldest thalline areas are likely to be at the fuzzy margins.

    examples: *Lasallia pustulata*
    Umbilicaria spp.

Combinations of (1), (2) and/or (3) are likely to occur.

Longevity in lichens needs to be critically interpreted. Pseudomeristems of macrolichens (see below) persist, as do meristems of perennial plants. However, in very old trees the metabolically active cells are not centuries old, and the same is true of old thalli of morphologically complex lichens. Senescent thalline areas are either overgrown or disintegrate. In the case of foliose taxa, the

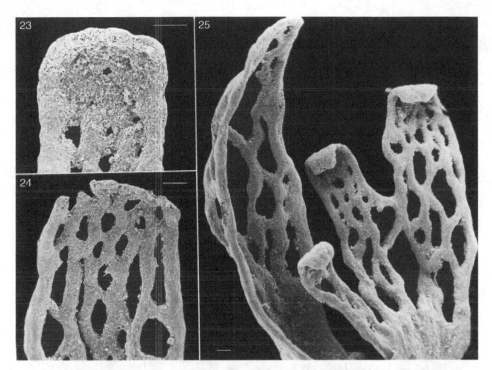

Figs. 5.23–5.25 Growth patterns in macrolichens (*continued*): *Ramalina menziesii* ("lace lichen") with combined apical/marginal and intercalary growth. When apical pseudomeristematic zones (Fig. 5.23) are lost by damage or arthropod grazing (Fig. 5.24) the band-shaped, perforate thallus lobe will grow by intercalary elongation, visible in the increasing length of the meshes (Fig. 5.25) (for further details see Sanders, 1989). Magnification bars represent 0.1 mm.

remaining thallus becomes strikingly ring-shaped. In very cold or very hot climates it may take centuries to achieve a wide ring diameter, but only the youngest portions, i.e. the growing marginal area and adjacent zones, are retained.

5.6.1   Growth patterns

Mycobionts of the morphologically less advanced crustose lichens grow more or less like molds: either on or within the substrate where they meet their compatible photobiont (see Chapter 3). The situation is very different in squamulose, foliose, or fruticose lichens with an internally stratified thallus. The coordinated growth of the dominant fungal exhabitant (the partner that lives outside) around the photoautotrophic inhabitant (the photobiont) is a most remarkable biological phenomenon. Three distinct growth patterns have been recognized in macrolichens (Table 5.3, Figs. 5.17–5.25) and intermediates are likely to occur. Foliose and fruticose lichens are more extensively investigated.

New cells are produced almost exclusively in terminal or marginal thalline areas with meristematic properties. Because these growing edges differ in many respects from meristems of plants (Fletcher 2002) and, moreover, because fungi have no real tissues, these growing edges or tips have been termed pseudomeristems (Honegger 1993). High cell turnover rates and small average cell sizes in both partners are typical features of such pseudomeristematic zones (Fig. 5.26a). Both mycobiont and photobiont cells achieve their full size in the elongation zone behind the pseudomeristem (Fig. 5.26b, c). Fully differentiated fungal and photobiont cells have low cell turnover rates (few or no cell divisions). A high

Xanthoria parietina (Teloschistales, Ascomycotina)

photobiont: *Trebouxia arboricola* (Chlorophyta)

a foliose, dorsiventrally organized macrolichen with mainly apical/marginal and limited intercalary growth

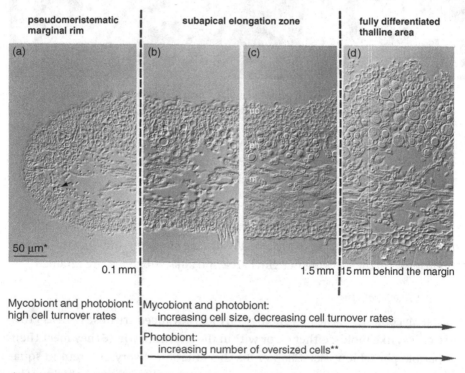

pseudomeristematic marginal rim | subapical elongation zone | fully differentiated thalline area

(a) (b) (c) (d)

50 μm*

0.1 mm   1.5 mm 15 mm behind the margin

Mycobiont and photobiont: high cell turnover rates

Mycobiont and photobiont: increasing cell size, decreasing cell turnover rates

Photobiont: increasing number of oversized cells**

*   same magnification in (a) – (d)
**   cells having exceeded the size required for autospore formation without undergoing mitosis, i.e. with arrested cell cycle (Hill, 1985, 1989). Arrow in (a) points to an autosporangium.

lc: lower cortex; uc: upper cortex; m: medullary layer; ph: photobiont layer;

Fig. 5.26 Internal thalline differentiation in *Xanthoria parietina* as seen in differential interference contrast light microscopy of semithin sections.

percentage of oversized photobiont cells (cells having exceeded the size required for autospore production without undergoing mitosis; Hill 1985, 1989; Fiechter 1990) are typically found in adult thalline areas (Fig. 5.26d; Honegger 1993). A large number of lichens do not grow exclusively at their margins or tips and adjacent elongation zones (Fig. 5.26a–c) but retain the capacity to expand and enlarge even in their older parts (Chapter 2). Such intercalary (inserted) growth processes have not yet been analyzed in detail, and it is unknown whether they are due primarily to elongation of cells that have been produced by the apical/ marginal pseudomeristem or to continuous cell division. The growth pattern of umbilicate lichens, which are fixed to the substrate by means of a central umbilicus (i.e. navel; Figs. 5.19, 5.22), are least understood. Umbilicariaceae have no distinct marginal pseudomeristem but reveal irregular, patchy intercalary growth (Honegger 1993, 2001; Figs. 5.19, 5.22). Patterns of productivity (photosynthetic activity) are also diffuse in such thalli (Larson 1983; Hestmark et al. 1997). The regulatory mechanisms behind all these differentiation processes in morphologically complex macrolichens are poorly understood. It is inferred from presently available data that the cell turnover in the photobiont is strictly controlled by the mycobiont. The cell cycle of the photobiont in nongrowing thalline areas is arrested (Hill 1985; Honegger 1993). As inhibitors of cell division are of general biological interest, especially with regard to tumor suppression in humans and vertebrates, the molecular basis of this inhibitory principle in mycobiont–photobiont relations of lichens merits thorough investigation.

6

Sexual reproduction in lichen-forming ascomycetes

R. HONEGGER AND S. SCHERRER

A high percentage of lichen-forming ascomycetes reproduce sexually and thus are assumed to disperse primarily via ascospores, which have to relichenize. However, one should keep in mind that even fertile lichens have options for vegetative dispersal in the symbiotic state, either via symbiotic propagules such as soredia, blastidia or isidia, or via thallus fragmentation. Viable fungal and algal cells were shown to be contained in fecal pellets of lichenivorous slugs (McCarthy and Healey 1978; Fröberg et al. 2001) and of the ever-present lichenivorous mites (Meier et al. 2002). Thus, it is not known how often relichenization occurs in natural habitats.

A detailed knowledge of sexual reproductive strategies is required for understanding evolutionary traits and population genetics. Zoller et al. (1999) were the first to recognize that lack of ascomata in strongly fragmented and geographically isolated populations of Lobaria pulmonaria ("lungwort") might be due to missing mating partners. As this species produces abundant isidiate soredia, one might conclude that ascospores are unnecessary. However, recombination as the centrally important element of sexual reproduction has an impact on genetic stability, whereas favorable and unfavorable mutations are transmitted to the offspring in clonal (vegetative) dispersal. As pointed out by Hestmark (1992), sexual reproduction may often be a mode of escape from old, severely parasitized thalli (Seymour et al. 2005b). It remains to be seen how often new thalli are formed from germinating ascospores in rarely fertile species with efficient dispersal via vegetative symbiotic propagules, such as Pseudevernia furfuracea, Hypogymnia physodes and others.

The majority of lichen-forming ascomycetes can be cultured in the aposymbiotic state, but they fail to differentiate sexual reproductive stages under these conditions; thus, classical genetic crossing experiments cannot be conducted.

Lichen Biology, ed. Thomas H. Nash III. Published by Cambridge University Press.
© Cambridge University Press.

This problem occurs also in many groups of nonlichenized fungi. In most lichens the vegetative mycelium is not hidden in the substrate or host, as is the case in nonlichenized taxa, but fully visible above ground. Therefore, the abundance of ascomata and their location on the vegetative mycelium can be evaluated. Some species of lichen-forming ascomycetes are always fertile and have many ascomata (e.g. *Xanthoria parietina* [Fig. 6.1], *Ramalina fastigiata*), others have no or few to many ascomata (e.g. *Xanthoria calcicola* [Fig. 6.4], *Parmelia tiliacea*, *P. sulcata*), and in a third group ascomata are very rare (e.g. *Pseudevernia furfuracea*, *Hypogymnia physodes*) or absent (*Thamnolia vermicularis*, *Lepraria, and Leprocaulon* spp.). In some species ascomata are formed in the subapical part of the thallus, which continues growth at its periphery and may reach large dimensions (Fig. 5.17); examples are *Xanthoria parietina*, *Parmelia tiliacea*, *Ramalina menziesii*, *R. fraxinea*, *Lobaria pulmonaria*. Others produce ascomata at or near the growing tip or margin which subsequently stops growth; thus, only a limited size can be achieved. Examples are *Teloschistes chrysopthalmus* (Fig. 5.18) and many other *Teloschistes* spp., *Ramalina fastigiata*, and *Peltigera venosa* (Fig. 5.21).

6.1   Mating systems

Nonlichenized and lichen-forming ascomycetes are not female or male; each haploid mycelium is theoretically capable of differentiating both gametangia (ascogonia) and gametes (microconidia = spermatia). Their sexual reproduction is regulated by mating type (*MAT*) genes (review: Debuchy and Turgeon 2006). In contrast, basidiomycetes have several *MAT* loci and thus very complex mating systems. Filamentous ascomycetes have one *MAT* locus, which is completely different in haploid mycelia carrying only one out of the two *MAT* alleles of the same heterothallic (cross-fertilized) species. They are referred to as *MAT 1-1* and *MAT 1-2* (Turgeon and Yoder 2000), but other terms are found in the literature as well (e.g. *MAT 1* and *MAT 2*, *MAT A* and *MAT a*, etc.). As *MAT 1-1* and *MAT 1-2* are completely different within the same species the term idiomorph is used instead of allele, one idiomorph carrying one to several genes. Homothallic (self-fertile) species have either elements of both *MAT* idiomorphs in one haploid mycelium, or of only one, the other being lost. Each haploid mycelium forms ascogonia, but in cross-fertilized species a dikaryon can only be formed with a mating partner carrying the complementary idiomorph. Self-fertile species do not need a mating partner. Homothallism is a derived character (Yun *et al.* 1999), which can be achieved in only one mutation (Pöggeler 1999). *MAT* genes evolve very rapidly, but need to be conserved within a species. Thus, mutations in *MAT* idiomorphs might play important roles in speciation (see below).

How can mating systems (homothallism or heterothallism) of lichen-forming ascomycetes be explored when sterile cultured mycelia do not form sexual reproductive stages? The progeny of meiosis can be analyzed with fingerprinting techniques. Genomic DNA derived from single ascospore isolates from one ascoma (Murtagh *et al.* 2000; Seymour *et al.* 2005a) or from one ascus (Honegger *et al.* 2004) can be subjected to RAPD-PCR (randomly amplified polymorphic DNA-polymerase chain reaction) or AFLP (amplified fragment length polymorphisms) and resulting products be screened for polymorphisms. In homothallic species, all sporelings from the same ascus and ascoma, and sterile cultured vegetative mycelium of the mother thallus, reveal identical fingerprints (Fig. 6.3). Due to recombination events, polymorphisms are found among RAPD or AFLP markers of sibling isolates in heterothallic species (Fig. 6.5; Murtagh *et al.* 2000; Honegger *et al.* 2004; Seymour *et al.* 2005a; Honegger and Zippler 2007). Already the sporeling phenotype may provide interesting information about the mating systems. In heterothallic species of numerous Teloschistaceae, Parmeliaceae and Physciaceae, all with 8-spored asci, a maximum of four distinct sporeling phenotypes per ascus were observed, which differed in growth rate, growth pattern and/or pigmentation (Fig. 6.5), but in the homothallic *Xanthoria parietina* (Fig. 6.2) and *X. elegans* all sporelings per ascus and ascoma grew equally fast and looked the same (Honegger *et al.* 2004; R. Honegger, unpublished). Heterothallism is easily detectable with fingerprinting techniques, but homothallism is more difficult to identify. Uniform fingerprints among sibling single spore isolates obtained with large numbers of markers (PCR with 10–30 primers, Murtagh *et al.* 2000; Honegger *et al.* 2004) suggest homothallism, but the situation can only be properly interpreted when *MAT* genes are characterized. As *MAT* genes evolve rapidly and reveal little sequence homology they are difficult to track down, especially when the genome of the species in question has not yet been fully sequenced (as is so far the case in all lichen-forming ascomycetes). Successful characterization of *MAT* genes in a range of *Xanthoria* spp. confirmed heterothallism in *Xanthoria polycarpa* and *X. flammea*, but homothallism in *Xanthoria parietina*, with *MAT 1-2* present in all sibling isolates, and in *X. elegans*, with *MAT 1-1* and *MAT 1-2* in all sibling isolates (Scherrer *et al.* 2005). *MAT 1-2* of *Cladonia galindezii* was characterized and identified in 40–60% of randomly selected siblings, a further proof of heterothallism in this species (Seymour *et al.* 2005a).

Does the abundance of ascomata per thallus correlate with mating systems? Most investigators wish to know more about mating systems in lichen-forming ascomycetes without investing in time-consuming laboratory experiments. In Teloschistaceae, Parmeliaceae, Ramalinaceae and Physciaceae all irregularly fertile species with no or few to many ascomata turned out to be heterothallic,

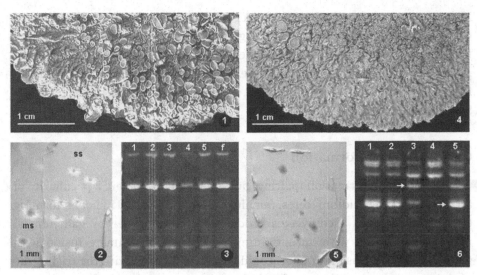

Figs. 6.1–6.6 Mating systems in lichen-forming ascomycetes visualized with RAPD-PCR fingerprinting techniques applied to genomic DNA derived from single asco-spore isolates. Figs. 6.1–6.3. Thalli of the self-fertile (homothallic) *Xanthoria parietina* (yellow wall lichen; Fig. 6.1) are always richly fertile, their older parts being covered by apothecial disks (fruiting bodies). All eight ascospores derived from one ascus were manually separated and allowed to grow as single spore isolates (ss). Multispore isolates (ms) comprise the contents of the whole ascus (Fig. 6.2). Ascospore germination rates were high (7–8 ascospores per ascus germinated). All sporelings revealed the same phenotype and grew equally fast. Fingerprints of genomic DNA derived from five out of eight single spore isolates (1–5) and from sterile cultured vegetative mycelium isolated from the fruiting body (f) revealed the same pattern (Fig. 6.3) with all primers tested. Figs. 6.4–6.6. Thalli of the cross-fertilized (heterothallic) *Xanthoria calcicola* are irregularly fertile, carrying no, few, to many fruiting bodies (apothecia). Ascospore germination rates were low to medium (0–6 ascospores per ascus germinated). The maximum 6 single spore isolates obtained per ascus revealed different phenotypes and growth rates (Fig. 6.5). Fingerprints of genomic DNA varied (arrows point to 2 polymorphic markers in Fig. 6.6).

but species with numerous ascomata were either homothallic (e.g. *X. parietina*, *X. elegans*), or heterothallic (e.g. *X. polycarpa*, *Physcia aipolia*, *Ramalina fastigiata*) (Honegger *et al.* 2004; Honegger and Zippler 2007).

Phylogenetic analyses combined with studies on mating systems give an insight into evolutionary and speciation processes. The widespread and common *Xanthoria parietina* was assumed to be the primary species, from which the European *X. calcicola* is derived (Purvis *et al.* 1992). *Xanthoria flammea*, a morphologically interesting South African endemic, was placed in the monotypic genus *Xanthomaculina* (Kärnefelt 1989). *Xanthoria calcicola* and *X. flammea* are heterothallic,

X. parietina is homothallic (Honegger *et al.* 2004; Scherrer *et al.* 2005). Comparative analyses of the noncoding rDNA region (ITS 1 and 2, 5.8 S) and of the hydrophobin gene sequences revealed a relatively close relationship of *X. flammea* with the *X. parietina* complex, and the homothallic *X. parietina* was found to be a derived, not a primary, species (Scherrer and Honegger 2003; Scherrer *et al.* 2005).

6.2   Dikaryon formation

Dikaryon formation (pairing, but not fusion of nuclei), a peculiarity of ascomycetes and basidiomycetes, remains largely unexplored in lichen-forming taxa. Nonlichenized ascomycetes produce dikaryons either by gametangial fusion (ascogonium with antheridium), spermatization (gamete fusion with gametangium, i.e. fusion of microconidia or macroconidia with the trichogyne of the ascogonium), or fusion of undifferentiated vegetative hyphae and subsequent pairing of nuclei (somatogamy).

In thalli of lichen-forming ascomycetes ascogonia are formed within or slightly underneath the algal layer. In many species trichogynes are seen protruding above the thallus surface (Figs. 6.8–6.9), which is often locally covered by mucilaginous material at this particular site (examples in Jahns 1970; Henssen and Jahns 1973; Honegger 1978a, b). The majority of lichen-forming ascomycetes produce large numbers of tiny, often bacteria-sized microconidia in flask-shaped conidiomata with narrow ostiole (pycnidia; Figs. 6.10–6.13), whose cavity is filled with a hydrophilic mucilage. In wet weather the swelling mucilage causes masses of microconidia to ooze out of the narrow apical ostiole (Fig. 6.11), whence they are most likely dispersed either by invertebrate vectors, as in many nonlichenized fungi such as *Claviceps purpurea*, and *Epichloë* spp., or by rusts (summarized by Naef *et al.* 2002) or by rain (splash dispersal). In many species pycnidia are well exposed, e.g. in projections along the thallus margin of *Cetraria islandica* ("Icelandic moss") and other *Cetraria* spp., or at the tip of podetia in *Cladonia* spp. (reindeer and cup lichens; Figs. 6.7, 6.10). In most foliose species (Parmeliaceae, Physciaceae, Teloschistaceae, etc.), pycnidia are immersed, but the ostiole opens at the surface of the lobes. Spermatization is assumed to be the mode of dikaryon formation in the numerous species of lichen-forming ascomycetes, in which microconidia have been seen adhering to protruding trichogynes (Jahns 1970; Henssen and Jahns 1973; Honegger 1978a, 1984a, b). In *Cladonia furcata* microconidia adhere, tip first, to the surface of trichogynes (Fig. 6.9), dissolve their own and the trichogyne wall at the contact site and leave a hole in it (Honegger 1984a). Trichogynes are short-living, ephemeral receptive hyphae which die off once that dikaryotization is completed.

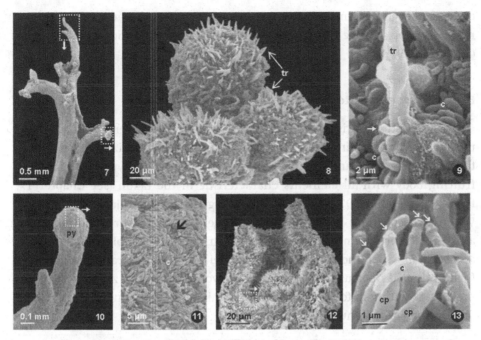

Figs. 6.7–6.13 Spermatization in *Cladonia furcata*. Fig. 6.7. Topmost fragment of a fertile thallus. Details of framed areas are shown in Figs. 6.8 and 6.10. Fig. 6.8. Group of three ascomatal primordia with numerous trichogynes (tr). Fig. 6.9. Detail of a trichogyne (tr) and adhering microconidia (c). Arrow points to a microconidium (= spermatium) that is attached, tip first, to the wall of the trichogyne. Fig. 6.10. Pycnidium (py) at the tip of the branch. Detail of framed area is in Fig. 6.11. Fig. 6.11. Arrow points to the pycnidial ostiole. Innumerable microconidia (c) were oozing out of the pycnidial cavity. Fig. 6.12. Longitudinally dissected pycnidium after removal of the hydrophilic mucilage that filled the pycnidial cavity. Conidiophores grow out of the pycnidial wall and release microconidia into the pycnidial cavity. Detail of framed area is in Fig. 6.13. Fig. 6.13. Mature microconidium (c) detached from the conidiophore (cp). Arrows point to conidiophores with developing microconidia. (Figures 6.8 and 6.13 are from Honegger 1984*b*, with permission of the publisher.)

Migration of conidial nuclei through the trichogyne and pairing with ascogonial nuclei likely occurs, but has never been documented in lichen-forming ascomycetes. In many species ascogonia were never found with protruding trichogynes (e.g. in the homothallic *Xanthoria parietina*; Janex-Favre and Ghaleb 1986; confirmed with SEM techniques by R. Honegger, unpublished.); others produce large numbers of trichogynes but neither macroconidia nor microconidia, which might serve as gametes (e.g. *Peltigera* spp.; R. Honegger, unpublished.).

In the vast majority of species ascomal primordia and microconidiomata (pycnidia) are differentiated in the lichenized part of the thallus, usually within the algal layer. However, in *Rhizocarpon* spp. and possibly in other taxa of crustose lichen-forming ascomycetes, ascomal primordia and microconidio-mata develop in the prothallus of thalli adjacent to lichenized areoles (Honegger 1978a).

6.3   Ascomal ontogeny

The stimuli triggering ascomal initiation and differentiation are unexplored in lichen-forming ascomycetes. As in most nonlichenized taxa the vegetative hyphae surrounding the dikaryotic, ascogenous hyphae build up a fruiting body with all elements characteristic of the taxon (see Chapter 17; Henssen and Jahns 1973; Henssen 1981; Parguey-Leduc and Janex-Favre 1981). When an ascogenous hypha has reached the hymenial layer it differentiates, after crozier formation and thus ultimate distribution of complementary nuclei, an ascus, in which the paired haploid nuclei fuse (i.e. undergo karyogamy) to form a diploid nucleus, the zygote. In the young ascus the zygote undergoes meiosis, resulting in four haploid nuclei, which may subsequently go through one or more mitotic nuclear divisions.

6.4   Ascosporogenesis

From a cell biological point of view, ascospore formation is a very interesting process, during which new cells are formed within the protoplast of an existing one (free cell formation). As in nonlichenized ascomycetes, the postmeiotic ascus of lichen-forming taxa comprises four haploid nuclei, which divide once, two times or even more, resulting in 8, 16, 32, or 64, etc. nuclei per ascus, 8 being found in the majority of species. A double membrane sac, often termed the peripheral membrane cylinder, is differentiated near the periphery of the protoplast. As in nonlichenized ascomycetes (Beckett 1981) different membrane systems were found to contribute to the formation of this membrane sac: the plasma membrane in *Peltigera* spp. (Fig. 6.14), *Baeomyces rufus* or *Chaenotheca chrysocephala* (Honegger 1982b, 1985), or the endoplasmatic reticulum in *Physcia stellaris*, *Pleurosticta acetabulum* or *Rhizocarpon geographicum* (Honegger 1982b). Guided by microtubules of the spindle apparatus, the double membrane sac invaginates around nucleate portions of cytoplasm and finally ruptures at the invagination fronts, thus becoming fragmented into small sacs, each of which is a future ascospore. Ascospore delimitation is completed upon closure of each of these small membrane sacs (Fig. 6.15). The inner membrane is the plasma membrane of the ascospore, the outer one the investing membrane, which is in contact with the cytoplasm of the ascus; it breaks down in the mature spore after wall completion.

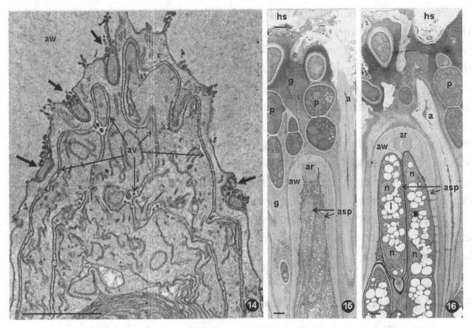

Figs. 6.14–6.16 Ascosporogenesis in *Peltigera canina*, as seen in transmission electron micrographs of longitudinally sectioned asci that had been chemically fixed with potassium permanganate. Fig. 6.14. Differentiation of a peripheral membrane sac, the ascus vesicle (av), in the premeiotic ascus via blebbing of the plasma membrane (bold arrows) near the ascus tip, where the ascus wall (aw) is thickened. In the postmeiotic ascus the ascus vesicle envelops nucleate portions of cytoplasm, the future ascospores. Figs. 6.15–16. Asci after ascospore delimitation (Fig. 6.15) and ascospore maturation (Fig. 6.16). Asci have an apically thickened wall (aw) and an eversible apical ring (ar) that stains blue with iodine in light microscopy preparations. Asci and paraphyses (p) are embedded in the hydrophilic hymenial gelatine (g) secreted by the paraphyses, which swells dramatically in the wet state, thus generating a high pressure on the flanks of the asci. Empty asci (a) after ascospore ejection are tangentially sectioned. The topmost part of expanding asci reached the hymenial surface (hs) during ascospore release. At maturity the fusiform, multicellular ascospores (asp) comprise one nucleus (n) per cell and numerous lipid-storing vesicles (l). Magnification bars represent 1 μm.

Large numbers of genera and species of lichen-forming ascomycetes have one-celled ascospores. However, within the freshly delimited ascospore nuclear divisions, followed by cell divisions, may occur according to species- or genus-specific parameters, leading to 2-, 4- or multicelled spores (Fig. 6.16). Under favorable conditions each cell of the ascospore germinates. Multicellularity increases the chance of survival under difficult conditions. During ascospore wall differentiation, at least two layers are deposited. The innermost corresponds in its structure and composition to the hyphal wall and will form a continuum with the wall of the germ tube. Outer wall layers may be locally thickened and hyaline or pigmented, e.g. gray to black due to inclusion of melanin granules (e.g. *Rhizocarpon* or *Physcia* spp.; Honegger 1978*b*, 1980). Several genera of lichen-forming ascomycetes from extreme climates (arctic, antarctic and high alpine ecosystems) such as rock tripes (*Lasallia* and *Umbilicaria* spp.) or map lichens (*Rhizocarpon* spp.) have muriform (multicellular) ascospores with strongly mela-nized walls. Melanin protects from UV radiation and possibly prevents loss of soluble compounds (Butler and Day 1998). Multicellularity enhances survival rates, one or few cells of the spore staying viable while others die under harsh conditions. An exceptional situation is found in *Pertusaria* spp. with either one or two large, multinucleate ascospores per ascus, which presumably comprise all, as in the unisporate *P. bryontha*, or half of the postmeiotic nuclei as in the bisporate *P. pertusa*.

6.5   Ascus structure and function

Most lichen-forming ascomycetes develop their asci in a hymenium, which is filled with a hydrophilic mucilage (Figs. 6.15, 6.16). This mucilage is produced and released by the paraphyses (Figs. 6.15, 6.16), vegetative, haploid hyphae with characteristic growth patterns. Tip cells of the paraphyses may also secrete secondary compounds, which crystallize at the hymenial surface and give the apothecial disk its characteristic color. Paraphyses of some species secrete the same secondary compounds as cortical vegetative hyphae in the thallus (e.g. yellow to orange anthraquinones in numerous Teloschistaceae) or different ones (e.g. the blood red naphtaquinone haemoventosin in the grayish *Ophioparma ventosa*, or the bright red anthraquinone bellidiflorin in the greenish gray *Cladonia bellidiflora*; Hunek and Yoshimura 1996). During hydration the ascomal mucilage swells substantially and generates a high pressure on the flanks of the asci. In the course of ascospore maturation the ascus wall under-goes a series of differentiation processes, which culminate in ascospore release. These differentiation processes include deposition of wall layers and, in species which eject ascospores, differentiation of a characteristic apical apparatus (Fig. 6.16; see Fig. 4.36 in Chapter 4). The apical apparatus of immature asci withstands the high pressure generated by swelling of the hydrophilic mucilage within the hymenial layer and stays closed. When ascospores are nearly mature, defined zones of the apical apparatus change their structural integrity and finally rupture, a prerequisite for successful ascospore discharge. In species with rostrate ascus dehiscence (see Fig. 4.36B–D in Chapter 4), the rupture of outer wall layers at the apex facilitates the tube-like expansion of inner ones, which reach the hymenial surface or expand even above. This rostrate dehiscence is typically found in Lecanoraceae, Cladoniaceae, Physciaceae, etc. (Honegger 1978b). Many Peltigeraceae carry an eversible amyloid ring at the tip of their expansible inner wall layer (Figs. 6.16, 4.36A; Honegger 1978b). Mazaediate ascomata are not filled with mucilage and their asci do not eject the spores, as the ascospore wall simply disintegrates during ascospore maturation (Honegger 1985), leading to a powdery ascospore mass on the surface of the fruiting body.

Ascus structure and function and features of ascomal ontogeny are widely used as taxonomic characters. Phylogenies based on molecular markers are largely in agreement with these sets of data, but partly show that structurally and functionally similar ascus types and ascomata have evolved independently in different groups. An example is the genus *Sphaerophorus*, formerly included in Caliciales on the basis of its mazaediate ascomata and deliquescent asci (Henssen and Jahns 1973), but now recognized as a family Sphaerophoraceae within the Lecanorales (Wedin and Döring 1999). The same applies for Caliciaceae, now included in Lecanorales, the Caliciales having been omitted (Eriksson *et al.* 2006b).

7

Biochemistry and secondary metabolites

J. A. ELIX AND E. STOCKER-WÖRGÖTTER

7.1 Intracellular and extracellular products

There are two main groups of lichen compounds: primary metabolites (intracellular) and secondary metabolites (extracellular). Common intracellular products occurring in lichens include proteins, amino acids, polyols, carotenoids, polysaccharides, and vitamins, which are bound in the cell walls and the protoplasts, are often water-soluble, and can be extracted with boiling water (Fahselt 1994b). Some of these products are synthesized by the fungus and some by the alga. Since the lichen thallus is a composite structure, it is not always possible to decide where a particular compound is biosynthesized. Most of the intracellular products isolated from lichens are nonspecific, and also occur in free-living fungi, algae and in higher green plants (Hale 1983).The majority of organic compounds found in lichens are secondary metabolites of the fungal component, which are deposited on the surface of the hyphae rather than within the cells. These products are usually insoluble in water and can only be extracted with organic solvents. Carbon for the lichen is furnished primarily by the photosynthetic activity of the algal partner. Mosbach (1969) summarized the overall carbon metabolic sequence as involving photosynthesis in the photobiont followed by transport of the carbohydrate to the fungus, metabolism of the carbohydrate and subsequent biosynthesis of lichen secondary metabolites. The type of carbohydrate released by the alga and supplied to the fungus is determined by the photobiont, while in lichens containing cyanobacteria, the carbohydrate released and transferred to the fungus is glucose. In lichens containing green algae, the carbohydrate released and transferred to the fungus is a polyol: ribitol, erythritol, or sorbitol (Section 10.2.1).

Lichen Biology, ed. Thomas H. Nash III. Published by Cambridge University Press.

7.2   The fungal origin of the secondary metabolites

All of the secondary substances which are so characteristic of lichens are of fungal origin. Consequently it seems rather surprising that with more than 700 secondary metabolites known from lichens (Huneck 1999; Dembitsky and Tolsikov 2005), most are unique to these organisms and only a small minority (c. 50–60) occur in other fungi or higher plants. As an example, the anthraquinone parietin, the orange pigment common in most Teloschistales, occurs in species of the non-lichenized fungal genera *Achaetomium, Alternaria, Aspergillus, Dermocybe, Penicillium,* as well as in the vascular plants *Rheum, Rumex,* and *Ventilago.* Similarly, the common *para*-depside lecanoric acid also occurs in the fungal genus *Pyricularia,* while the typical higher plant sterol, brassicasterol, has also been detected in the lichens.

7.3   Biosynthetic pathways to lichen secondary metabolites

Direct evidence from biosynthetic investigations on intact lichens using labeled compounds is meager, but hypothetical pathways are often proposed on the basis of what is known for the biosynthesis of analogous fungal products (Turner and Aldridge 1983). In addition, further circumstantial evidence may be forthcoming from observed joint occurrence of compounds and laboratory interconversions and biomimetic syntheses (C. Culberson and Elix 1989). In the past nearly all the chemical data came from studies of natural lichens because cultures of lichen fungi grow very slowly and failed to show all products characteristic of mature thalli in nature. However, recent advances in the controlled growth of recombined species promise to open new areas of research whereby the biosynthetic sequence to various lichen acids can possibly be confirmed (Section 7.5; Hamada *et al.* 2001). Most of the secondary metabolites present in lichens are derived from the acetyl-polymalonyl pathway, but some come from the shikimic acid and mevalonic acid pathways (C. Culberson) and Elix 1989; Huneck 2001). An overview of the probable biosynthetic pathways to the major classes of lichen products is illustrated in Fig. 7.1. One of the more interesting developments in recent years is the recognition of the key role played by *para*-depsides as potential precursors (or biosynthetic intermediates) to *meta*-depsides, depsones, diphenyl ethers, depsidones and dibenzofurans (C. Culberson and Elix 1989). Very recent experimental evidence obtained with cultured lichens is consistent with these suggestions.

7.4   Major categories of lichen products

The first classification of lichen substances based on known structures and biosynthetic pathways was constructed by Asahina and Shibata (1954). This

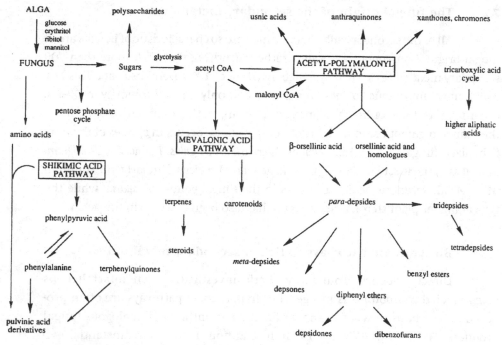

Fig. 7.1 Probable pathways leading to the major groups of lichen products.

system is modified from time to time, as more information has become available, most recently by C. Culberson and Elix (1989). The principal classes of lichen secondary metabolites are listed in Table 7.1 according to their probable biosynthetic origin, with the approximate number of compounds of known structure indicated (in brackets) and the structure of typical representative compounds (Figs. 7.2–7.5).

Of the acetyl-polymalonyl derived compounds, aromatic products are especially well represented (Figs. 7.2, 7.3), the most characteristic being formed by the bonding of two or three orcinol or β-orcinol-type phenolic units through ester, ether and carbon–carbon linkages (Fig. 7.3). The large majority of depsides, depsidones, dibenzofurans, usnic acids and depsones all appear to be produced by such mechanisms and all are peculiar to lichens. Other aromatic compounds of acetate-polymalonate origin, such as the chromones, xanthones, and anthraquinones, are probably formed by internal cyclization of a single, folded polyketide chain (Table 7.1) and are often identical or analogous to products of non-lichen-forming fungi or higher plants. In addition to the compounds of known chemical structure, many of unknown structure are given common names and assigned to compound classes, because they are frequently encountered and easily recognized by microchemical methods.

Table 7.1. *Major classes of secondary metabolites in lichens*

1. Acetyl-polymalonyl pathway
   1.1 Secondary aliphatic acids, esters and related derivatives (45)
   1.2 Polyketide derived aromatic compounds
      1.2.1 Mononuclear phenolic compounds (19)
      1.2.2 Di- and tri-aryl derivatives of simple phenolic units
         1.2.2a Depsides, tridepsides and benzyl esters (185)
         1.2.2b Depsidones and diphenyl ethers (112)
         1.2.2c Depsones (6)
         1.2.2d Dibenzofurans, usnic acids and derivatives (23)
      1.2.3 Anthraquinones and biogenetically related xanthones (56)
      1.2.4 Chromones (13)
      1.2.5 Naphthaquinones (4)
      1.2.6 Xanthones (44)

2. Mevalonic acid pathway
   2.1 Di-, sester- and triterpenes (70)
   2.2 Steroids (41)

3. Shikimic acid pathway
   3.1 Terphenylquinones (2)
   3.2 Pulvinic acid derivatives (12)

7.5   Molecular studies on polyketides and secondary metabolites of lichens (polyketides of lichen-forming fungi)

7.5.1   *Origin and distribution of polyketides*

Polyketides are a class of naturally occurring metabolites found in bacteria (prokaryotes), fungi (lichen-forming fungi; Fig. 7.6), algae, and higher plants, as well as in the animal kingdom (e.g. in dinoflagellates, insects, mollusks, and sponges). Polyketides are usually categorized on the basis of their chemical structures. An immensely rich diversity of polyketide structural moieties have been detected and structurally elucidated, and more await discovery.

7.5.2   *Biosynthesis and assembly of polyketides*

Polyketides are biosynthesized by sequential reactions catalyzed by an array of polyketide synthase (PKS) enzymes. PKSs are large multienzyme protein complexes that contain a typical core of coordinated active sites. The biosynthesis of polyketides occurs stepwise from 2-, 3- and 4-carbon building

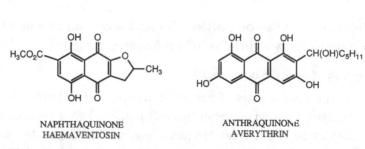

Fig. 7.2 Structures of typical acetyl–polymalonyl lichen products derived from a single polyketide chain.

blocks, such as acetyl-CoA, propionyl-CoA, butyryl-CoA and their activated derivatives malonyl, methylmalonyl, and ethylmalonyl-CoA (Fig. 7.7). The major polyketide chain building step is a decarboxylative condensation (closely related to the chain elongation step in fatty acid biosynthesis). By chemical and biochemical comparisons, a mechanistic relationship between polyketide and fatty acid biosynthesis has been recognized, whereby the carbon backbones of the respective molecules are assembled by successive condensation of acyl

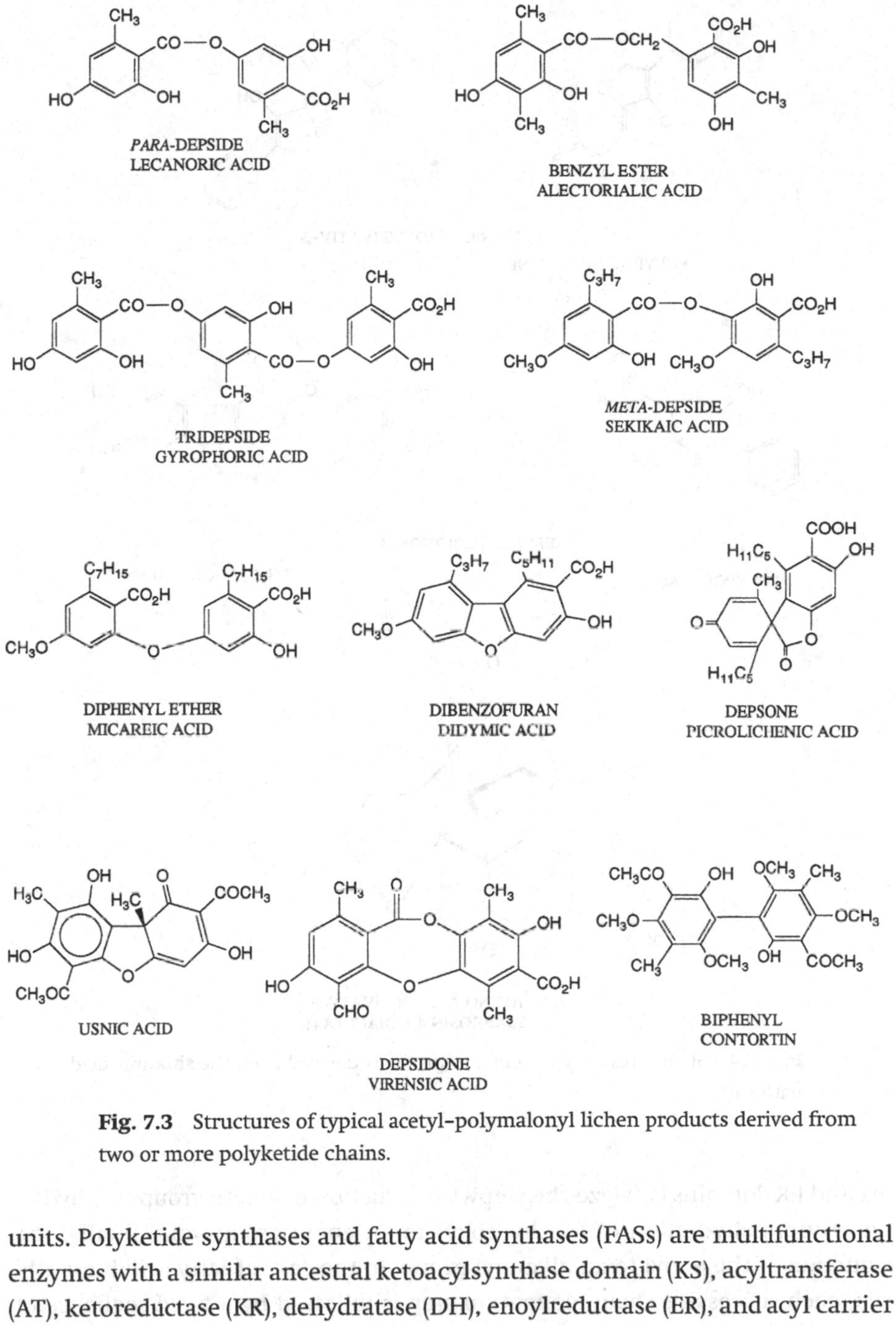

Fig. 7.3 Structures of typical acetyl–polymalonyl lichen products derived from two or more polyketide chains.

units. Polyketide synthases and fatty acid synthases (FASs) are multifunctional enzymes with a similar ancestral ketoacylsynthase domain (KS), acyltransferase (AT), ketoreductase (KR), dehydratase (DH), enoylreductase (ER), and acyl carrier protein (known as a phosphopantetheine attachment site or PP domain).

The KS, AT and PP domains are essential for both FASs and PKSs. Although the KR, DH, and ER are found in all FASs, some or all are absent in PKSs. Ketoreductase,

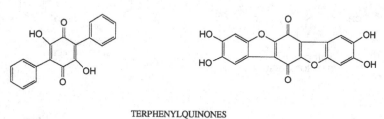

PULVINIC ACID DERIVATIVES

PULVINIC DILACTONE   CALYCIN

TERPHENYLQUINONES

POLYPORIC ACID   THELEPHORIC ACID

AMINO ACID DERIVATIVE
SCABROSIN 4,4'-DIACETATE

Fig. 7.4   Structures of typical lichen products derived from the shikimic acid pathway.

DH and ER domains catalyze the stepwise reduction of a keto group to a hydroxyl group, dehydration of the hydroxyl to an enoyl group, and, finally, the reduction of the enoyl to an alkanoyl group. In the case of fatty acid biosynthesis, each successive chain elongation step is followed by a fixed sequence of ketoreduction, dehydration and enoylreduction; whereas the individual chain elongation intermediates of polyketide biosynthesis undergo all, some, or none of the functional group modifications. This results in a remarkable diversity of

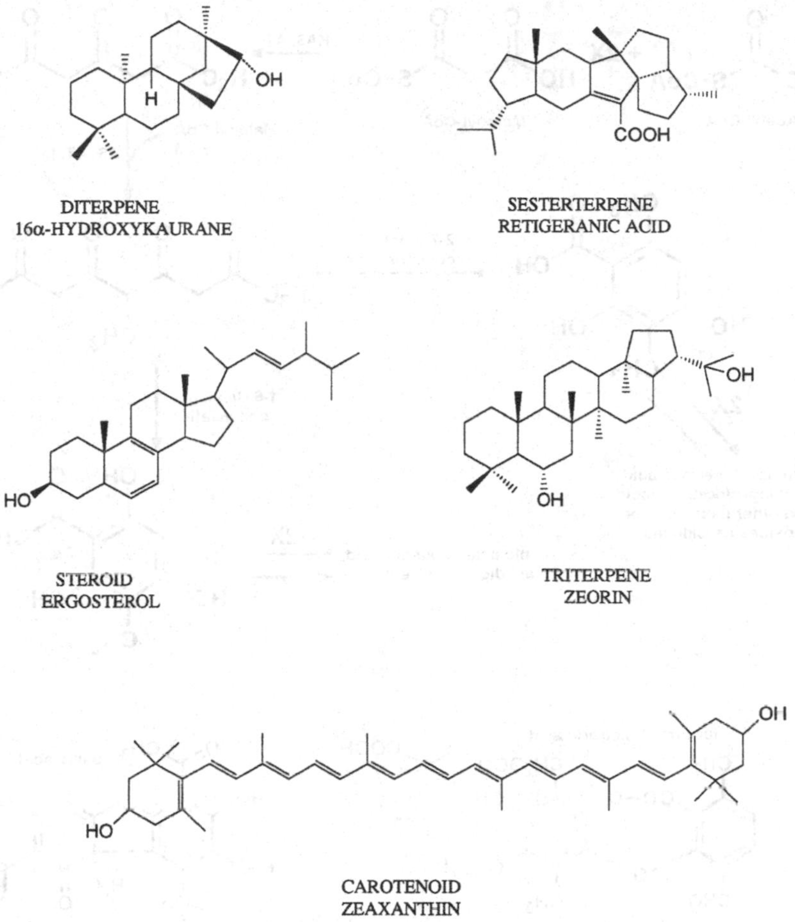

DITERPENE
16α-HYDROXYKAURANE

SESTERTERPENE
RETIGERANIC ACID

STEROID
ERGOSTEROL

TRITERPENE
ZEORIN

CAROTENOID
ZEAXANTHIN

Fig. 7.5 Structures of typical lichen products derived from the mevalonic acid pathway.

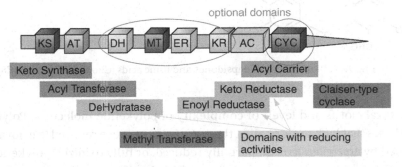

Fig. 7.6 Model of fungal type I PKSs.

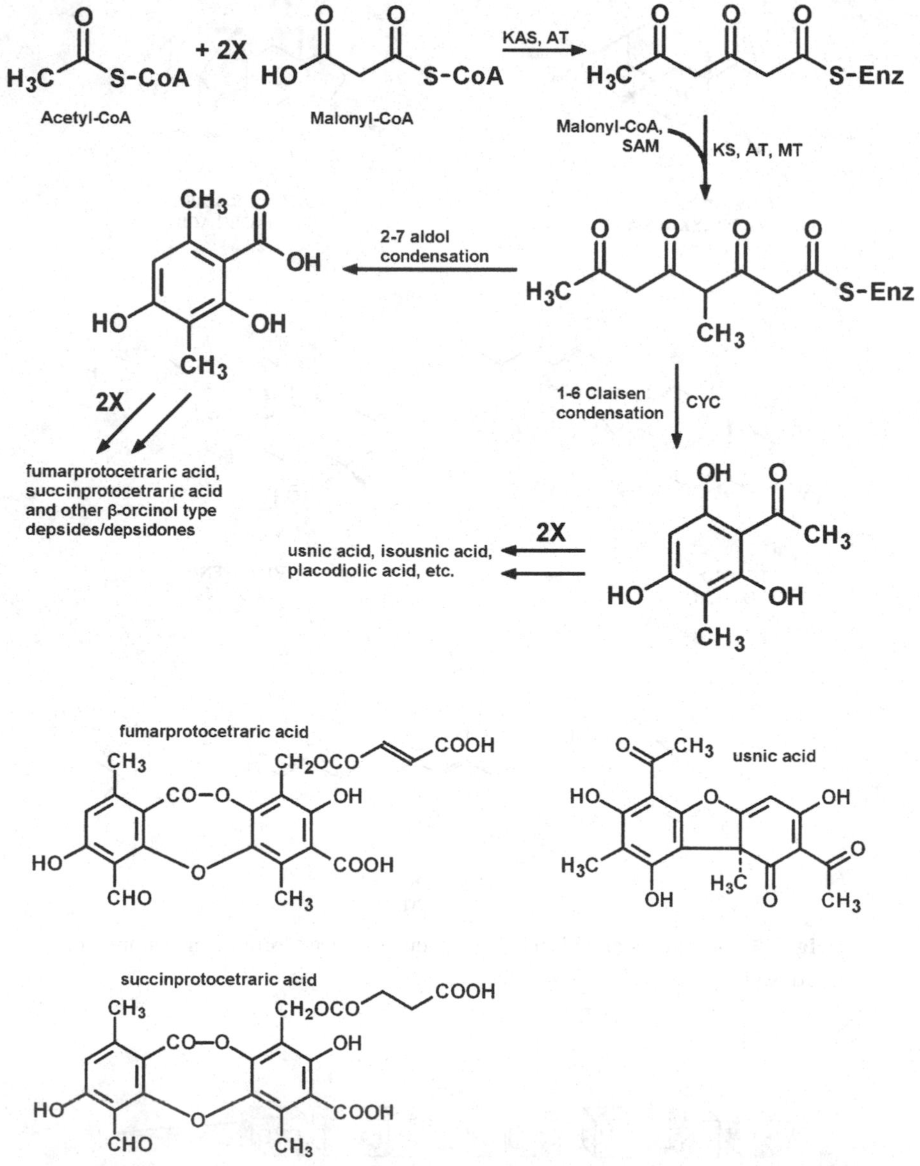

Fig. 7.7 Biosynthesis of depsidones and usnic acids (Chooi *et al.* 2006, IMC8).

structural motifs and levels of complexity of polyketide molecules. Polyketide synthases that lack some or all of these domains produce reduced (e.g. lovastatin formed by *Aspergillus terreus*), partially reduced, or fully oxidized polyketides.

Both types of polyketides are found in lichen-forming fungi, e.g. reduced polyketides, such as bourgeanic acid and anthrones (reduced anthraquinones), and fully oxidized polyketides, such as depsides, depsidones, β-orcinol depsidones, and dibenzofurans (Section 7.4). The formation of oxidized polyketides (most of the well-known and common lichen polyketides) is controlled by nonreducing PKS genes (Schmitt *et al.* 2005a).

7.5.3 Types of PKSs, biosynthesis of lichen polyketides, and PKS genes from lichens

Recent research has shown that three architecturally different polyketide synthases (PKSs) occur in the prokariotic and eukaryotic organismal world (Bedford *et al.* 1995; Cox *et al.* 1997; Hopwood 1997; Cane *et al.* 1998; Bingle *et al.* 1999; Nicholson *et al.* 2001). Types I and II are present in bacteria and fungi, have multifunctional enzymes or aggregates of monofunctional enzymes, which operate upon substrates bound by thioester linkages to an acyl carrier protein. Type III PKSs, found in higher plants, lack the ACP moiety, and instead use coenzyme A esters.

Type I systems consist of large multifunctional proteins, which can be either noniterative (e.g. modular systems responsible for biosynthesis of macrolides, large ring compounds such as erythromycin, rifamycin, etc.) or iterative (Cane *et al.* 1998). The iterative type I PKSs are single protein complexes (single modules), that contain all the necessary domains and use their active sites repeatedly (iteratively) to produce a particular polyketide. They add a C_2 molecule (e.g. a CoA ester) to the growing chain with each condensation and cycle repeat. The products of an iterative and noniterative PKS can be joined; and, in this case, result in the formation of a branched PK. The diversity of PKSs results from the use of the three optional PKS reducing domains as described above (Kroken *et al.* 2003).

Iterative type I polyketide synthases, analogous to vertebrate FASs, are typical for the biosynthesis of fungal polyketides, e.g. 6-methylsalicylic acid and aflatoxins. The former has been identified as a precursor of aflatoxins, as well as norsolorinic acid, and other anthraquinones. Anthraquinones are common, polyketide-derived pigments in lichens, and they also occur in nonsymbiotic fungi and higher plants (e.g. *Rumex* spp.). Another common fungal metabolite is orsellinic acid. In *Penicillium griseum*, penicillic acid is formed by a gross structural modification of orsellinic acid. Orsellinic acid is also a common precursor of many lichen substances, including depsides and depsidones.

To date, PKS genes have been found in clusters (genes adjacent along one stretch of a chromosome). Fungal secondary metabolites are encoded by clusters of sequentially arranged genes (Keller and Hohn 1997). Fungal PKS genes encode multifunctional proteins (fungal type I PKSs) with only one single, reiteratively used ketoacyl synthesis domain, which sequentially condenses C_2 units. The gene fragment encoding the ketoacyl domains are highly conserved and can be easily targeted with PKS primers.

Grube and Blaha (2003) and Schmitt *et al.* (2005a) have utilized a phylogenetic approach to elucidate the relationships between fungal genes by using amino acid sequences of KS (ketoacyl domains) of fungal PKSs, which putatively produce nonreduced (oxidized) polyketides (lichen substances). Such molecular analyses can be useful in reconstructing the evolutionary history of PKS genes in general, but more particularly, in identifying subgroups of type I PKS genes. In particular, they hope to elucidate the evolution of metabolic diversity within selected lichen orders or families that have been chemically characterized.

The location and identification of putative PKS genes and FASs are sometimes complicated by the fact that both genes have a strong sequence similarity, reflecting the roles of the enzymes coded by the genes. Other genes, e.g. those coding for cyclases, which catalyze the formation of aromatic polyketides (typical for lichens) have no counterparts among the genes coding for FAS (forming molecules arranged in long chains).

7.5.4   Heterologous expression of a lichen PKS in filamentous fungi

Although filamentous fungi produce an immense variety of polyketides (Simpson 1995), only a few PKS genes have been isolated (Sinnemann *et al.* 2000). The first PKS gene from a lichen has recently been isolated and sequenced by O. Andrésson and S. P. Davidsson (pers. comm., EUKETIDES Meeting, 2006). This PKS gene was obtained from *Solorina crocea*, then cloned and expressed in several filamentous fungi (*Aspergillus nidulans, A. niger, A. oryzae,* and *Fusarium venenatum*) by standard cloning and recombination technique. A 16 kb plasmid with a marker mediating hygromycin resistance was constructed, and together with a strong fungal promoter, the transcription of the lichen PKS gene was achieved. Further genetic transformation of *A. niger* with this plasmid construct yielded transformants that were able to produce a pigment of yet unknown chemical structure (from Gagunashvili and Andrésson, 2006). Such experiments hope to produce polyketides and polyketide-type pigments of considerable potential for drug discovery and novel biological activity.

7.5.5   Fatty acids and polyketides in lichens and cultured lichen fungi (mycobionts)

In many of the earlier investigations (Yamamoto 1990; Ahmadjian 1993; Kinoshita 1993), the majority of cultured mycobionts did not produce polyketides, i.e. typical lichen metabolites. Interestingly, when polyketides were produced in culture, alternative substances were often formed rather than those present in the original lichen or voucher specimens. The observed results were often difficult to interpret, and factors which favored the production of lichen substances remained unrecognized for several decades. More recently, it was shown that lichen mycobionts, which do not produce polyketides, may biosynthesize fatty acids instead. Molina *et al.* (2003) found that axenic cultures of *Physconia distorta* grown on nutrient-rich media produced mainly fatty acids (oleic, linoleic, and stearic acids) and their triglyceride derivatives, substances which were deposited on the surface of the mycelia as fat droplets. These experiments showed that FAS (fatty acid synthase) was switched on and activated, whereas PKS was obviously inhibited.

In another study (Adler *et al.* 2004), an aposymbiotically grown mycobiont, cultured under stable culture conditions, did not produce the typical medullary polyketide gyrophoric acid, but instead generated hydrocarbons, monoacylglycerides, and triacylglycerides. Such metabolic switching has also been observed in filamentous fungi, such as *Aspergillus nidulans* (Archer *et al.* 1999). Heterologous expression and cloning yielded both PKS and FAS genes. If PKS and FAS genes do not form separate gene clusters, the search for the location of putative PKS genes can be very difficult and needs advanced molecular genetic methods. In this case, the research challenge is to locate the genes that control the biosynthesis, modification, and in some cases secretion and resistance of polyketides. Then one could clone the PKS genes so that they can be moved to and be expressed in well-defined cell factories, like *Aspergillus niger* or *A. oryzae*.

A recent approach (Brunauer and Stocker-Wörgötter 2005; Stocker-Wörgötter 2005, 2008; Brunauer *et al.* 2006) was undertaken to sequence a cDNA of the PKS for anthraquinone production. The lichen and especially the cultured mycobiont of *Xanthoria elegans* were found to be excellent model organisms, as the aposymbiotically grown mycobiont readily produced anthraquinones, like parietin, teloschistin, etc. This meant that the mRNA was actually transcribed. From the axenically grown mycobiont, clean RNA was isolated, and then used for synthesis of cDNA by utilizing the SMART RACE cDNA synthesis technique. The SMART™ technology provided an efficient method for producing a cDNA pool, enriched in full-length cDNA and incorporating primer binding sites, at the 5′- and the 3′-ends of the cDNA, following 5′- and 3′-RACE (Rapid Amplification of cDNA Ends)-PCRs (Zhu *et al.* 2001). To obtain specific amplification of the PKS cDNA, gene specific primers (GSP) were designed based on the known sequence of the KS domain of the enzyme and used together with the oligos for the incorporated primer binding sites at the 5′- and 3′-ends of the cDNA. The resulting amplicons of the 5′- and 3′-RACE-PCRs were cloned into a T-vector and sequenced. The cDNA sequence was then analyzed using the ORF prediction program implemented in VestorNTI. The resulting amino acid sequence was then subjected to a Blast search against the NCBI database. The Blast search revealed high homology to other known PKS enzymes, especially to the wA gene product (Accession: Q03149; Fujii *et al.* 2001) of *Aspergillus nidulans*, which produces a polyketide structurally homologous to the polyketides of *X. elegans*.

Several catalytic domains on the enzyme could be identified and, finally, a gene bank for PKS of the *Xanthoria elegans* fungus was established.

Chooi *et al.* (2006) searched for a PKS gene responsible for production of β-orsellinic acid and methyl-phloroacetophenone, as precursors of typical lichen polyketides (depsidones and usnic acids) in the Australian lichen *Chondropsis semiviridis*. They found that both genes, either controlling depsidone and/or usnic acid production, were probably coregulated and were part of a common, larger gene cluster. In this case, the polyketide gene was identified by heterologous expression in a surrogate host, e.g. *Aspergillus nidulans*. For this procedure, the PKS gene clones of *Chondropsis semiviridis* were transformed and a strong promoter in the chosen transformation host was found.

Today heterologous expression of PKS genes and successive production of larger quantities of biologically active polyketides (e.g. actinomycetous compounds and fungal metabolites) are becoming common strategies to obtain and design pharmaceutically useful molecules. Probably in the near future similar molecules will also be found in lichens (Boustie and Grube 2005).

7.6   Detection and identification of secondary lichen substances

7.6.1   History

The application of chemical discriminators to lichen taxonomy began inadvertently when thalline color was accepted as a generic or specific character. Hence the gray-green genus *Physcia* (containing the colorless substance atranorin in the cortex) was segregated from the superficially similar but yellow-orange genus *Xanthoria* (containing cortical parietin, an orange pigment). Similarly *Parmeliopsis ambigua* (with a yellow thallus due to the presence of usnic acid) was separated from the morphologically similar *P. hyperopta* (gray with atranorin). Nevertheless, most lichen substances are colorless and can be detected only by indirect means. The first chemical tests conducted on lichen thalli for taxonomic purposes were carried out by Nylander in the 1860s (Nylander 1866; Vitikainen 2001). He detected the presence of various colorless lichen substances by spotting chemical reagents directly on the lichen thallus (spot tests) to produce characteristic color changes. He used solutions of iodine, potassium hydroxide (K), and calcium hypochlorite (C). Further test reagents followed – KC (K solution followed by C) and CK (with reverse addition) – but the origin of these characteristic color reactions remained unknown. The first extensive chemical investigations on lichens were conducted by Hesse and Zopf, culminating in Zopf's (1907) publication of *Die Flechtenstoffe*, in which descriptions of over 150 lichen compounds appeared. The ultimate structural elucidation of

Table 7.2. *Reagents for thalline spot tests*

K = 10% aqueous KOH solution
a. Turns yellow then red with most o-hydroxyl aromatic aldehydes.
b. Turns bright red to deep purple with anthraquinone pigments.

C = saturated aqueous Ca(OCl)$_2$ or common bleach (NaOCl) solution
a. Turns red with m-dihydroxy phenols, except for those substituted between the hydroxy groups with a -CHO or -CO$_2$H.
b. Turns green with dihydroxy dibenzofurans.

KC = 10% aqueous KOH solution followed by saturated aqueous Ca(OCl)$_2$ or common bleach (NaOCl) solution
a. Turns yellow with usnic acid.
b. Turns blue with dihydroxy dibenzofurans.
c. Turns red with C- depsides and depsidones which undergo rapid hydrolysis to yield a m-dihydroxy phenolic moiety.

PD = 5% alcoholic p-phenylenediamine solution
a. Turns yellow, orange or red with aromatic aldehydes.

many common lichen metabolites was due to the subsequent meticulous work of Asahina and coworkers in Japan during the 1930s (Asahina and Shibata 1954). This laid the foundation for further research on these compounds in recent times. Methods were recently summarized by Huneck and Yoshimura (1996).

7.6.2   Localization of secondary products

Thalline spot tests and the distribution of pigments provided the first evidence that the lichen substances were not distributed evenly throughout the thallus. In some species the striking red or orange anthraquinone derivatives and the yellow pigment usnic acid were obviously restricted to the upper cortex. Similarly spot tests demonstrated that many of the colorless depsides and depsidones were restricted to the medullary layer. The common spot test reagents (summarized in Table 7.2) not only indicate where particular compounds are located in sectioned thalli, but may also give a clue to the chemical nature of the substance.

In recent times scanning electron microscopy and laser microprobe mass spectrometry have been utilized to identify particular crystals present on or in the thallus. For instance, scanning electron microscopy (SEM) of the cortex of *Lecanora cerebellina* Poelt showed crystals with two morphologies. This species is known to contain the chlorinated xanthone, vinetorin. Luminescent crystals showing a signal for chlorine by energy-dispersive X-ray spectrometry (EDX) were identified as vinetorin, while nonluminescent crystals of different morphology gave a strong signal for calcium but little chlorine, and were tentatively identified as calcium oxalate. X-ray diffraction analysis of the crystals picked from the surface of *Pyxine caesiopruinosa* (Nyl.) Imsh. confirmed that the abundant bipyramidal crystals were calcium oxalate dihydrate and not the secondary product lichexanthone, known to occur in this species.

Crystals have also been identified by mass spectrometry using an instrument that combines a light microscope and a microprobe mass spectrometer with a lateral resolution of about $1\,\mu m$. Because ionization is laser induced, this method is also potentially applicable to thermally labile compounds. Laser microprobe mass spectrometry (LMMS) of solid inclusions in a cross section of the thallus of *Lauerera benguelensis* (Müll. Arg.) Zahlbr. confirmed the presence of lichexanthone by the prominent $(M + H^+)$ peak at m/z 287 in the positive ion spectrum. LMMS has also been coupled with fluorescence microscopy and transmission electron microscopy (TEM) to locate compounds in semithin sections of several species. The point analyzed on a laser microprobe is then examined in detail by TEM. For example, lichexanthone and russulone, a new tetracyclic anthraquinone, were located in different zones of the fruiting body of *Lecidea russula* Ach. (C. Culberson and Elix 1989).

7.6.3   Microchemical methods

Although the structural elucidation of lichen compounds results from a combination of classical chemical methods and modern spectroscopic techniques, the extensive data on the natural occurrence of these compounds are primarily based on microchemical methods of analysis. Extensive surveys based on extracts from herbarium specimens began in the 1930s when Asahina developed a simple microcrystallization technique to identify particular compounds, most of which were of known chemical structure and were located in particular histological regions of the thallus.

7.6.4   Microcrystallization

Asahina's microcrystallization technique allowed definitive recognition of individual lichen acids on a routine basis (Orange *et al.* 2001). This simple technique required no special equipment, and with experience generally yielded accurate analyses of major products. It involved extraction of a lichen fragment with acetone, evaporation of the solvent and recrystallization of the remaining residue from a suitable solvent – all conducted on a microscope slide (Asahina and Shibata 1954). A particular lichen substance crystallized in a distinctive shape and color and was identified by comparison with photographs of authentic materials (Hale 1974). Nevertheless, it soon became evident that this method could not detect minor components and was inadequate for the study of mixtures. This method is now superseded by more accurate and sensitive chromatographic methods. Even so, using Asahina's method, botanists discovered extensive correlations between chemistry, morphology, and the geographic distribution of lichens.

7.6.5   Paper and thin layer chromatography

In the period 1952–56 the Swedish chemist Wachtmeister introduced paper chromatography to identify lichen acids and their hydrolysis products (Wachtmeister 1956; Elix 1999). This method established that the chemistry of many species was more complex than was indicated by microcrystallization techniques. Experimental problems, poor spot resolution, low sensitivities, and long analysis times were subsequently overcome by the development of thin layer chromatography (TLC), which is now the most widely used method for identifying lichen products. This technique improved vastly the speed and certainty of recognition of lichen substances by means which are simple to use and relatively inexpensive.

7.6.6   Standardized TLC methodology

A standardized method developed by C. Culberson and coworkers remains in general use. It uses commercially available silica gel TLC plates and employs three solvent systems (designated A, B and C) and two internal controls (atranorin and norstictic acid), to which all R_f data are compared (C. Culberson 1972; C. Culberson and Amman 1979; C. Culberson et al. 1981). An acetone extract of the lichen is spotted on the plate and subsequently eluted in each solvent system. For each solvent system, a spot is assigned to an R_f class determined by its position relative to the controls. Data on punched cards or computer are then sorted to find all the compounds with the same R_f classes. Of these possibilities, those with similar spot characteristics (color, fluorescence, etc.) are compared chromatographically with the unknown. Additional solvent systems and visualizing agents are available for compounds that do not separate well in the initial analysis, and two dimensional TLC exhibits considerably improved R_f discrimination of structurally similar compounds and enables the identification of minor constituents present in complex mixtures (C. Culberson and Johnson 1976).

The solvent systems

Solvent A Toluene-dioxane-acetic acid (180:45:5) is reputed to owe its distinctive characteristics to the ability of dioxane to associate with phenolic hydroxy groups.

Solvent B Hexane-methyl *tert*.-butyl ether-formic acid (140:72:18) gives good separation of compounds that differ only slightly due to the length of side chains or the number of *C*-methyl substituents.

Solvent C Toluene-acetic acid (170:30) is an excellent general solvent for a wide variety of different compounds.

Solvent E Cyclohexane-ethyl acetate (75:25) is recommended for less acidic compounds that have high R_f values in solvents A, B, and C (e.g. many pigments, esters, triterpenes; Elix *et al.* 1988; Elix and Ernst-Russell 1993).

Solvent G Toluene-ethyl acetate-formic acid (139:83:8) is particularly useful in separating compounds with relatively low R_f values in solvents A, B and C (e.g. β-orcinol depsidones, secalonic acids).

Visualization of spots

One of the most useful features of TLC is the broad range of spot characteristics that can be used in addition to R_f data. Before spraying, the dried plates are examined in daylight for pigments and for fluorescence or quenching under short and long wavelength ultraviolet (UV) light. Subsequently the plates are sprayed with 10% sulfuric acid and then heated at 110 °C in an oven or hotplate for 10 minutes to develop the spots. Sulfuric acid charring detects the broadest range of compound types, including virtually all terpenes and phenolic derivatives (Fig. 7.8). After the plates are charred, the compounds give a range of characteristic visible colors and some even have a characteristic fluorescence (White and James 1985).

More recently a standardized TLC analysis procedure was developed to take advantage of computer technology (Elix *et al.* 1988; Mietzsch *et al.* 1993). This method utilizes six solvent systems and eight control compounds. Measured relative R_f values are used to sort within a computerized database. The programs also list biosynthetically related compounds as an aid to the identification of minor satellite substances.

High performance thin layer chromatography (HPTLC)

This modification of the standard TLC method utilizes TLC plates comprising a thinner layer of smaller grained silica particles (average 5–6 μm compared with 10–12 μm for ordinary TLC plates). It is reported to be a more sensitive method, requiring shorter run times and less solvent but is much more sensitive to humidity than the standard method (Arup *et al.* 1993).

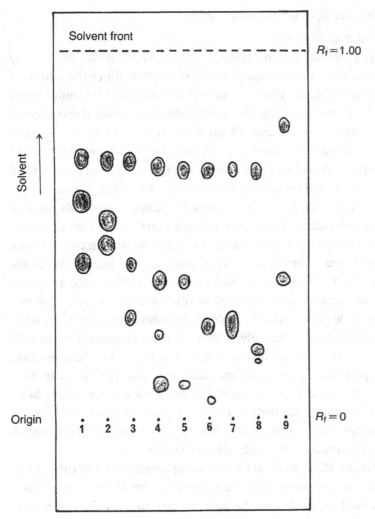

Fig. 7.8 Tracing of a TLC plate of extracts from *Xanthoparmelia* species in TA (toluene, 200: acetic acid, 30; solvent C). Compounds listed in decreasing R_f. 1, *X. barbatica* (usnic acid, barbatic acid, 4-*O*-demethylbarbatic acid); 2, *X. notata* (usnic acid, 4-*O*-methylhypoprotocetraric acid, notatic acid); 3, *X. scabrosa* (usnic acid, loxodin, norlobaridone); 4, *X. metastrigosa* (usnic acid, hypostictic acid, hypoprotocetraric acid, hypoconstictic acid); 5, *X. terrestris* (usnic acid, norstictic acid, salazinic acid); 6, *X. tegeta* (usnic acid, stictic acid, constictic acid); 7, *X. hypoprotocetarica* (usnic acid, hypoprotocetraric acid); 8, *X. pertinax* (usnic acid, succinprotocetaric acid, fumarprotocetraric acid); standard mixture (atranorin, norstictic acid).

7.6.7   High performance liquid chromatography (HPLC)

Isocratic and gradient elution

All of the aromatic lichen products are ideally suited for analysis by high performance liquid chromatography (Elix *et al.* 2003). Since the advent of bonded reverse phase columns, this technique provides a powerful complement to the established TLC methods. Samples are dissolved in methanol and injected into the appropriate partition column, through which an appropriate solvent or sequence of solvents is passed under high pressure. The substances separate and are detected using a UV detector. The retention time (Rt, or time of passage) and peak intensity are recorded by a chart recorder (Fig. 7.9). HPLC is also used to measure either absolute or relative concentrations of lichen compounds, because the peak intensity (area under the curve) is proportional to the concentration. Earlier applications utilized isocratic elution (an eluant of constant composition) to achieve excellent separations of a variety of depsides, depsidones, and dibenzofuran derivatives (C. Culberson *et al.* 1979; Lumbsch and Elix 1985). However, gradient elution methods (using a sequence of solvent mixtures) are more efficient in analyzing crude lichen extracts, which often contain compounds of wide-ranging hydrophobicities. For example, a 30-minute linear gradient from 0.5% acetic acid in water to 100% methanol was used to separate six known (constictic, stictic, norstictic, psoromic, gyrophoric, and rhizocarpic acids) and seven unidentified components in the *Rhizocarpon superficiale* group (Geyer *et al.* 1984). More recently Archer and Elix (1993) used a 35-minute gradient from 30% aqueous methanol containing 0.7% *ortho*-phosphoric acid to 100% methanol to distinguish ten components in an undescribed *Pertusaria* species (Fig. 7.9).

Most workers using HPLC to detect lichen compounds combine this technique with TLC and/or mass spectrometry to verify the identification of the peaks. This has often proved to be a bonus, because the unique chemistry of reverse phase separations and the high sensitivity of UV detectors led to the discovery of many compounds new to science. Nevertheless, verification of the identity and purity of peaks remains a problem in screening large numbers of specimens.

Retention indices

New standardized methods that use retention indices relative to two internal standards (one of low and one of high retention) rather than retention times, avoid problems caused by column age and minor variations in solvent composition (Huovinen *et al.* 1985; Feige *et al.* 1993). Retention indices also provide structural information; for example, C. Culberson *et al.* (1984) showed that there was a linear relationship between the retention index and the number of side-chain carbon atoms in homologous series of naturally occurring orcinol depsides and their hydrolysis products. Thus, this method can provide

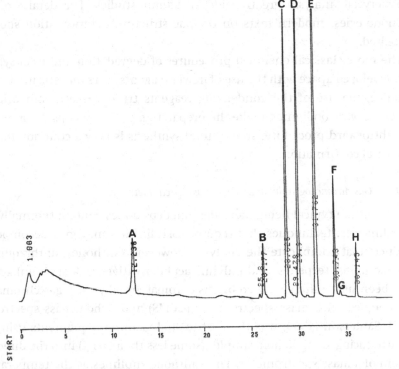

Fig. 7.9 Trace of HPLC of methanol extract of *Pertusaria*. sp. (horizontal scale in minutes). A, benzoic acid (internal standard); B, 4,5-dichloronorlichexanthose; C, arthothelin; D, asemone; E, thiophanic acid; F, 3-0-methylasemone; G, 6-0-methyl-asemone; H, superlatolic acid (internal standard).

the first clue to the identity of new products (particularly satellite compounds) in lichens. Because of the expense and technical complexity of HPLC, TLC will probably continue to be more widely used in routine identifications. Even so, HPLC is the method of choice for detecting trace satellite compounds, analyzing very small samples, quantifying the lichen products present, and for providing structural information from retention characteristics.

7.6.8   Chemical methods

As with many areas of natural product chemistry, new impetus in the chemistry of lichen substances is provided by the more rapid and improved methods for detecting, isolating, and purifying these compounds and in determining their structure. The techniques of preparative TLC, radial chromatography and preparative HPLC provide rapid and efficient methods for the purification of lichen substances, and developments in mass spectrometry, proton and carbon-13 NMR (nuclear magnetic resonance) spectroscopy and

X-ray crystal analysis greatly aide structural studies. For details of these methodologies, modern texts on organic structure determination should be consulted.

The more classical chemical procedures of degradation and total synthesis also developed apace with the use of newer reagents and synthetic methods. For instance, the use of the condensing reagents trifluoroacetic anhydride and dicyclohexylcarbodiimide make the preparation of lichen depsides a relatively straightforward procedure, so that total synthesis is now a common means of structural confirmation.

7.6.9   Gas chromatography and lichen mass spectrometry

As the typical lichen depsides and depsidones contain thermally labile ester linkages, techniques that require volatilization may give decomposition products that complicate the analysis. However, xanthones, anthraquinones, dibenzofurans, terpenes, and pulvinic acid derivatives lack such linkages and have been successfully studied by gas chromatography (GC), gas chromatography coupled with mass spectrometry (GCMS) and lichen mass spectrometry (LMS). Santesson (1969) was the first to study xanthone pigments using LMS, by introducing small lichen samples (some less than 50 ng) into the direct inlet system of a mass spectrometer. The xanthone sublimes as the temperature is raised (100–150 °C) under very low pressure and the mass spectrum is recorded. Xanthones generally give prominent molecular ions and the spectra of mixtures can often be seen as additive of the individual components (Fig. 7.10; Elix 1999). Aptroot (1987) was able to resolve and identify the main terpenoid components of many lichens of the Pyxinaceae by GCMS, and found this technique to be far superior to standardized TLC for these particular compounds.

7.7   Application to systematics

The secondary metabolites in over 5000 lichens, approximately 33% of the known species, have now been studied and metabolite data are used more extensively in the routine identification of lichens than in any other group of organisms. These data are not only used extensively in lichen systematics but also in discussions of origins and relationships.

7.7.1   Cortical chemistry

For many years taxonomists consistently underestimated the ecological importance and possible evolutionary significance of the chemical

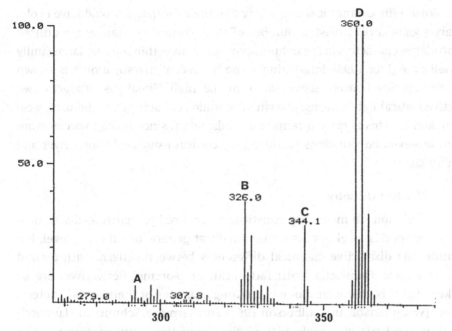

Fig. 7.10 Lichen mass spectrum of *Lecanora* sp. (high mass region). Horizontal scale = m/z values. Vertical scale = % abundance. Molecular ion peaks: A, 4-chloronorlichexanthone; B, 4,5-dichloronorlichexanthone; C, usnic acid; D, arthothelin.

components of the upper cortex in lichens. Certainly it was recognized that some cortical substances are correlated with higher taxonomic ranks–for example, at generic level: vulpinic acid in *Letharia*; or at the family level: anthraquinones and particularly parietin in the Teloschistaceae. These cortical compounds appear to have systematic significance because of their vital ecological roles.

In lichens growing on exposed substrates, various light-absorbing compounds are located in the upper cortical tissue of the vegetative and generative parts of the thallus, and these cortical lichen substances commonly show variation in concentration along light gradients. Clear evidence suggests that these pigmented compounds have a primary biological role as light-screens, regulating the solar irradiation reaching the algal zone in the upper cortex. In addition to the general filtering effect (*Trebouxia* grows best at relatively low light intensity), these compounds may have a secondary value in protecting the lichen thallus from excessive ultraviolet irradiation (Solhaug and Gauslaa 1996; Rancan *et al.* 2002; Rubio *et al.* 2002). The major groups of substances involved include the β-orcinol *para*-depsides atranorin and chloroatranorin, the usnic acids, anthraquinones, xanthones and pulvinic acid derivatives. Given their apparent physiological importance it seems likely that their formation would have evolutionary significance. Indeed, a number of these cortical substances are utilized as correlative characters in the delimitation of genera within the very large family Parmeliaceae (Elix 1993). Interestingly another recent investigation has shown that the species *Lecanora somervellii* from the high Himalayas produces two effective cortical light-screens, calycin (a pulvinic acid derivative) and usnic acid (a polyketide). This is quite a remarkable adaptation since related species from lower, less-exposed situations produce only cortical usnic acid (Obermayer and Poelt 1992).

7.7.2   Medullary chemistry

Variations in medullary constituents are used primarily as discriminators at the species level but also occasionally at generic or suborder level. For example, the distinctive chemical differences between *Cetrelia* (with orcinol derivatives) and *Platismatia* (with fatty acids or β-orcinol derivatives) are so marked that this character alone is a strong indicator of their generic heterogeneity (W. Culberson and Culberson 1968). Furthermore, Schmitt and Lumbsch (2004) on the basis of a molecular phylogeny of the Pertusariaceae support secondary chemistry as important systematic characters for the family.

At suborder level some chemical substances have restricted distributions. For instance, the hopane triterpenes appear to be distributed in the Lecanorineae (e.g. in *Physcia*, *Heterodermia*), the Cladonineae (e.g. in some *Cladonia*), the Teloschistineae (e.g. some *Xanthoria*) and the Peltigerineae (in *Pseudocyphellaria* and *Peltigera*), whereas sesterterpenes are found only in the Peltigerineae (in *Pseudocyphellaria*) and the dammarane triterpenes in the Lecanorineae (some *Pyxine*).

The use of chemical discriminators at species level has been a controversial topic (Lumbsch 1998). The discovery of chemical differences often led to an appreciation of the importance of previously overlooked morphological features, as in *Punctelia subrudecta* with lecanoric acid and a pale tan lower surface, and *P. borreri* with the related tridepside gyrophoric acid and a black lower surface; these species also have different geographic distributions. In some species complexes the morpholological variations may be more subtle, as is observed with the *Cladonia chlorophaea* group, within which there are correlations of the different chemical races with color and soredia size (at least in parts of their range).

Fortunately, most morphologically defined species have a constant chemistry, irrespective of their geographic origin, substrate or ecology, and this justifies the use of chemistry in lichen taxonomy. Within a complex of morphologically similar species, three common patterns of chemical variation are observed: replacement compounds, chemosyndromic variation, and accessory type compounds. With replacement type compounds, congeneric chemotypes show simple replacement of one substance by another. Morphologically these lichen populations are sometimes indistinguishable, but they have well-defined, constant variations in chemical composition. One of the classical examples is that of *Pseudevernia furfuracea*, which has three chemical races: an olivetoric acid-containing race from northern Europe; a physodic acid-containing race from southern Europe and north Africa; and a lecanoric acid-containing race from North America. Biogenetically the first two races appear closely related. The metabolites can be considered biosequential, because one can be derived from the other by a single biosynthetic step. But the third race is not related, because lecanoric acid is biosynthetically remote from the other two compounds. It is now generally accepted that, when there is a biogenetic demarcation allied with a biogeographical separation, such taxa should be recognized as species and the North American taxon is distinguished as *P. consocians*. More recent studies of the European races revealed that they possess distinctive but overlapping chemistries and show no significant correlation with habitat ecology, a result which convinced some lichenologists that these races represent a single species that shows some chemical variation (Dahl and Krog 1973; W. Culberson *et al.* 1977).

In summary, most lichenologists, who recognize chemically distinct races as species, support their decision primarily on the basis of the different geographic distributions that such races usually show. However, it is suggested by the Culbersons that the best evidence for chemical variation being under genetic control, rather than being environmentally determined, is the fact that chemical races, where sympatric, maintain their integrity even when growing side by side (W. Culberson 1967; W. Culberson and Culberson 1967). As a corollary, the occurrence of chemical intermediates in areas of sympatry either indicates that such races belong to a single species or that hybridization is occurring between the races.

The existence of chemosyndromic variation in some lichen groups may make the recognition of chemical intermediates more difficult (C. Culberson and Culberson 1976). A chemosyndrome refers to a group of biosynthetically related metabolites and in this pattern of chemical variation the major metabolite (or metabolites) in any one taxon is invariably accompanied by minor quantities of several biosequentially related substances (Table 7.3). Further, the compounds that are the major constituents of some species may be minor constituents of related taxa and vice versa. Hence a true chemical intermediate cannot simply be defined as containing both of two replacement compounds, but would have to contain both chemical constellations in comparable concentrations.

Table 7.3. *Chemosyndromic variation in the* Relicina samoensis *complex*

Species (distribution)	Echinocarpic	Conechinocarpic	Hirtifructic	Gyrophoric	Fatty acids	Distribution
R. samoensis	major	minor	–	–	–	Pan-Pacific
R. amphithrix	major	minor	–	–	–	Australia/ Indonesia
R. terricrocodila	major	minor	trace	–	–	Australia
R. fijiensis	–	–	major	–	–	Fiji
R. niuginiensis	–	–	major	trace	minor	Papua New Guinea
R. relicinula	–	–	–	–	major	Indonesia

Source: After Elix (1991).

The biogenetic relationships between secondary products in lichens is used in cladistic analysis of evolutionary relationships amongst taxa. Thus C. Culberson (1986) has used the *Cladonia chlorophaea* group as an example of how preliminary biogenetic hypotheses can lead to a cladogram for the 14 chemotypes presently known in this complex.

A further important feature of chemical races is their ecology. In several cases that are studied in detail, different chemical races were found to be ecologically sorted into distinct habitats in their range of sympatry. Although the underlying physiological causes of this sorting or the related phytogeographically significant distributions remains unknown, they do indicate that the chemical races have a more than superficial genetic basis.

W. Culberson (1986) recently presented a very convincing case for interpreting the ecological and biological characteristics of the major chemotypes, into which many Linnean species of lichens are divided, as indicating that these chemotypes are better considered as sibling species rather than as components of traditional morphological species. In fact sibling speciation, where the reproductive isolation of populations is often accompanied by ecological but little or no morphological differentiation, is a common product of evolution and is well documented amongst animals and in the vascular plants.

Even so, caution must be exercised because of the occurrence of accessory metabolites. These substances occur sporadically in a species, usually in addition to the constant constituents, and have no correlation with any morphological or distributional variations and hence are accorded no taxonomic significance. Such compounds commonly occur as accessory compounds in more than one species and often vary in quantity from deficiency to abundance.

In summary, most chemotypes (i.e. disregarding accessory chemical variations) appear to have subtle morphological, ecological or distributional tendencies and consequently should be afforded some taxonomic recognition.

7.7.3   Cell wall polysaccharides

These polymeric storage products of lichens require different techniques for their detection, study, and structure determination (Gorin et al. 1988; Common 1991). The best known lichen polysaccharides are lichenan, isolichenan, and galactomannan, each of which has a range of different, but related, chemical structures depending on the parent lichen.

While secondary products are often useful taxonomically at the specific and generic level in lichens, polysaccharide content is often diagnostic for larger phylogenetic units (Shibata 1973; Common 1991). Polysaccharides have a fundamental role in the biochemistry of fungi and tend to be conservative features in their evolution. Some polysaccharides are taxonomically significant at the highest levels of classification. For example, the presence of chitin, chitosan, or cellulose in the cell wall is a feature which helps define the classes of fungi. Within the class Oomycetes, Aronson and coworkers showed that a biochemical dichotomy exists with respect to hyphal wall composition between Rhipidiaceae and Leptomitaceae, the two families comprising the order Leptomitales. Furthermore, these biochemical differences paralleled the traditionally accepted morphological and anatomical differences between these families (Aronson 1977). In a similar vein, Shibata and coworkers (Shibata 1973) showed that pustulan is a characteristic polysaccharide in the Umbilicariaceae (Umbilicaria and Lasallia), and glycopeptides are important cell-wall components of the Lobariaceae. The taxonomic utility of such chemical characters in the Parmeliaceae was developed by Common (1991) who recognized four major groups: (1) isolichenan, (2) a Xanthoparmelia-type lichenan, (3) a Cetraria-type lichenan, or (4) an intermediate-type lichenan. Chemically these polysaccharides differ primarily in the stereochemistry of the glycosidic bonds, being largely b in lichenan and a in isolichenan. Xanthoparmelia-type lichenan, Cetraria-type lichenan and the intermediate-type lichenan differ primarily in their staining properties with various iodine reagents (Table 7.4); the structural features responsible for these differences have yet to be elucidated. The utility of this character is readily demonstrated by application to related but well-accepted genera (for example, Hypogymnia contains Cetraria-type lichenan whereas Menegazzia contains isolichenan). It is also used as one of the primary discriminators to differentiate the following yellow-green parmelioid genera: Psiloparmelia and Flavoparmelia (containing isolichenan) from Arctoparmelia (Cetraria-type lichenan) and Xanthoparmelia (Xanthoparmelia-type lichenan).

Table 7.4. *A comparison of the staining properties of isolichenan and various types of lichenan*

Polysaccharide	20–0.15% IKI	0.15% LPIKI	1.5% IKI	CaIKI	ZnIKI	SIKI	Meltzers
Isolichenan	blue	pale blue	bluish	bluish	bluish	–	bluish
Cetraria-type lichenan	–	–	red	deep red	–	red ppt.	orange
Xanthoparmelia-type lichenan	intense blue	–	red	deep red	purple	red ppt.	deep red
Intermediate-type lichenan	pale blue	–	red	deep red	–	red ppt.	red

Source: IKI, iodine, potassium iodide solution; LP, lactophenol; S, 10% sulfuric acid; Ca, calcium chloride; Zn, zinc chloride; Meltzers Reagent, chloral hydrate + iodine potassium iodide. Summarized from Common (1991).

7.8   Application to pharmacology and medicine

The use of lichens in folk medicines persists to the present day (Richardson 1988). Both the Seminole Indians in Florida and the Chinese herbal doctors employ various lichens in medicines, especially as expectorants. *Usnea* species were most commonly utilized. *Cetraria islandica* ("Icelandic Moss") is claimed to be effective in treating lung diseases and catarrh, and preparations from this species are still sold in Europe, usually as pastilles. *Peltigera canina* is eaten in India as a remedy for liver ailments and its high content of the amino acid methionine may be the basis for its alleged curative power (Hale 1983).

Today a wide range of secondary metabolites are recognized as having medicinal value (Pearce 1997). For example, Burkholder *et al.* (1944; Burkholder and Evans 1945) discovered that extracts from 52 different species of lichen in eastern North America inhibited growth of several kinds of bacteria. This led to a feverish race to identify the antibiotic components of such lichens. The antibiotic effect of a number of lichen metabolites was found to be significant for gram-positive bacteria, but ineffective against gram-negative bacteria. Thus gram-positive bacteria were significantly inhibited by usnic acid, protolichesterinic acid, and a variety of orcinol derivatives.

The usnic acids (Ingólfsdóttir 2002) have also been found to exhibit antihistamine, spasmolytic, and antiviral properties as well as being active against gram-positive bacteria and streptomycetes. Indeed they are used in commercially available antiseptic creams including "Usno" and "Evosin." Usnic acid is reported to be more effective than penicillin salves in the treatment of external wounds and burns and is also used to combat tuberculosis. The active centers of the usnic acid molecule seem to be the benzofuran or dihydrodibenzofuran nucleus, the phenolic hydroxy groups and the 4,4a-double bond in the dihydroaromatic ring (Asahina and Shibata 1954). The antibiotic action of usnic acid is due to the inhibition of oxidative phosphorylation, an effect similar to that shown by dinitrophenol. More recently Shibuya *et al.* (1983) showed that 4-0-methylcryptochlorophaeic acid was a powerful inhibitor of prostaglandin biosynthesis and a potentially useful anti-inflammatory drug.

Lichen substances are also known to exhibit antitumour activity. Usnic acid has low level activity against lung carcinoma. However, the most active antitumor lichen substances are water soluble polysaccharides which appear to be partially 0-acetyled homo-D-glucans. The lichen polysaccharide GE-3 was shown to be a host-mediated antitumor active substance, effective because of its stimulation of the immune system. The administration of GE-3 to mice caused inflammatory changes in the liver, a temporary increase in leukocytes followed by excretion of an a1-acid glycoprotein. Purification of the latter led to the isolation of a1-AG-1, which inhibited growth of cancer cells (Shibata 1992). Sulfated GE-3 (GE-3-S) was the most promising agent for HIV suppression. GE-3-S (S content 13.8%, mol. wt. 200 000) was prepared by sulfonation of GE-3 by $ClSO_3H$. Dextran sulfate and heparin are also suppressive against HIV infection. None of these polysaccharide sulfates expresses an inhibitory effect on cell free HIV reverse transcriptase activity. Therefore the polysaccharide sulfates appear to interfere with the adsorption of HIV particles onto the surface of T4 cells (Shibata 1992).

Antifungal activity is also found among lichen substances. Growth of the mould *Neurospora crassa* is strongly inhibited by usnic acid, as well as by haematommic acid, a monocyclic phenol derivative present in some lichen depsides (Hale 1974). A number of lichen metabolites also act as plant-growth regulators, with usnic acid being particularly active (Huneck and Schreiber 1972).

Although there are few commercial applications of lichen substances, the variety of antibiotic properties they exhibit obviously encourages further investigations (Pearce 1997; Huneck 1999).

7.9   Harmful properties of lichen substances

In northern Europe the lichen *Letharia vulpina* was used traditionally as a poison for foxes and wolves. The toxic principle is the pulvinic acid derivative vulpinic acid, which is not only poisonous to all meat eaters but also to insects and mollusks. Surprisingly this compound is ineffective against rabbits and mice. The secalonic acid derivatives are also highly poisonous. These substances are mycotoxins and, like vulpinic acid, may have evolved to serve a twofold ecological role. Thus, in addition to screening incoming light, they are highly poisonous to grazing herbivores.

Contact dermatitis, a severe skin rash, is well known among forestry and horticultural workers in North America, forming part of a syndrome known as "woodcutter's eczema" or "cedar poisoning." These complaints are an allergic response resulting from exposure to various lichen substances. Among the lichen substances responsible are usnic acid, evernic acid, fumarprotocetraric acid, stictic acid, and atranorin. Usnic acid, for instance, is a common lichen substance in the corticolous species of *Alectoria*, *Evernia*, and *Usnea*, which are widespread in the forests of North America. A dusting of soredia on clothing causes allergic reactions in the wives of lumbermen not directly exposed in the forests. Atranorin and stictic acid are also capable of photosensitizing human skin as well as being contact allergens. This can lead to photocontact dermatitis, where the allergic reactions become much more acute when the persons are exposed to the lichen substances in combination with light (Hale 1983; Richardson 1988).

Periodically hundreds of elks die in western North America when these large ruminants are forced out of their normal winter habitats by excessive snows and at lower elevations primarily find *Xanthoparmelia chlorochroa* to eat (Durrell and Newsom 1939; MacCracken *et al.* 1983; Anonymous 2004). Although the toxin is fully resolved, the abundance of salazinic acid is suspected. In contrast, these animals eat other epiphytic lichens without apparent ill effects.

7.10   Lichens in perfume

One of the more important economic uses of lichens today is in the perfume industry. The two most important species, *Evernia prunastri* ("oak moss") and *Pseudevernia furfuracea* ("tree moss") are harvested in southern France, Morocco and the former Yugoslavia in large quantities, with a harvest in the range of 8000–10 000 tonnes annually. The combined lichen material and tree bark is subsequently extracted with an organic solvent and treated with ethanol. The concentrate of this solution contains a mixture of essential oils and depside derivatives (degradation products). The final extract with its sweet "mossy" smell is used in some perfumes to ensure persistence on the skin, as the major ingredients do not evaporate readily. The lichen extract may amount to 1–12% of the finished perfume. The precise identity of the scented component remains a trade secret but comprises a very small proportion (*c.* 0.04%) of the total extract, the majority of which comprises borneol, cineole, geraniol, citronellol, camphor, naphthalene, orcinol, orsellinate esters and their homologues (Moxham 1980; Richardson 1988; Hiserodt *et al.* 2000).

7.11   Lichens in dyeing

Lichens were used as a source of dyestuff from the time of the ancient Greeks and probably earlier (Henderson 1999), but are of little economic importance today. Historically *Roccella montagnei*, a common fruticose lichen on rocks, provided valuable red or purple dyes in the Mediterranean region. These dyes were produced by "fermenting" the *Roccella* or chemically equivalent species (*Ochrolechia tartarea*, *O. androgyna*, or *Parmotrema tinctorum*) with dilute ammonia solution. The macerated lichen and dilute ammonia were sealed in a container containing twice the volume of air. The purple color developed after a week and was used as a direct dye (orchil) for protein fibres (wool and silk). The simple *para*-depsides erythrin (*Roccella*) and lecanoric acid (*Ochrolechia* and *Parmotrema tinctorum*) present in these lichens are responsible for these colors. Rapid base hydrolysis of the lecanoric acid or erythrin by ammonia gives ammonium orsellinate and then orcinol (by decarboxylation). Subsequent oxidative coupling in the presence of ammonia gives rise to the dyestuff, orcein, which comprises a mixture of three major chromophores, 7-hydroxyphenoxazone, 7-aminophenoxazone and 7-aminophenoxazine (Hale 1983). The common acid-base indicator litmus, formerly widely used in chemistry laboratories, is closely related to orcein but represents a more complex mixture of polymeric compounds with the 7-hydroxyphenoxazone chromophore and its anion being responsible for the sensitivity of the color to pH.

Some Harris tweeds manufactured in Scotland are still dyed with lichen dyestuffs (Richardson 1988). All of these dyes are quite colorfast, and impart a unique musty odor to the fabric. For instance, *Parmelia omphalodes* is utilized to provide a rich brown dye for dyeing protein fibres, particularly wool. Subsequent investigations showed that this was due to the salazinic acid present, and in fact most of the tawny yellow-brown to reddish-brown colors produced on wool by lichen dyes are produced by lichen substances with *o*-hydroxyaldehyde functionalities. The aldehyde functional group condenses with free amino groups present in the wool proteins to form a stable Schiff base (azomethine) linkage (Hale 1974). For practical applications, Casselman (2001) is a useful reference.

8

Stress physiology and the symbiosis

R. P. BECKETT, I. KRANNER, AND F. V. MINIBAYEVA

Lichens are the dominant life forms in about 8% of the land surface of the Earth (Ahmadjian 1995), mainly in polar regions and on the tops of mountains. These places are characterized by severe abiotic stresses such as desiccation, temperature extremes, and high light intensities. Arguably, what really makes lichens special, and what separates them from most other eukaryotic organisms, is their ability to tolerate extreme stresses. For this reason, some have called lichens "extremophiles," organisms that can thrive in conditions that would kill other, less specialized organisms. Scientists have found that hardy lichens can survive a trip into space, and now the list of natural astronauts includes lichens. During a recent experiment by the European Space Agency, lichen astronauts were placed on board a rocket and launched into space, where they were exposed to vacuum, extreme temperatures, and ultraviolet radiation for two weeks. Upon analysis, it appeared that the lichens handled their space-flight just fine (Young 2005)!

In the typical environments that many lichens inhabit, stresses such as low thallus water content and temperature extremes can develop within just a few minutes. However, others, such as a nutrient deficiency, can take months to develop. The stressfulness of a particular habitat is the result of the interaction of climate and substrate. It plays a major role in determining lichen distribution. Understanding the physiological processes that lie behind stress injury, and how lichens tolerate environmental stress, is therefore of great importance in lichen biology.

A "stress" factor can be defined as any external influence that has a harmful effect on an organism. This chapter will discuss environmental or "abiotic" factors that produce stress in lichens, although biotic factors, such as competing higher plants or other lichens, pathogens, and insect predation can also result in

Lichen Biology, ed. Thomas H. Nash III. Published by Cambridge University Press.
© Cambridge University Press.

stress. It is rarely possible to see at a glance whether a lichen is alive or dead. Furthermore, because of their slow growth rates, it is difficult to use growth to assess stress. Instead, lichenologists tend to measure parameters such as the inhibition by stress of net photosynthesis or chlorophyll fluorescence to quantify stress effects on the photobiont. If researchers are more interested in the mycobiont, they may study the stress-induced inhibition of respiration or leakage of intracellular soluble potassium through membranes.

The concept of stress is closely linked to that of stress tolerance, which is the ability of an organism to cope with an unfavorable environment. If tolerance increases as a result of exposure to prior stress, the plant is said to be acclimated (or hardened). Acclimation can be distinguished from adaptation, which usually refers to a genetically determined level of resistance acquired by a process of selection over many generations. Adaptation and acclimation to environmental stresses result from changes that occur at all levels of organization, from the anatomical and morphological level to the cellular, biochemical, and molecular level. The specific stresses we will consider here include desiccation, temperature extremes, and high light intensities. Air pollution is an important source of stress, and is discussed in Chapter 15, while Chapters 11 and 12 review nutrient stresses. Although it is convenient to examine each of these stresses separately, most are interrelated, and a common set of cellular, biochemical, and molecular responses accompanies many of the individual acclimation and adaptation processes. For example, desiccation and light stress often accompany high temperatures. Furthermore, at the cellular level, many stresses may have the same effects, for example the production of reactive oxygen species and damage to the cytoskeleton. We shall see that while each stress causes particular problems for lichens, the underlying mechanisms of resistance probably share many common features with each other, and with those in other organisms.

8.1   Stress tolerance – protection or repair?

We are still far from understanding fully the mechanisms of stress tolerance in lichens. In particular, we are not sure if tolerance is largely a matter of reassembly and reactivation of components conserved intact through a time of stress, or whether a more or less extensive "repair" process is involved. For example, during rehydration following desiccation some parameters, for example chlorophyll fluorescence, recover almost immediately following rehydration (e.g. Beckett *et al.* 2005*b*). This suggests that the integrity of the thylakoids is preserved throughout the events of desiccation and remoistening. On the other hand, the fact that *complete* recovery of carbon fixation can take much longer suggests the involvement in recovery of other cellular mechanisms, as yet unknown. Furthermore, during rehydration following desiccation, membranes are initially leaky to ions and metabolites, but later regain their integrity (Weismann *et al.* 2005*a*). Therefore, it is possible that some form of repair-based desiccation tolerance mechanism exists. In bryophytes, initial studies on moss ultrastructure during and after desiccation seemed to imply that some form of repair takes place. Freshly rehydrated plants appear to display swelling of chloroplasts and mitochondria, and major changes in the endomembrane domains and microtubular cytoskeleton – damage that is repaired gradually. However, recent more careful investigations using improved procedures for preparing material for microscopy (e.g. Pressel *et al.* 2006; Proctor *et al.* 2006) no longer support a simple damage-repair hypothesis of desiccation tolerance. The latest view is that tolerance involves a suite of protective mechanisms, including scavenging reactive oxygen species, or preventing their formation, and probably synthesizing sugars and dehydrins. The latter are LEA (late embryogenesis abundant)-like proteins that share some features with LEA proteins in seeds. In bryophytes, recovery of the essential systems, such as respiration, light capture and carbon dioxide fixation, and protein synthesis, now looks to be largely physical, and probably not metabolically costly in terms of either energy or materials. We need to do more work to test if the same is true for lichens. If so, then the "cost" of a lichen being so stress tolerant may well be mainly in producing "protective mechanisms" that enable lichens to survive the next stress event.

8.1.1   *Limits to stress tolerance*

Early workers determined the limits of stress tolerance in lichens, and Kappen (1974) provides an excellent review of this work. The main conclusions of this review are summarized here. For water supply, the great majority of lichens are highly desiccation tolerant. Providing that desiccation occurs reasonably slowly (over hours rather than minutes), most lichens can withstand drying to water contents of 5% or less, and most can remain viable for months, if stored at low relative humidities. Even aquatic species such as *Dermatocarpon fluviatile* can survive desiccation for four weeks. Conversely, most lichens are highly intolerant of submergence, or in many cases even moist storage, for more than a few days. Such lichens appear to become overrun with pathogenic fungi, or dissociate into separate symbionts. For temperature, lichens are tolerant of extremely low temperatures (e.g. liquid nitrogen, $-196\,^{\circ}C$) when dry. Hydrated thalli can also tolerate these temperatures, as can even tropical species, providing that cooling is slow enough. Interestingly, the heat tolerance of hydrated lichens is lower than that of higher plants, and most lichens, including tropical species, die when the thallus temperature exceeds 35 to 43 $^{\circ}C$. Normally, in the field, lichens will dry before they reach these temperatures. However, even in temperate climates, the temperature of dry thalli can reach 60 °C – not surprisingly, as the heat tolerance of dry thalli is high. For example, *Teloschistes flavicans* easily survived three days at 60 °C. Generally, lichens growing in open habitats have a higher tolerance to high temperatures.

More recently, high light stress has received much attention. Although some species possess carbon concentrating mechanisms (see Chapters 9 and 10), lichens display classical C3 carbon fixation and therefore light intensities will often exceed those that saturate photosynthesis. A significant decrease in the levels of stratospheric ozone has been observed during the last few decades, with accompanying increases of UV radiation. It is unclear if hydrated lichens are more tolerant to UV radiation than higher plants. However, some lichens can grow in very low light intensities in caves, where apparently they must be deriving carbon saprophytically. Finally, because lichens grow in exposed sites, they must be subjected to considerable mechanical damage from forces such as wind and abrasion by sand particles. The ability of lichens to propagate by "thallus fragmentation" (see Chapter 4) suggests that they have strong tolerance to mechanical stress.

8.2   Harmful effects of stress

As already mentioned, each type of stress can potentially cause specific problems for lichens. For example, desiccation damages the cytoskeleton, makes membranes leaky, and changes the structure of proteins so that, for example, the activity of some enzymes is reduced. Chilling is responsible for a different kind of membrane damage, a large increase in viscosity called "gelling", while heat can denature proteins. It seems unlikely therefore that we will ever have a concept of a "general resistance of the protoplasm" that we can apply to lichens. However, one feature shared by almost all stresses is that they cause the formation of reactive oxygen species (ROS). Intracellularly produced ROS can cause considerable damage to cells by attacking nucleic acids, lipids, and proteins. To survive stress, lichens must be able to either reduce the formation of ROS, or detoxify them once formed.

8.2.1   Formation of ROS

Many ROS are free radicals, atoms, or molecules with unpaired electrons. This unpaired electron is readily donated, and, as a result, most free radicals are highly reactive. Oxygen radicals include superoxide ($O_2^{\bullet-}$), the hydroxyl radical ($^{\bullet}OH$), hydroperoxyl (protonated superoxide, $HO_2^{\bullet}$) and the nitric oxide radical ($NO^{\bullet}$). Hydrogen peroxide (H_2O_2), singlet oxygen ($^1O_2$), ozone (O_3) or peroxynitrate (O_2NOO^-) are not free radicals but are, nevertheless, reactive and potentially harmful, and together with the above-mentioned radicals are usually classified as "ROS." It is important to realize that, while small quantities of ROS are produced as by-products of oxygen-dependent reactions during normal metabolism, almost all stresses enhance their production. The brief outline below summarizes the most important pathways that produce ROS, and the general reactivity of the various species. Full details of all these reactions can be found in Kranner and Lutzoni (1999), and Kranner and Birtić (2005), and an excellent review on reactive species and redox biology was recently published by Halliwell (2006).

- **Singlet oxygen** is formed inside cells, particularly in the photosynthetic apparatus. Light energy trapped by chlorophyll molecules can be transferred to $^3O_2$ (triplet oxygen, ground state oxygen), forming singlet oxygen ($^1O_2$). Singlet oxygen can react directly with polyunsaturated fatty acid side chains to form lipid peroxides.

- **Superoxide anions** are formed by the capture of an electron by $^3O_2$, and its formation is an unavoidable consequence of aerobic respiration (Møller 2001). Normally, when the terminal oxidases, cytochrome *c* oxidase and the alternative oxidase, react with oxygen, four electrons are transferred, and water is the product. However, $^3O_2$ can also react with other electron transport components, from which only one electron is transferred, thus forming $O_2{}^{\bullet-}$. In addition, enzymes such as nitropropane dioxygenase, galactose oxidase, and xanthine oxidase catalyze oxidation reactions in which a single electron is transferred from the substrate onto oxygen to produce $O_2{}^{\bullet-}$. Autoxidation of some reduced compounds (e.g. flavins, pteridines, diphenols, and ferredoxin) can also transfer a single electron to $^3O_2$ to produce $O_2{}^{\bullet-}$. In the cell wall and plasma membrane, NAD(P)H oxidases, peroxidases, poly- and diamino-oxidases and laccases can form extracellular $O_2{}^{\bullet-}$. Some groups of lichens were recently demonstrated to contain strong extracellular laccase activity (Laufer *et al.* 2006*a*, *b*; Zavarzina and Zavarzin 2006), and almost certainly produce $O_2{}^{\cdot-}$ in and around the cell wall. Compared with $^1O_2$ and the $^{\bullet}OH$ radical, $O_2{}^{\bullet-}$ is less reactive, having a half-life of 2–4 μs, and a low cellular concentration ($<10^{-11}$ M). It cannot react directly with membrane lipids to cause peroxidation, and cannot cross biological membranes. Most $O_2{}^{\cdot-}$ formed in biochemical systems reacts with itself nonenzymatically or enzymatically (catalyzed by superoxide dismutase) to form H_2O_2.

- **Hydroxyl radicals** are formed when hydrogen peroxide (formed by the dismutation of $O_2{}^{\bullet-}$ or other reactions) reacts with $O_2{}^{\bullet-}$ in the iron-catalyzed Haber–Weiss reaction. In the cell wall this reaction may be catalyzed by apoplastic peroxidases. The $^\bullet OH$ radical is the most reactive and aggressive species known to chemistry, having a half-life of 1 ns. Through electron transfer from $^\bullet OH$, $^1O_2$ can be formed.

- **Hydrogen peroxide**, as noted, can be formed by $O_2^{\bullet-}$ dismutation, but also by oxidases such as glycolate oxidase, glucose oxidase, urate oxidase, oxalate oxidase (the so-called "germin-like" oxidase), and amino acid oxidases.

8.3   Reducing ROS during stress is a key part of stress tolerance

The effects of ROS are not all negative, as they also play important roles in signaling processes. However, stress will arise if pro-oxidative processes prevail. Therefore, stress tolerance requires mechanisms that keep ROS under control. Measuring the production of the above molecules in lichen tissues is difficult, but experiments using dyes that fluoresce, following reaction with specific ROS, clearly show that stresses increase ROS production in thalli (Weissman *et al.* 2005*a*). However, lichens must reduce ROS formation during stress, or vigilantly scavenge them once formed.

8.3.1   Prevention of ROS formation

In mitochondria, stress can disrupt normal function and increase superoxide production. Controlled uncoupling of electron flow from phosphorylation protects cells by reducing the formation of harmful ROS (Skulachev 1998). Uncoupling can occur, firstly, via the alternative oxidase that dissipates the redox potential, and, secondly, via uncoupling proteins that dissipate the proton motive force (for reviews, see Jarmuszkiewicz 2001; Borecký and Vercesi 2005). Both of these processes will result in heat formation, and can be monitored using microcalorimetry. Beckett *et al.* (2005*b*) used this method to measure heat production in *Peltigera polydactylon* (Fig. 8.1). Freshly collected lichens produced heat at rates of *c.* 2 mW g^{-1} dry mass. After desiccation for *c.* 2.5 hours, heat production dropped to almost zero. Following rehydration, heat production increased to *c.* 8 mW g^{-1} dry mass and then gradually declined, although production was still *c.* 6 mW g^{-1} dry mass 3 hours after rehydration. These preliminary results suggest that the ability to dissipate energy may be an important component of desiccation tolerance mechanisms in lichens.

In chloroplasts "photophosphorylation" is the mechanism used by lichen photobionts to trap the energy of the sun and make ATP and NADPH. The molecules are later used in the Calvin cycle to fix carbon dioxide. Both processes are described in detail in Chapter 10. However, as discussed above, situations

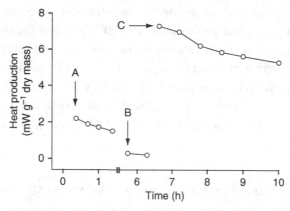

Fig. 8.1 Heat production by *Peltigera polydactylon* (A) after collection from the field, (B) after desiccation for 3 h to a water content of *c.* 10% that at full turgor, and (C) after rehydration (error bars smaller than symbols). (Modified from Beckett *et al.* 2005*b.*)

often exist when the light a lichen is absorbing contains more energy than it can use to fix carbon dioxide. Furthermore, during recovery from stress, carbon fixation takes longer to recover than photophosphorylation (Beckett *et al.* 2005*b*). Under these conditions, ROS can form in the photosynthetic apparatus (McKersie and Lesham 1994). Excitation energy can be transferred from chlorophyll molecules onto ground state oxygen ($^3O_2$), which is then converted into the highly toxic $^1O_2$. As we discussed above, $^1O_2$ can cause considerable damage; it can, for example, initiate lipid peroxidation. When high light stress is accompanied by another stress, such as dehydration, the problem worsens due to a further restriction of photosynthesis. In higher plants, it is well known that ROS production is partly prevented by dissipating the excess energy harmlessly as heat, in a process known as "nonphotochemical quenching" or NPQ (Szabo *et al.* 2005). To carry out NPQ, plants probably use a variety of reactions, mostly involving carotenoids. For example, β-carotene can quench $^1O_2$. Xanthophylls, the oxygenated derivatives of carotenes, can also quench $^1O_2$, e.g. lutein and neoxanthin. In particular, in the xanthophyll cycle (Frank *et al.* 1999; Demmig-Adams 2006) solar radiation is dissipated as heat while violaxanthin undergoes de-epoxidation to antheraxanthin and then zeaxanthin, partly preventing the formation of $^1O_2$ (Fig. 8.2). With the development of readily portable chlorophyll fluorescence devices, it is now quite easy to measure NPQ in the field (Jensen 2002). Careful analysis of seasonal variation of NPQ in the chlorophycean lichens *Lobaria pulmonaria* (MacKenzie *et al.* 2001, 2002) and *Xanthoria parietina* (Vráblíková *et al.* 2006) have shown that NPQ does indeed track the amount of solar radiation (Fig. 8.3). Cyanobacterial lichens, with some exceptions, e.g. the

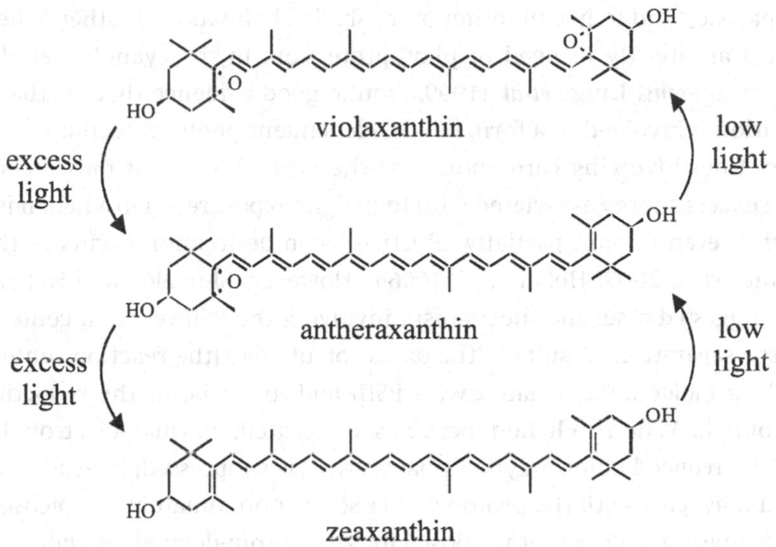

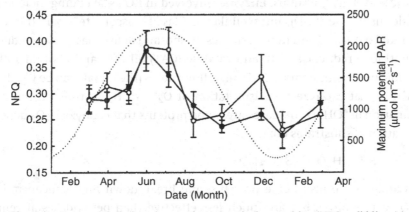

Fig. 8.2 Dissipation of excess light energy as heat in the xanthophyll cycle. In the xanthophyll cycle, violaxanthin is converted to antheraxanthin and then zeaxanthin. This pathway uses solar energy for the removal of two epoxy-groups in violaxanthin. Thus, some solar radiation is dissipated as heat rather than being involved in the excitation of chlorophyll, which can cause singlet oxygen formation due to transfer of excitation energy from chlorophyll to ground state oxygen. De-epoxidation is catalyzed by the enzyme violaxanthin de-epoxidase. In higher plants, ascorbate may be involved in the reduction of the epoxy group, but which compounds contribute to this reaction in the absence of ascorbate in cryptogams is not known. Violaxanthin can be recycled by zeaxanthin epoxidase in an NADPH dependent reaction.

Fig. 8.3 Seasonal variations in nonphotochemical quenching (NPQ, solid lines) in *Xanthoria parietina* track maximum potential photosynthetically available radiation (PAR, dotted line). In this experiment, removal of lichen substances from lichens using acetone (open symbols) had little effect on NPQ, suggesting that in this case parietin may not have been acting as a block to PAR. (Modified from Vráblíková *et al.* 2006.)

Lichinaceae, tend to live in rather more shaded habitats than other lichens, but are still apparently in need of photoprotection. In the cyanobacterial lichen *Peltigera rufescens*, Lange *et al.* (1999*a*) found good evidence that canthaxanthin formation is involved in a form of NPQ-dependent photoprotection.

Reactions involving carotenoids are the normal way that most lichens dissipate excess energy associated with high light exposure when a lichen is moist, and they even remain partially effective when desiccation occurs in the light (Kranner *et al.* 2005; Heber *et al.* 2006*b*). However, Shuvalov and Heber (2003) have suggested a second mechanism involving the PSII reaction center where "charge separation" results in the oxidation of P680 (the reaction center chlorophyll, a molecule associated with PSII) and, initially, in the reduction of a chlorophyll. When a lichen becomes desiccated, normal electron transfer from the reduced chlorophyll to phaeophytin is suppressed; instead, the chlorophyll may react with the protonated P680 in approximately one picosecond, a time frame as rapid as that proposed for zeaxanthin-dependent energy dissipation (Holt *et al.* 2005). Apparently, desiccation causes a change in the conformation of the PSII reaction center, such that it switches from being an energy conservation center to an energy-dissipating center. Both mechanisms provide photoprotective pathways in green-algal lichens (Kopecky *et al.* 2005); in contrast, cyanolichens appear to be primarily protected by the second, desiccation-activated mechanism (Heber *et al.* 2006*a*,*b*).

8.3.2   *ROS scavenging*

Once formed, ROS can be removed by either enzymatic or nonenzymatic scavenging systems. Enzymes involved in ROS scavenging include superoxide dismutase (SOD), intracellular or Class I ("ascorbate") peroxidases (AP), secreted or Class II and III peroxidases, mono- and dehydroascorbate reductases, glutathione reductase (GR), and catalase (CAT) (Elstner and Oßwald 1994). All aerobic organisms contain SOD, metalloproteins (the metals can be Fe, Mn or Cu and Zn) that catalyze the dismutation of $O_2^{\bullet-}$ to H_2O_2, thus preventing the formation of $^\bullet OH$. Peroxidases are heme proteins that catalyze H_2O_2-dependent oxidation of substrates (S):

$$SH_2 + H_2O_2 \rightarrow S + 2H_2O$$

Catalases are also heme proteins, and break down high concentrations of H_2O_2 very rapidly, but are much less effective than peroxidases at removing H_2O_2 present in low concentrations because of their low affinity (high K_m) for this substrate.

$$2H_2O_2 \rightarrow 2H_2O + 3O_2$$

It seems very likely that antioxidants and enzymes actually work together to scavenge ROS, as first suggested by Foyer and Halliwell (1976). They postulated an ascorbate-glutathione cycle for the scavenging of the H_2O_2 produced from $O_2^{\bullet-}$ by SOD. This cycle involves reactions of glutathione, ascorbic acid, GR, AP, and mono- and dehydroascorbate reductases. However, it is presently unclear if lichens produce ascorbate at all – a thorough investigation of *Cladonia vulcani* (discussed in more detail below) showed that neither the mycobiont nor the photobiont contained this antioxidant (Kranner *et al.* 2005). Although some fungi contain homologues of erythroascorbate (Loewus 1999), it is not clear if these erythroascorbate homologues occur in lichenized fungi and whether they can substitute for ascorbate in the above pathway.

Work on the role of antioxidant enzymes in stress tolerance in lichens is just starting. First indications are that no simple relationship exists between the actual levels of free radical scavenging enzymes and stress tolerance (Mayaba and Beckett 2001; Kranner *et al.* 2003). Furthermore, even moderate stress appears to decrease the activity of these enzymes. For example, Mayaba and Beckett (2001) measured the activities of AP, CAT and SOD during wetting and drying cycles in *Peltigera polydactyla*, *Ramalina celastri*, and *Teloschistes capensis*. These species normally grow in moist, xeric, and extremely xeric microhabitats, respectively. Enzyme activity was measured shortly after collection, after slow rehydration, after desiccation for 14 d and 28 d, and during the first 30 min of subsequent rehydration with liquid water. In all species, enzyme activities tended to rise or stay the same during slow rehydration. After desiccation for 14 d, enzyme activities decreased, and then decreased further to very low values after desiccation for 28 d. In all species, including the *T. capensis* from an extremely xeric habitat, the activities of all enzymes remained at very low values during the 30 min following rehydration, and were therefore unavailable to remove any reactive oxygen species accumulating in lichen tissues as a result of desiccation stress. Weissman *et al.* (2005*b*) later showed that sudden rehydration of even the desiccation-tolerant *Ramalina lacera* collected dry from the field decreased SOD activity by 50–70% and caused a transient decrease in total catalase activities. The study of Weissman *et al.* (2005*b*) was particularly interesting, because these workers carefully identified different SOD and CAT using gel electrophoresis. Each symbiont produced one catalase isoform. The alga contained four Fe-SOD and four Mn-SOD isoforms, while the fungus contained a Cu/Zn-SOD and a Mn-SOD. The activity of all SOD isoforms changed in a similar way in response to desiccation, although the algal CAT was more strongly inhibited than the fungal form. The apparent high sensitivity to stress of ROS scavenging enzymes in lichens suggests that enzymatic antioxidants are more likely to be involved in removing ROS produced during moderate stress or the normal metabolic processes of lichens rather than severe desiccation.

ROS do not only occur intracellularly. H_2O_2 can freely diffuse across the plasma membrane into the apoplast (Allan and Fluhr 1997; Henzler and Steudle 2000), and here they can be harmful to plasma-membrane-bound and cell-wall-bound enzymes, and indeed the plasma membrane itself. Interestingly, Beckett and Minibayeva (2007) showed that all tested lichens had the ability to rapidly break down exogenously supplied H_2O_2, but apparently lacked extracellular peroxidases and catalases. However, in one group of lichens, the Peltigerineae, extracellular tyrosinase activity could be readily detected, and the ability to break down H_2O_2 was directly correlated with tyrosinase activity. Tyrosinases have been shown to break down H_2O_2 using a catalase-like mechanism (Garcia-Molina *et al.* 2005). Peltigeralean lichens are generally more sensitive to desiccation than other lichens (Beckett *et al.* 2003). They are likely to produce more ROS during desiccation stress, and therefore need more protection from oxidative stress. Thus for some lichens, the breakdown of H_2O_2 by tyrosinases may defend lichens against the harmful effects of desiccation-induced ROS.

It is worth noting that, in addition to the ROS described above, stress can also cause the formation of other kinds of radicals, e.g. semiquinones (Gutteridge and Halliwell 1999). Laufer *et al.* (2006*a*) recently showed that stress strongly activates cell wall tyrosinases in lichens. This enzyme is directly involved in quinone metabolism, and possibly detoxifies harmful quinone radicals.

Turning now to the nonenzymatic antioxidants, for the cells of many life forms, the major water-soluble low-molecular-weight antioxidants are glutathione (γ glutamyl-cysteinyl-glycine; GSH) and ascorbate (Noctor and Foyer 1998), although as discussed above, many cryptogams apparently lack ascorbate (Kranner *et al.* 2005; Kranner and Birtić 2005). GSH and ascorbate are hydrophilic and their major function is cellular protection from oxidative damage in liquid phases, particularly in the cytoplasm. Tocopherols and β-carotene (Munne-Bosch and Alegre 2002) are the main lipid-soluble antioxidants, and are therefore the key antioxidants in membranes (see Kranner and Birtić 2005 and Kranner *et al.* 2008 for an overview). For example, GSH can scavenge $^\bullet OH$ in the following way:

$$GS^- + {}^\bullet OH \rightarrow {}^- OH + GS^\bullet$$

$GS^\bullet$, the glutathiolate anion, can react with many other molecules; in the simplest case, it reacts with another $GS^\bullet$, forming glutathione disulphide (GSSG):

$$GS^\bullet + GS^\bullet \rightarrow GSSG$$

GSSG can then be recycled by the NADPH-dependent enzyme glutathione reductase.

$$GSSG + NADPH + H^+ \rightarrow 2GSH + NADP$$

Other antioxidants such as ascorbate and tocopherol also react rapidly with ROS (Halliwell and Gutteridge 1999). However, as discussed above, antioxidants do not necessarily work in isolation. The antioxidant defenses of lichens should be understood as a network of free radical scavengers that includes both low-molecular-weight antioxidants and enzymes.

8.4   Other mechanisms of stress tolerance

8.4.1   *Other adaptations to desiccation*

Desiccation tolerance in lichens was recently reviewed by Nash *et al.* (2007). Apart from reducing ROS formation, work carried out on other organisms suggests that lichens could tolerate desiccation using a combination of sugars, amphiphilic substances, and dehydrins. Nonreducing sugars such as trehalose and sucrose are thought to promote vitrification, i.e. the formation of a "glass-phase" in the cytoplasm. Vitrification is a phenomenon that has been studied extensively in the so-called orthodox, i.e. desiccation tolerant, seeds (Black and Pritchard 2002). In the glassy state, the cytoplasm has the properties of a liquid with the viscosity of a solid. Due to the high viscosity of the cytoplasm, possibly harmful chemical reactions proceed much more slowly, and alterations in ionic strength and pH, and solute crystallization, are prevented. The glass phase also fills space, thus preventing cellular collapse following desiccation. Additionally, nonreducing sugars may substitute for water by forming hydrogen bonds, maintaining hydrophilic structures in their hydrated orientation (Crowe *et al.* 1984), and thus helping to stabilize proteins and membranes under dry conditions (for review see Leprince *et al.* 1993). "Polyols" are polyhydric alcohols such as sorbitol and mannitol, and can also accumulate in plants during water deficit (Popp and Smirnoff 1995). It seems likely that sugars and polyols play a role in desiccation tolerance in lichens, because they certainly contain high concentrations of these compounds (Roser *et al.* 1992). Interestingly, the proportion of the osmotic potential of lichens attributable to ions (mostly K) varies from about 10% in desiccation-tolerant species to 75% in highly sensitive species (Beckett 1995). The implication is that sugars take over the role of potassium in generating turgor (needed for cell division) in more tolerant species, and are therefore available to protect lichens from desiccation.

In seeds, pollen, and resurrection higher plants, "amphiphilic substances" (those with some solubility in both water and lipids) are believed to play an important role in desiccation tolerance (Oliver *et al.* 2001, 2002). Many types of amphiphilic substances exist in plants, including alkaloids, flavonoids, and other phenolic compounds. The distribution of these compounds within a cell differs, depending on hydration level, with the amphiphiles relocating from the aqueous cytoplasm to the membrane bilayer as water is removed. Here, they can act as powerful antioxidants, and strongly stabilize membranes. It is tempting to predict that future research will reveal the presence of amphiphiles in lichens.

"Dehydrins" are a type of "late embryogenesis abundant" or "LEA" proteins, which are characterized by high glycine content and high hydrophilicity index. As their name suggests, they are abundant in many seeds (Rorat 2006), but also occur in many bacteria and fungi (Abba' *et al.* 2006). From their amino acid sequences, it has been predicted that some classes of LEA proteins exist as random coil structures and others as amphipathic alpha helical structures. In comparison with a pure sucrose glass, the presence of LEA proteins increases both the glass transition temperature and the average strength of hydrogen bonding of the amorphous sugar matrix. LEA proteins could play a structural role as anchors in a tight molecular network to provide stability to macromolecular and cellular structures in the cytoplasm in the dry state. In the highly viscous or vitreous cytoplasmic matrix, this network would inhibit fusion of cellular membranes, denaturation of cytoplasmic proteins, and effects of harmful free radical reactions. It is worth noting that dehydrins have been suggested to play various other roles in the tolerance of organisms to stresses other than desiccation (Wise and Tunnacliffe 2004). Very preliminary evidence suggests that dehydrins may play a role in desiccation tolerance of the lichen *Peltigera horizontalis* (Schulz 1995), but much more work is needed to assess the role of dehydrins in the desiccation tolerance of lichens.

Finally, we still have very little idea how lichens can tolerate the huge reduction in cell volume that accompanies desiccation. Recent work on bryophytes has suggested that a key feature of desiccation tolerance is the depolymerization of the microtubule cytoskeleton (Pressel *et al.* 2006), and it would be interesting to test if the same thing happens in lichens. Honegger *et al.* (1996) in their elegant SEM and TEM investigations using cryotechniques have demonstrated that dramatic shrinkage of fungal and photobiont protoplasts occur with drying, but that the protoplasts maintain close contact with the cell walls. As a thallus dries, large cytoplasmic gas bubbles are formed, and therefore the relatively thick fungal cell walls are only deformed to a limited degree. These bubbles help the protoplast to remain in contact with the wall during desiccation. Upon rehydration these gas bubbles rapidly disappear and the protoplast refills the cells. Although some vertical shrinkage of whole lichen thallus occurs during drying, the overall shape of most lichens remains relatively unchanged because of the rigidity of the fungal cell walls.

8.4.2   Other adaptations to high light stress

Apart from NPQ, nonphotosynthetic pigments also help to protect lichens from excessive light. In *Lobaria pulmonaria*, melanins are synthesized in spring as solar radiation increases (Gauslaa and Solhaag 2001). In *Xanthoria parietina*, parietin may have the same role (but see Fig. 8.3). Parietin is an orange-colored anthraquinone pigment that occurs as tiny extracellular crystals in the top layer of the upper cortex of members of the lichenized fungal order Teloschistales, to which *X. parietina* belongs. Parietin efficiently absorbs solar radiation in the 400–500 nm PAR band. Distinct seasonal variations occur, with lichens having *c.* 50% of the parietin content in midwinter compared with midsummer (Gauslaa and McEvoy 2005). Apart from generally reducing solar radiation reaching the photobionts, recent studies have shown that UV light alone can induce the synthesis of melanic compounds and parietin (Solhaug *et al.* 2003). These pigments may therefore protect the lichens from potentially harmful UV radiation.

8.4.3   Other adaptations to extreme cold

In very cold habitats, such as mainland Antarctica, the metabolic activity of lichens is severely limited by water availability and low temperatures. Here, it is very important for lichens to be able to carry out photosynthetic activity and rehydrate from snow at subzero temperatures. Interestingly, desiccated lichens can take up enough water to photosynthesize from the sublimation (vapor phase) of snow (e.g. Schroeter and Scheidegger 1995). The lowest temperature measured in active lichens was $-17\,°C$ for *Umbilicaria aprina* at a continental antarctic site. If water-saturated thalli of *U. aprina* were slowly cooled at subzero temperatures, ice nucleation activity could be detected at $-5\,°C$, indicating extracellular freezing of water. Extracellular ice formation leads to cytorrhysis or cellular collapse of the photobiont cells and to cavitation in the mycobiont cells. However, both processes were reversible if the lichen thallus was rewarmed. Furthermore, even if lichens were frozen in a hydrated state, they could still photosynthesize at subzero temperatures. Later studies in Antarctica (Pannewitz *et al.* 2003) showed that, although the snow cover represented the major water supply for the lichens, they only became photosynthetically active for a brief time at or close to the time that the snow disappeared. The snow did not provide a protective environment, as occurs in some alpine habitats, but appeared to limit lichen activity.

Several physiological mechanisms of cold tolerance by animals, plants, and free-living fungi have been proposed, and it is probable that these also apply to lichens (Zachariassen and Kristiansen 2000; Robinson 2001). As noted above, lichens are naturally rich in trehalose and sugar alcohols (Roser *et al.* 1992), and these compounds have been proposed to act as general cryoprotectants in free-living fungi. The recently discovered antifreeze proteins (AFPs; Griffith and Yaish 2004) are unusual proteins: they have multiple, hydrophilic ice-binding domains that appear to function as inhibitors of ice recrystallization and ice nucleation. These proteins are known from bacteria, fungi, plants, and invertebrates. Although there are no published reports from lichens, intriguingly, a United States patent was recently registered claiming that an AFP derived from a lichen, with an apparent molecular weight of from 20 to 28 kDa, may be used in preventing food from freezing! Curiously, at the same time, the lichens that have AFPs also contain extracellular proteinaceous ice nucleators that trigger freezing at high subzero temperatures (Kieft and Ruscetti 1990). These have much higher molecular masses than the AFPs, and either provide cold protection from released heat of fusion or alternatively establish a protective sheath of extracellular ice in freeze-tolerant species. Clearly, lichens are able to regulate ice formation on and in their thalli very well.

8.5   Constitutive and inducible stress tolerance

As discussed above, if tolerance increases as a result of exposure to prior stress, the plant is said to be acclimated (or hardened). In theory, both "protection" and an "ability to repair" could be induced. It is often assumed that lichens rely mainly on constitutive mechanisms, as they frequently grow in highly stressful environments, where stresses such as desiccation may be sudden and severe. In this respect, they may differ from higher plants. A lichen exposed to dry air will dry to the point where metabolism stops within a few minutes, while a drying shoot of a vascular resurrection plant may remain metabolically active for many hours. If the lichen is to survive, its desiccation tolerance *must* be constitutive; the vascular plant has time to put a protective mechanism in place when drought threatens. However, the disadvantage of constitutive mechanisms is that they are present even when not needed, and at these times divert energy away from growth and reproduction. Selection may therefore favor inducible tolerance mechanisms in environments that are usually moist, and in which lichens are predictably (and probably slowly) desiccated. The latter conditions are typical for the habitats where many of the larger cyanobacterial lichens grow. While we know that lichens can acclimate their photosynthetic and respiratory apparatus to work optimally under changing environmental conditions (Kershaw 1985; Lange and Green 2005), work on the acclimation of lichens to stresses such as desiccation is just beginning. Recently, Beckett *et al.* (2005*b*) successfully hardened the lichen *Peltigera polydactylon* to desiccation

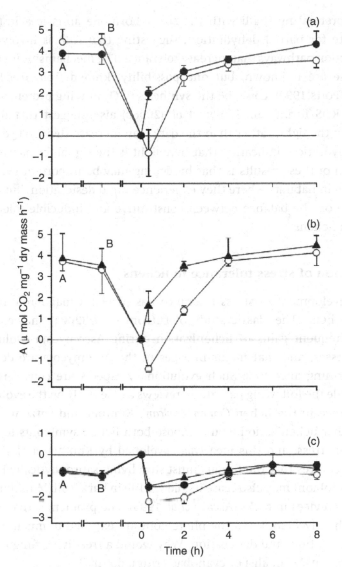

Fig. 8.4 Both partial dehydration (a) and treatment with 100 μM ABA (b) improve
the recovery of net photosynthesis (A) during rehydration following desiccation
for 15 days in *Peltigera polydactylon*. Neither pretreatment had a significant effect
on respiration (c). A, after collection from field; B, following 3 days pretreatment.
Time 0 indicates the start of rehydration. Open circles, control material; solid
circles, material pretreated by partial dehydration (losing *c.* 35% of its water);
solid triangles, ABA pretreated material. (Modified from Beckett *et al.* 2005b.)

stress by slowly dehydrating thalli to a water content of *c.* 65% that of full turgor
for three days, and then storing them fully hydrated for a further one day. This
treatment significantly improved the ability of thalli to recover net photosynthesis during rehydration after desiccation for 15 but not 30 days (Fig. 8.4).

Interestingly, pretreating thalli with the stress hormone abscisic acid (ABA) could substitute for partial dehydration, suggesting that ABA is involved in signal transduction pathways that increase tolerance. The mechanisms involved in this increase are unknown, but one possibility, known from free-living cyanobacteria (Potts 1994), could be the synthesis of O_2 binding protein which would reduce ROS formation. Heber *et al.* (2006*b*) also suggest that drying chlorolichens in the light, rather than the dark, can increase their NPQ during subsequent rehydration, indicating that here light is the signal for hardening. The implication of these results is that hardening may be important even for lichens growing in habitats where they experience rapid desiccation, but more work is needed on the balance between constitutive and inducible tolerance mechanisms in lichens.

8.6   Evolution of stress tolerance in lichens

The development of stress tolerance has played a major role in the evolution of lichens. The classic study of Lutzoni *et al.* (2001) showed that there were infrequent gains of lichenization during Ascomycota evolution, but multiple losses, and that major lineages in the Ascomycota are derived from lichen-forming ancestors. Such evolutionary aspects are considered in Chapter 16, while the following paragraph reviews a case study on the evolution of stress tolerance in the lichen *Cladonia vulcani*. Kranner and Lutzoni (1999) hypothesized that lichenization would expose both lichen symbionts to additional oxidative stress, and this idea was developed by Kranner *et al.* (2005). Lichenization requires that the fungus must stop living saprophytically below ground; the photobiont may also cease its hidden life in bark, soil (Mukhtar *et al.* 1994), or small crevices in rocks (Ascaso *et al.* 1995). The problem is that above ground, both the mycobiont and the photobiont are exposed to much higher levels of solar radiation, and desiccation. Why would a free-living fungus, and even more surprisingly, an alga or cyanobacterium, do this?

At this stage, it is worth noting that lichenization involves the evolution of a complex above-ground structure, which neither a fungus nor an alga can form alone. Kranner *et al.* (2005) conducted a detailed investigation into the mechanisms involved in protection from oxidative stress in the intact lichen, *Cladonia vulcani*, and in its symbionts when isolated and grown in axenic culture. This study showed that antioxidant and photoprotective mechanisms in the lichen *Cladonia vulcani* are more effective by two orders of magnitude than those of its isolated partners. The major low-molecular-weight antioxidant in the fungus was GSH. Neither the *Cladonia vulcani* mycobiont nor the photobiont contained ascorbate; tocopherol was only found in the photobiont. In isolation, both alga and fungus suffered oxidative damage during desiccation. On its own, the alga only tolerated very dim light and its photoprotective systems were only partially effective. Without the alga, the fungus's GSH-based antioxidant system was slow and ineffective. However, in the lichen, the two symbionts appeared to induce up-regulation of protective systems in each other. In the lichen, where it is exposed to higher light intensities, the alga has lower chlorophyll concentrations, which helps to avoid $^1O_2$ formation. In addition, it has higher concentrations of photoprotective pigments involved in NPQ, and of the antioxidant α-tocopherol. In addition, total glutathione (GSH plus GSSG) is present in the lichen at a level 30% greater than the sum of the contents in isolated alga and fungus. The lichen therefore appears to be better adjusted to cope with oxidative stress than its isolated partners. The authors noted that for the lichen, this mutually enhanced resistance of its symbionts to oxidative stress, and in particular the lichen's enhanced desiccation tolerance, are required for life above ground with the major benefit of the new lifestyle being the increased chance of dispersal of reproductive propagules. Overall, the enhanced antioxidant capacity apparently contributes to the evolutionary success of the lichen symbiosis (Kranner et al. 2005).

8.7   Conclusions

According to the "competitor-ruderal-stress tolerator" model of plant life histories proposed by Grime (1979), lichens would be classified as classic stress tolerators. High stress habitats should favor species with inherently low growth rates, and indeed many lichens grow very slowly (see Chapter 9 for more details). Slow growth is probably a result of first, low resource availability (including water, which limits the time lichens can be active), and second, metabolic "costs" associated with being stress tolerant. As we discussed earlier, it is not always easy to know what these metabolic costs are. However, we know that in mosses, for example, increased NPQ can reduce photosynthetic efficiency (Beckett et al. 2005a). Similarly, mitochondria with strong alternative oxidase or uncoupling protein activity will respire less efficiently. The need to synthesize sugars, dehydrins, or antifreeze proteins to protect cells during desiccation or freezing will divert reserves away from growth and respiration. Despite their undoubted inefficiency, it is their extraordinarily high stress tolerance that enables lichens to live in places that no other plants can; the only competition lichens face in many of their habitats is from other lichens!

9

Physiological ecology of carbon dioxide exchange

T. G. A. GREEN, T. H. NASH III, AND O. L. LANGE

Photosynthesis is used by autotrophic organisms to convert light energy into chemical energy for maintenance, growth, and reproduction. Heterotrophs have shown considerable agility in forming symbiotic relationships with autotrophs so that they obtain a reliable carbon source. The endosymbiont theory proposes that the chloroplast of eukaryotic cells evolved from photosynthetic cyanobacterium-like organisms which were engulfed by nonphotosynthetic cells, leading eventually to the evolution of algae and plants. Fungi have also developed symbioses, one being the lichen, an extrasymbiosis with photosynthetic algae and/or cyanobacteria, which, in many respects, appears to function like a single autotrophic "organism." Its photosynthesis and respiration are complex biophysical and biochemical processes that will not be discussed here in any detail. We shall restrict our analysis to those aspects that are of ecological relevance, in particular carbon dioxide (CO_2) exchange, which is the subject of this chapter.

On average, approximately 40 to 50% of a lichen's dry mass consists of carbon which is almost exclusively fixed by photobiont photosynthesis. Photosynthetic processes are vital for the existence, survival, and growth of the lichen. Energy-producing respiratory processes that release CO_2 occur in both the mycobiont and photobiont, although the individual contributions of the symbionts to the respiration of the whole lichen thallus are not yet known. However, it is most probable that total thallus respiration mainly reflects the metabolic activity of the fungal partner (Quispel 1960). This is suggested by the much larger proportion of the mycobiont in a heteromerous lichen thallus; the photobiont partner may constitute as little as 10% of the total lichen biomass (Sundberg et al. 1999a). When measuring CO_2 exchange, net photosynthesis (NP) rates are monitored and they are the difference between gross photosynthetic CO_2 assimilation (GP)

Lichen Biology, ed. Thomas H. Nash III. Published by Cambridge University Press.
© Cambridge University Press.

and concomitant respiratory CO_2 release. Gas exchange measurements do not allow gross photosynthesis per se to be monitored directly, and it is often approximated by adding dark respiration (DR) to NP under similar conditions. Net photosynthetic CO_2 gain, or carbon income, less the dark respiratory CO_2, or carbon loss, integrate CO_2 exchange over any selected time period. The value represents the net primary production of the lichen. Lichen productivity and production is of considerable general interest because they dominate, and are sometimes the only autotrophs, in a number of ecosystems (e.g. coastal deserts and many polar systems) and are major components of other ecosystems (e.g. epiphytes in some forests).

Rates of lichen CO_2 exchange are highly dependent on the individual species or lichen type involved, on its internal status and on the actual external environmental conditions. In this chapter, we first summarize techniques for measuring CO_2 exchange and then we discuss inherent thallus variability which impacts rates of photosynthesis and respiration. We then analyze the physiological response patterns due to the effects of major external factors. In nature, the environment is always a complex mixture of influences, and we acknowledge Kershaw's (1985) major contribution to this topic. He stated that lichen physiological behavior needs to be understood within a multivariate framework of environmentally important factors. We discuss the interrelationships between environmental factors and evaluate the transference of laboratory results to the interpretation of lichen performance in nature. In several case studies, we use this information to analyze CO_2 exchange of lichens growing in different habitats.

Chapter 10 covers important aspects of carbon economy and lichen growth.

9.1   Techniques for measuring gas exchange

Instantaneous CO_2 gross assimilation can be measured over seconds with radioactive ^{14}C techniques (Moser et al. 1983a) and this method is suited for special studies. However, the most frequently used technique is to measure net CO_2 exchange of a lichen sample within a cuvette by using infrared gas analyzers in an open, differential system. In the laboratory, internal cuvette conditions (such as temperature, light, humidity, CO_2 concentration) can be precisely controlled, allowing the response patterns of the lichen to be analyzed. In the field, precautions have to be taken to prevent conditions in a cuvette deviating strongly from those for the free-living lichen. A commonly applied solution to this problem is the manually operated "CO_2 porometer" with brief enclosure times, of only one or a few minutes, that minimize change and allow the samples to be freely exposed to ambient conditions between the gas exchange measurements (Lange et al. 1985). Long-term measurements of CO_2 exchange are

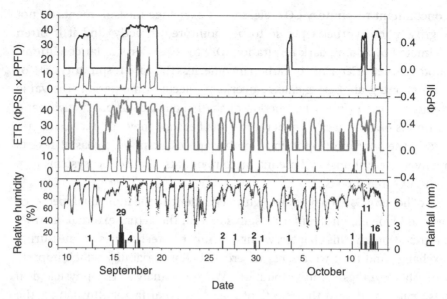

Fig. 9.1 Monitoring of green-algal and cyanobacterial sectors of a photosymbiodeme in a New Zealand temperate rain forest. (Upper panel) Cyanobacterial sector, upper line ETR (relative electron transport rate; PPFD, photosynthetically active photon flux density), lower line Φ_{PSII} (quantum efficiency of photosystem II).(Middle panel) Green-algal sector, lines as for cyanobacterial sector. (Lower panel) Relative humidity and precipitation (bars with rainfall in mm). Note the coincidence of activity of the cyanobacterial sector with rainfall and of the green-algal sector with high humidity. (From Green *et al.* 2002.)

possible with a "klapp cuvette" that briefly encloses the lichen sample every 30 minutes and can be operated over months and years with minimal disturbance to the measured lichen (Lange *et al.* 1997).

In the past two decades, monitoring of chlorophyll fluorescence with modulated fluorometers using the saturation pulse concept has become an important tool in photosynthesis research on both plant leaves and lichens (Jensen 2002). It is a noncontact system and fluorescence signals can be measured at some distance from the lichen sample as nonintrusive indicators for rapid assessment of a variety of photosynthesis parameters (see Schreiber *et al.* 1994 for definitions, calculations, and symbols). The efficiency of photosystem II (Φ, quantum yield) is a sensitive measure of the core behavior of the photosynthetic light reaction. It has become an elegant signal to detect and monitor metabolic activity of lichens in the field and has proved to be a good indicator of when they are potentially photosynthetically active (Fig. 9.1), for instance during long-term monitoring under snow in Antarctica (Schroeter *et al.* 2000; Pannewitz *et al.* 2003). Maximal Φ (after dark adaptation) is often used to detect damage (photoinhibition or photodestruction) to the photosynthetic apparatus. Quantum yield of illuminated samples allows relative electron transport rate (ETR) to be calculated and this can be used for an indicator of *in situ* photosynthesis for higher plants. However, the photobionts of lichens utilize alternative electron transport processes which do not contribute to CO_2 fixation (such as photorespiration and Mehler reaction), and a marked deviation between actual CO_2 uptake and ETR can occur under some conditions (Leisner *et al.* 1997; Green *et al.* 1998). Other fluorescence signals are nonphotochemical or photochemical quench parameters, and imaging fluorometers allow depiction of the spatial distribution of signals over a lichen thallus (Barták *et al.* 2006).

9.2 CO_2 exchange and its relation to inherent thallus variability

Large differences in photosynthetic capacity, i.e. maximal rates under optimal conditions, occur amongst different species. Many, in particular slow-growing lichens, have very low maximal rates and this is especially the case when net photosynthesis is related to thallus dry weight because photobionts can comprise only a very small proportion of total thallus biomass. In many textbooks lichens are quoted for their extremely low photosynthetic rates (e.g. Larcher 2003) but there are many species that have remarkably high rates of area-related photosynthesis and chlorophyll contents similar to those of phanerogamous leaves. Soil crust lichens from fog desert habitats reach NP between 5 and 7 μmol CO_2 m^{-2} s^{-1}, and almost 9 μmol CO_2 m^{-2} s^{-1} was measured for the tropical *Dictyonema glabratum*. These are much lower than maximal values for C3 crop plants of 20–40 μmol CO_2 m^{-2} s^{-1}, but compare well with sun (10–50 μmol CO_2 m^{-2} s^{-1}) and shade (3–6 μmol CO_2 m^{-2} s^{-1}) leaves of deciduous trees and the 8–10 CO_2 m^{-2} s^{-1} for evergreen coniferous trees. However, in nature it appears that lichens very seldom, if at all, and then only for very short periods, experience optimal conditions for maximal photosynthesis (Section 9.4.2).

Within one single species, differences in photosynthetic capacity occur amongst thalli, within thalli, and with time. Separate thalli from the same environment can differ greatly in maximal NP, for instance the order of magnitude range found for *Umbilicaria aprina* (Sancho *et al.* 2003) with no obvious explanation for the variability. Substantial variation of photosynthetic capacity can also occur within one individual lichen thallus. For the two *Cladonia* species in Fig. 9.2 c. 50% of the dry-weight-related photosynthetic activity occurs in the upper 10% of the podetia and almost no activity was measurable in the lower half (Nash *et al.* 1980). Maximal photosynthetic activity is concentrated in the actively growing portion of the thalli where the high total chlorophyll content indicates a much greater concentration of photobionts. Linear relationships

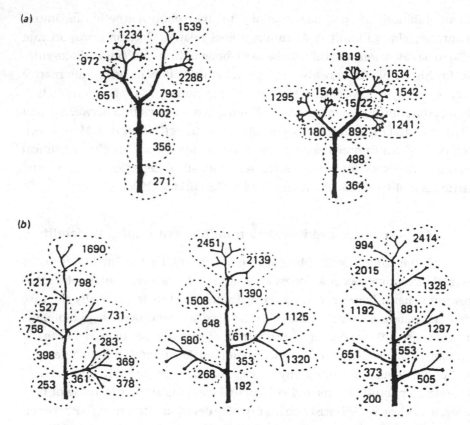

Fig. 9.2 Photosynthetic assimilation patterns (^{14}C technique, radioactive counts per minute) within the upper one fourth (2.5 cm) from the podetia of (a) *Cladonia stellaris* and (b) *C. rangiferina*. (From Nash *et al.* 1980.)

have been often found between photosynthetic capacity, thallus nitrogen concentration and chlorophyll concentration, indicating the close metabolic relationships of these parameters (Palmqvist *et al.* 2002). Such correlations do not suggest control of photosynthetic rate by chlorophyll content but, more likely, an increased number of photosynthetic centers, each with an associated number of chlorophyll molecules (Kershaw 1985). Photosynthetic rates are most often given on a dry weight basis for ease of determination. Changes in thallus thickness with age can alter these rates as the ratio of mycobiont to photobiont increases (Larson 1984). This can give the impression that NP rates also change with age but, in the case quoted above, the rates remain constant when presented on an area basis (Green *et al.* 1985). Seasonal acclimation of lichen photosynthesis is another possible complexity (Kershaw 1985). Variability also occurs in DR, and in the lower half of the two *Cladonia* species (Fig. 9.2) respiration was approximately one tenth that of the upper 10 mm. Almost complete acclimation of mitochondrial respiration to seasonal changes in temperature has also been found (Lange and Green 2005).

The high intrathalline, individual, and temporal variability of CO_2 exchange capacity makes it difficult to determine representative measurements especially *in situ*. A single CO_2 gas exchange measurement is simply a spot check for the sample enclosed in the cuvette, a sample which often consists of several lobes of different individuals of the same population. Repeated measurements can reveal diel and seasonal courses and allow interpretations to be made. The high variability has also often led researchers to normalize data, for example light response curves, as a percentage of the highest maximal rate for each sample (Green *et al.* 1999), and, in general, this reveals relatively stable response patterns.

9.3 CO_2 gas exchange in relation to environmental variables

9.3.1   Moisture

Poikilohydry

Lichens are poikilohydric (Section 1.4) and this dominates their CO_2 exchange behavior. Metabolically there will be a strong link to water status with negative effects when water potentials are below zero. Physically there are also complex effects. First, there is the interaction between water storage and CO_2 diffusion in the capillary spaces. Second, water is generally taken up and lost over the entire thallus surface and this is a stark contrast to the situation in homoiohydric higher plants where water vapor and CO_2 follow the identical diffusion pathway, the stomata. Lichen water use efficiencies (LWUE; Máguas *et al.* 1997) are in no way comparable to similar ratios for higher plants.

Thallus water content

Gas exchange changes dramatically as a function of thallus water content, WC (Fig. 9.3). Four zones can be defined as thallus water content increases from minimal values. In the highly desiccated thallus no apparent CO_2 exchange can be measured and the thalli are dormant. The first metabolic switching point occurs at 5–10% WC (about −90 MPa, in equilibrium with 50% relative humidity at 20 °C) with the activation of enzymes of the TCA cycle (use of tritiated water vapor; Cowan *et al.* 1979a, b) and indicates a change from localized to a mobile absorbed water layer on the enzymes (Koga *et al.* 1966). It should be noted that so-called "air-dried" lichens can fall within these conditions and cannot, therefore, be considered inactive. A second switching point occurs at about 20% water content (−40 MPa in solutions or as matric potential through air humidity; see Nash *et al.* 1990; Fig. 9.4) with the transition from gel

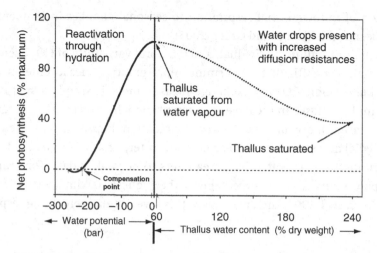

Fig. 9.3 Idealized response curve of net photosynthesis (ordinate, normalized as % maximal net photosynthesis) to thallus water content (abscissa) from data obtained for *Ramalina maciformis*. The water content (WC) axis is given as water potential (bar) at water contents below that optimal for net photosynthesis and as % dry weight at higher WC when potential is at zero bar. Net photosynthesis is decreased by desiccation at the low WC and by increased CO_2 diffusion resistances at higher WC. (From Schulze *et al.* 2002.)

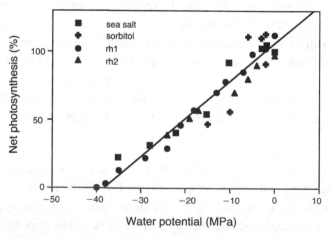

Fig. 9.4 Dependence of net photosynthesis expressed as a percentage of pretreatment measurements, as a function of different water potentials (-MPa) established in salt or sorbitol solutions, or by varying relative humidity, in *Dendrographa minor*. (From Nash *et al.* 1990.)

to solution within the cells (see Koga *et al.* 1966) and photosynthetic metabolism is detectable. Cyanolichens need substantially higher hydration, in the range of 85–100% to show the same activity level (Büdel and Lange 1991). At approximately 22% water content the moisture compensation point is reached and photosynthesis is sufficient to balance respiration (Fig. 9.3). At higher WC, NP increase rapidly to a maximum at WC of 70–150% (c. 80% in Fig. 9.3) for chloro-lichens, and a substantially higher 300–600% for cyanolichens. At this WC it is probable that the algal cells and fungal hyphae are fully turgid and further increases in water content represent free water somewhere in the lichen thallus. At WC above the NP maximum the NP response becomes species specific (Lange et al. 1993b) and some species maintain near maximal NP to the highest water contents while others show various degrees of depression (suprasaturation). The depressed NP is normally a result of increased diffusion resistances at the high water contents, as demonstrated by the ameliorating effect of increased CO_2 concentrations (Lange and Tenhunen 1981; Lange et al. 1999b). On rare occasions, it can also be due to increasing dark respiration with water content and this may be a result of selective rehydration within the thallus (Sancho et al. 2000a). Further aspects of this depression are covered in the next section on diffusion; an example of its impact on ecological performance is given in Section 9.4.2, temperate climate. A somewhat surprising development is the demonstration that maximal NP and maximal WC appear to be positively and highly significantly linearly correlated (Sancho et al. 2004) with two separate relationships, one for foliose and one for epiphytic fruticose lichens. There is no obvious explanation for these correlations, although the relationship for foliose lichens is from low values for both parameters for lichens which favor survival, to the opposite extreme that emphasizes growth. Further research is obviously needed to clarify these results.

Dark respiration exhibits a somewhat different relationship with thallus water content. At lower water contents CO_2 release increases markedly with increasing WC until a saturation level is reached. The initial WC at which maximal dark respiration is reached is approximately the same as that at which photosynthetic maxima occur.

Diffusion pathways and resistances in the lichen thallus

The diffusion of CO_2 and water vapor within, and exchanging with, lichen thalli is governed by Fick's Law and the rate of diffusion is influenced by the driving force (approximating the difference in concentration) across a structure, and the diffusion resistance (inverse of conductance) provided by the structure. Resistances are present in two main locations, outside and inside the lichen thallus.

Outside the thallus is the boundary layer resistance, r_a. The thicker the layer is, then the slower the diffusion rate. Boundary layer resistances are related directly to the square roots of the width of the thallus and inversely to the wind speed. Narrow lichens that hang free in the air from their substrates (e.g. Usnea

Table 9.1. *Estimates of resistance to diffusion (ms) in* Ramalina maciformis *at different thallus water contents, which, in increasing order, were achieved by equilibrium with an atmosphere at 97% relative humidity (at 10 °C), lightly wetting, copiously wetting, and infiltration. Resistances refer to gaseous (*r_g*) and aqueous (*r_w*) phases*

	Hydration levels (%)			
Resistances	50	120	200	600
r_g	10.5	19.3	60.5	0.0
r_w	0.0	6.8	142.3	556.8

spp., *Ramalina* spp.) will have thin boundary layers and low resistances. This will result in rapid loss of water and rapid energy exchange so that thallus temperature will be effectively nearly identical to air temperature while wet or dry (Coxson *et al.* 1984). Foliose lichens will have greater r_a and this can slow water loss; r_a will increase with thallus width, and the associated decreased water loss may explain the faster growth of *Pseudocyphellaria* spp. in New Zealand evergreen forests (Snelgar and Green 1982) and *Degelia plumbea* in Norway (Gauslaa and Solhaug 1998). Lichens on the ground or bark of trees are within the surface boundary layer. This effectively means that water loss and thallus temperature will be strongly linked to irradiance. The greater the radiation load the more rapid the water loss and the higher the thallus temperature.

Water evaporates from the entire surface of a lichen and does not follow the same pathway as CO_2, which must diffuse through internal pathways. Lichens are, therefore, very different from higher plant leaves where the two substances follow the same route between the photosynthetic cell and the atmosphere, and calculations used in higher plants to calculate internal CO_2 concentration are not applicable to lichens. This has considerably slowed our understanding of the situation at the algal surface within a lichen.

Cowan *et al.* (1992) estimated diffusion resistances in lichens by using CO_2 exchange comparisons for different CO_2 concentrations in air (79% N_2; 21% O_2) versus helox (79% He; 21% O_2) (Table 9.1). This is possible because CO_2 diffuses 2.3 times faster in helium than in nitrogen so that equivalent CO_2 assimilation rates are achieved at lower CO_2 concentrations in helox. Their analysis allowed them to make several statements about the conditions within a lichen thallus. First, even when there is no liquid water in the diffusion pathways (pores) the gas phase resistance is still 2–3 times greater than the liquid phase within the photobiont. Second, the gas phase resistance is mainly in the cortex of the thallus and, even when NP is nearly maximal, the required porosity is only

0.1 to 0.2%; i.e. between 1/1000th and 1/500th of the surface is composed of air-filled pores. Third, increase in diffusion resistance occurs predominantly in the cortex due to the narrowing and lengthening of pores with swelling of the hyphae and cortex overall, and with blocking by extracellular water (Stocker 1927). The cellular interstices in the algal and medulla layer remain air-filled, except when the lichen is infiltrated under vacuum. Fourth, the origin of the CO_2 is produced by DR models as if it is close to the photobionts (i.e. it is released into the medulla and photobiont zones because the cortex resistances are so high). This situation means that NP only become negative under favorable light when an extra water phase resistance forms between the photobiont and the source of the dark respiration at very high WC.

Cowan et al. (1992) confirmed that the medulla and the photobiont cells do not have liquid water around them, but are in the gas phase (Zukal 1895; Stocker 1927; Snelgar et al. 1981a, b; Honegger 1998). Zukal (1895) makes the point that the internal diffusion pathways of the thallus will remain air-filled unless water is pushed in under pressure; on potential grounds water cannot pass by capillarity from narrow to wider pores. It is now clear that the photobiont and also some mycobiont cells are covered with hydrophobins, water repellent proteins (Honegger 1998). These proteins will help prevent water films forming on the photobiont cell surfaces and will also assist with keeping the medulla and other sites free of water. The presence of the hydrophobins causes any water inside the medulla and photobiont zones to form droplets, and the high water potential of the droplet will ensure that the water is rapidly transferred through the gas phase to the lower potential photobiont and fungal cells. The hydrophobins do not waterproof the cells but simply make them water repellent (Chapter 5); vapor exchange can still occur rapidly, rather like the situation in the artificial Teflon-based Goretex® fabrics.

Thallus internal CO₂ concentration

One major benefit of the resistance analysis by Cowan et al. (1992) is that the values obtained can be used to calculate other parameters such as internal CO_2 concentration inside the thallus, and the CO_2 gradient between the outside of the lichen and the surface of the photobiont cells. Such data are presented for *Ramalina maciformis* in Table 9.2. The internal CO_2 concentration is higher and the gradient lower when the light is lower (i.e. lower net photosynthetic rate). Unexpectedly low internal CO_2 concentrations are calculated under conditions of high light and higher resistances; as low as 132 ppm for a lightly wetted thallus and a very low 16 ppm when copiously wetted. These values confirmed the low compensation points already reported for a variety of lichens (Snelgar and Green 1980, 1981; Bauer 1984). Cowen et al. (1992) confirmed that there was

Table 9.2. *Carbon dioxide diffusion parameters in the thallus of* Ramalina maciformis *calculated at low (105 μmol m^{-2} s^{-1}) and high (300 μmol m^{-2} s^{-1}) PPFD and three different thallus water contents (from Cowan et al. 1992). Parameters are: CO$_2$ concentration (ppm) at the surface of the photobiont cells; CO$_2$ gradient (ppm) from thallus surface to photobiont cells with external CO$_2$ concentration set at 350 ppm; calculated ^{13}C discrimination, parts per thousand*

		Hydration status		
		97% RH	Lightly wetted	Copiously wetted
PPFD (μmol m^{-2} s^{-1})	Parameter	WC = 50%	WC = 120%	WC = 200%
105	CO$_2$ (ppm) at photobiont surface	290	238	63
	CO$_2$ gradient (ppm)	60	112	287
	Discrimination (‰)	−23.1	−19.8	−8.5
300	CO$_2$ (ppm) at photobiont surface	244	132	16
	CO$_2$ gradient (ppm)	106	218	334
	Discrimination (‰)	−20.2	−12.9	−5.4

no apparent photorespiration in some lichens and that the CO$_2$ fixation pathway must differ from that of C3 higher plants. Since then it has become clear that all cyanobacteria and some green algae possess carbon concentrating mechanisms (CCMs) which allow them to reduce internal CO$_2$ to very low levels in the presence of moderate and greater diffusion resistances (Badger *et al.* 1993; Palmqvist 1993; Chapter 10). The occurrence of CCMs is not specific to lichens as they are primarily used to counteract CO$_2$ gradients formed in liquid boundary layers. However, their occurrence in lichens is advantageous in that it allows higher net photosynthetic rates by generating lower internal CO$_2$ and greater CO$_2$ gradients to drive diffusive uptake.

^{13}C isotopic discrimination

The occurrence of a CCM in some lichens also helps explain the disparity of ^{13}C discrimination values ($^{13}\partial$) reported in the literature. For phanerogams $^{13}\partial$ values generally range from −23 to −34‰ for C3 plants and from −14 to −23‰ in C4 plants. Although there is no evidence of C4 metabolism in lichens, reported $^{13}\partial$ values range from −17 to −34‰ in green-algal lichens, from −20 to −26‰ in heteromerous cyanolichens, and from −14 to −23‰ in homoiomerous cyanolichens (Máguas and Griffiths 2003). The occurrence of variable internal CO$_2$ concentrations is probably sufficient to explain most of the observed $^{13}\partial$ variability in lichens. For example, when ^{13}C discrimination values are calculated for the internal CO$_2$ concentrations in *Ramalina maciformis* (Table 9.2), then

$^{13}\partial$ values from $-23.1‰$ to $-5.4‰$ are possible. Interestingly, many lichens have discrimination values in the -20 to $-23‰$ range, very similar to those of higher plants (Lange et al. 1988; Palmqvist 2000), and this implies that low internal CO_2 is not common and that other limitations like low light or low water content influence the results. Such discrimination values suggest that lichens rarely operate at optimal, high net photosynthesis (Section 9.5).

9.3.2   Modes of water uptake

Rundel (1988) reviewed lichen water uptake under field conditions and showed that lichens can rapidly rehydrate within a few minutes when placed in water, such as becoming moistened by rain. Optical changes occur a few seconds after spraying of the dry thalli. Then, for green algal *Ramalina maciformis* (Lange et al. 1989), Fo (minimal fluorescence measured in the dark after a period of darkness) first increased rapidly within 2 to 3 minutes, followed by a slower increase to 75% of maximal Fo after about 4 minutes, and variable fluorescence was detectable as soon as 15 to 30 seconds after spraying. The rise in Fo was much slower in cyanobacterial *Peltigera rufescens*, taking about 20 minutes when variable fluorescence was also detectable. Rehydration does not necessarily mean quick reactivation of NP; in fact, there are examples of NP increasing slowly, for instance, over 12 hours (Schlensog et al. 2004). It appears lichens may resemble bryophytes and that the speed of reactivation is related to the moisture status and activity duration in their normal environments. If the lichens are wet and active for long periods, then reactivation is slower, and vice versa (Schlensog et al. 2004). Lichens, of course, predominate in drier habitats and reactivate rapidly. It is impressive that, even after about eight months of a harsh winter in Antarctica, lichens can reactivate and achieve near normal photosynthesis within minutes (Kappen et al. 1998a).

In addition to rain, lichens can utilize a variety of water sources, such as fog, dew, and even elevated water vapor, to activate gas exchange (Chapter 10). There is an abundance of lichens in coastal deserts where rainfall is minimal and sporadic, but fog, dew, and/or elevated humidities occur very frequently (Kappen 1988; Figs. 1.1 and 9.8; Section 9.4.2). The utilization of water vapor alone as a moisture source is remarkable (Butin 1954; Lange and Bertsch 1965). Photosynthesis reactivates at water potentials above -40 to -30 MPa, and -30 MPa is equivalent to a relative humidity of about 80%. Air temperature has little effect on rehydration at this required humidity but, in any typical day with a constant water vapor pressure, humidity will decline in the warmer daytime and rise in the cooler night. Rehydration by sources other than rain is, therefore, usually a nocturnal event. Some chlorolichens grow in habitats like overhanging ledges, where they almost exclusively rely on hydration through uptake of water vapor (Pintado and Sancho 2002). Matthes-Sears and Nash (1986) built a photosynthesis model based on *in situ* measurements to predict photosynthetic gain by moisture source over a 3-year period for *Ramalina menziesii*.

Chlorolichens and cyanolichens differ in their ability to rehydrate from water vapor alone (Lange *et al.* 1986, 1988, 1993a). At relative humidities around 90–95%, initially dry chlorolichens are able to resume high levels of NP in a few hours while cyanolichens do not, an effect that is ecologically significant (Fig. 9.1; Green *et al.* 1993, 2002). Investigation of several photosymbiodeme pairs found no difference in the WC achieved by the green-algal and cyanobacterial sectors at high relative humidities (Schlensog *et al.* 2000). However, cyanolichens need much higher WC than chlorolichens to commence positive NP (around 70–150% versus around 20%), so, because WC in equilibration with water vapor reaches only about 80%, this means that cyanobacterial lichens will show little activity. They appear unable to reach full turgor (Büdel and Lange 1991) through water vapor uptake. There is no satisfying explanation for this difference in performance between lichens with the two different types of photobionts.

Thallus structure seems to affect water vapor uptake (Sancho and Kappen 1989; Palmer and Friedmann 1990). It seems that further complexities in this story, possibly relating to thallus structural changes, still await discovery.

9.3.3   Light (photosynthetically active photon flux density)

Lichens show the typical saturation-type photosynthetic response to incident PFD (photon flux density or quantum flux density; see Chapter 10). Photon flux density required to saturate net photosynthesis is very variable. In general, lichens occurring in sheltered habitats exhibit saturation at lower PFD than lichens from exposed habitats. In a survey of lichens covering a range of habitats within a New Zealand evergreen forest to exposed soil, Green *et al.* (1997) measured saturation PFD from 82 to 766 μmol m^{-2} s^{-1} and light compensation for 4 to 136 μmol m^{-2} s^{-1}. Saturating PFD was positively correlated with compensation value and negatively with quantum efficiency. However, although *Pseudocyphellaria* species saturated at about 120 to 150 μmol m^{-2} s^{-1}, PFD values in the forest rarely exceed 10 μmol m^{-2} s^{-1} (Green *et al.* 1997). Lichens are apparently rarely active, and only briefly, when light saturated (Section 9.4.2). Lichens also adjust photosynthetically when low light conditions develop, as occurs with the closure of the canopy of deciduous forest (Kershaw 1985).

Because the photobiont is usually found in a layer beneath cortical hyphal tissue, the light reaching the photobiont is reduced compared with surficial light measurements by 54–79% when dry and 24–54% when fully hydrated (Ertl

1951; Dietz *et al.* 2000), and even less in some cases (Büdel and Lange 1994). The presence of the fungal cortex and included compounds provides protection against excess PFD and UV radiation. Although some early studies using chlorophyll fluorescence suggested that photoinhibition occurs (Demmig-Adams *et al.* 1990; Kappen *et al.* 1991), later studies have shown this to be a problem under natural conditions only when lichens are exposed to PFD greater than their normal environment (Schlensog *et al.* 1997; Kappen *et al.* 1998a; Gauslaa *et al.* 2001). Damage to the photobiont by high light can also occur in the "air-dried" state of lichens (Gauslaa and Solhaug 1999), and special mechanisms are involved to protect lichens against this danger (Heber *et al.* 2006a, b). There are a few lichens in which the photobiont is the dominant partner (e.g. *Coenogonium*), and hence the photobiont occurs as the principal surficial tissue. These lichens are mainly found in the low light environments of tropical and subtropical forest interiors. In many ways lichen photobionts behave as shade organisms protected from excess light by the fungal tissues and compounds (Green and Lange 1995). Physiological and biochemical aspects of lichen adaptation to light stress are covered in Chapter 8.

9.3.4   *Temperature*

Lichens are remarkable in their ability to survive a wide range of temperatures in nature, surviving temperatures as low as $-60\,°C$ in Antarctica and as high as $70\,°C$ on soil under temperate conditions (Lange 1953). In tests, cold resistance of many species seems to be almost unlimited, while heat resistance of dry thalli exceeds $100\,°C$ (30 minutes exposure). The photosynthetic apparatus was not damaged when two species were exposed for 14 days to space conditions in satellite experiments (Sancho *et al.* 2007). The biochemical and biophysical prerequisites for such tolerances are discussed in Chapter 8.

Many lichens are also highly resistant to high and low temperatures when hydrated and active. Small rates of apparent CO_2 uptake were found for *Cladonia alcicornis* in the laboratory at $-24\,°C$ (Lange 1965), and carbon fixation at such low temperatures was also confirmed (Lange and Metzner 1965). In Antarctica, *Umbilicaria aprina* maintained positive net photosynthesis to temperatures as low as $-17\,°C$ (Schroeter *et al.* 1994). However, it seems that this characteristic is restricted to chlorolichens, as cyanolichens were not able to photosynthesize at temperatures below about $-2\,°C$ (Lange 1965). Cyanobacterial lichens appear to be excluded from continental Antarctica (Schroeter *et al.* 1994) but, in contrast, cyanolichens appear to be better adapted to higher temperatures and can dominate in the tropics. Upper temperature compensation points for CO_2 exchange can approach $50\,°C$; gross photosynthesis for the tropical basidiolichen *Dictyonema glabratum* reaches maximal rates at *c.* $23\,°C$ and then remains unchanged until

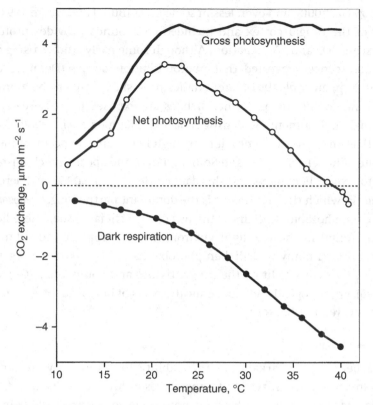

Fig. 9.5. Net photosynthesis, dark respiration, and estimated gross photosynthesis in relation to temperature for *Dictyonema glabratum*, Panama, measured under controlled conditions. (From Lange *et al.* 1994.)

40 °C (Fig. 9.5). In contrast, net photosynthesis has a clear optimum at about 22 °C and subsequently decreases due to increasing dark respiration so that the compensation point is reached near 40 °C. Even so, tropical lichens can still have positive NP at temperatures that are fatal for species in cooler habitats.

Lichens from different habitats exhibit different temperature optima (Lechowicz 1982), but it is remarkable that divergence between polar lichens and hot desert species is not more extreme, given the huge difference in macroclimates between these habitats (Kappen 1988). Monitoring of lichen activity at sites as different as the continental Antarctic to a desert in Spain has confirmed that the microclimate when the lichens are active is much more similar than the macroclimate (Green *et al.* 2007). Mean daily light inputs when active are almost identical, with the exception of the continental Antarctic location (Table 9.3). The mean temperature difference when active between the high Antarctic site and the Spanish desert is small compared with the much larger difference between the macroclimates.

Table 9.3. *Mean values for incident radiation and thallus temperature for three species of lichens in Spain, the maritime Antarctic and the continental Antarctic. The mean values have been calculated in each case either for the entire measuring period (1, 14, and 3 years respectively) and for times when the lichens were active (Active only)*

		Spain, Guadarrama Summit, Madrid, (2000 m) 41° N *Lasallia hispanica*	Antarctica Livingston I. (10 m) 62° S *Usnea aurantiaco-atra*	Antarctica Granite Harbour (10 m) 77° S *Umbilicaria aprina*
Radiation mean PAR (mol m^{-2} day^{-1})	All values	17.5	9.8	38
	Active only	**2.7**	**3.8**	**77**
Mean temperature (°C)	All values	9.7	−2.2	−9.74
	Active only	**4.5**	**1.1**	**2.9**

Temperatures vary markedly over the course of a year outside tropical regions, and some lichens are able to acclimate their gas exchange responses (Stålfelt 1939). Kershaw and coworkers (Larson and Kershaw 1974; Kershaw 1985) documented such seasonal photosynthetic changes in lichens using multivariate experiments where photosynthetic response was measured across different water contents, temperatures, and light levels. Lange and Green (2005) showed that temperate lichen species can acclimate their respiration to seasonal changes in temperature to such an extent that they had near identical respiration under very different nocturnal temperature conditions during the course of the year. Thus, in comparison with a nonacclimating response, more energy from respiration is available at low winter temperatures and increased loss of assimilates is avoided at higher temperatures in summer.

9.3.5   Carbon dioxide

Carbon dioxide is the substrate for photosynthesis, and CO_2 response of lichen net photosynthesis follows a saturation type function with an almost linear initial slope, the carboxylation efficiency. Under conditions of optimal hydration and saturating light, lichens usually saturate between 1000–1200 ppm external CO_2 partial pressure (Nash *et al.* 1983; Lange *et al.* 1999b). Suprasaturation of the thallus by water impeding CO_2 diffusion drastically increases the concentration required for CO_2 saturation, in some cases up to several thousand ppm. Also unusual are the low CO_2 compensation points achieved by many species. Values as low as 5 ppm are reported compared with the typical 30–50 ppm of homoiohydric C3 plants (Snelgar and Green 1980; Bauer 1984). Such low values are due to functioning of CCMs (Chapter 10).

As photosynthesis is not saturated by present ambient natural CO_2 ($\sim$370 ppm) that occurs around most lichen thalli, lichens profit from short-term increases in ambient CO_2 with an almost proportional increase in their NP, if not limited by other factors. For example, low light under a closed snow cover reduces carbon gain even when average CO_2 concentrations are 450–500 ppm with peaks to 1641 ppm (Sommerkorn 2000).

Responses of lichens to long-term experimental increases in external CO_2 have been inconsistent (Tuba *et al.* 1999). Exposure of *Cladonia convoluta* for five months to 700 ppm CO_2 increased net CO_2 uptake by 50%, a gain that was especially beneficial during drying cycles (Tuba *et al.* 1998). In contrast, *Parmelia sulcata* acclimated after only 30 days to 700 ppm CO_2 (Balaguer *et al.* 1996) by reducing photosynthetic capacity and less investment in RubisCO present in the pyrenoid of the algal chloroplasts. In contrast, no evidence of a down-regulation was found compared with individuals growing at 355 ppm CO_2 for *Parmelia caperata* individuals around a natural CO_2 spring, where daytime CO_2 concentrations averaged 729 ppm (Balaguer *et al.* 1999). No increase in lichen primary production under elevated CO_2 could be found, although enhanced accumulation of lichen secondary metabolites was noted. It is not known if the photobiont and/or mycobiont from the two stands remained genetically identical or if selection has taken place under the high CO_2 that has existed for over 200 years. Because of inconsistent and scarce information, firm conclusions about the performance of lichens under elevated CO_2 are not yet possible.

9.4 CO_2 exchange in nature: response to the complex environment

9.4.1   *Interrelationship of environmental variables*

Lange *et al.* (2001) compared the results of laboratory investigations with field responses for several lichens and found that the controlled studies were good predictors of the performance in the natural habitat. At any chosen moderate temperature, the net photosynthetic rate of any lichen is primarily controlled by thallus water content and PFD. An example of the interaction between these two factors is given for *Lecanora muralis* (Fig. 9.6). This particular lichen shows extreme suprasaturation with strongly depressed NP at low WC, due to desiccation, and at high WC, due to increased diffusion resistances. The

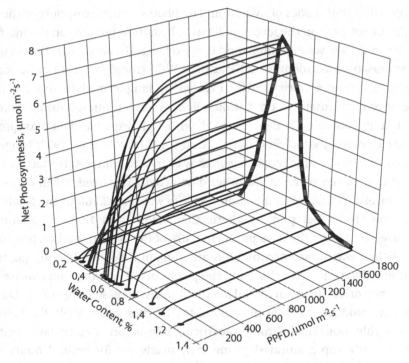

Fig. 9.6 Response of net photosynthesis of *Lecanora muralis* to thallus water content (% dw) and light (PPFD, μmol m^{-2} s^{-1}); the data were generated under controlled conditions in the laboratory, 12 °C. (From Lange 2002.)

response surface is, therefore, a ridge with its crest following saturating PFD at optimal WC and with declines on both sides of the ridge due to too much water at higher WC and too little at lower WC. A homoiohydric higher plant would remain on the crest of the ridge with NP effectively tracking changes in PFD. The lichen, however, can be at any point on the response surface.

Of interest is where the activity of lichens in the field would fall on the multifactorial, laboratory-generated response surface. Are conditions primarily optimal so that the lichen will behave like the higher plant, or are they mainly suboptimal, which would raise questions as to which is the major controlling environmental factor? Lack of optimality would also impact the modeling of photosynthetic productivity of lichens. These questions are investigated for several important lichen habitats in the following sections and synthesized in the conclusion.

9.4.2   Patterns of CO_2 exchange in lichen-rich habitats

Temperate climate

Lichens growing under temperate climate conditions are exposed to weather that changes extensively both seasonally and daily and which provides several different modes of hydration. The photosynthetic response of the lichen depends not only on the level and form of hydration but also on the interaction between thallus water content and CO_2 exchange. In particular, the occurrence of suprasaturation (depressed NP at high WC) strongly affects net carbon gain. These differences were clearly revealed by long-term data sets (30 minute measurement interval) on various lichens under quasi-natural conditions (Würzburg Botanic Garden) coupled with full response data sets produced under laboratory conditions (see Section 9.4.1; Leisner *et al.* 1997; Lange 2002, 2003*a*, *b*). The epilithic *Lecanora muralis* was dry and completely inactive for 25% of all days of the year. On the other days activity occurred, albeit occasionally only briefly, so that total metabolic activity was only 35.6% of the total time of the year (16.7% net photosynthetic CO_2 uptake and 18.9% respiratory CO_2 release). On 29% of the days producing 40% of the annual total carbon balance, the lichen was moistened by high air humidity, dew and/or fog (or frost in winter) which enabled a short and steep peak of CO_2 uptake after sunrise before the lichen dried out again and all metabolic activity ceased (Fig. 9.7, A). On 16% of the days producing 29% of the annual total carbon balance (B), the lichen was thoroughly moistened through nocturnal rain before a clear day. The thallus was initially suprasaturated giving low, steady NP for several hours until a strong and sudden increase in CO_2 uptake occurred as water loss resulted in water contents near optimal hydration. This lasted only for a very short time until full desiccation terminated all photosynthetic activity. These were the rare occasions when *L. muralis* achieved its highest rates of net photosynthesis, albeit only transiently. Heavy rain and long-lasting strong hydration usually resulted in an unfavorable diel carbon balance of the lichen (20% of the days, 4% of the annual total carbon balance). Nocturnal respiration was high and suprasaturation allowed only very low rates of CO_2 uptake even under favorable light conditions (D). Extended moistening by rain resulted in higher rates of net photosynthesis only when precipitation was light and/or when evaporative water loss was so rapid that suprasaturation was overcome at least for parts of the day (C). High WC seemed to be the most adverse feature for productivity of *L. muralis* and, over the whole year, it heavily depressed NP 38.5% of the time, when other conditions were suitable for photosynthetic carbon fixation and resulted in a negative carbon balance on as much as 25% of these days.

In contrast, *Cladonia convoluta* shows no suprasaturation so that maximal NP occurs at all WC above saturation (Lange and Green 2003). On days with substantial rain, when *L. muralis* had depressed NP, *C. convoluta* had its best carbon gain (35% of days, 78% of total carbon gain).

Nevertheless, annual net primary production of the *L. muralis* sample amounted to 15.8–20.7% carbon (related to thallus carbon content), which

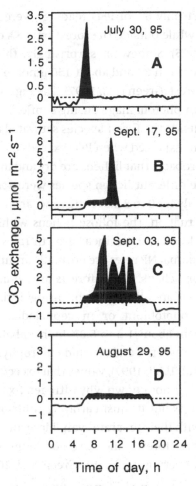

Fig. 9.7 Four different daily patterns of CO_2 exchange for *Lecanora muralis* (Lange 2003a); respiration negative and the period of carbon gain filled in black. (A) Thallus lightly moistened by overnight dew, with a brief but high peak of net photosynthesis (NP) after the sun rose and the lichen dried. (B) Thallus well hydrated by nocturnal rain, initially low NP from high water content (WC) and suprasaturation, a brief but strong peak as the thallus dried through the optimal WC range. (C) Several peaks of high NP as the WC remains in the optimal range for NP with regular wetting and drying. (D) Heavy rain gives a continually high WC and depressed NP from suprasaturation.

apparently enables this lichen species to be one of the most successful saxico-lous colonizers of nonnatural substrates of the temperate northern hemisphere.

Temperate rain forests
    New Zealand rain forests are an example of this habitat and, because they are evergreen, light conditions are relatively stable, but low, throughout

the year. In an upland *Nothofagus* forest only 6% of PFD readings were above 50 μmol m^{-2} s^{-1} over a 15-day period, while outside the forest only 8% of PFD were below that value (Green *et al.* 1995). Somewhat surprisingly, the light saturation levels for the lichens were between 62 and about 150 μmol m^{-2} s^{-1}, much higher than the PFD within the forest (Green *et al.* 1997). The high saturation values may protect the lichens from photodamage during sunflecks when the PFD can be considerably higher. Outside the forest species saturated at 550 to 766 μmol m^{-2} s^{-1}, which is lower than expected when PFD exceeds 800 μmol m^{-2} s^{-1} for about 50% of the time. It is probable that lichens are dry and inactive when PFD is high. Within the forest the different lichen species were active at optimal combinations of PFD and WC only for 0 to 8.4% of the time; and outside the forest, 3.3 to 13.3%. Despite this situation, the foliose lichens within the forest are some of the fastest growing known, with linear growth rates up to 27 mm y^{-1} (Snelgar and Green 1982). Maximal NP rates are normally around 1 to 2 nmol m^{-2} s^{-1} which are not high for lichens. These forests are dominated by *Sticta* and *Pseudocyphellaria* species that are all capable of nitrogen fixation by cyanobionts, either as the main photobiont or in cephalodia. These species have high nitrogen contents from about 1.2 to 5.2% (noncephalodiate chlorolichens <0.8% and cyanolichens between 4 and 5%) and chlorophyll contents up to around 5 mg gdw^{-1} (Green *et al.* 1980, 1997), values that exceed those for most other lichens (see Chapter 10). It is not known why nitrogen fixation is such a dominant life form in these forests but it must certainly contribute to high productivity. Photosymbiodemes with their separate cyanobiont and chlorobiont sectors occur on the margin of these forests, and the chlorobiont sectors regularly activate during high humidity overnight (Fig. 9.1; Green *et al.* 2002).

Tropical rain forests

Tropical rain forests are well known for their abundant lichens with high species diversity and biomass. A surprising gradient in predominant growth form also exists. In the lowland forests crustose species are common on bark and leaf surfaces, whereas foliose and fruticose species dominate in premontane and montane tropical rain forests. Differences in the environmental parameters that determine rates of CO_2 exchange are considered to be responsible for this situation (Zotz 1999).

Species from tropical montane forests, especially from well-illuminated edge habitats, have the highest chlorophyll, dry weight, or area-related net photosynthesis rates reported for lichens (see Section 9.2). Studies in Panama (Lange *et al.* 1994, 2000; Zotz *et al.* 1998) show that they benefit from extended periods of photosynthetic activity following moistening with only relatively short periods of desiccation between rain showers. Many of the dominant foliose species are adapted to high light with light saturation for NP occurring around 500 μmol m^{-2} s^{-1} PFD. Unexpectedly for lichens, the species in these habitats, especially the cyanolichens, are well adapted to high temperatures: the upper compensation point of homoiomerous *Leptogium phyllocarpon* is higher than 50 °C, and that of *Dictyonema glabratum* almost 40 °C. Net carbon gain is limited by two factors. First, most of the species show suprasaturation during and after heavy rain so that CO_2 exchange is heavily depressed; for example, *L. azureum* was estimated to lose more than one third of its potential diurnal carbon gain due to suboptimal hydration. Second, high respiratory CO_2 losses occur overnight as thalli are often hydrated and metabolically active at temperatures between 17 and 20 °C. The high rates of DR are often not compensated by diurnal carbon fixation, resulting in a negative daily balance. On average over a measuring period of 14 days, *D. glabratum* lost 72% of diurnal carbon gain by respiratory processes. However, their carbon balance is extremely sensitive to any change in temperature. Modeling shows that an increase in nocturnal temperature by 4 °C would be enough to generate a negative carbon balance for *D. glabratum* if no acclimation took place. Nevertheless, in spite of such disadvantages, there is a lush lichen vegetation.

The low altitude tropics are poor in macrolichens and differ from montane tropical rain forests by having higher temperatures, especially during the night. In the Panamanian lowland forest, *Leptogium azureum* had more days with negative than positive carbon balance (15 randomly chosen measurement days; Zotz and Winter 1994); and, for *Parmotrema endosulphureum*, cumulative dark respiration was almost 90% of diurnal carbon gain (Zotz et al. 2003). Higher precipitation might also have caused suprasaturation. Net carbon gain for lowland macrolichens is precariously close to zero and probably limits their ecological success.

In contrast, corticolous crustose chlorolichens are abundant and highly diverse in tropical lowland forests, as shown by Lakatos (2002) and Lakatos et al. (2006) in French Guiana. Their existence is due to a combination of specific morphological and physiological adaptations; suprasaturation, for instance, is reduced by the presence of water-repellant mycobiont hyphae at the thallus surface. These lichens can also achieve adequate carbon balances in spite of the very low light environment of their understory habitat, below 10 μmol photons m^{-2} s^{-1} for at least 95–99 % of the daytime. An extreme example is *Thelotrema alboolivaceum*, which is distinctly shade adapted with light compensation point at 7 μmol photons m^{-2} s^{-1}, and net photosynthesis becoming saturated at 39 μmol photons m^{-2} s^{-1} PFD. All of these shade lichens are also able to rapidly activate photosynthesis in order to effectively use shorter or longer lasting sunflecks.

Polar and alpine environments

These environments are linked by the low importance or complete absence of higher plants and a high importance of lichens. However, the environments are dissimilar and influence lichen CO_2 exchange differently. *Alpine* environments have a normal day/night cycle and cold nights can occur even in summer. The environment is wet with considerable additional water from mist and cloud, in particular, and, if not under snow, lichens can remain photosynthetically active during the entire day when light levels are low because of clouds. In a summer study in the Alps, Reiter and Türk (2000a, b) found long photosynthetic periods, 34% of study time for *Umbilicaria cylindrica*, 43% for *Cladonia mitis*, and 46% for *Thamnolia vermicularis*, but maximal photosynthetic capacity was reversed (6.9, 8.0 and 10.9 μmol kg^{-1} s^{-1}, respectively). Low temperatures appeared to have little effect on NP. High light levels can occur during the day and most lichens have very high (>2000 μmol m^{-2} s^{-1}) light saturation levels. Alpine populations of *Nephroma arcticum* are more "sun" adapted than at subalpine sites (Sonesson *et al.* 1992).

Polar sites all face long periods of light and dark during their annual cycle. In the *Arctic* the subpolar/polar tundra areas have been extensively studied (Moser *et al.* 1983a, b; Kershaw 1985, Hahn *et al.* 1993; Lange *et al.* 1998). Kershaw (1985) and coworkers carried out numerous studies of lichens in the Canadian Arctic. In these ground-breaking studies the importance of the microhabitat was clearly demonstrated, with strong effects of hummocks and hollows, lichen form and color, and wind on hydration duration and activity (Kershaw and Larson 1974; Larson and Kershaw 1975). In an Alaskan study (Hahn *et al.* 1993) the lichens could be active through the day (38% of days) and face being hydrated at high light. On average, lichens were dry and inactive for 41.5% of the measurement period (day and night). Carbon uptake was positively correlated with maximal NP (Lange *et al.* 1998). The cyanolichen *Peltigera malacea* was active for the shortest time but had the largest carbon gain due to its high photosynthetic capacity. Although there were large differences in thallus water content between the species, three times in mean water content, and four times in water contents during activity, there was only a difference of 6.5% across all species in length of active periods.

Antarctica has two major climatic zones: the maritime and continental Antarctic. In the maritime Antarctic *Usnea aurantiaco-atra* has been monitored on Livingston Island for several years using chlorophyll fluorescence combined with modeling from CO_2 response data (Schroeter *et al.* 2000). The maritime is warmer, summer monthly mean temperatures reach 6.2 °C, and the wet climate means that lichens can be active through the whole year except when snow covered. Although some positive photosynthesis can occur in winter, the major gains are in spring. Warm winters cause substantial respiratory losses due to low light and very short day length and may even produce a negative annual carbon budget. Lichens face the possibility of being hydrated and active under both high (warmer) and low (colder) light and lichens respond with a large change in optimal temperature for NP with change in PFD. Optimal temperatures range from 0 to about 12 °C as PFD changes from 50–1200 μmol m^{-2} s^{-1} for *Leptogium puberulum* and *Parmelia saxatilis* (Sancho et al. 1997; Schlensog et al. 1997).

Continental Antarctica is a cold desert with low precipitation. Lichens have been found as far south as 85° S, practically the limit of exposed rock, but are confined to areas with favorable microclimate and reliable water supply, and can have much shorter active periods. The crustose species *Buellia frigida* was hydrated mainly by snowmelt from small snow drifts and was active for only a few days in the early summer (Kappen et al. 1998b). A sample of *Umbilicaria aprina*, monitored at 3-hour intervals for a full year, was active on only 37 measurement points at midsummer. Exceptions are lichens like *Xanthoria mawsonia* in an unusual habitat growing along the sides of meltwater channels at Cape Hallett, 77° S latitude, which was active for 70% of the entire period of an investigation lasting 40 days (Pannewitz et al. 2006). This species had an optimal temperature of 10 °C, a very high maximal photosynthetic rate (11.0 mg g^{-1} h^{-1}) and was not light saturated at 2000 μmol m^{-2} s^{-1}. These lichens showed a reverse diel pattern of photosynthesis, in which they are not desiccated under high light but remain active because that is when the water flows in the high latitudes (Green et al. 2007).

In general, in comparison with species from the arctic and alpine areas, most lichens from Antarctica have low chlorophyll contents and low maximal NP. Calculated from a summary list (Kappen 1988) the average maximal net photosynthetic rates (mg g^{-1} h^{-1}) were: alpine, 1.37; arctic, 0.38; antarctic, 0.21. In comparisons between temperate and antarctic sites, *Parmelia saxatilis* and *Umbilicaria nylanderiana* had one seventh of the chlorophyll content and considerably lower NP in antarctic specimens (Sancho et al. 1997). There appears to be no explanation for this situation. However, it is interesting that a lowered photosynthetic rate with no change in respiration rate would lower the optimal temperature for NP in the same manner found for low light (Friedmann and Sun 2005).

Snow deposition is particularly important in all habitats. In Swedish Lapland lichens benefit from photosynthesis when snow covered despite the low light because the consistently near zero temperatures resulted in very low light compensation points <10 μmol m^{-2} s^{-1} (Sommerkorn 2000). Similar situations are known from the maritime antarctic and alpine regions (Schroeter et al. 1997, 2000). In contrast, snow banks in continental Antarctica trapped the winter cold and no photosynthesis occurred until immediately before complete snowmelt (Pannewitz *et al.* 2003). Extended snow cover can be detrimental due to extended respiration losses at low light in the maritime antarctic, arctic, and alpine regions (Kappen *et al.* 1995) and from decreased activity in continental Antarctica.

Deserts

Lichens are successful colonizers of the deserts of the world, because they do not rely on rain but can become active after moistening by fog, dew, or even high air humidity alone (see Section 9.3.2). A special epigeic plant formation in arid and semi-arid lands are biological soil crusts, an intimate association between soil particles and cyanobacteria, algae, microfungi, and bryophytes, in which lichens often play a dominant role. They form a coherent layer that resists wind and water erosion, influences water infiltration, and contributes to carbon and nitrogen budgets of the ecosystems (Belnap and Lange 2003). Crusts dominated by *Lecidella crystallina* (Namib coastal fog desert) have total chlorophyll contents of $500\,mg\;m^{-2}$, covering the ground with a continuous sheet of photosynthetic machinery equivalent to a phanerogamous leaf. However, time periods of activity are extremely limited. Apart from a few erratic rains, the thalli are dry during most of the day, and moistening from fog and/or dew allows only a short period of photosynthetic activity. Figure 9.8 shows a diel time course of CO_2 exchange for the crustose *Caloplaca volkii* on a favorable summer day with the characteristic early morning peak of photosynthesis. Overnight, the lichen crust absorbed water from dew and fog, and dark respiration was activated and increased with water content until sunrise when the well-hydrated lichen started photosynthesis. CO_2 uptake rose steeply to a maximum until high radiation, increasing temperatures, and decreasing air humidity rapidly dried the lichen and first limited, and then halted, all activity. Photosynthetic activity of desert lichens differs from day to day according to the specific hydration conditions (Lange *et al.* 2006).

This short morning peak of net photosynthesis, the "early breakfast response" after nocturnal hydration by dew or fog, is a regular photosynthetic response pattern of lichens in arid and semi-arid areas and also in many other regions including the drier seasons in the Mediterranean and temperate regions (Fig. 9.7A; Kappen 1988).

9.5   Lichen CO_2 exchange performance and their ecological niche – a synthesis

Lichens have a worldwide distribution and grow in a high diversity of sites. Nevertheless, the majority of lichens prefer certain habitats, which align with their most important general features: i.e. poikilohydric character, slow

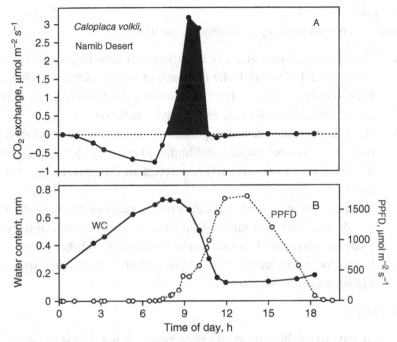

Fig. 9.8 Morning "early breakfast response" of CO_2 exchange of *Caloplaca volkii* measured in the field in the Namib Desert. (Upper panel, A) CO_2 exchange, respiration negative and the period of carbon gain filled in black. (Lower panel, B) Thallus water content, (WC) in mm rainfall equivalent (dark symbols); light intensity, PPFD, in $\mu mol\ m^{-2}\ s^{-1}$ (open symbols). (From Belnap and Lange 2003.)

growth, usually small habit (which result in low competitiveness), and ability to colonize areas which are marginal for most plants with respect to dryness, cold, and space, and are often called "extreme." What role does carbon dioxide exchange play in defining such a lichen niche? We can now try to answer this question, at least in a generalized and introductory form, using information presented in the previous sections. First we need to try to define what constrains the lichen niche: that is, what advantages and what disadvantages their occurrence.

Excluded

a. Lichens are excluded from all habitats that favor higher plants. Simply, if higher plants can establish sufficiently then lichens are competed out (of course, higher plants also provide a new possible environment for lichens, for example as epiphytes).

b. Cyanolichens, only, are excluded by cold temperatures.

Advantaged

Lichens are favored and higher plants are excluded by:

a.   Habitats with low water storage; these can have high or low absolute water availability, but if the presence of water is transient then the poikilohydric nature of lichens is favored. Such environments include arid zone soils, rock surfaces, epiphytic habitats.

b.   Habitats with low mean temperatures, when trees are excluded from the polar and alpine zones, and higher plants completely from very cold localities such as continental Antarctica and the nival zone of the high mountains.

c.   Habitats where water is not delivered by rain but rather in the form of fog, dew or high air humidity. Lichens are excellently adapted to use these sources, which are almost inaccessible for higher plants.

d.   In epiphytic habitats that are physiologically dry (too low storage for higher plants to exploit).

Disadvantaged

a.   At least macrolichens are disadvantaged by hot, moist nights (tropical lowland rain forest), and extended wet nights seem to damage lichens also in maritime Antarctica.

b.   Most of the lichens are sensitive to air pollution.

c.   Transient habitats, such as agricultural areas and managed forests, tend to change too rapidly to support major lichen populations.

Carbon dioxide exchange, i.e. the need to secure enough photosynthetic carbon production, plays an important role in all of the above. In most *Excluded* habitats the lichens are simply outcompeted when there is sufficient and con-sistent water for growth of higher plants (again, note the new habitat formed on the higher plants). The same is true when continuous availability of water enables luxuriant growth of mosses, such as in *Sphagnum* bogs or rain-rich forests. Cyanolichens, generally, appear to be unable to photosynthesize below about $-2.5\,°C$ and this is the probable reason why they are excluded from continental Antarctica. When *Advantaged*, the poikilohydric nature of lichens allows (a) rapid recovery of photosynthetic productivity after desiccation (Chapter 8), and con-siderable survival ability when dry. These are both excellent characteristics for habitats with transient water availability, such as epiphytic, lithic, and polar habitats. In (b) the lack of higher plants provides opportunities for lichens in these cold environments. Temperate and polar lichens retain high net photosyn-thetic rates at low temperatures and chlorolichens show positive net photosynth-esis well below freezing. Lush lichen growth in the fog deserts of the world, often

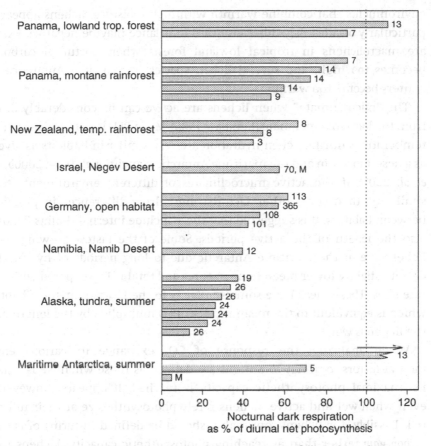

Fig. 9.9 Nocturnal respiratory carbon losses as percent of diurnal net photosynthetic carbon gain under natural conditions for a selection of lichen species from different habitats (figures at the end of the bars indicate number of diel courses used for integration; M means integration through a model). Dark respiration > 100% indicates that averaged carbon balance has been negative during time period of measurements. The summary was drawn from data in publications of many authors listed in Lange *et al.* (2000); additional data from Zotz *et al.* (2003).

in combination with a lack of any phanerogamous vegetation, is characteristic for *Advantaged* (c). However, dew and fog favors lichen productivity also in other climatic regions, such as in the Alps, in Mediterranean areas, and in the temperate zones. Lichens are *disadvantaged* in otherwise suitable habitats where respiration is too high. Net primary production is the result of diurnal photosynthetic carbon gain minus nocturnal respiratory carbon loss. The high proportion of mycobiont in lichen thalli results in proportionately high respiration at higher temperatures. For many species in many environments, about 40 to 60% of the carbon income is lost through nocturnal respiration (Fig. 9.9). However, in environments that combine warmth with high moisture, lichens appear to be particularly disadvantaged because carbon balance may be negative. Examples are macrolichens in tropical lowland forests when nocturnal carbon loss becomes too high, and several lichen species in maritime Antarctica when winters become too warm.

The "microclimate" when lichens are active can be considerably different from the local macroclimate (Kershaw 1985). Mean PFD inputs and mean thallus temperatures during lichen hydration are very similar in habitats as divergent as a desert in Spain and the maritime Antarctic (Sancho *et al.* 1997, 2000*b*; Green *et al.* 2007). If the active microclimates of different environments tend to similarity, then some other factor(s) must drive differences in productivity between habitats. It is suggested that these include internal thallus factors and also the length of the active periods. Some of the fastest growing crustose lichens are in the maritime Antarctic due to long periods of hydration and despite slightly lower mean temperatures and total PFD compared with tempe-rate sites. These ideas have some similarity to the I_{wet} parameter (Chapter 10) which is equivalent to the mean incident PFD multiplied by the length of time the lichen is wet.

When analyzing the response of CO_2 exchange to various environ-mental factors, optimal conditions are defined as those when NP is maximal, i.e. maximal photosynthetic capacity is reached. It appears, however, that even when wet and active, lichens rarely photosynthesize at their full poten-tial. Possibly, for lichens, optimality should be defined in terms of *maximized carbon gain* rather than as reaching photosynthetic capacity. Lichens may be hydrated for only brief intervals, and their problem is to make as much use as possible of these short periods under conditions of progressing desiccation. This is achieved by an apparently excessive (i.e. very high, photosynthetic) capacity, which allows maximal use of any short period of moistening. Examples would include the exceptional maximal photosynthetic rates in dry open areas, such as soil crusts, where lichen photosynthetic capacities approach those of C3 plants and allow the lichens to make good use of the brief active periods; and also the high, in comparison with normal light envir-onment, light saturation values for rain forest lichens, a situation that allows them to maximize use of the brighter sunflecks without risk of photosystem damage.

If lichens are not functioning optimally with respect to CO_2 exchange, then this makes the effects of global climate change even more difficult to predict. Increased growth rates probably require longer active times. However, more active time means more water availability and a greater likelihood, outside cold areas, that higher plants can invade. Changes in temperature will probably have little direct effect, as the active microclimates are very similar. However, these are broad conclusions from a small database, and there are many opportunities and needs for further research. Carbon dioxide exchange does appear to be crucially involved in defining the lichen niches and in interpreting lichen habitat selection.

The carbon economy of lichens

K. PALMQVIST, L. DAHLMAN, A. JONSSON, AND T. H. NASH III

Growth and survival of photosynthetic carbon autotrophs, such as lichens, are primarily limited by their photosynthetic carbon assimilation minus carbon dioxide (CO_2) losses related to growth and maintenance respiration. The absolute, as well as the relative, rates of these two processes will hence determine their capacity to grow. In lichens, both photosynthesis (P) and respiration (R) are strongly constrained by prevailing environmental conditions, particularly water and light. There is also a variation in inherent P and R capacities among species and individuals. Significant progress has been made during the last two decades in understanding how such variations in external conditions and internal capacities affect lichen growth (Boucher and Nash 1990a; Muir et al. 1997; Sundberg et al. 1997, 2001; Hyvärinen and Crittenden 1998b; Palmqvist and Sundberg 2000; Hilmo and Holien 2002; Dahlman and Palmqvist 2003; Hyvärinen et al. 2003; Gaio-Oliveira et al. 2004a, 2006; Gauslaa 2006; Gauslaa et al. 2006b; Palmqvist and Dahlman 2006), including re-establishment of vegetative propagules (Hilmo and Sastad 2001; Hilmo and Ott 2002). This progress has been driven by the development of transplantation techniques in combination with more mechanistically oriented studies, knowledge that has further been used to formulate both conceptual (Palmqvist 2000) and more mechanistically oriented models (Link et al. 1985; Palmqvist and Sundberg 2000; Dahlman and Palmqvist 2003). Such models are useful for the direction of future research towards more explicit hypothesis testing, and could also be adopted to predict how environmental changes may affect particular lichen species, or compare species' abilities to utilize and acclimate to varying environmental conditions.

Besides providing the quantitative data per se, transplantation experiments have also proven that lichens display measurable growth rates within a few

Lichen Biology, ed. Thomas H. Nash III. Published by Cambridge University Press.
© Cambridge University Press.

months under field conditions, ranging from a few percent to over 100% increase in dry weight, depending on the species and the particular environmental conditions during the studied period. The inherent difficulty in cultivating lichens under controlled conditions (Dibben 1971) has thus been circumvented by instead growing them under the stochastically varying light, water, and temperature regimes occurring in the field. The transplantation experiments and the subsequent monitoring of environmental conditions for a wide range of habitats and species (e.g. Palmqvist and Sundberg 2000; Gaio-Oliveira et al. 2004a; Gauslaa et al. 2006b) has further provided the necessary information to manipulate light, water, and nutrient supplies during the growth experiments, as exemplified in Dahlman and Palmqvist (2003). In this way it has even been possible to increase the growth rate of lichens, allowing compilation of more data much faster, so that one can compare species and individual responses with altered conditions.

In addition to quantifying lichen growth in relation to prevailing environmental conditions when lichens are metabolically active, a mechanistic approach also allows direct measurements of metabolic activity (i.e. photosynthetic and respiratory gas exchange) in situ (e.g. Lange et al. 1993a). However, such measurements are technically advanced and restrict measurements to very few thalli (Leisner et al. 1997), thus reducing the generality of the results. An alternative approach is to combine laboratory gas exchange measurements with field experiments, allowing measurements of significantly more samples (e.g. Sundberg et al. 1997; Gaio-Oliveira et al. 2004a). Nevertheless, laboratory gas exchange measurements, as well as environmental monitoring, particularly the continuous thallus water content (WC) measurements of individual thalli (Coxson 1991), are also laborious and restrict measurements to only a few individuals (e.g. Gaio-Oliveira et al. 2004a, 2006). A model allowing prediction of thallus WC from more easily measured environmental variables (A. Jonsson, J. Moen and K. Palmqvist, unpublished data) and the possibility of using indirect and more easily measured markers for metabolic capacity of the lichens (Palmqvist et al. 2002; Palmqvist and Dahlman 2006) are therefore approaches that may be adopted for more broad-scale surveys of species and habitats.

The above-outlined progress is summarized in this chapter, initially discussing the allocation of resources between the photobiont and the mycobiont and how this may affect their carbon economy (Section 10.1). Carbon flow in the thallus is also surveyed, emphasizing the need for more data and techniques (Section 10.2). Environmental factors affecting lichens and their water relations in particular are assessed (Section 10.3), followed by a summary of the regulation of photosynthesis by environmental as well as internal factors (Section 10.4), and a discussion of what we know and what we need to study regarding lichen respiration (Section 10.5). Finally, lichen growth is analyzed in a mechanistic context (Section 10.6), comparing the capacity of lichens to convert assimilated environmental resources into new biomass, and discussing how these findings may help us understand the lichen symbiosis better.

10.1   The requirement of controlled resource allocation in a symbiotic system

Growth of an individual plant or lichen is the result of resource gain and subsequent biosynthesis of cellular compounds minus losses related to dispersal, fragmentation, grazing, or necrosis. Since the dominant part of both lichen and plant biomass is made up of carbohydrate [(CH$_2$O(n)] equivalents (Palmqvist and Sundberg 2000), their growth will primarily be dependent on photosynthetic CO$_2$ assimilation minus respiratory CO$_2$ losses. In addition to carbon, the plant or lichen must also acquire mineral nutrients such as nitrogen and phosphorous for the synthesis of new proteins, membranes, and DNA. To maintain a balanced growth, the acquisition of carbon must therefore be balanced in relation to mineral availability and vice versa (Chapin 1991). Resources must further be allocated to different cells and organs in a regulated manner to secure future development (Grace 1997). According to functional equilibrium models (Brouwer 1962), it is assumed that acquired resources are allocated so that pool sizes of key elements remain constant within and among organs (Schulze and Chapin 1987), and that environmental limitations or excesses that reduce resource use will also reduce uptake (Chapin 1991). This model has been summarized for vascular plants as: if the carbon to nutrient ratio of your tissue is too high, grow roots; otherwise, shoots (Grace 1997).

For lichens, carbohydrate acquisition is directly related to photobiont photosynthesis, while minerals are acquired through wet or dry deposition across the whole thallus. Since mycobiont hyphae almost always make up the dominant part of the lichen biomass (Chapters 1 and 3), we can assume that the fungus acquires most of the minerals, similar to roots in plants, with the exception of those lichens that have cyanobacterial N$_2$ fixation. So, even if we have little knowledge of which metabolites should be regarded as key elements for lichens, we may still adopt a carbon-based functional equilibrium model for these organisms, replacing shoots with photobiont cells and roots with fungal hyphae. According to this view, a positive carbon gain will then require that resources are balanced between the photobiont and mycobiont so that the summed carbon gain exceeds respiratory CO$_2$ losses over a defined timespan, which is probably weeks and months rather than single days (Lange and Green 2006). So, even though the mycobiont appears to dominate in terms of biomass, the metabolic demands of this fungal tissue must be lower than that which can be supported by the photobiont. In this context, and as for plants, nitrogen (N) is likely to be a key component, because increased investments of N in photobiont cells may potentially increase the carbon (C) gain capacity whereas increased investments of N in respiratory fungal tissue will increase the rate of maintenance respiration (Lambers 1985; Reich *et al.* 1998). So, even though lichen mycobionts may control the photobiont through some means of N starvation (Richardson 1999), the mycobiont must refrain from overinvestments in its own tissue.

10.1.1   *Resource allocation trends across lichen species*

To test whether a functional equilibrium model might be applicable also to lichens, data on C gain capacities and maintenance respiration was gathered for 75 lichen associations with various morphology, photobiont affiliation, and habitat preferences (Palmqvist *et al.* 2002). Similar broad-scale surveys had previously been made for plants and bryophytes (Enriquez *et al.* 1996; Reich *et al.* 1998, 1999), but were lacking for lichens. In addition to measuring CO_2 gas exchange characteristics and thallus N concentrations, the potential usefulness of the indirect metabolic markers chlorophyll (Chl) *a* (for C gain capacity), and chitin and ergosterol (for mycobiont mass and activity) were also evaluated. Chl *a* had previously been shown to be well correlated to both photosynthetic capacity and RubisCO concentration in a smaller set of these lichen species (Palmqvist *et al.* 1998), and ergosterol with maintenance respiration (Sundberg *et al.* 1999a).

The thallus N concentration ranged from 1 to 50 mg g^{-1} dry weight (Fig. 10.1), net photosynthesis (NP) varied 50-fold, and R 10-fold (Fig. 10.2). Green-algal lichens had the lowest N concentrations; cyanobacterial lichens had the highest; and tripartite lichens were intermediate (Fig. 10.1). All the three markers increased with thallus N concentration. Lichens with the highest Chl *a* and N concentrations had the highest rates of both P and R. Chlorophyll *a* alone accounted for *c.* 30% of the variation in NP and R across species (Fig. 10.2), which may seem surprising since less than 1 or 2% of the thallus N is actually invested in this pigment (Palmqvist *et al.* 1998, 2002; Sundberg *et al.* 2001). However, as will be discussed later (Section 10.4), this emphasizes that the ratio of Chl *a* in relation to the other proteins of the photosynthetic machinery may be relatively constant in lichen photobionts. As a result, Chl *a* appears to be a relatively useful surrogate variable for photosynthetic unit (PSU) or even photobiont concentration in lichen thalli (Palmqvist 2000), but this needs to be established more firmly, as discussed in Palmqvist and Dahlman (2006).

In general, lichens had invested their N resources so that their maximal C input capacity matched their respiratory C demand around a similar (positive)

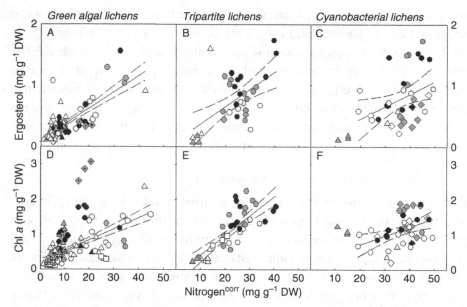

Fig. 10.1   Relationships between the concentrations of thallus nitrogen (N) and ergosterol (A–C) or Chl *a* (D–F) for 75 different lichen species using three individuals of each, collected from pristine Arctic/Antarctic (white symbols), boreal (gray symbols) and temperate/subtropical (black symbols), extracted from Palmqvist *et al.* (2002) covering 43 mycobiont genera, 8–10 photobiont genera and different morphologies; crustose (squares), fruticose (triangles), foliose (circles) and homoiomerous (diamonds). Thalli that had been N fertilized either by birds or anthropogenically (symbols with cross) are not included in the regressions. The N content of Chl *a* (6.27% of molecular weight) was subtracted from total thallus N ($= N^{corr}$) to avoid autocorrelation. The regression lines ($\pm$ 95% confidence interval) were obtained by pooling all thalli in the respective plots excluding the N-fertilized thalli. Parameter details are presented in Table 10.1.

equilibrium across species. Taken together, this indeed emphasizes that lichens are able to optimize their resource investments between carbohydrate input and expenditure tissue, suggesting that a carbon economy view might be a fruitful way to compare and understand the performance of different lichens. As we will see later in this chapter (Section 10.6), the carbon balance of lichens is, moreover, in the same range as that of higher plants (Palmqvist and Sundberg 2000).

However, even though the carbon equilibrium appeared to be similar across the species, there were also some conspicuous differences among species or groups of lichens, with some apparently being more carbon efficient compared with others. The green-algal fruticose lichens had, for instance, invested

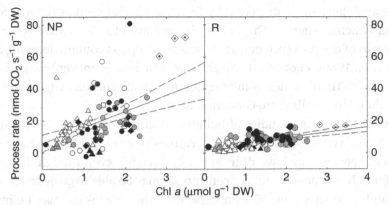

Fig. 10.2   Light-saturated net CO_2 fixation (NP) and dark CO_2 efflux (R) rates, as a function of thallus Chl a concentration of the boreal and temperate/subtropical lichens presented in Fig. 10.1 (adapted from Palmqvist *et al.* 2002). Measurements were made at 15°C and optimal thallus water contents awaiting steady-state rates. The regression lines (± 95% confidence intervals) were obtained by pooling all data in the respective plot.

proportionally more of their thallus nitrogen in Chl a compared with the foliose, cyanobacterial lichens, a difference that was manifested as a higher photosynthetic efficiency quotient $[K_F = (NP_{max} + R)/R)]$ in the former group. This indicates that P capacity increases more than maintenance R when a larger fraction of the thallus N is invested in the photobiont. Such differences in resource allocation may then mechanistically explain the large differences in inherent growth capacity that exist among species as summarized by Rogers (1990). In addition to differences in resource allocation patterns that may be linked to photobiont and/or morphology groups, we should also expect there to be an acclimation of resource allocation related to environmental conditions. An increased N supply or decreased light supply would for instance require increased investments in the photobiont simply to maintain energy equilibrium as discussed in Dahlman *et al.* (2003). For the same reason, more photobiont cells in relation to mycobiont hyphae would also be required at higher temperatures (Friedmann and Sun 2005; Sun and Friedmann 2005).

10.1.2   *Intraspecific resource investment trends*

The above broad-scale survey presents data for whole thalli or lobes, although the whole thallus is seldom homogeneous with respect to N concentration or photobiont to mycobiont ratio. For example, foliose species, such as *Nephroma arcticum* and *Peltigera aphthosa*, have thinner thallus margins compared with the older interior parts, with a higher photobiont to mycobiont ratio in the margins (Sundberg *et al.* 2001; Dahlman *et al.* 2002). A similar difference between young and old tissue is displayed by fruticose species such as the mat-forming *Cladina* lichens, where both photobiont and mycobiont hyphae occur in the upper part of the podetia whereas the lowermost parts contain decaying hyphae (Nash *et al.* 1980). Photosynthetic activity is therefore unevenly distributed in these *Cladina* thalli, which is in agreement with an uneven distribution also of their photobiont cells (Gaio-Oliveira *et al.* 2006).

Similar to previous studies of lichen N relations (Rai 1988, 2002; Palmqvist *et al.* 1998), the broad-scale survey (Palmqvist *et al.* 2002) emphasized that different species may have different N optima. So, even though lichens with the higher N and photobiont concentrations may be able to grow faster because of a higher inherent net C gain capacity, other trade-offs may define their optimal N concentration range. The data presented in Fig. 10.3 is in agreement

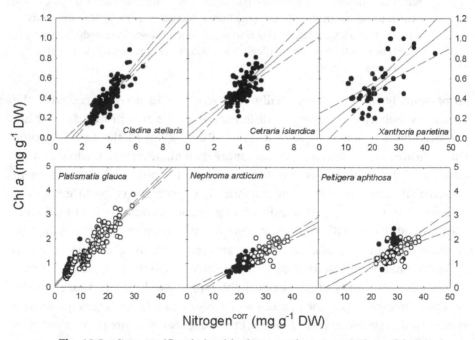

Fig. 10.3   Intraspecific relationships between the concentrations of thallus nitrogen and Chl *a* of the terricolous fruticose green-algal lichens, *Cetraria islandica* and *Cladina stellaris* (A. Jonsson and K. Palmqvist, unpublished data), the foliose green-algal lichens *Platismatia glauca* (Palmqvist and Dahlman 2006) and *Xanthoria parietina* (Gaio-Oliveira *et al.* 2005b), and the foliose tripartite lichens *Nephroma arcticum* and *Peltigera aphthosa* (Dahlman *et al.* 2002; Dahlman and Palmqvist 2003). Black symbols represent untreated thalli collected from their natural habitat, or thalli treated as control in manipulative experiments; open symbols, thalli that were experimentally deprived of or fertilized with nitrogen. Regression lines (± 95% confidence interval) were obtained by pooling all thalli of the respective species; equations and parameter details are presented in Table 10.1.

with the idea that each lichen may have a relatively fixed N concentration range, because the data presented here have been compiled from different populations, habitats, and experimental conditions, and thus, as far as possible, represent each species range. The N range was lowest in the green-algal lichens from nutrient-poor habitats (*Cetraria islandica* and *Cladina stellaris*) and significantly higher in the two N_2-fixing species (*Nephroma arcticum* and *Peltigera aphthosa*) (Fig. 10.3). The most scattered Chl a to N relation occurs in the nitrophilic lichen *Xanthoria parietina* (Fig. 10.3), probably reflecting the fact that, in contrast to the other five species, a significant and largely varying part of the thallus N can be stored in soluble pools (Gaio-Oliveira *et al.* 2005a) or as chitin (Crittenden *et al.* 1994). The widest N concentration range was displayed by *Platismatia glauca*, a species that can tolerate both nutrient-poor conditions and high loads of anthropogenic N deposition (Dahlman *et al.* 2003). The increased Chl a with increasing N concentration (Fig. 10.3) also increased the growth rate of the latter species (Palmqvist and Dahlman 2006). We infer that an increased photosynthetic efficiency quotient occurs with increasing Chl a, similar to the observation for fruticose green-algal lichens (see Section 10.1.1). However, the necessary CO_2 gas exchange data are lacking to substantiate this conclusion.

  In the broad-scale survey, there was, moreover, a correlation between ergosterol concentration and thallus N concentration (Fig. 10.1), emphasizing an increased amount of metabolically active hyphae per unit dry weight of the lichen with higher N status. In the few species investigated so far, however, the ergosterol concentration instead seems to be insensitive to thallus N concentration (Sundberg *et al.* 2001; Dahlman *et al.* 2002; Dahlman and Palmqvist 2003; Gaio-Oliveira *et al.* 2005a, b; Palmqvist and Dahlman 2006), suggesting that the concentration of active fungal hyphae in relation to thallus mass may be quite fixed for each species. The significant increase in Chl a with increasing thallus N concentration (Table 10.1, Fig. 10.3) without a concomitant increase in ergosterol concentration within species thus is consistent with the inference that individuals may display higher growth rates with higher N status.

10.2   Carbon sinks and translocation

  As indicated in Fig. 10.2, up to 50% of the CO_2 assimilated in photobiont photosynthesis may be consumed in lichen respiration. This is similar to estimates that have been made for whole plants. So, in terms of their C demands, lichen thalli have maintenance and construction costs comparable to those of roots and translocation tissue of vascular plants. A large fraction of the remaining C compounds make up the thallus structure, with cell walls being the largest sink. However, reduced C compounds may also serve additional functions, apart

Table 10.1. *Regression parameters for the data presented in Figs. 10.1 and 10.3. Data were fitted to a linear equation of the form* y = ax + b, *where* y *is component concentration (mg g^{-1} DW),* a *is slope of the line (dimensionless),* x *is thallus nitrogen concentration (mg g^{-1} DW) and* b *is the y-axis intercept. Thallus N concentrations were corrected for the N content of chlorophyll. All slopes were significantly different from zero for* p < 0.001 *if not otherwise stated after the adjusted* r^2 *value.*

Group or species	Ergosterol – nitrogen			Chlorophyll *a* – nitrogen		
	a	*b*	*r^2* [*n*]	*a*	*b*	*r^2* [*n*]
Green-algal lichens	0.026	0.1	0.61 [92]	0.031	0.2	0.43 [137]
Tripartite lichens	0.028	0.1	0.31 [34]	0.056	−0.2	0.64 [40]
Cyanobacterial lichens	0.020	0.05	0.11; $p = 0.02$ [36]	0.025	0.2	0.16; $p = 0.004$ [42]
Cladina stellaris				0.15	−0.18	0.68 [114]
Cetraria islandica				0.12	0.01	0.40 [110]
Xanthoria parietina				0.02	0.04	0.38 [67]
Platismatia glauca				0.11	0.06	0.88 [306]
Nephroma arcticum				0.05	−0.02	0.25 [98]
Peltigera aphthosa				0.06	−0.14	0.28 [97]

from providing the mere structure of the thallus. The soluble polyols may, for instance, function both as short-term storage or contribute to the osmotic pressure of the thallus, thereby lowering its water potential and enabling absorption of water from the air (see Section 9.3.2). The more complex secondary C compounds (Chapter 7) may further protect the lichen from drought (Chapter 8), high light stress (Chapter 8), and predation (Section 14.7.2). A comprehensive understanding of the C economy of lichens would hence require a quantitative analysis also of these various C sinks and their turnover rates, something that remains to be more thoroughly investigated.

10.2.1   Carbon translocation

Lichens lack specific cells or tissues for the translocation of metabolites, water, and nutrients, between photobiont and mycobiont. Translocation patterns vary among species, and are in part dependent on the chemical composition of the biont cell walls and how the bionts are integrated (Honegger 1991a). The mechanisms behind the induction of carbohydrate export and mass transfer from photobiont to mycobiont have not been elucidated and so far, no specific polyol or glucose transporter has been isolated from lichens, even though such a carrier has been postulated (Collins and Farrar 1978). The exported C varies depending on the photobiont. Green algae release a polyhydric sugar alcohol (polyol) to the mycobiont, and cyanobacteria release glucose, in the latter case possibly mediated through the gelatinous polysaccharide sheath surrounding the cyanobiont (Richardson and Smith 1966, 1968; Hill and Smith 1972). The exported carbohydrate is ribitol in lichenized *Coccomyxa*, *Myrmecia* and *Trebouxia*; erythritol in *Trentepohlia*; and sorbitol in *Hyalococcus* (Honegger 1991a). Once taken up by the mycobiont, the carbohydrate is rapidly and irreversibly metabolized into mannitol, via the pentose phosphate pathway (Lines *et al.* 1989), and thereby made unavailable to the photobiont (Galun 1988).

Variation in thallus hydration status may affect the efficiency of carbohydrate translocation in different ways, depending on the species. For instance, in the *Nostoc* lichen *Peltigera polydactyla*, mannitol formation was significantly enhanced when water contents were increased (MacFarlane and Kershaw 1982); whereas alternating drying and wetting was required for polyol metabolism and translocation in the *Trebouxia* lichen *Hypogymnia physodes* (Farrar 1976b). Another study, using labeling of photosynthates with $H^{13}CO_3^-$ followed by in vivo NMR measurements, detected ribitol, glucose, mannitol, and polysaccharides (Sundberg *et al.* 1999b), and showed similar differences in labeling and transfer rates with rapid ^{13}C labeling of mannitol in the *Nostoc* lichen *Peltigera canina*. In contrast, in the *Trebouxia* lichen, *Platismatia glauca*, the ^{13}C label was only accumulated as ribitol and glucose in the photobiont, and no detectable labeling of mannitol was detected during a 12-hour chase period of wet thalli. Since no photosynthesis occurs in desiccated lichens, we are accustomed to thinking negatively about these periods: that is, as lost opportunities for carbon assimilation. However, if one considers carbon translocation, the alternating wetting and drying may apparently be a prerequisite for a functioning symbiosis in green-algal lichens. Although more experiments of the above type are needed, particularly with green-algal lichens, this hypothesis is consistent with the observation of Ahmadjian and Heikkilä (1970) that relichenization of axenically separated symbionts also produced better results in cultures subjected to wetting and drying cycles.

10.2.2   Carbon sinks

Polyols, secondary carbon metabolites, and cell walls are the major carbon sinks that have been most intensively investigated, and, even if quantitative measures are rare, it appears that these three sinks are also the largest.

The polyol content varies between 2% and 10% of the thallus dry weight, depending on species and season, with the highest content recorded for the *Nostoc* lichen *Peltigera polydactyla* in late summer (Lewis and Smith 1967). Apart from mannitol, lichen fungi also contain arabitol. Arabitol is depleted more rapidly than mannitol under conditions of stress (Farrar 1976b). We infer that this polyol may function as a short-term carbohydrate reserve. In contrast, mannitol was proposed to serve as a substrate for respiration during prolonged dark periods (Drew 1966). When ribitol and mannitol pools were separately quantified in three lichens collected in late summer (Sundberg *et al.* 1999*b*), it was found that mannitol concentrations were 4–5% and ribitol concentrations *c.* 1% of the dry weight in the two tripartite lichens *Lobaria pulmonaria* and *Peltigera aphthosa*. Most of the soluble carbohydrates may then be found in the mycobiont. This conclusion is supported by a more recent study of the green-algal lichen *Platismatia glauca*, where the ribitol concentration was consistently lower than 0.5% and the mycobiont arabitol and mannitol pools were above 2% of the dry weight across a large range of thalli, with significant differences in thallus N and Chl *a* concentrations and thereby growth capacities (Palmqvist and Dahlman 2006). The sizes of all the three polyol pools were moreover similar both before and after the 73-days growth experiment, although significant turnover could be envisaged because the weight gain exceeded the pool sizes significantly. We infer that lichens may maintain relatively fixed polyol concentrations in their thalli. This would be consistent with the interpretation that the physiological function of these compounds includes protecting enzymes during drought stress, and regulating turgor and osmotic pressure (Bewley and Krochko 1982; Farrar 1988).

Cell walls may constitute a significant fraction of the dry weight of a lichen (Boissière 1987), particularly in species with high C to N ratios such as *Lasallia pustulata* (Palmqvist *et al.* 1998), where 68% of the dry weight could be attributed to the cell wall (Boissière 1987). In this species, chitin constituted 5.5% (w:w) of all wall components. In contrast, in the *Nostoc* lichen, *Peltigera canina*, which has a significantly lower thallus C to N ratio, only 36% of the dry weight could be attributed to cell wall compounds, with chitin constituting 13% (w:w). This difference in chitin concentration in relation to other cell wall compounds further emphasizes that the relative C to N requirements of different tissues may vary significantly depending on the lichen species.

The majority of organic compounds found in lichens are secondary metabolites of the fungus. Often they are unique to lichens and are deposited on the surface of the hyphae (Chapter 7). These products can amount to between 0.1% and 10% of the dry weight of the thallus, sometimes up to 30% (Galun and Shomer-Ilan 1988; Chapter 7). These *c.* 800 different low molecular weight metabolites (Huneck 2001), referred to as lichen substances and including lichen acids, are among the more intensively investigated aspects of lichenology (Galun and Shomer-Ilan 1988; Chapter 7). They belong to a variety of chemical groups, and are synthesized from side branches of the pentose-phosphate cycle, glycolysis, or the mitochondrial TCA cycle (Mosbach 1969; Section 7.3).

Secondary carbon metabolites are usually absent in lichens with cyanobacterial photobionts (Galun and Shomer-Ilan 1988), which can fix N_2 and so have a larger access to nitrogen for biosynthesis. In agreement with the carbon-nutrient balance (Bryant *et al.* 1983) or the growth-differentiation balance hypotheses (Stamp 2004), synthesis of complex secondary carbon compounds may then simply be a way to make use of excessive carbon when nitrogen is a limiting resource.

The more dynamic aspects of secondary metabolism in relation to growth rates and environmental limitations were recently studied by McEvoy (2006), who showed that, although the synthesis of some secondary compounds may respond to varying levels of irradiance, other compounds in the same thallus were produced with a more fixed ratio in relation to other structural compounds in the thallus. Many secondary lichen compounds strongly absorb ultraviolet radiation, and consequently numerous field and laboratory experiments have been carried out to study their possible solar radiation protective role (e.g. Gauslaa and Ustvedt 2003; Nybakken *et al.* 2004; Solhaug and Gauslaa 2004a, b; Bjerke *et al.* 2005). Solar protection by carbon-based pigments may be particularly beneficial in exposed habitats and for those species that rely on nitrogen supply through wet or dry deposition. For these species, it may be impossible to make use of the excess irradiance by increasing the photobiont density, which will require the input of nitrogen, a limiting resource in many lichen environments. The lichen acids may also serve a biological function other than solar protection, such as a defensive role towards parasitizing fungi or bacteria, or browsing animals, or as rock-weathering agents (Lawrey 1986; Fahselt 1994a).

A better understanding of the regulation of all the above major carbon sinks in lichens requires refined tools to quantify them in relation to environmental resource supplies and in relation to each other (McEvoy 2006). The recently explored technique of excising thallus parts prior to more longer term manipulative experiments to enable quantitative measurements of initial and final concentrations of various metabolite pools should also be explored further (Palmqvist and Dahlman 2006).

10.3   Carbon economy in relation to a poikilohydric lifestyle

The ecological success of lichens can in part be explained by their poikilohydric nature, and their ability to resist desiccation and low temperatures (Kappen 1988; Kappen and Valladares 1999). They may thereby colonize habitats that are inaccessible to more homoiohydric plants. Desiccation tolerance involves tolerance of both mycobiont and photobiont, a trait that the lichen symbionts share with their free-living relatives (Raven 1992; Qui and Gao 2001; Chapter 8). The extent to which lichens can tolerate drought stress is, however, also partly related to the moisture conditions to which they are adapted in their natural habitat. For example, xeric species recover more quickly and from longer periods of drying than mesic species (Bewley 1979). In terms of productivity, lichens are still limited by their poikilohydric nature, because their metabolism is constantly switched on and off when they alter between a hydrated and a desiccated state. Their growth is therefore strongly constrained by length and frequency of their wet and metabolically active periods and by light availability during these events (Palmqvist and Sundberg 2000). In contrast and as already discussed in the previous section (10.2), the translocation of assimilated carbon from the photobiont to the mycobiont may in some species still be dependent on these alternating desiccation and rehydration cycles.

10.3.1   Loss and uptake of water

Both the uptake and the loss of water are physical processes without metabolic control, with the water content (WC) of the thallus constantly changing, as it strives to equilibrate with the WC of the surrounding atmosphere (Blum 1973; Rundel 1982; A. Jonsson, J. Moen and K. Palmqvist, unpublished data). However, rates of water movement between lichen thalli and their environment vary among species, in relation to thallus morphology, anatomy, and color, and the amount of water the thallus can hold at saturation (Rundel 1982, 1988; Lange *et al.* 1993*b*; Valladares *et al.* 1997; Dahlman and Palmqvist 2003; Jonsson, Moen and Palmqvist, unpublished data). For instance, due to their larger surface area to volume ratios, filamentous and fruticose species take up and lose water more rapidly than flat, foliose species. For the same reason, a thick foliose lichen will equilibrate more slowly with the surrounding air than a thinner one (Gauslaa and Solhaug 1998). Furthermore, the lichens and the environmental conditions depicted in Fig. 10.4 emphasize that species living relatively closely together in the same habitat can display quite variable lengths and frequencies of their hydration–desiccation cycles. In this case, the terricolous and mat-forming lichen *Cladina rangiferina* remained wet for considerably longer periods compared with the epiphytic fruticose lichen *Alectoria sarmentosa*, reflecting a higher exposure to the air of the latter species and the absence of a large boundary layer reducing water diffusion rates (Jonsson, Moen and Palmqvist, unpublished data). The mat-forming species was, moreover, exposed to larger fluctuations in temperature resulting in significant condensation of morning dew (see day numbers 252 and 253 in Fig. 10.4).

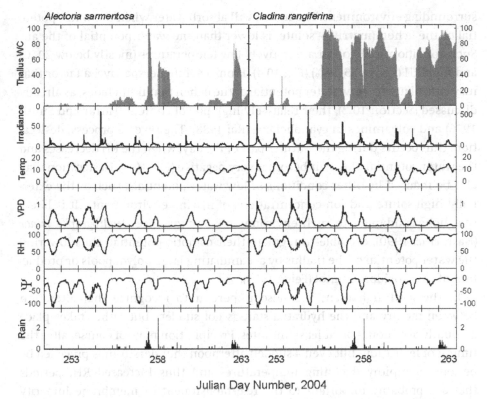

Fig. 10.4   Variation in relative thallus water content (WC) of the epiphytic pendulous lichen *Alectoria sarmentosa* and the mat-forming, terricolous *Cladina rangiferina*, with irradiance measured close to the lichen, thallus temperature, air relative humidity, vapor pressure deficit of the air (VPD), air water potential (Ψ), and precipitation, in a spruce forest stand close to Umeå, Sweden, during c. 2 weeks in September 2004. Data were obtained using the techniques and equations described in Jonsson *et al.* (2007).

Lichens can utilize a variety of water sources to activate their metabolism, such as precipitation, fog, and dew, and even high relative humidity (RH) (Matthes-Sears *et al.* 1986). The importance of nonprecipitation sources of water is well demonstrated by the abundance of lichens in coastal deserts where precipitation is minimal, but fog, dew, or elevated RH occur almost daily (Kappen 1988; Fig. 1.1). The uptake of water from nonsaturated atmospheres is extraordinary, as it represents essentially the reverse of evaporation and is a process that does not occur in the relatively homoiohydric vascular plants. The water moves between the lichen and the air along a decreasing water potential gradient (Rundel 1982, 1988; Nash *et al.* 1990; Jonsson, Moen and Palmqvist, unpublished data) until the lichen WC has equilibrated with the surrounding environment. The lichen will absorb water when the water potential of the lichen (matrix + solute) is lower than the water potential of the air. Such conditions only occur at relatively low temperatures (mostly below 20 °C) and high RH (above 75–95%) (Fig. 10.4). From a solute perspective, a factor that may contribute to a low water potential of the lichen thallus includes, as already discussed (Section 10.2), their relatively high polyol content (Farrar and Smith 1976) and polyamines in cyanolichens (Rai 1988). The reverse process, desiccation, thus occurs during the warmer part of the day, when the RH decreases and the water potential of the air becomes very negative (Fig. 10.4). Desiccation may also be induced by other conditions of low water potentials, such as the externally high solute and ion concentrations of marine environments. It is likely that lichens adapted to such environments, for instance *Dendrographa minor* (Nash *et al.* 1990), are able to regulate their osmotic potential, thus lowering the water potential of the thallus by accumulating large polyol pools or production of mannosidomannitol (Feige 1973).

   In the natural habitat, it is also important to recognize that alteration between the dry and the hydrated state is not sudden, but rather takes place gradually over hours or, at least, minutes. Precipitation may, of course, alter the timing of hydration, but even a sudden afternoon thunderstorm is preceded by periods of rapidly declining temperatures and thus increased RH, periods that are probably important to the re-establishment of membrane integrity (D. H. Brown, personal communication). This may explain why a "respiratory burst," termed resaturation respiration (Farrar 1976*a*; Link and Nash 1984*b*), is seldom recognized during field measurements (Section 10.5).

10.3.2   *Activation of metabolism*

   In lichens with green-algal photobionts, photosynthesis can be recovered if the thallus is allowed to equilibrate with air of water potentials above *c.* −10 MPa (Lange *et al.* 1986; Scheidegger *et al.* 1995*a*). However, this generally requires equilibration periods of up to 30–60 h, and also fails to induce maximal photosynthetic rates (Scheidegger *et al.* 1995*a*). In contrast, addition of liquid water, such as that from precipitation, fog, or dew, may activate metabolism to maximal rates within a few minutes, or at least within one hour, as discussed in Palmqvist (2000). Photosynthesis in lichens with cyanobacterial photobionts always requires the addition of liquid water to recover from desiccation (Lange *et al.* 1986). The reason for this has not been elucidated, but photosynthetic light absorption and the subsequent electron transport reactions may be uncoupled until liquid water is added (Bilger *et al.* 1989; Lange *et al.* 1989). The *Nostoc* cells of cyanobacterial lichens further form small colonies surrounded by a gelatinous, aqueous-polysaccharide sheath in the thallus with the fungal hyphae protruding into the sheath (Honegger 1991*a*), whereby a somewhat longer rehydration period may be required simply to imbibe this sheath and the *Nostoc* cells with water. This may then be a part of the 10–15 minute lag after the addition of liquid water before photosynthesis is induced in these lichens (Lange *et al.* 1989).

10.3.3   *The water content dependence of photosynthesis and respiration*

Once hydrated, lichen metabolism will be reactivated more or less rapidly, depending on the species. The total water holding capacity and relative WC required for optimal metabolism varies widely depending on lichen species. For instance, different species living under apparently identical rain-forest conditions may contain between 3 g and 30 g water per g dry weight at maximal hydration (Lange *et al.* 1993*a*). Among these, WC at optimal photosynthesis also varies, so that species with the highest maximal water contents also require more water. The photosynthetic WC response of a particular lichen species (exemplified in Fig. 10.5) is apparently determined by a range of different factors, with morphological constraints being most important (Collins and Farrar 1978; Lange *et al.* 1993*b*, 1999*b*). For instance, the structural composition of the lichen will determine the relative distribution of water among

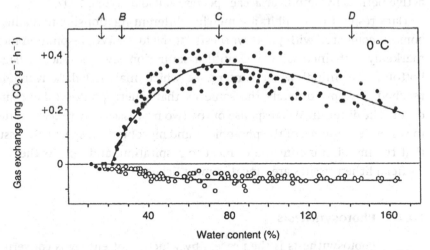

Fig. 10.5 Population estimates for the dependence of net photosynthesis (solid symbols) and dark respiration (open symbols) on thallus water content in *Ramalina maciformis* at 0 °C. Each point is a separate piece of thallus (Lange 1980). The letters across the top correspond to the water contents at which gas exchange is initiated (*A*), the water compensation point for CO_2 exchange in the light (*B*), net photosynthetic optimum (*C*) and region within which increasing CO_2 diffusion resistance depresses net photosynthesis (*D*).

intracellular, apoplastic, and extracellular compartments within the thallus (Beckett 1995, 1997). Moreover, fungal cell wall composition and deposition of water-repellent secondary metabolites on the hyphal surface are important in determining the degree of hyphal swelling upon hydration and the subsequent risk of blockage of gaseous pores. A depression of photosynthesis at high water contents is a general phenomenon among lichens, and is caused by increased thallus resistance for CO_2 diffusion (Collins and Farrar 1978; Lange and Tenhunen 1981; Cowan *et al.* 1992). The quantitative effect of depressed CO_2 acquisition at high thallus water contents must, however, be considered in relation to the beneficial effect of being able to remain metabolically active for longer periods, if this is coupled to a relatively large water-holding capacity and reduced desiccation rate. Such a difference in water-holding capacity could, for instance, partly explain the higher growth rate of *Peltigera aphthosa* growing under identical environmental conditions as *Nephroma arcticum* (Dahlman and Palmqvist 2003). To study trade-offs like this, the recently proposed parameter lichen water use efficiency (LWUE) is useful (Máguas *et al.* 1997). Similar to the water use efficiency (WUE) parameter of higher plants, LWUE reflects carbon gain versus water loss. Indeed, as found by Máguas *et al.* (1997), variations in morphology and structure may lead to altered LWUE, where features such as thallus density seem to play a crucial role in prolonging photosynthetically active periods in Umbilicariaceae species (Valladares *et al.* 1997).

Dark respiration exhibits a somewhat different relationship to thallus water content compared with photosynthesis. At the lower WC, respiration increases markedly with increasing WC until a saturation level is reached (Fig. 10.5, bottom). The initial water content, at which maximal dark respiration is reached, is approximately the same as that where photosynthetic maxima occur. The different WC response of the two processes may reflect a difference in water dependence of the photobiont and mycobiont, respectively, assuming that the mycobiont contributes most to respiration, and reflects the fact that surficial lichen tissues are fungal in most lichens.

10.4   Photosynthesis

Photosynthesis is the process by which light energy is converted into chemical energy used by autotrophic organisms for maintenance, growth, and reproduction. This complex process, which traps radiant energy and uses this to reduce CO_2 into carbohydrates, is one of the most conserved processes in the biosphere, essentially being similar across the plant kingdom, including cyano-bacteria and unicellular algae. Chlorophylls and certain other pigments in association with protein complexes absorb the radiant energy, and their energy state is thereby raised. In oxygen-evolving organisms, as the energy passes through paired reaction centers (photosystems I and II), electrons are generated that are used to reduce NADP (photosystem I) and to split water (photosystem II), a process that results in the evolution of O_2 as a by-product. Through a series of subsequent steps, the energy level of the photosynthetic protein–pigment complexes return to a baseline level and concomitantly ATP and $NADPH + H^+$ are generated. Incorporation of CO_2 is mediated by a series of enzymes in the Calvin cycle, with ribulose 1,5 bisphosphate carboxylase/oxygenase (RubisCO) being the first. RubisCO is further the most abundant protein in all photosynthetic cells, and, as a result, is also the major nitrogen sink of the biosphere (Chapin et al. 1987; Evans 1989). The photosynthetic performance of lichens and their photobionts has been comprehensively reviewed before (Kershaw 1985, Kappen 1988; Nash 1996; Campbell et al. 1998; Palmqvist 2000, 2002; Chapter 9), so the following presentation is only a brief summary.

Literature data on lichen photosynthesis in relation to varying water, light, temperature, and CO_2 supplies is vast, covering both in situ (Section 9.4.2) and more controlled laboratory (e.g. Kershaw 1985; Lange et al. 1995; Lange 2002) CO_2 gas exchange measurements. In situ measurements have been made in largely contrasting habitats ranging from Antarctica and the Arctic to deserts and temperate and tropical rain forests, also covering a wide spectrum of species (Section 9.4.2). These studies have emphasized that, similar to other photosynthetic carbon autotrophs, lichen photosynthesis displays a curvilinear relationship to light (Link et al. 1985; Palmqvist 2000) and is able to adapt to prevailing temperature conditions in the habitat (Kappen 1988). One must acknowledge, however, that lichen photobionts are unicellular algae or cyanobacteria. Thus, in some major respects they are different from the cells of higher plants. Without going into details, this includes the absence of a vacuole in the photobionts, generally only one large chloroplast in the algae as opposed to the 30–50 in plant cells, and the sharing of electron transport components between photosynthesis and respiration in cyanobacteria (Campbell et al. 1998). Similar to their free-living relatives, many photobionts moreover possess a photosynthetic CO_2 concentrating mechanism (CCM), different from the C4 and CAM metabolism of some higher plants, as discussed elsewhere (Palmqvist 2000).

In nature, quantum flux densities in the 400–700 nm spectral range, the action spectrum of the photosynthetic light reactions, may differ by at least two orders of magnitude among habitats. Within a habitat, quantum flux densities also vary seasonally, diurnally, and spatially. However, a remarkable feature of plant and algal chloroplasts, as well as cyanobacteria, is their ability to adjust their photosynthetic apparatus in relation to available irradiance. Short-term acclimation is achieved through so-called State I to State II transitions, whereby electron transport through the two photosystems can be adjusted in relation to light quality and/or relative ATP to NADPH demand (Manodori and Melis 1984; Allen *et al.* 1989; Guenther and Melis 1990). Longer-term acclimation involves regulation of the relative abundance of the various proteins of the photosynthetic apparatus in relation to irradiance during growth (Larson and Kershaw 1975). Cells grown under low irradiances thereby invest relatively more of their nitrogen into light-harvesting and thylakoid proteins; whereas high light grown cells invest relatively more into RubisCO and Calvin cycle proteins (Boardman 1977; Björkman 1981). Such acclimation in relation to seasonal variations in irradiance has also been documented for lichens (Stålfelt 1939; MacKenzie *et al.* 2001), and is discussed by Kershaw (1985).

10.4.1   *Photosynthetic capacity*

As shown in Fig. 10.2, maximal net CO_2 assimilation capacity (P_{max}), measured at optimal WC, light saturation, 15 °C and ambient CO_2, can vary up to 80 nmol g^{-1} DW s^{-1}, or 8.5 μmol m^{-2} s^{-1} (Palmqvist *et al.* 2002; K. Palmqvist, unpublished). This range is in agreement with many other studies of lichens but is well below those for higher plant leaves (Chapin *et al.* 1987; Green and Lange 1995). In contrast, when related to chlorophyll, P_{max} of lichens and their photobionts is more similar to the rates in higher plants (Table 10.2). Thus, at the cellular level, P_{max} seems to be determined by the same factors in lichen photobionts, including cyanobacteria, as in plants. The relatively lower photosynthetic capacities per thallus area, or dry weight, of lichens can then be explained by their relatively low concentration of photosynthetic cells or photosynthetic units (PSU) (Palmqvist 2000), a trait that may be explained by the following four characteristics. First, in contrast to plants and bryophytes, lichens do not construct two-dimensional surfaces entirely made of photosynthetic tissue, and all photobiont cells are surrounded by fungal tissue that, even though maintaining structural integrity, probably limits photobiont expansion within the thallus (Honegger 1991*a*). Second, photobiont development may be constrained by lack of sufficient nitrogen required for the proteins of the photosynthetic apparatus. The latter is supported by the strong correlation between chlorophyll *a* and thallus nitrogen concentration as depicted in Figs. 10.1 and 10.3. Third, even if cyanobacterial lichens are characterized by high thallus nitrogen concentrations, these species still have low PSU densities in comparison with plant leaves with similar nitrogen status (Palmqvist *et al.* 1998). This may in part be explained by the chitin content of the fungal cell wall and thus a relatively higher requirement of nitrogen for fungal growth and biosynthesis compared with plants and bryophytes (Duchesne and Larson 1989). Fourth, the poikilohydric nature of lichens restricts photosynthetic

Table 10.2. *Maximum rates of net photosynthesis in relation to tissue chlorophyll concentration of some photosynthetic organisms*

Chlorophyll is a + b for all except the cyanobacteria, which have only chlorophyll a. Measurements were made at light saturation, at temperatures that are typical for the respective species, and unless otherwise stated at ambient (35 Pa) CO_2. Photosynthesis was measured as CO_2 fixation for plants, lichens and bryophytes, and as O_2 evolution for algae and cyanobacteria

Species	Photosynthetic rate		Comment	Reference
	$\mu mol\ g\ Chl^{-1}\ s^{-1}$	$\mu mol\ m^{-2}\ s^{-1}$		
Plants				
Triticum aestivum (flag leaf)	50	28	Nitrate fertilized	Evans, 1983
Triticum aestivum (flag leaf)	75	24	Without nitrate treatment	Evans, 1983
Atriplex triangularis	62	35	High light grown	Björkman et al., 1972
Atriplex triangularis	15	7	Low light grown	Björkman et al., 1972
Betula verrucosa	29	11	High light grown	Öquist et al. 1982
Betula verrucosa	21	5	Low light grown	Öquist et al., 1982
Cordyline rubra	4.7	3.3	Shade-plant	Boardman et al., 1972
Green-algal lichens				
Parmelia caperata (Trebouxioid)	25	–		Tretiach and Carpanelli 1992
Hypogymnia physodes (Trebouxioid)	19	9	Low light, nitrogen fertilized	Palmqvist et al. 2002
Hypogymnia physodes	14	1.8	Exposed natural habitat	Palmqvist et al. 2002
Peltigera aphthosa (Coccomyxa PA)	18	3.9	Saturating CO_2	Palmqvist 1993
Peltigera aphthosa	10	1.2		Palmqvist 1993; Palmqvist et al. 1994
Umbilicaria deusta (Trebouxioid)	6	–		Sundberg, B., unpublished,

Table 10.2. (cont.)

Species	Photosynthetic rate		Comment	Reference
	µmol g Chl^{-1} s^{-1}	µmol m^{-2} s^{-1}		
Dermatocarpon miniatum (*Stichococcus*)	2.5	0.6		Smith and Griffiths 1996
Cyanobacterial lichens				
Peltigera praetextata (*Nostoc*)	115	–		Hawksworth and Hill 1984
Peltigera canina (*Nostoc*)	50	5.4	Saturating CO_2	Palmqvist 1993
Peltigera canina	13–42	–		Palmqvist, K., unpublished
Pseudocyphellaria dissimilis (*Nostoc*)	35	–		Snelgar 1981
Leptogium saturninum (*Nostoc*)	8	3.5		Palmqvist, K., unpublished
Peltigera leucophlebia (*Nostoc*)	3.6	1		Smith and Griffiths 1996
Lichina pygmea (*Calothrix*)	1	–	O_2-evolution, 5 °C	Raven *et al.* 1990
Bryophytes				
Marchantia sp.	27	4.5		Palmqvist, unpublished
Thuidium delicatulum	25	–		McCall and Martin 1991
Dicranum scoparium	22	–		McCall and Martin 1991
Grimmia laevigata	9–12	–		Alpert 1988
Free-living green algae				
Chlamydomonas reinhardtii	70	–	Saturating CO_2	Leverenz *et al.* 1990
Scenedesmus obliquus	70	–	Induced CCM, ambient CO_2	Palmqvist *et al.* 1997

Green-algal photobionts

Coccomyxa PA	–	170	High light grown, saturating CO_2	Ögren 1993; Ögren, unpublished data
Coccomyxa PA	–	36–72	Low light grown, saturating CO_2	Ögren 1993; Ögren, unpublished;
Coccomyxa PA	–	15	Low light grown, ambient CO_2	Palmqvist 1993
Trebouxia erici	–	15	Low light grown, saturating CO_2	Palmqvist et al., 1997
Trebouxia erici	–	10	Low light grown, ambient CO_2	Palmqvist et al., 1997

Cyanobacteria

Synechococcus sp. PCC 7942	–	60–85	Induced CCM, saturating CO_2	Palmqvist 1993; Clarke et al. 1995
Calothrix 7601	–	50–55	High CO_2 grown, saturatingCO_2	Campbell 1996
Nostoc PC	–	50	Induced CCM, saturating CO_2	Palmqvist 1993

–, Data lacking.

activity to occasions when the thallus is able to maintain sufficient hydration. Because exposure to high irradiances enhances the evaporative losses of water, lichen photosynthesis is most often restricted to periods when irradiances are relatively low, such as during rainfall and early morning hours following dew-fall. The beneficial effects on lichen productivity by a high photobiont density may then be limited by increased self-shading.

10.4.2   *The light response curve and the colimitation of photosynthesis by light and nitrogen*

Photosynthesis always displays a characteristic response to irradiance levels (Fig. 10.6). Typically, the relation between photosynthesis and irradiance has three different phases: the light-limited part where photosynthesis is limited by irradiance; the light-saturated part where carboxylation efficiency is the limiting factor; and the transition zone between these two phases (denoted convexity) (Fig. 10.6A). The efficiency of photosynthesis will evidently

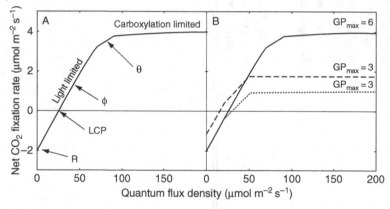

Fig. 10.6   A typical response of net photosynthesis to quantum flux density using the nonrectangular hyperbola equation described by Leverenz and Jarvis (1979) and later reworked by Lambers *et al.* (1998); see Palmqvist (2000). The intercept with the *x*-axis is the light-compensation point (LCP), the initial slope gives the quantum yield (ϕ) and the intercept with the *y*-axis is the rate of respiration in darkness (R). The rate of bending of the curve from the light limited to the light saturated part is described by the convexity value θ. CO_2 fixation is light-limited at low quantum flux densities and carboxylation limited at the higher densities. In (A) GP_{max} was set to 6 µmol m^{-2} s^{-1}, R to –2 µmol m^{-2} s^{-1}, ϕ to 0.08 mol CO_2 (mol quanta)$^{-1}$, and θ to 0.99. In B, θ was set to 0.99, and ϕ to 0.08 in all three curves. In the uppermost curve (solid line), GP_{max} was set to 6 and R to –2. In the lowermost curve (dotted line), GP_{max} was set to 3 and R to –2, and in the middle curve (dashed line) GP_{max} was set to 3 and R to –1.2. Note that the light saturation value is doubled when carboxylation capacity (GP_{max}) is doubled, and that the LCP is increased and NP decreased when R is increased.

be highest if the photosynthetic apparatus has been optimized to operate close to the light saturation value, which should then match the irradiance level to which the cell, leaf, or lichen is most often exposed (see Palmqvist, 2000, and references therein). The concentration of pigments and the other proteins of the photosynthetic tissue, and thus the nitrogen concentration of this tissue, will therefore vary depending on prevailing light conditions (Evans 1989). As a result, photosynthesis becomes saturated at lower irradiances in low-light acclimated compared with high-light acclimated cells. Due to the higher PSU densities and thus also RubisCO concentration of high-light acclimated cells, these also display higher P_{max} rates (Björkman 1981). However, because of the higher nitrogen concentrations of high-light acclimated cells, dark respiration rates will also be increased (Fig. 10.6B), and subsequently the irradiance required to reach light compensation (Björkman 1981). In addition to the above-mentioned risk of increased self-shading with increased PSU density (Section 10.4.1), it will then also be too expensive to maintain such "over-investments" in a low light environment. This again emphasizes the necessity for a tight acclimation of the photosynthetic apparatus to varying light conditions, for instance among the seasons. Few attempts have been made to study the cellular acclimation of photosynthesis in lichens in its mere mechanistic details (Kershaw 1985), but recent advances using molecular techniques to quantify *de novo* RubisCO synthesis, in combination with fluorescence analysis of electron transport capacity, have proved that such studies are possible for lichens (MacKenzie *et al.* 2001). However, even though it appears that the ratio between Chl *a* and RubisCO concentrations may be relatively fixed in many lichens (Palmqvist *et al.* 1998; Sundberg *et al.* 2001), and thus explains the relatively high correlation between Chl *a* concentration and photosynthetic capacity (Fig. 10.2), our knowledge of photosynthetic acclimation to light and nitrogen in these organisms is still incomplete. The relatively tight correlation between photosynthetic pigment concentration and RubisCO in lichens may well be an effect of the low light environment of many lichens when they are wet and active. Alternatively, this may also be a result of N limitation, because RubisCO is the largest N sink of the photosynthetic proteins (Chapin *et al.* 1987).

10.4.3 *Temperature dependence*

Laboratory measurements of net photosynthesis demonstrate that lichens have fairly broad thermal optima that shift to somewhat higher temperatures as light increases (Lange 1969; Section 9.3.4). In general, enzyme activity increases as temperatures increase from low values, and the overall temperature response curve can be modeled using an Arrhenius equation (Link *et al.* 1985). At higher temperatures, dark respiration increases disproportionally, and, as a consequence, net photosynthesis declines, which is also an effect of increased oxygenase activity of RubisCO (photorespiration) at higher temperatures (Björkman 1981). Although lichens from different habitats exhibit different temperature optima (Lechowicz 1982), it is probably more remarkable that the differences between antarctic lichens and hot desert species are not more extreme, given the huge difference in macroclimates between these habitats. Review of the extensive literature dealing with field measurements of lichen photosynthesis demonstrates that many lichens are photosynthetically active primarily within the 5–20 °C range, although 0–10 °C may better characterize some antarctic species (Kappen 1988). Convergence of photosynthetic temperature optima is not overly surprising, because principal photosynthetic gain may be constrained to times when temperatures are similar in different habitats, as for instance exemplified in Gaio-Oliveira *et al.* (2004*a*). Moreover, lichens from polar regions are photosynthetically most active in summer periods, when temperatures are warmer; whereas lichens from hot deserts are most active in winter periods or in early morning hours, when temperatures are cooler (Nash and Moser 1982; Kappen 1988).

10.5   Respiration

In plants, a significant portion of the photoassimilated carbohydrates becomes the main substrate for respiration (Amthor 1995). This is apparently also the case for lichens where up to 50% of the photoassimilated carbon may be "lost" in respiration (Fig. 10.2). However, respiration should not be viewed as a futile carbon loss process, because the function of respiration is to convert photoassimilates into substances used for growth and maintenance, and to energize this, as well as transport and nutrient assimilation processes (Amthor 1995; Lambers *et al.* 1998). During respiration and growth, CO_2 is then released as a by-product. The fraction of carbon that is "lost" during respiratory metabolism is dependent on the pathways of respiration, the mitochondrial ADP:O ratio, and substrate composition (Amthor 1995). A comprehensive review on plant respiration was presented by Lambers *et al.* (1998), and, since plant and fungal respiration appears to be fundamentally similar (Fahselt 1994*b*), only a few specific characteristics of lichen respiration are given here.

10.5.1   *Variation in respiration related to "internal" factors*

Lichen respiration varies about 4–5 fold across species (Palmqvist 2000), both when related to area and dry weight (Fig. 10.2). Since mycobiont biomass dominates in most lichens, one may assume that mycobiont respiration should dominate photobiont respiration but this has never been established, and may vary among species. It is still clear that both bionts contribute to the dark respiration of the thallus. Lichens with low thallus N, ergosterol and Chl *a* concentrations have the lowest rates of respiration (Sundberg *et al.* 1999a; Palmqvist *et al.* 2002). This implies that species with low respiration may be relatively rich in carbon compounds that contribute to biomass and have low metabolic turnover and energetic requirements. The increase in lichen respiration with increased thallus nitrogen concentration is then most probably an effect of increased energy demand related to protein turnover (Lambers 1985), as also discussed by Lange and Green (2006).

As indicated in Fig. 10.2, a large part of the variation in both photosynthetic capacity and maintenance respiration across species can be attributed to variation in Chl *a* concentration. Thus, we infer that the two processes are somehow coregulated. In plants it has long been established that the magnitude of P and R covaries with each other (Enriquez *et al.* 1996; Lambers *et al.* 1998), where the general view is that P is feedback inhibited by carbohydrate demands rather than vice versa (Lambers *et al.* 1998). Assuming that the mycobiont is the largest C sink in lichens, we may then speculate that the fungus controls the C availability by regulating the size of its photobiont population. A mycobiont with high carbohydrate requirements must then also ensure sufficient nitrogen investments in photobiont cells to achieve efficient photosynthesis, as discussed in previous sections. This further emphasizes that lichen mycobionts that require high levels of both carbohydrates and nitrogen, such as the fast-growing members of the Peltigerales, must form a thallus providing space for a large enough photobiont population to meet the concomitantly high demands (Hyvärinen *et al.* 2002). Even if we know very little about this, these speculations again emphasize that we might be able to understand the lichen symbiosis better, if we knew more about their overall C demands and the mechanisms regulating their C economy.

10.5.2   *Respiration in relation to environmental factors*

Several environmental factors also affect respiration rates in lichens (Kershaw 1985; Green and Lange 1995). As discussed above, respiration in lichens increases significantly with increasing temperature. A 10 °C increase in temperature may result in a 2–3-fold increase in respiration (Lambers 1985). This is mainly related to an overall increased metabolism at higher temperatures, including increased growth respiration and maintenance costs (Lambers *et al.* 1998). However, as in plants, respiration can acclimate to increased temperatures so that individuals adapted to a higher temperature display relatively lower increases in respiration with increasing temperature compared with a low temperature adapted population (Sancho *et al.* 2000b). Lichens may also adapt their respiration to seasonal variations in temperature (Lange and Green 2005). As already discussed, respiration also increases with increased water content, probably reflecting raised metabolism in the fungus (MacFarlane and Kershaw 1982). Respiration may be particularly high when dry lichens are rehydrated, resulting in a burst of respiration (Brown *et al.* 1983). The underlying biochemical mechanism for this so-called resaturation respiration is not known, although there are several suggestions, such as an increased energy demand for repair of damaged membranes (Smith and Molesworth 1973), and/or a burst in respirable substrates related to drought damage of the membranes (Farrar and Smith 1976). Both amplitude of the burst as well as time required to reach steady state are probably dependent on the time the thallus was active during its previous active period, and the rate of drying, with fast drying resulting in a larger burst (Brown *et al.* 1983). It also appears that the burst is less prominent when CO_2 fluxes are followed *in situ* (Lange *et al.* 1994; Zotz *et al.* 1998), probably due to the slower transition between the desiccated and the hydrated state in the field situation (Section 10.3.1) compared with the laboratory experimental procedures detecting this burst.

Because respiration is composed of so many subprocesses, simply measuring CO_2 efflux at steady state, as in Palmqvist *et al.* (2002) (shown in Fig. 10.2), will not tell us how all these subprocesses are being regulated, or which subprocess is dominant. Aiming to investigate lichen respiration in some more detail, Lange and Green (2006) analyzed a large data set of *in situ* measurements of CO_2 gas exchange for several species. Based on the hypothesis that respiration might also be more directly regulated by the concentration of photosynthates from the preceding wet period in the light, they made regression analyses of the various R to NP relations that could be extracted from the data set. However, no clear correlations were found; so, respiration in lichens does not seem to be driven by the substrates from the immediately preceding NP under natural conditions. This, however, may not be so surprising considering the relatively large arabitol and mannitol pools that lichen mycobionts apparently possess (e.g. Gaio-Oliveira *et al.* 2005*a*; Palmqvist and Dahlman 2006), thus reducing their dependence on current input of photosynthates for respiration. We also know too little about C translocation rates from photobiont to mycobiont or the R requirements of the various catabolic reactions in the mycobiont to evaluate this "negative" finding. Nevertheless, this recent reanalysis of previously obtained *in situ* data to test a new hypothesis emphasizes that much can be learned without having to redo technically advanced experiments.

10.6   Lichen growth

Lichens need to expand and increase in biomass in order to exploit additional resources, reach maturation, and reproduce. To understand lichens we must then understand how internal and external factors coregulate and limit their growth. Lichen growth has frequently been expressed as a linear measure (mm y^{-1}), i.e. as increased radius for foliose and crustose lichens and as increased tip length for fruticose species, where foliose species may grow 0.5–4 mm y^{-1}; fruticose species 1.5–5 mm y^{-1} and crustose species 0.5–2 mm y^{-1} (Hale 1973), but many lichens fall outside these ranges. However, even though these linear measures may be useful for the special application of lichen growth rate known as lichenometry (Section 10.7), they have a somewhat limited value when one attempts to understand lichen growth processes mechanistically. For instance, the rate of linear expansion of a thallus will be dependent on its already existing area and its specific weight; so, it would be better to express lichen growth as area or weight increases relative to initial area or weight: for a discussion of using relative growth rate (RGR) for lichens, see Kytöviita and Crittenden (2002). More recent investigators have done this and expressed lichen growth as biomass and/or expansion changes. It has then been found that prominent epiphytic species such as *Lobaria oregana* (Rhoades 1977), *Lobaria pulmonaria* (Muir *et al.* 1997) and *Ramalina menziesii* (Boucher and Nash 1990*a*), may increase in biomass by some 30–50% in a year. Similar high growth rates have also been recorded for terricolous *Peltigera* and *Nephroma* species (Sundberg *et al.* 2001; Dahlman and Palmqvist 2003), and for the cushion-forming species *Cladonia portentosa* (Hyvärinen and Crittenden 1998*a*) and *Cladina stellaris* (Gaio-Oliveira *et al.* 2006). However, annual biomass increases are not always as high, sometimes being below 2–3% y^{-1} as in *Hypogymnia physodes*, *Lobaria pulmonaria* and *Platismatia glauca*, when grown in relatively dark or dry habitats (e.g. Renhorn *et al.* 1997; Palmqvist and Sundberg 2000; Gaio-Oliveira *et al.* 2004*a*). Growth rates also vary among years (Matthes-Sears and Nash 1986) and sites, and even among individual lobes of the same thallus (Armstrong 1993). A mechanistic approach must therefore also include an analysis of growth rates in relation to environmental resource supply, and the species' or individual's capacity to assimilate these resources and convert them into new biomass.

10.6.1   *Limitations of lichen growth – the I_{wet} concept*

It is reasonable to assume that lichen growth is the net result of photosynthetic carbon gain minus the respiratory losses associated with growth and maintenance respiration, with additional losses due to dispersal (Gauslaa 2006), death and necrosis, and grazing. Because both P and R are so strongly constrained by the water status of the thallus (see Sections 9.3.1 and 10.3), lichen growth may therefore be primarily limited by water availability (Armstrong 1974; Muir *et al.* 1997; Renhorn *et al.* 1997; Hyvärinen and Crittenden 1998*b*). Once being wet and metabolically active, lichen growth will possibly be limited

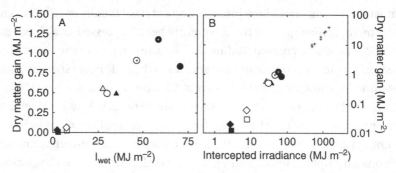

Fig. 10.7 (A) Dry matter gain per unit area as a function of received irradiance during their wet and active periods (I_{wet}) for the five lichens: *Cetraria islandica* (triangles), *Hypogymnia physodes* (squares), *Lobaria pulmonaria* (diamond), *Peltigera aphthosa* (circle) and *Peltigera canina* (circle with cross) during one growing season (spring to autumn) in two contrasting habitats (open symbol – southern Sweden; black symbol – northern Sweden) (adapted from Palmqvist and Sundberg 2000). Dry matter gain was converted to its energy equivalent using the energy content of sugar and the irradiance was converted using the mean energy content of the photons within the action spectrum for photosynthesis, as described in the original study. The slope of the regression line is a measure of the light use efficiency (*e*). (B) Dry matter gain of the same lichens is compared with that of higher plants (crosses) (barley, potatoes, sugar beet and apples) (extracted from Montieth 1977).

by light because of the above-discussed dependence of P on light (Fig. 10.6). This is also supported by observations in the literature; *Lobaria pulmonaria*, for instance, grew faster in the spring than in late autumn (Muir *et al.* 1997), and faster close to a clear-cut edge than in the interior of a forest (Renhorn *et al.* 1997; Hilmo and Holien, 2002; Gauslaa *et al.* 2006*b*).

The effect of water and light on lichen growth was investigated by comparing growth of five lichen species with different life-history traits at two contrasting macro- and microclimate sites in Sweden (Palmqvist and Sundberg 2000). The two parameters were combined in the parameter I_{wet}, i.e. the summed irradiance received during wet and metabolically active periods. Despite significant microclimatic differences between the two sites the relationship between lichen growth and I_{wet} was indeed quite similar across species and sites (Fig. 10.7A). The slope of the line that describes this relationship is a measure of the light use efficiency (*e*) of lichen growth, which was further surprisingly high (0.5–2%), in fact being in the same range as for vascular plants (Montieth 1977) (Fig. 10.7B). In contrast to many earlier claims that lichens might be particularly inefficient because of their symbiotic and poikilohydric nature, this latter observation instead emphasizes that energy losses along the way from light absorption to final biomass accumulation are of the same magnitude

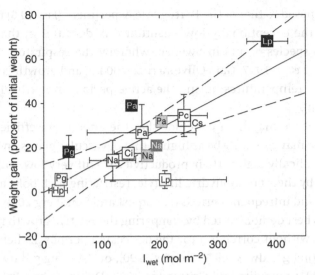

Fig. 10.8   Relative weight gain as a function of I_{wet} for the lichens: *Cetraria islandica* (Ci), *Cladina stellaris* (Cs), *Hypogymnia physodes* (Hp), *Lobaria pulmonaria* (Lp), *Nephroma arcticum* (Na), *Peltigera aphthosa* (Pa), *Peltigera canina* (Pc), and *Platismatia glauca* (Pg). Data have been pooled from several studies to obtain mean values of weight gain and I_{wet}, respectively. Lichens studied in Sweden have been pooled to a separate average (white symbol), and lichens either deprived of (gray symbol) or fertilized with nitrogen (black symbol) have been pooled separately. *Lobaria pulmonaria* has been studied both in Sweden (white symbol) and in Portugal (black symbol) but were not nitrogen manipulated. Data are extracted from Sundberg *et al.* (1997), Palmqvist and Sundberg (2000), Sundberg *et al.* (2001), Dahlman and Palmqvist (2003), Gaio-Oliveira *et al.* (2004a), Palmqvist and Dahlman (2006). The regression lines ($\pm$ 95% confidence intervals) were obtained by pooling all the data.

in lichens and vascular plants. This observation is, however, not so surprising in view of the similar respiratory load of lichens compared with plants (see Sections 10.2 and 10.5), further emphasizing that lichen photosynthesis can acclimate to prevailing light conditions to operate at or close to light saturation during most of their active time (Fig. 10.6).

A range of subsequent studies have now explored this mechanistic approach further, covering additional species, both spatial and temporal variations in habitat conditions, and in some cases manipulated water, light, and nitrogen supplies (Dahlman and Palmqvist 2003; Gaio-Oliveira *et al.* 2004a, 2006; Gauslaa *et al.* 2006b; Palmqvist and Dahlman 2006). Figure 10.8 summarizes those studies where raw data were available, showing that the I_{wet} to weight gain relation is quite conservative across species and environmental conditions. The significantly lower weight gain in relation to I_{wet} of *Lobaria pulmonaria* in the Swedish experiments as opposed to that of the Portuguese experiment (Fig. 10.8) may be a combination of the exceptionally low quantum flux densities in the dense forests where this species occurs in Sweden, whereby the respiratory load is increased (Sundberg *et al.* 1997; Gaio-Oliveira *et al.* 2004*a*), and growth is stimulated by the higher temperatures during the active periods in Portugal (Gaio-Oliveira *et al.* 2004*a*).

However, even if we may then conclude that the energy use efficiency of lichen biomass gain may generally be as high as that of vascular plants once they are wet and metabolically active, their productivity is still very low. This may then be explained by their poikilohydric lifestyle, restricting metabolic activity to relatively short and infrequent periods (Palmqvist and Sundberg 2000). Their low productivity is best demonstrated by comparing the relative growth rates of lichens and plants without correction for the desiccated periods. Lichens then range from 0.5–69 mg g^{-1} dw week^{-1} (Rogers 1990), or 0.4–4 mg g^{-1} dw day^{-1} (Sundberg *et al.* 2001; Kytöviita and Crittenden 2002; Dahlman and Palmqvist 2003; Gaio-Oliveira *et al.* 2006), depending on species, season, and variation in hydration status; while plants may display rates up to 100–500 mg g^{-1} dw day^{-1} (Grime *et al.*, 1988).

10.6.2   *Energy use efficiency of lichen growth – the role of photobiont density*

The discussion in the last section implies that lichen growth may be more limited by the environmental conditions, such as light and water availability, than variation in internal capacity to assimilate and convert environmental resources into new biomass. However, the observation that species exposed to identical environmental conditions for extended periods can display largely different growth rates as a result of different capacities to remain wet and metabolically active contradicts this idea (Renhorn *et al.* 1997; Sundberg *et al.* 1997; Palmqvist and Sundberg 2000; Dahlman and Palmqvist 2003). Moreover, lichen species and individuals with higher photobiont concentrations also display higher growth rates compared with those with a lower photobiont density when exposed to the same environmental conditions (Sundberg *et al.* 2001; Dahlman and Palmqvist 2003; Gaio-Oliveira *et al.* 2006; Palmqvist and Dahlman 2006), manifested as an increased light use efficiency (e) with increasing Chl *a* concentration in the thallus (Fig. 10.9). This is then in agreement with the increased overall metabolism with increasing thallus N and Chl *a* concentration (Fig. 10.1, Palmqvist *et al.* 2002), and the increased photosynthetic efficiency quotient (K_F), when the proportion of the thallus N invested in the photobiont is increased (see Section 10.2).

It may then be concluded that lichen growth is limited by a combination of environmental conditions and variation in the capacity of the thallus to convert

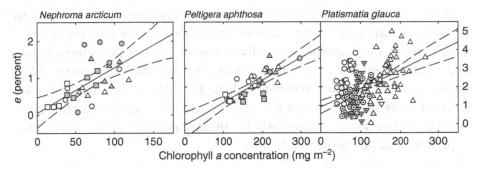

Fig. 10.9   Light use efficiency (*e*) as a function of the thallus Chl *a* concentration for the tripartite lichens *Nephroma arcticum* and *Peltigera aphthosa* (Sundberg *et al.* 2001), and the green-algal lichen *Platismatia glauca* (Palmqvist and Dahlman 2006). The tripartite lichens were grown during two (white symbols) or four (gray symbols) summer months in northern Sweden and either watered weekly with artificial rainwater (circles), nitrogen fertilized (triangles) or nitrogen deprived by removing the cephalodia (squares). *Platismatia glauca* was treated daily during 73 days either with artificial rainwater (circles), ammonium (triangles), nitrate (upside-down triangle with cross) or glutamine (hexagon with cross). The regression lines (± 95% confidence intervals) were obtained by pooling all data in the respective plot.

environmental resources into new biomass. So far, however, there are few studies that have been designed to quantify the separate effects of these two groups of factors. Even fewer investigations have followed whether individual thalli can optimize their resource allocation between the photobiont and the mycobiont when environmental conditions are altered, as attempted in Palmqvist and Dahlman (2006). Nevertheless, lichenologists have now gathered significant data, both from the field and laboratory, on environmental conditions, CO_2 gas exchange, and major resource allocation trends across species and habitats to allow formulation of more general and mechanistic models for lichen functioning. This may provide the necessary theoretical framework, which is still lacking, for addressing questions concerning the regulation of the lichen symbiosis. Such a theory would then be useful to direct future research towards more explicit hypothesis testing. We have here argued that we might be able to understand lichens better if we relate their performance to the necessity to maintain a positive C balance, and their ability to convert some 0.5 to 5% of the light energy received during wet and metabolically active periods into new biomass (Figs. 10.7, 10.9). The ability of a species to acclimate to altered environmental conditions might then be assessed by comparing its light use efficiency when resource supplies have been altered, concomitant with quantification of their major C pools and the relative concentration of photobiont cells in the thallus.

Finally, in addition to the above-discussed limitations of lichen growth imposed by water, light and nitrogen, phosphorous may also significantly inhibit thallus development, particularly species with cyanobacterial photobionts (Crittenden *et al.* 1994; Knops *et al.* 1996; Kurina and Vitousek 1999; Benner *et al.* 2007). The mycobiont and the photobiont may further display different concentration optima for various nutrients whereby the photobiont might be relatively more limited by N, but the mycobiont might be more limited by phosphorous (Makkonen *et al.* 2007).

10.7   Lichenometry

A special application of lichen growth rate studies is dating surfaces using procedures known as lichenometry (Easton 1994). Although accuracy of date estimates is probably no better than 10% (Innes 1988), it has proved useful in fields such as archaeology and geomorphology in regions where alternative dating techniques (e.g. dendrochronology in forested areas) are not applicable. For example, the procedures allow dating of moraines and glacial movement (Beschel 1961; Palacios *et al.* 1999; Harrison and Winchester 2000; Casely and Dugmore 2004; Koch *et al.* 2005; Matthews 2005), rockfall and rock slide frequency (McCarroll *et al.* 1998; Ferber 2002; Gupta 2005), earthquake frequency (Nikonov and Shebalina 1979; Bull 1996; Bull and Brandon 1998), sediment transport (Gob *et al.* 2003), water level fluctuations (Timoney and Marsh 2004), petroglyph age (Sedelnikova and Cheremisin 2001), and monuments on Easter Island or Isla Pasqua (Richardson 1974).

Lichenometry has been limited primarily to studies with crustose lichens, with special emphasis in alpine environments on the genus *Rhizocarpon* on open rock surfaces. Obviously dating is limited to the potential age of the lichen, which in many cases may be decades but in others extends to centuries. A few arctic/alpine lichens may even live to over 1000 years, perhaps even up to *c.* 4500 years (Beschel 1961), an age rivaling the oldest vascular plants. The technique works best in regions where growth can be calibrated against substrates of known age (Larocque and Smith 2004), such as that provided by old cemeteries in Europe, or in combination with other dating techniques (Evans *et al.* 1999). This is particularly necessary in situations where lichen radial growth rate varies as a function of thallus size. In places where dated substrates are not available, greater uncertainty with estimated dates necessarily occurs. For example, attempts to establish growth curves on Baffin Island in the Canadian high Arctic were made by studying growth of different sized thalli of *Pseudephebe* over shorter time intervals (Miller 1973), but such curves make the implicit assumption that climatic control of growth for the period of measurement is representative of climatic effects over the life of the species being studied. In light of the recorded variability of climate over the last several centuries, the assumption probably does not hold. Given the known influence of such factors as substrate variability and exposure of rock surface on growth, it is obviously necessary to stratify one's sampling design as much as possible to maximize the equivalency of different sites on which lichens are measured. Once study sites are selected, then usually the largest lichen is selected within an area of c. 0.1 ha and its longest axis is measured (Innes 1988). Other measurements have been tried, such as the largest inscribed circle or measuring the largest five thalli to obtain a better statistical estimate. The biological basis for lichenometry has been recently addressed by McCarthy (1999), Sancho and Pintado (2004), and Loso and Doak (2006).

11

Nitrogen, its metabolism and potential contribution to ecosystems

T. H. NASH III

Nitrogen is a macronutrient essential for life in the formation of proteins and nucleic acids. Its limited availability frequently constrains growth and productivity, both of individual organisms, such as lichens (Chapter 13 and Crittenden *et al.* 1994), and of ecosystems as a whole. On the other hand, nitrogen may also be present in excessive amounts in some regions, such as those affected by high deposition of ammonia associated with fertilizer use and/ or animal husbandry. Because nitrogen is not part of the Earth's crust's material, rock weathering and subsequent soil formation do not provide a nitrogen source, as they do for many other elements. Rather the principal pool of nitrogen is atmospheric nitrogen N_2, which is not readily utilized by most organisms. Nitrate (NO_3^-) and ammonia (NH_3, or ammonium ions, NH_4^+) are the inorganic forms of nitrogen that are universally processed by organisms, and their availability is critical to growth and survival of green-algal lichens (Chapter 10). With chlorolichens, nitrogen concentrations can often be related to atmospheric deposition of these ions (Hyvärinen and Crittenden 1998a). At least some cyanolichens are part of the small group of organisms, including some bacteria and actinomycetes, capable of utilizing atmospheric N_2 directly. As a result cyanolichens have higher total N concentrations (2.2–4.7%) than chlorolichens (0.4–0.85%) (Rai 1988). Because cyanolichens are abundant in a number of ecosystems, their potential contribution to nitrogen fixation in these ecosystems has led to attempts to quantify the rate of nitrogen fixation by these lichens and to understand the environmental and physiological factors limiting their fixation rates.

Approximately 10% of the lichen species contain cyanobacteria. They are primary symbionts of about 50 genera and 1000 species of lichens and secondary symbionts of about 20 genera and 500 species (Olafsen 1989; Rai 2002). They often contain species of *Calothrix*, *Fischerella* (= *Stigonema*), *Gloeocapsa*, *Nostoc*, or *Scytonema* (Chapter 2), all of which are capable of fixing N_2. When they are secondary to chlorophotobionts (i.e. in tripartite lichens), these cyanobacteria are often contained within specialized structures called cephalodia (James and Henssen 1976), which are gall-like and a unique product of the symbiosis. In many species, the cephalodia are external, scattered over the surface of the thallus, but in others (e.g. *Lobaria* and *Sticta*) they occur internally in small packets. The question of whether the ubiquitous *Azotobacter*, a nitrogen-fixing bacterium, has a functional role in lichens has not been answered unequivocally, but at least in the *Lobaria*-dominated *Pseudotsuga* forests of Oregon, these bacteria do not appear to contribute to fixation (Denison 1973). More recently some members of the Lichinaceae (i.e. *Thyrea* spp., a group with *Chroococcidiopsis*; Chapter 2), which are very common in arid habitats, have also been shown to fix nitrogen (Crittenden *et al.* 2004).

11.1   Nitrogen metabolism

Nitrogen fixation is an energy-dependent process and overall it involves the utilization of 8 electrons and 16 MgATPs to convert one N_2 to two NH_3 (Salisbury and Ross 1992). The process involves an enzyme complex called nitrogenase, which involves both a Fe-protein and a Fe-Mo protein. Ferredoxin provides the electrons to the Fe-protein, which, when bound with ATP, becomes a strong reducing agent. The electrons are transferred to the Fe-Mo protein, which is subsequently responsible for the transfer of electrons and protons to N_2 to form the two NH_3. Nitrogenase activity can be readily inhibited by O_2 and this requires separation of nitrogen fixation activity from the photolysis of water during photosynthesis, even though the electrons may originate from the photosynthetic apparatus. The solution in the case of the common lichen photobiont *Nostoc*, is to form specialized cells called heterocysts, in which the phycocyanin accessory pigments and the water-splitting magnesium center of photosystem PS II are lost (Tel-Or and Stewart 1976). This limitation to PS I photosynthesis maintains internal anaerobiosis or microaerobic conditions, either of which protect nitrogenase from O_2 inactivation, and is thought to provide reductant and ATP for nitrogenase in the light without competition from CO_2 (Tel-Or and Stewart 1977). Because only glutamine synthase (GS) is present within the heterocyst, glutamine is formed and is exported to the vegetative cyanobiont cells, where further processing by glutamate synthase (GOGAT) occurs. In free-living cyanobacteria heterocysts constitute 5–10% of total cells (Weissman and Benemann 1977; Stewart 1980; Rai 2002) and similar percentages of heterocysts are found in the cyanobiont when it is the primary photobiont (Hitch and Millbank 1976). But a much higher percentage of hetero-cysts (15–36%) may be found in the cyanobiont of tripartite lichens (Hitch and Millbank 1975; Millbank 1976). Concomitant with higher heterocyst occurrence is proportionally higher N-fixation activity (Rai 2002).

Although reduced inorganic nitrogen incorporation occurs in N_2-fixing spe-cies, nitrogen assimilation in most other organisms begins with uptake of either nitrate (NO_3^-) or ammonium (NH_4^+) ions. Because all amino acids contain an amine group (-NH_2), the uptake of nitrate requires subsequent conversion to ammonium ions, an energy-dependent process, which overall requires 8 elec-trons, as N oxidation goes from $+5$ to -3, and $10\,H^+$ ions (Salisbury and Ross 1992). The process involves initial conversion to nitrite (NO_2^-), a step in which NADH provides an electron, before conversion to NH_4^+, which in algae involves 6 electrons from H_2O as mediated by the chloroplast noncyclic electron trans-port system. The two steps are catalyzed respectively by the enzymes nitrate reductase and nitrite reductase, both of which are widely distributed among organisms, including lichens.

Ammonium ions are quite toxic (Salisbury and Ross 1992), and, as demon-strated by pulse chase experiments with $^{15}N_2$ followed by $^{14}N_2$, are rapidly converted to other nitrogen products in lichens (Rai 1988, 2002). In the case of the cyanobiont, NH_4^+ is catalyzed by GS to glutamine, an amide that may accumulate without being toxic. Subsequently, glutamine is catalyzed by GOGAT to glutamate. In contrast, ammonium assimilation in the mycobiont is catalyzed by glutamate dehydrogenase (GDH) to glutamate (Rai 1988, 2002). There is evidence of a tight coupling of metabolic activity in heterocystic cyanobionts and adjacent mycobiont hyphae. For example, GS activity is sup-pressed in the heterocyst; thereby allowing NH_4^+ export, and simultaneously GSH activity in the adjacent mycobiont is stimulated. In fact ^{15}N tracer studies have documented c. 55% N transfer to the mycobiont of the dipartite *Peltigera canina* and over 95% transfer in the tripartitite *Peltigera aphthosa* (Stewart and Rowell 1977; Rai 2002).

In *Peltigera* species the transferred NH_4^+ is converted to glutamate by the mycobiont (Rai 1988, 2002). Initial evidence pointed to peptides being secreted (Millbank 1974; Stewart and Rowell 1977), but subsequent work has not con-firmed this, and it is now assumed that the peptides were a fungal product and not from the cyanobiont. Simultaneous studies with $^{15}N_2$ and $H^{14}CO_3^-$ and digitonin (which disintegrates the mycobiont), have demonstrated only the release of $^{15}NH_4^+$ from the heterocysts and no detectable organic fraction (Rai et al. 1980; Rai 1988). The massive export of NH_4^+ explains why cyanophycin granules, the common form of nitrogen storage of both N_2- and NH_3-grown *Anabaena*, were absent from the cyanobiont of *Peltigera canina*, when it was grown on a nitrogen-free medium (Stewart and Rodgers 1978). Similarly Boissière (1982) found that the free-living form of *Nostoc commune* with 3% heterocyst frequency always contained numerous cyanophycin granules, but the majority of the cyanobiont cells of *Nostoc* were found to contain few or none.

Subsequent coupling with the Kreb cycle results in production of glutamic acid (glutamate in the ionic form), aspartic acid, and asparagine. Pulse chase experiments involving initial exposure to $^{15}N_2$ followed by $^{14}N_2$ and the use of a mass spectrometer to quantify ^{15}N have specifically confirmed this sequence and show the gradual transfer of ^{15}N throughout the lichen thallus over hours (Stewart and Rowell 1975; Rai 1988, 2002). The amine group on glutamic acid can be readily transferred by transamination to a variety of α-keto acids and thereby form other amino acids. Twenty amino acids are essential to protein formation, and most of these are specifically known from lichens (Rai 1988; Nash *et al.* 2001). Glutamate, glutamine, alanine, and aspartate are the most common, soluble amino acids (Rai 2002). In addition, many other amino acids occur in plants, and a number are reported from lichens (Rai 1988). Protein formation results from a series of condensations, initially of amino acids to form polypeptides, and subsequently into larger units called peptones, proteoses and then proteins themselves. For details of protein and nucleic acid synthesis the reader is referred to basic plant biochemistry texts.

11.2   The cephalodial-thallus relationship: case studies with *Peltigera aphthosa*

In lichens with cephalodia, such as *Peltigera aphthosa*, a closely integrated interrelationship occurs among the mycobiont, the dominant green-algal photobiont, and the cyanobiont found in the cephalodia. Increased heterocyst frequency up to 10–55% are found in the cyanobiont (Hitch and Millbank 1976). In digitonin pretreated thalli of *Peltigera aphthosa*, 90–95% of the fixed $^{15}N_2$ is released as NH_4^+ (Rai *et al.* 1980). Apparently most of this ammonium is incorporated by the cephalodial mycobiont, because very high GDH activity is found there, and concurrently GS and GOGAT activity in the cyanobiont is reduced by over 90% compared with free-living cyanobacteria (Rai 1988). Also the addition of MSX (L-methionine SR-sulfoximine), an amino acid analog that inactivates glutamine synthase in cyanobacteria (Thomas *et al.* 1975), to the cephalodia caused no change in NH_3 excretion or acetylene reduction activity (ARA; Section 11.4). This indicates that no GS-mediated enzymatic feedback control on the cyanobiont is present (Englund 1977), and further substantiates that the principal nitrogen sink is the mycobiont. Initially, accumulation in the mycobiont occurs as glutamate, and subsequently most of the glutamate is transformed to alanine (Rai 1988). Alanine appears to be the principal transfer compound from the cephalodial mycobiont to the main thallus. Ammonia assimilation, principally through the GDH pathway, avoids accumulation of glutamine and the potential inhibitory effect it would have on nitrogenase.

Although cephalodia are intimately associated with the main thallus, their nitrogenase activity exhibits a degree of independence, as excised cephalodia continue to fix and transfer nitrogen for up to three days (Rai *et al.* 1981). However, decreased rates of ARA activity observed in some excised cephalodia (Englund 1978; Huss-Danell 1979) have led to the suggestion that the cephalodia may be dependent on photosynthate from the algal photobiont. In fact, Wolk *et al.* (1994) demonstrated fixed carbon movement from vegetative cells into the heterocysts of tripartite lichens. Furthermore, the dominant chlorobiont, *Coccomyxa*, receives over 60% of its N-requirement from the heterocysts of the cyanobiont (Rai *et al.* 1983).

However, when exogenous sources of NH_4^+ are available, negative feedback control on nitrogenase activity apparently occurs via the algal photobiont *Coccomyxa* (Rai *et al.* 1980). The NH_4^+ stimulates high glutamine synthase activity, which is almost exclusively located in *Coccomyxa*. This results in production of glutamine, the release of which is demonstrable by ^{14}C procedures in freshly isolated *Coccomyxa*. Blocking glutamine synthase with MSX results in restoration of nitrogenase activity (Rai 1988).

11.3   Excessive nitrogen

In regions associated with high fertilizer use, intense animal husbandry, and some industrial activity, atmospheric N deposition may easily be one to two orders of magnitude higher than in rural regions with minimal agriculture, and lichens may be used to monitor these deposition patterns (Søchting 1995; Hyvärinen and Crittenden 1998a, b; Walker *et al.* 2003; Jovan and McCune 2006; Section 11.7). In fact, research today is often focused on ecosystems experiencing N saturation (Curtis *et al.* 2005; Mitchell *et al.* 2005) rather than N-limited ecosystems. For example, Houdijk and Roelofs (1991) reported bulk deposition of 80 kg N ha^{-1} yr^{-1} for forests in the Netherlands, but in contrast Knops *et al.* (1996) reported bulk deposition of less than 1 kg N ha^{-1} yr^{-1} for a relatively pristine oak woodland in central coastal California.

The frequent occurrence of a few lichen species in microhabitats high in nitrogen compounds, such as tree bases and around bark wounds, has long been recognized, and these species are called nitrophilous and are predominantly members of the Xanthorion parietinae alliance (Barkman 1958). With decreased SO_2 pollution in the Netherlands, species belonging to this alliance have increased their abundance (van Dobben 1996; van Dobben and Bakker 1996; van Dobben and ter Braak 1998). Furthermore, there is now widespread evidence that lichens can be usefully used to monitor ammonia deposition (van der Eerden *et al.* 1998; Jovan and McCune 2005, 2006; Wolseley *et al.* 2006; Jovan and Carlberg 2007).

Lichens occurring in high deposition areas may thus require mechanisms for dealing with excessive nitrogen, particularly as rapid (minutes to saturation) uptake of NH_4^+ or NH_3 to extracellular exchange sites is well demonstrated, and subsequent intracellular uptake is inferred (Miller and Brown 1999). There is now experimental evidence documenting that one of the common nitrophilous species, *Xanthoria parietina*, is able to tolerate high NH_3 or NH_4^+ levels (Gaio-Oliveira *et al.* 2004b, 2005a, b). It is inferred that it is more efficient than most species in metabolizing NH_4^+ to a less toxic storage form of nitrogen, but the mechanism is not yet understood fully. In contrast, the disappearance of a number of *Cladonia* species in the course of fertilizer experiments (van Dobben 1993) is consistent with the inference that NH_4^+ may be toxic to these species, although competition may have played a role as well. Similarly, in a 10-year, elevated N experiment Fremstad *et al.* (2005) documented that a number of alpine chlorolichens, such as *Alectoria nigricans* and *Cetraria ericetorum*, exhibited decline in abundance (cover) and developed extensive discoloration (chlorosis?).

In addition to gaseous release of reduced N gases, there are many oxidized N gases, such as NO and NO_2 released from combustion processes (vehicles, power plants, etc.). Consequently, a strong correlation between traffic volume and N-status of lichens is reported (Gombert *et al.* 2003). Furthermore, atmospheric interactions of these gases contribute to the development of acid rain, which, at least at low pH, has proved detrimental to lichen N-fixation capability (Kytöviita and Crittenden 1994). Only recently has it become documented that HNO_3 and to a lesser extent HNO_2, both as gases, are widely present in significant quantities in arid and semi-arid environments (Fenn *et al.* 2003). Because of its extremely high deposition velocity, HNO_3 plays a major role in N deposition in such environments. The interaction of lichens with these gases is just beginning (Nash and Riddell 2006), and appears to cause significant detrimental effects as dry deposition.

A special case of high nitrogen levels involve guano deposits of birds, and a few lichen species are well known to commonly associate with these deposits (Barkman 1958). These lichens, such as *Candelariella coralliza*, *Lecanora muralis* and *Xanthoria parietina*, have the highest range of nitrogen content (4.2% to over 13%) reported in the literature (Rai 1988, 2002). Bird deposits are high in urea, a highly toxic nitrogen product that generally inactivates proteins by affecting hydrogen bonds. The toxicity of urea may well account for the lack of most lichen species, and the ability of a few lichens to tolerate these habitats is reflected in their very high urease activities (Rai 1988), by which urea is converted to CO_2 and NH_4^+

(Salisbury and Ross 1992). In fact Gaio-Oliveira *et al.* (2004*b*) recently demonstrated that *Xanthoria parietina* can tolerate up to *c.* 1000 kg N ha^{-1} y^{-1}, as aqueous solutions of NH$_4$Cl applied weekly over 10 months.

11.4   Methods of determination of nitrogen fixation

The enzymatic reduction of gaseous N$_2$ to NH$_3$ by the enzyme nitrogenase in organisms can only be conclusively demonstrated by ^{15}N$_2$ enrichment studies, which require the use of a mass spectrometer. This technique, however, is costly and the results are often difficult to analyze. The acetylene reduction technique (Stewart *et al.* 1967) provides an alternative tool for estimating nitrogenase activity, because nitrogenase also catalyzes the reduction of acetylene (C$_2$H$_2$) to ethylene (C$_2$H$_4$). These gases are readily measured with gas chromatographic techniques, and the amount of ethylene produced over a period of time is an indication of nitrogenase activity. The ethylene produced from this "acetylene reduction activity" (ARA) may be correlated with nitrogenase activity (Hardy *et al.* 1973), but not necessarily with ammonia production. The method is so deceptively simple that often pitfalls are ignored, such as the possible C$_2$H$_2$ inhibition of photosynthesis (David and Fay 1977), although Crittenden and Kershaw (1978) found no such inhibition. Long-term exposure (6 to 24 hours) causes an apparent stimulation of nitrogenase activity (Kershaw 1985) by interfering with the normal ammonia metabolism of the cell. Ethylene, in contrast to ammonia, cannot be metabolized (Masterson and Murphy 1984), and thus may accumulate. In addition, nitrogenase functions as a hydrogenase (Kerfin and Boger 1982), and this hydrogen production, which varies with species and cultural conditions (Bothe *et al.* 1978), is inhibited by acetylene.

Nevertheless, ARA measurements do have much to recommend for field studies, if one uses an internal standard (Kaiser and Debbrecht 1977) to correct for losses and leakages (McNabb and Geist 1979). Any estimates of nitrogen production should have parallel ^{15}N$_2$ uptake to provide a valid C$_2$H$_4$:N$_2$ conversion ratio (e.g. Kurina and Vitousek 2001), since the theoretical stoichiometric (originally thought to be 3:2, Hardy *et al.* 1973, but now modified to 2:1) is rarely attained (Belnap 2001). Empirical measurements of the C$_2$H$_4$:N$_2$ ratios have varied with organism and situation from 4:1 to 24:1 (Peterson and Burris 1976; Millbank 1981, 1982*a*).

11.5   Daily nitrogen fixation patterns: an example
     from Anaktuvuk Pass, Alaska

Nitrogen fixation patterns, as measured by ARA, in relation to photosynthesis and environmental parameters over two 24-hour periods are given in Fig. 11.1

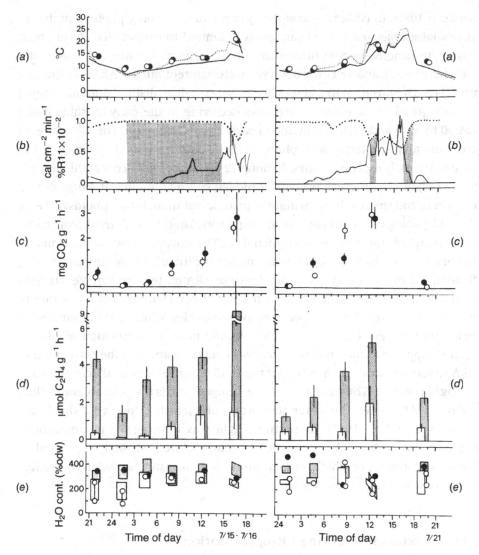

Fig. 11.1 Environmental variables, gross photosynthesis and C_2H_2 reduction
activity (ARA) of *Peltigera canina* (solid circles) and *Stereocaulon tomentosum* (open circles)
within a flood plain, lichen mat for two days (15–16 July and 21 July) in 1981. The
horizontal panels correspond to: (a) temperature (°C): air (solid line) at 8 cm, mat
(dotted line), during photosynthetic measurement (circles), during C_2H_2 assay if
different (solid line); (b) solar radiation (solid line), relative humidity (dotted line),
precipitation (shaded region); (c) mean gross photosynthesis ($n = 4$) and 95% confidence interval based on the *t*-distribution (vertical lines); (d) mean acetylene
reduction activity ($n = 5$) and 95% confidence interval based on the *t*-distribution
(vertical lines); (e) thallus water contents during CO_2 assay (circles) and C_2H_2 assay
(open and shaded bars).

(Olafsen 1989). In *Peltigera canina* the cyanobiont is the only photobiont, but in *Stereocaulon tomentosum*, the cyanobiont is limited to cephalodia. During both days, the lichens were continuously moist, generally above 200% oven dry weight (panel a), and, over the observed water content ranges, ARA is essentially unaffected by variation in water content. Nevertheless, distinct diurnal patterns in ARA (panel d) are evident with lows occurring in the early morning hours (24:00 to 3:00) and highs occurring in early to late afternoon. The ARA patterns correlate almost exactly with photosynthetic patterns (panel c), which fall to zero in the early morning hours. Although 24 hours of light occur at this latitude (68° 30′ N) in July, the sun is actually hidden behind a mountain range during this time, and the resulting twilight is insufficient to maintain photosynthesis. Both physiological responses are strongly correlated with diurnal solar radiation (panel b) and temperature (panel a). The covariance of environmental factors such as these in field studies makes it difficult to identify controlling factors, but in this study the dependence of ARA on temperature versus light could be distinguished by considering partial correlations that allow one to document the relationship between two variables when a third variable is held constant. For the study as a whole, the putative correlation of ARA to light disappeared when temperature was held constant, but the correlation of ARA to temperature (primarily over the 0–15 °C range) was unaffected by holding light constant. Thus in this example, temperature is the primary controlling factor on ARA when the water content is sufficient to support maximal ARA (Nash and Olafsen 1995). In contrast, light was limiting for photosynthesis (Olafsen 1989). The ARA dependence on temperature is strongly supported by laboratory studies, in which temperatures are deliberately varied in controlled, systematic ways (Kershaw 1985).

11.6   Factors influencing nitrogen fixation

11.6.1   Moisture

Water is of paramount importance to ARA (Kershaw 1985; Fritz-Sheridan and Coxson 1988*a*). Although respiration and photosynthesis can occur at very low moisture levels in some lichens, nitrogen fixation does not begin until at least *c.* 80% ODW (oven dry weight) is reached. Minimum values for a species remain almost constant with different temperatures, light intensities, season, or habitat (MacFarlane and Kershaw 1977). For species with saturation occurring at water contents of 500–800% ODW, maximum nitrogen fixation rates are found at 200–400% ODW and in most species maintain high, fairly constant ARA rates up to saturation (Kershaw 1985; Fritz-Sheridan and Coxson

1988*b*). This ARA to water content relationship applies in general to both cephalodial species and those with a cyanobiont as the principal photobiont. However, a few species (e.g. *Peltigera polydactyla*) exhibit marked depression in ARA near saturation (Kershaw 1985).

After desiccation over several days, ARA recovers slowly over hours upon rewetting in most species (Kershaw 1985), although Huss-Danell (1978) found no effects of desiccation on *Stereocaulon paschale*. The effects in desiccation-sensitive species are more pronounced with longer periods of desiccation. Following a 14-day drought, *Stereocaulon* from Canadian woodlands, monitored *in situ*, exhibited a 12-hour lag after rainfall before ARA resumed, which then slowly climbed (Crittenden and Kershaw 1979). Under extreme drought conditions ARA may become negligible, as illustrated in *Stereocaulon virgatum*, a tropical species growing on recent volcanic lava flows in the cloud zone of Guadeloupe Island, where annual precipitation is *c.* 10 m and cloudy moist conditions usually prevail (Fritz-Sheridan and Coxson 1988*b*). At least part of the explanation for reduced ARA following drought allieviation is related to constraints on available energy. In free-living *Nostoc commune*, Potts and Bowman (1985) found that ATP generation was immediate in vegetative cells, but exhibited an 8–10 hour lag in the heterocysts, with a corresponding nitrogen fixation lag.

11.6.2   *Temperature*

In general ARA increases with increasing temperatures up to 25–30 °C, but decreases at higher temperatures (Kershaw 1985; Fritz-Sheridan and Coxson 1988*b*; Kurina and Vitousek 2001). Measurable ARA is reported from as low as − 5 °C (Kallio *et al.* 1972) and up to 35–39 °C (MacFarlane and Kershaw 1977; Fritz-Williams and Coxson 1988*b*). To a limited degree optimum temperatures for ARA vary among species, ecological habitat, and pretreatment. Kallio *et al.* (1972) found that maximal rates ranged from 15 to 25 °C for two arctic species, depending on the length of pretreatment (assays being carried out at 15 °C). In the field in polar and many temperate locations, the most important temperatures for nitrogen fixation may actually be lower than implied by laboratory measurements because hydrated conditions are not sustainable. For example, Huss-Danell (1977) saw maximal field activity in *Stereocaulon paschale* at 16–20 °C. At higher ambient temperatures, the lichen dry too quickly for measurements to be meaningful. In addition, ARA frequently declines substantially over several hours, if hydrated conditions are maintained at higher temperatures (Kershaw 1985).

Temperature is doubtlessly important in observed changes in ARA across seasons, but it is difficult to distinguish precisely from other factors affecting ARA simultaneously, such as drought, which is more common and intense during summer periods. Englund and Myerson (1974) found higher field ARA in December than March or May for equivalent light and temperature in *Peltigera canina*, indicating either a winter depletion of necessary metabolites, or a time lag in nitrogenase mobilization after a long period of freezing. Low ARA in soil lichens following melting of the snow pack is well established (Kershaw 1985), and is probably more related to winter depletion of energy and carbon reserves than freezing temperatures. When Huss-Danell (1978) subjected summer collections to simulated nighttime frost, ARA was stimulated. Both Huss-Danell (1978) and MacFarlane and Kershaw (1980) reported higher ARA in the fall than in the summer, and this may be due to an indirect affect of temperature on respiration. With lower temperatures, respiration is reduced, and a closer coupling of the Kreb cycle with nitrogen fixation is possible. Overall the dependence of nitrogenase activity on photosynthetically generated products makes it difficult to separate the temperature optimum for the enzyme complex from the conditions which provide the maximum reductant and energy (Englund 1978).

Environmental variables often covary in the field, and consequently teasing out the relative importance of individual variables is often difficult. For example, ARA often varies in concert with both temperature and light variation in the field (e.g. Fig. 11.1). The use of partial correlation (or partial regressions where appropriate) is an effective procedure for assessing which variable is most important (Nash and Olafsen 1995). In the case of a summer season data set from Anaktuvuk Pass, Alaska (68.5° N), they demonstrated that when *Peltigera canina* and *Stereocaulon tomentosum* were sufficiently moist to exhibit positive ARA, temperature was the most important variable controlling ARA, because the partial correlation of ARA with light virtually disappeared when temperature was held constant, and the partial correlation of ARA with temperature remained highly significant when light was held constant. This positive relationship between warmer temperature and higher rates of nitrogen fixation might lead one to assume that global warming will be beneficial. However, Bjerke *et al.* (2003) reported no change in nitrogen fixation activity after a 5-year simulated climate-change experiment. Critically no ARA measurements were apparently made across the duration of the experiment, and consequently whether or not increased periods of N-fixation occurred is unknown. Community change in polar regions as temperatures increase is likely to be most important, and at any given location the periodicity of hydration events will be very important for controlling lichen ARA.

11.6.3   Light

With free-living cyanobacteria, a positive relationship between ARA and light occurs up to saturation at the very low light value of *c.* 30 μmol photons m^{-2} s^{-1}

(Lex *et al.* 1972), and is dependent only on far red (> 695 nm) for cyclic photo-phosphorylation (Bothe and Loos 1972). For lichens, Denison (1979) found that in *Lobaria oregana*, the entire thalli monitored *in situ*, the same ARA was reported day and night, but noted that *Peltigera membranacea* showed a dark depression of ARA. Differential responses among species to the ability to maintain ARA during dark conditions is now widely demonstrated (Fritz-Sheridan and Coxson 1988*b*). Dark depression of ARA can be correlated with the disappearance of glycogen reserves and may be relieved by added glucose (Bottomley and Stewart 1977) or fructose (Ernst *et al.* 1984). However, it is difficult to separate, in the dark, the requirements for (1) carbon skeletons to accept newly fixed NH_3, (2) carbon for respiratory protection of O_2, and (3) carbon for ATP/reductant generation. In the lichen system this is further com-plicated by the mycobiont's contribution to any of those requirements. In lichens stored in the dark up to 20 days, Kelly and Becker (1975) reported no light saturation for the forest species *Lobaria pulmonaria* or *Sticta weigelii* at irradiance levels as high as 1200 μE, a value far higher than documented by Lex *et al.* (1972). The apparent relationship to light may have been spurious, because the carbon-starved cyanobionts may have fixed nitrogen only as photo-synthetic products became available, thus appearing dependent on elevated light levels.

Under more normal conditions where carbon starvation is unlikely, high light (*c.* 1600 μmol photons m^{-2} s^{-1}) may be detrimental. Fritz-Williams and Coxson (1988b) demonstrated at optimal water contents a 45% ARA reduction in *Stereocaulon virgatum* at 20 °C, a relatively normal temperature in this tropical cloud zone. At higher temperatures (upper 30s°C) ARA was reduced to nearly zero (over 95% decline) regardless of light levels. Even at intermediate light levels (200–800 μmol photons m^{-2} s^{-1}) ARA may be reduced compared with low light (50–100 μmol photons m^{-2} s^{-1}), as demonstrated by Kurina and Vitousek (2001) for *Stereocaulon vulcani* in Hawaii. High light is well known to be detrimental to lichen photosynthesis (Demmig-Adams *et al.* 1990) and in fact net photosynthesis was reduced at high light in *S. vulcani* (Kurina and Vitousek 2001). The requirement of ARA for photosynthetic products (Rai 2002) implies a potential causal link.

11.6.4   *Partial pressures of oxygen*

The initial reports of high ARA in lichens (Huss-Danell 1979), coupled with the knowledge that the cyanobionts are tightly surrounded by actively respiring fungal hyphae, caused initial speculation that cyanobionts might fix nitrogen vegetatively within an anaerobic medulla (Millbank 1972). However, subsequent measurements of oxygen tensions in the light of *c.* 20% in the primary photobiont layers of *Peltigera polydacyla*, *P. canina* and *P. aphthosa* and 30% in *P. aphthosa* cephalodia (Rai 1988) invalided this suggestion. In the dark O_2 levels fell to below 20% in thick thalli, but remained near atmospheric in thin thalli. These measurements point to the necessity of heterocysts in the *Nostoc* of these species in order to avoid oxygen inactivation of the nitrogenase complex.

11.6.5   Nutrient limitations and pH

Acetylene reduction activity (ARA) is potentially limited by several nutrients. For example, potential phosphorous limitation, suggested by Crittenden *et al.* (1994), was recently confirmed by Kurina and Vitousek (1999) for *Stereocaulon vulcani* in Hawaii. In late successional forests in Hawaii, Benner *et al.* (2007) have demonstrated that P fertilizer additions results in a luxurious development of epiphytic cyanolichens (Section 12.2). Also in Hawaii, Crews *et al.* (2001) showed that a suite of nutrients in fertilizer experiments where N and P were not included resulted in a doubling of ARA in *S. vulcani*. Lastly, Horstmann *et al.* (1982) demonstrated enhanced ARA in *Lobaria* spp. in the Pacific Northwest (USA) with the addition of Mb, which is a component of the nitrogenase enzyme.

In the context of concerns regarding acid rain associated with industrialization and ever-increasing populations, effects of acidity on ARA have been investigated by several researchers. For example, Fritz-Sheridan (1985) found an 82% reduction in ARA of *Peltigera aphthosa* after 24 days of exposure to simulated acid rain of only pH 4.0. However, Gunther (1988) reported much less dramatic effects, possibly as a result of buffering from the lichen's substrate. Hallingbäck and Kellner (1992) worked with lichens separated from their substrates and also found reductions in ARA due to acidity, but the effects were modified by relative concentrations of nitrate and/or ammonium. In fact, Hallingbäck (1991) inferred that the widespread disappearance of *P. aphthosa* from southern Sweden was due to acid rain.

11.6.6   Exogenous nitrogen effects

Many free-living cyanobacteria routinely form heterocysts only when there is insufficient available nitrogen (Fogg *et al.* 1973). Those from nutrient-rich environments form heterocysts only after a transfer to a nitrogen-free medium. Thus, the presence of heterocysts in cyanobionts highlights their nitrogen stress. Stewart and Rowell (1977) reported that a 24-hour incubation with NH_3 completely inhibited ARA in *Nostoc* isolated from *Peltigera canina*, but only 50% in intact lichen disks. In contrast, nitrate reduced ARA by 50% in the isolated *Nostoc*, but it had no effect on that of intact lichen disks. The lack of complete inhibition in the disks exposed to NH_4Cl may be partially attributed to the slower growth rate of the intact cyanobiont rather than to the lack of any inhibition mechanism. The lack of any inhibition of disks in the KNO_3 implies that the mycobiont may exert a control over nitrogenase inhibition (i.e. by the mycobiont selectively taking up the N ions such that the photobiont does not see them; P. D. Crittenden, personal communication), and studies with isolated mycobionts show considerable growth on NH_3 and NH_4NO_3 (Ahmadjian 1993).

Nitrogen fertilizers exert differential affects on ARA. Nitrogen fixation by *Stereocaulon paschale* and *Peltigera aphthosa* (Kallio 1978) and *P. praetextata* (Hallbom and Bergman 1979) was inhibited by most fertilizers containing any N, but to a greater degree by those containing NH_4^+. Even the low concentration of 3 mM NH_4NO_3 seriously disrupted the symbiosis of *P. praetextata*. Of all species with all fertilizers, *S. paschale* experienced the greatest inhibition of ARA. In contrast, Kurina and Vitousek (1999) found no change in ARA with N-fertilizer treatments for *Stereocaulon vulcanii* in Hawaii. Furthermore, Davis *et al.* (2000) documented long-term (over months) beneficial effects of 1.0–2.0 mM NO_3^- concentrations on an aquatic lichen *Hydrothyria venosa* (= *Peltigera hydrothyria* now), but concentations of 4–30 mM were detrimental, with the effects becoming manifest at shorter time intervals for the higher concentrations.

11.6.7   Species of cyanobiont

Relatively little is known about taxonomic differences of cyanobionts (Chapter 2). Some lichen genera always contain a particular genus of cyanobiont, but the degree to which variation (if any) may occur within a genus is not well documented. For example, *Peltigera* always contains *Nostoc*, as do *Collema*, *Leptogium* and a number of other genera. While no studies have compared the activities of respective cultured cyanobionts from several species to ascertain the degree to which the strain of cyanobacterium may contribute to variability, one would expect the same diversity of response in cyanolichens as reported in free-living cyanobacteria. Members of the genus *Peltigera*, for example, show differing ARA maxima (Kershaw 1974), yet the cyanobiont is reported as *Nostoc* and differences tacitly attributed to lichen species. At the very least, the ability to form heterocysts, as occurs in *Nostoc*, potentially results in major increases in ARA.

11.7   The role of lichens in the nitrogen economy of ecosystems where they are a dominant component

Lichens may potentially affect the nitrogen cycling of ecosystems in a number of ways. Ecosystems where cyanolichens are dominant components will contribute through nitrogen fixation. Examples of such ecosystems include

Pseudotsuga forests of the Pacific Northwest in North America where *Lobaria* spp. occur in abundance as epiphytes, *Nothofagus* forests in temperate rain forests of the southern hemisphere where *Pseudocyphellaria* spp. occur in abundance as epiphytes, and subarctic woodlands where *Stereocaulon* spp. occur as extensive mats among the scattered spruce trees. In addition, chlorolichens are dominant components of many other ecosystems and they may capture significant quantities of nitrogen, particularly by dry deposition. Leaching and decomposition are effective means of transferring captured N to the soil nutrient pool.

11.7.1   *Chlorolichen enhancement of nitrogen retention*

Chlorolichens are in fact very efficient accumulators of NH_4^+ and NO_3^- (Lang *et al.* 1976; Reiners and Olsen 1984), even under cold polar conditions where *c.* 90% retention from snow meltwater is demonstrated (Crittenden 1996, 1998). Furthermore, Crittenden (1983, 1989) has demonstrated 80% retention of these ions from rainwater by both *Cladonia stellaris* and *Stereocaulon paschale* mats occurring in the subarctic spruce woodlands. In addition to accumulation from wet deposition, lichens also accumulate significant quantities of N from dry deposition. For example, Knops *et al.* (1996) have demonstrated that the epiphyte *Ramalina menziesii* is capable of augmenting N input to a nitrogen-poor oak woodland by $5\,kg\,ha^{-1}\,y^{-1}$ for closed canopies ($3\,kg\,ha^{-1}\,y^{-1}$ for woodlands with 60% tree cover). Because bulk deposition was less than $1\,kg\,N\,ha^{-1}\,y^{-1}$, it was inferred that most of the N gain came from dry deposition by the process of impaction.

11.7.2   *Cyanolichens and nitrogen fixation in the context of nitrogen dynamics*

To ascertain the importance of lichens, information is needed on more than biomass and accurate nitrogen fixation rates. In addition, one needs to ascertain how much of the fixed nitrogen enters the intrasystem nutrient cycle (Section 12.10), as opposed to being directly lost by such processes as denitrification (see below). Only the nitrogen entering the intrasystem cycle, as NO_3^-, NH_4^+ or organic N, can be utilized by other organisms. Also, it is important to compare the contributions by lichens with those from other sources. For example, Vitousek *et al.* (1987) studied a volcanic sequence in Hawaii and estimated that *Stereocaulon vulcani*, a common lichen colonizing lava flows, contributed $0.2\,kg\,N\,ha^{-1}\,y^{-1}$, but this contribution was small when compared with the subsequent contribution of $23.5\,kg\,N\,ha^{-1}\,y^{-1}$ by *Myrica faya*, a successional, nitrogen-fixing tree. Nevertheless, there are some ecosystems in which lichens probably contribute a much higher percentage of nitrogen. One of the more intriguing ecosystems is the commercially important *Pseudotsuga menziesii* forests of Oregon, in which Denison (1973) asserts that *Lobaria oregana* contributes approximately 50% of the total N input. Unfortunately, this estimate, as well as many others in the literature, has not been subjected to critical analysis, and information on the determination methods listed in Section 11.4 is generally not available for most of the studies discussed hereafter.

Decomposition of nitrogen in lichen litter obviously provides transfer to the intrasystem cycle (Holub and Lajtha 2003). By multiplying the lichen nitrogen concentration times the amount of lichen litter one can estimate how much nitrogen may become released, but this may be an overestimate as part of the nitrogen may remain in a recalcitrant fraction that is not fully decomposed. Nevertheless, almost all the N may be released in time frames varying from almost a year to a few years (McCune and Daly 1994; Knops et al. 1996; Esseen and Renhorn 1998; Coxson and Curteanu 2002; Holub and Lajtha 2004). But in warmer habitats such as deserts, not all of the released N may become available, as part may be revolatilized as N_2 to the atmosphere through denitrification. For example, Rychert and Skujins (1974) at first esti-mated the N_2 fixation by cyanolichen crusts in the Great Basin Desert of the western USA to be 10–100 kg N $ha^{-1} y^{-1}$, with an average of 16 kg N $ha^{-1} y^{-1}$, but in follow-up studies Skujins and Klubek (1978) concluded that 80% of the available NH_4^+ may be lost via denitrification and may never enter the soil pool. Decomposition will also release NO_3^- and NH_4^+, both of which may be absorbed by other organisms, but estimating their fluxes may require ^{15}N tracer studies. In most systems direct mineralization studies are difficult because the amount of N from lichens is usually small compared with other sources. Alternatively, the use of resin bags with lichen litter provides an index of ionic release, but such studies may not be realistic in that secondary consumption is excluded (McCune and Daly 1994).

Lichens affect nitrogen cycling in other ways. The most important may be leaching, particularly of organic forms of N, while the lichens are in place (Millbank 1982b; Millbank and Olsen 1986); elevated soil N is often found within a meter or more of cyanolichen mats (Knowles 2006). In succession sequences in the Pacific Northwest (USA), L. H. Pike (unpublished) calculated that leaching from cyanolichens (Peltigera spp.) may provide as much as 16 kg N $ha^{-1} y^{-1}$. Even green-algal lichens are subject to leaching. Knops et al. (1996) estimated that the green-algal lichen Ramalina menziesii increased N deposition by 3.0 kg N $ha^{-1} y^{-1}$, and of this total c. 10% was due to leaching. In addition, lichens with their high surface area/biomass ratios may be important in absorbing other nitrogen gases, such as the greenhouse gas N_2O, which is commonly released from soils, but unfortunately there are no data thus far.

In the case of cyanolichens, direct fixation of N_2 from the atmosphere occurs, and as a result N levels in these lichens are generally higher (generally 2.2–4.7%)

Table 11.1. *Estimations of lichen nitrogen fixation contributions to the nitrogen economy in various ecosystems*

Region	Input range	Major lichens	Reference
Glacial drift, Iceland	6.2 kg N ha^{-1} y^{-1}	*Stereocaulon* spp., *Peltigera* spp.	Crittenden 1975
Sand dunes and quarry	2.4–5.8 kg N ha^{-1} y^{-1}	*Peltigera* spp.	Millbank 1981
Boreal forest, Sweden	1.0 kg N ha^{-1} summer^{-1}	*Stereocaulon paschale*	Huss-Danell 1977
Pine–birch forest, Sweden	10.0–40.0 kg N ha^{-1} y^{-1}	*Stereocaulon paschale*, *Nephroma arcticum*	Kallio 1974
Subarctic Alaska	0.04–0.21 kg N ha^{-1} y^{-1}	*Peltigera* spp. + others	Gunther 1989
Pseudotsuga and *Picea* forest, New Mexico, USA[a]	0.04–3.3 kg N ha^{-1} y^{-1}	*Peltigera canina*, *Peltigera aphthosa*	Forman and Dowden 1977
Oak forest, North Carolina, USA	0.22–1.23 kg N ha^{-1} y^{-1}	*Leptogium cyanescens*, *Lobaria quercizans*	Becker *et al.* 1977
Gray beech (*Fagus*) forest, North Carolina, USA	0.8 kg N ha^{-1} y^{-1} (gaps) 0.17 kg N ha^{-1} y^{-1} (forest)	*Lobaria quercizans*, *Lobaria pulmonaria*	Becker 1980
Volcanic succession, Hawaii, USA[b]	0.2–0.45 (− 1.5) kg N ha^{-1} y^{-1}	*Stereocaulon vulcani*	Kurina and Vitousek 2001; Crews *et al.* 2001
Cloud forest, Colombia[a]	1.5–8 kg N ha^{-1} y^{-1}	*Sticta* spp., *Leptogium* spp.	Forman 1975
Nothofagus forest, New Zealand	1–10 kg N ha^{-1} y^{-1}	*Sticta* spp., *Pseudocyphellaria* spp.	Green *et al.* 1980
Pseudotsuga old growth forest, NW USA	1.5–16.5 kg N ha^{-1} y^{-1}	*Lobaria oregana*	Antoine 2004; Denison 1979
Cold desert, Utah, USA	2–20 kg N ha^{-1} y^{-1}	*Collema* spp.	Rychert and Skujins 1974 but modified by Skujins and Klubek 1978
Canyonlands National Park, Utah, USA	13 kg N ha^{-1} y^{-1}	*Collema* sp.	Belnap 2002

[a] No field measurement of C_2H_2 reduction or ^{15}N uptake made, estimations based on published laboratory ARA rates and conversion of 3:2 to nitrogen.

[b] Kurina and Vitousek's (2001) model estimates are based on daytime hours only for a site at 1500 m. At lower elevations with warmer temperature may potentially approach 9 kg N ha^{-1} y^{-1} based on their model. Crews *et al.* (2001) conducted their study at 1130 m on the same mountain and obtained the 1.5 kg N ha^{-1} y^{-1}. These are the only investigations to calibrate their ARA measurements with ^{15}N procedures.

than in green-algal lichens (generally 0.4–1.9%), which do not occur on nitrogen-rich substrates (Rai 1988). As a consequence, interest has focused on nitrogen inputs in ecosystems where cyanolichens are abundant, such as temperate rain forests and some subarctic ecosystems. In a number of these ecosystems estimates of annual nitrogen input by lichen fixation from ARA data are available (Table 11.1), but there are a number of potentially serious errors associated with some of these calculations. The frequent use of the theoretical 3:2 ratio for C_2H_4 produced to N_2 fixed, may thus lead to serious overestimates of the actual N_2 fixed, as ratios as high as 25:1 have been measured in free-living cyanobacteria (Peterson and Burris 1976). Because of these losses through hydrogen production and the differential diffusibility of C_2H_2 and N_2, the conversion for each system must be determined by parallel $^{15}N_2$ uptake (e.g. Crews et al. 2001; Kurina and Vitousek 2001), but this is rarely done. All these factors, combined with the necessity of establishing that the fixation rate is linear with time (David and Fay 1977), and the need for reasonable estimates of biomass, potentially make the jump from nanomoles of C_2H_4 g^{-1} y^{-1} to kilograms of $N\,ha^{-1}$ y^{-1} a great leap into the unknown. Unfortunately some authors fail even to specify conversion factors used.

Thus, the numbers given in Table 11.1 must be considered provisional at best. More accurate estimates include those by Huss-Danell (1977), who coupled biomass, cover, and environmental parameter measurements with field ARA measurements. More recently the Hawaiian ARA measurements have been calibrated against ^{15}N uptake measurements (Crews et al. 2001; Kurina and Vitousek 2001). On the other hand, estimates by Forman (1975) for cyanolichens in a Colombian cloud forest ($1.5–8\,kg\,N\,ha^{-1}\,y^{-1}$) are among the most problematic, as no ARA measurements were made in the field. The estimates are based on biomass estimates and extrapolations of ARA based on published cephalodial rates for temperate lichens under laboratory conditions. In addition, no correction was made for dark respiration, which at high temperatures may decrease daytime fixation (MacFarlane and Kershaw 1977). Gunther (1989) has cogently argued that some of the estimates presented in Table 11.1 are too high due in part to environmental complexities (especially drought and prehistory effects) and the experiments of Millbank (1981). Where possible, Gunther presented alternative estimates (1989, table 5). Unfortunately assumptions are often unstated and the lack of other data being presented made alternative estimates impossible. Nevertheless, optimistically high estimates are still being published in the literature.

12

Nutrients, elemental accumulation, and mineral cycling

T. H. NASH III

As with any organisms the accumulation and processing of both macro-nutrients and micronutrients essential for life's physiological functions are critical to the growth and development of lichens. The fact that lichens do not possess roots, the efficient nutrient absorption system of vascular plants, has led to major dependence on atmospheric sources of nutrients instead of the soil pool exploited by vascular plants (Nieboer *et al.* 1978), although some soil uptake by terricolous *Peltigera* species has been demonstrated (Goyal and Seaward 1982). Because atmospheric sources of nutrients are relatively meager compared with soil nutrient pools, nutrient concentrating mechanisms are critical for lichen survival. The fact that such mechanisms exist has led to more general scientific interest in lichens as surrogate receptors for atmospheric deposition. For example, the extremely high body burdens of radio-nuclides in indigenous human populations of arctic regions in the 1950s and 1960s resulted from their high consumption of caribou and reindeer, which in turn ate mostly lichens for 6–8 months of the year (Palmer *et al.* 1963; Lidén and Gustafsson 1967). Although most areas were remote from areas of surface nuclear testing, the lichens of these areas were efficient accumulators of radio-nuclides in this food chain. In more recent years, interest in such phenomena has led to using lichens for studies of regional atmospheric deposition of metals and other atmospheric contaminants (Puckett 1988). From such studies it has become evident that some lichens are quite tolerant of high metal concentrations, and mechanisms for such tolerance are now being elucidated (Nash 1989). The natural occurrence of high metallic concentrations has also been used to some extent in geologic prospecting. In addition to these aspects, the role of lichens in initial pedogenetic processes affecting the Earth's crust has been recognized for some time (Jones 1988). Such processes influence mineral

Lichen Biology, ed. Thomas H. Nash III. Published by Cambridge University Press.
© Cambridge University Press.

cycling processes on a geological timescale and more recently it is becoming recognized that lichens may play important roles in more rapid mineral cycling processes, such as the capture of allogenic nutrients that would not otherwise be retained in the ecosystem (Knops *et al.* 1996). Before reviewing these topics, it is important to consider the chemical and physical properties of nutrients and metals to be discussed.

12.1   Chemical and physical properties of nutrients and metals

Nutrients of importance to plant metabolism exist primarily as ions, either positively charged cations or negatively charged anions. Negatively charged anionic binding sites, such as carboxylic and hydroxycarboxylic acid moieties within the structural polysaccharides of the cell walls, and positively charged anionic binding sites are assumed to be responsible for the cell wall exchange capacity in lichens. Absorption from these sites into the cytoplasm may occur as well as directly from external solutions. Metabolism of the nutrients within the cytoplasm is covered in basic physiology and biochemistry texts and is not considered here.

In atmospheric deposition studies the term "heavy metals" has received widespread, popular usage for metals that are potentially toxic. However, the term "heavy metals" lacks rigorous definition, conveys little chemical meaning, and is used for different groups of metals by different authors. Nieboer and Richardson (1980) argued convincingly that the term is better replaced by their chemically and biologically meaningful groups designated as Class A, borderline, and Class B metals. These fall within natural groups within the periodic table and may be divided by a covalent index that reflects different ligand binding affinities. Class A metals prefer ONS donors; whereas Class B metals prefer SNO donors, and borderline metals fall in-between. At cation binding sites, Class A ions will displace borderline ions, which in turn will displace Class B ions.

Although potentially all elements could be considered, there is within lichenological literature limited information available on many of them. For example, studies dealing with metallic tolerance have frequently examined populations occurring on old mine tailings, where the most common metals include Cd, Cu, Fe, Ni, and Zn. In the context of atmospheric deposition, information on Al, As, Ba, Be, Ca, Ce, Co, Cr, Cs, Ga, Hg, K, La, Li, Mg, Mn, Na, Nd, Pb, Rb, Sc, Sn, Sr, Ti, V, and Y is available (Nash 1989). Of these, the following elements belong to Class A: Al, Ba, Ca, Ce, Cs, K, La, Li, Mg, Na, Nd, Rb, Sc, Sr, and Y; the following belong to borderline: As, Cd, Co, Cr, Fe, Ga, Mn, Ni, Sn, Ti, V, and Zn; and the following belong to Class B: Cu, Hg, and Pb. In contrast, ecosystem studies are primarily focused on the macronutrients: C, N, P, S, K, Mg, and Mn.

12.2   Nutrient requirements

Because whole lichens cannot be readily grown in culture, precise concentrations of nutrients necessary for lichen growth remain undefined. Apart from the obvious necessity of having both macro- and micronutrients necessary for normal metabolic functioning, evidence for nutrient require-ments can be assessed indirectly. For example, some lichens exhibit a high degree of substrate specificity (Brodo 1973) and this may at least partially be related to nutrient status. In addition, the nutrient requirements for growth of separated symbionts is fairly well defined (Ahmadjian 1993). Interestingly, *Trebouxia*, the most common lichen photobiont, is a facultative heterotroph in that it can show some growth in the dark when an organic source of nutrients is available. Additional evidence of nutrient requirements can be gleaned from field studies in which the dynamics of nutrient exchange are monitored. For example, Knops *et al.* (1996) in a canopy manipulation study documented that canopy lichens removed sulfate and phosphate from throughfall and hence inferred that these nutrients were limiting their growth at his Californian oak woodland site.

The addition of nutrients often stimulates growth and various metabolic processes, and thereby demonstrates nutrient requirements. For example, con-siderable literature focuses on potentially limiting effects of nitrogen (N) on lichen productivity (Chapter 10), but phosphorus (P) limitation deserves much more consideration. Unlike N, P has no gas phase, and frequently P may be lost from ecosystems by sedimentation and secondary mineral formation. Benner and Vitousek (2007) recently conducted separate, long-term (years) N and P fertilizer experiments in Hawaiian *Metrosideros* forests. Essentially no lichen response was found for the N-fertilized plots, but dramatic increases in the lichens occurred on the P-fertilized plots. Particularly for cyanolichens, lichen cover of trees increased from 3.2% in the controls up to 62.7% in the P-fertilized plots, and species richness increased from 2.9 species/tree to 18.8 species/tree. Chlorolichens did not increase as dramatically; cover increased from 0.3% to 3.3%, and species richness increased from 0.5 species/tree to 2.9 species/tree. Further evidence of P limitation was obtained by Hyvärinen and Crittenden (2000), who demonstrated probable P recycling within the podetia of *Cladonia portentosa* based on ^{33}P investigations. They demonstrated similar recycling of N with ^{15}N investigations (Ellis *et al.* 2005), and conclude that the obvious ecologi-cal success of these extensive *Cladonia* mats in subarctic regions is due to their ability to recycle potentially limiting P and N.

In addition, it is conceivable that nutrients in too high a level may cause a breakdown in the symbiosis (Section 12.7). For example, nitrate deposition in western Europe vastly exceeds deposition in natural ecosystems, and in regions such as the Netherlands, lichen impoverishment is well documented (van Dobben 1993).

12.3 Sources of nutrients

12.3.1 Atmospheric sources

Processes by which atmospheric deposition to lichens occurs include precipitation and occult precipitation (principally fog and dew) in the case of wet deposition, and sedimentation, impaction, and gaseous absorption in the case of dry deposition (Knops et al. 1991). Occult precipitation is very important to lichens both for nutrients and as a moisture source. Concentrations of nutrients and contaminants in occult precipitation may be substantially higher than in rainfall because more dilution occurs in the formation of rain. Sedimentation of large contaminant aerosols and their incorporation into lichen thalli is well demonstrated by comparisons of chemical profiles, based on X-ray diffraction data, of atmospheric aerosols sampled directly from the air and particles trapped within lichens (Garty et al. 1979). Impaction rates have not specifically been measured in lichens, but the small dissected form of many lichens is consistent with the inference that impaction rates should be high (Chamberlain 1970). Gaseous uptake measurements are limited, but in the case of sulfur dioxide (SO_2), uptake by the arctic lichen *Cladonia rangiferina* is at least an order of magnitude greater than typical vascular plants (Winner et al. 1988). Both the nitrogen and sulfur cycles have significant gaseous phases involving several different gases, and lichens may play significant roles in the exchange of these gases because of the high surface areas of lichens. Also lichens lack stomata, and thus they exchange gases across their entire surface.

12.3.2 Substrate sources

Many lichens occur on soils or rocks and hence are in intimate contact with lithic sources of nutrients. Lichens can affect weathering of rock surfaces by both mechanical and chemical means (Syers and Iskandar 1973), and, once nutrients are soluble, uptake into the lichen may occur by the mechanisms discussed below (Section 12.4). The greater availability of calcium (Ca) in limestone and the higher pH of limestone are obvious differences compared with acidic rocks. Solubility of many nutrients is affected by pH and this implies that availability of nutrients may be very different between limestone and acidic substrates. As a consequence, it is not surprising that very different lichen communities occur on limestone versus acidic rocks (Wirth 1972; Roux 1981).

Furthermore, on gypsum, sulfate, as well as Ca, is present in much higher concentrations, and there are a few lichens, such as *Acarospora clauzadeana*, that are restricted to gypsum.

Most lichens, especially those occurring on soils and rocks, are affected by blowing dust, most of which originates from exposed soil surfaces. These soil particles can readily become incorporated into intracellular spaces within lichens and result in relatively high concentrations of Al, Fe, Sc, Ti, and other elements of lithic origin within the thallus. Solubilization of these particles is a potential source of nutrients, but this process is slow, and most elements in these particles probably remain unavailable. To understand the nutrient status of a lichen, it is thus important to distinguish between total elemental concentrations versus ionic or solubilized concentrations. Currently we have no easy way to assess this, but the relative contribution of soil particles to total elemental loadings of lichens can be assessed by comparing ratios of macro- or micronutrients to inert elements, such as Sc or Ti (Nieboer *et al.* 1978 and Section 12.6 below).

Epiphytic lichens are also affected by dynamics of nutrient processing within the canopy as well as the nature of the bark on which they occur. Elements such as potassium readily leach from foliage and subsequently may be taken up by epiphytes. On some trees stemflow occurs readily during precipitation and well-developed nutrient streaks may develop on one side of the stem. Some lichens are restricted to these high nutrient areas and a rich terminology has been developed to describe this phenomenon (Barkman 1958). In addition, there are major differences in nutrient composition of different types of bark and the availability of these nutrients is affected by pH. This results in very different lichen communities occurring on trees with acidic bark (e.g. conifers, *Betula* spp., etc.) in contrast to trees with more neutral bark (e.g. *Fraxinus*, *Tilia*, etc.). Pollution sources may modify bark nutrient status, both by adding elements and affecting pH. Acidic deposition frequently results in the lowering of pH and the change of lichen communities occurring on previously neutral substrates (e.g. Skye 1968), but Ca deposition associated with cement plants may increase pH and result in a community shift in the opposite direction.

12.4   Accumulation mechanisms

In addition to lacking a vascular root system, the surface of lichens does not have the waxy cuticle found in vascular plant leaves, although most lichens do have an upper cortex layer of fairly tightly interwoven fungal hyphae. Lichens are in fact very efficient at scavenging ions from very dilute solutions, such as melted snow (Crittenden 1998). As a consequence, elemental exchange

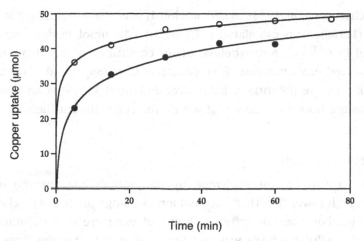

Fig. 12.1 Uptake of copper by 1 g samples of *Cladonia mitis* (open symbols) and *Umbilicaria muhlenbergii* (solid symbols) from solutions containing 5000 μmol Cu. (From Puckett *et al.* 1973.)

in lichens occurs across the entire lichen surface. The principal mechanisms by which cation accumulation occurs are discussed below.

12.4.1   Ion exchange

Metallic ions exist primarily as cations, although some anions of elements such as U are known. The kinetics and thermodynamics (as defined by uptake capacities, binding constants, and mass and charge balances) of cation uptake are well known (Nieboer and Richardson 1981). Initially cation uptake is a rapid, passive, physiochemical process occurring extracellularly in lichens (Nieboer *et al.* 1978). Saturated levels are reached in a matter of minutes (Fig. 12.1) and the capacity to retain cations within the cell wall is estimated to vary between 6 and 77 μmol g^{-1} depending on species (Nash 1989). These cations are primarily retained external to the photobiont's or mycobiont's cytoplasm at cation exchange sites. On the basis of competition experiments, affinity of ions for exchange sites varies in the sequence monovalent Class A < divalent Class A < borderline divalent < divalent Class B (Nieboer and Richardson 1980).

Anion uptake by lichens has been investigated to a more limited degree. In general it is both a slower process (hours) than cation exchange and results in less total accumulation. For example, the anionic uranyl complex ($UO_2L_2{}^{2-}$), accumulated much more slowly than uranyl cations and total uptake of anions was much less (1.6 μmol g^{-1} versus 49 μmol g^{-1}) than cations. In contrast, the uptake of the anionic form of arsenic (H_2AsO_4) was greater (10 μmol g^{-1}) and followed Michaelis–Menten kinetics, a fact that implies active uptake (Richardson *et al.* 1984), as had previously been established for phosphate by Farrar (1976*d*), who calculated a K_m of 3.3×10^4 nmol Pi dm^{-3} and V_{max} of 0.5 nmol Pi g^{-1} s^{-1}. Nevertheless, under conditions of dry deposition in heavily polluted environments, Boonpragob *et al.* (1989) found that external anion loadings were potentially huge (over 200 µmol g^{-1}). It is assumed that anion exchange sites must also exist within the cell wall, but their identity is unknown.

12.4.2   Intracellular uptake

In contrast to ion exchange uptake, intracellular uptake usually involves a much lower flux than external ion exchange processes (Beckett and Brown 1984*a*). For example, after 2.5 hours of exposure to a cadmium (Cd) solution, intracellular uptake was less than 10% of total uptake (Brown and Beckett 1985). Intracellular uptake increases with time (over hours) frequently following Michaelis–Menten kinetic patterns (Brown and Beckett 1984). Because intracellular uptake shows saturation kinetics with response to increasing external concentration of Cd, uptake is assumed to be a carrier-mediated process and hence to involve energy expenditure. Such energy dependency is demonstrated for the intracellular uptake of phosphate (Farrar 1976*d*).

12.4.3   Particulate entrapment

Within lichens considerable intercellular space exists (e.g. estimated as 18% for *Xanthoria parietina* by Collins and Farrar 1978). There is good direct and indirect evidence that particles may be trapped within these spaces (Nieboer *et al.* 1978). Among many lichens a linear relationship exists between the elemental concentrations of Ti and Al. Because the slope of this line approximates the known ratios of these elements in 12 of the 13 major minerals found in the Earth's crust, Nieboer *et al.* (1978) inferred that soil dust trapping frequently occurs within lichens. Around pollution sources the occurrence of particulates within lichens has frequently been demonstrated with scanning electron microscope procedures (Garty *et al.* 1979; Jones *et al.* 1982).

Such entrapment of metallic-rich particles is doubtlessly responsible for some of the extremely high elemental loadings reported in the literature. For example, Tomassini *et al.* (1976) measured Ni concentrations (µg g^{-1}) from 220 to 310; Fe, from 1700 to 5200; and Cu, up to 250. Similarly, in the vicinity of a steel complex in England, Seaward (1973) reports concentrations (µg g^{-1}) in *Peltigera rufescens* as high as 90 000 Fe, 5000 Mn, 91 Cu, 127 Cr, 454 Pb, and 38 Ni. In the city of Leeds, Seaward also reported concentrations (µg g^{-1}) up to 35 800 Fe, 349 Mn, 159 Cu, 97 Cr, 3124 Pb, and 183 Ni for *Lecanora muralis*. Selected other studies are incorporated into Table 12.1, but more extensive references are cited in Nash (1989).

Table 12.1. *Selected elemental contents for lichens from sites near industrial/urban (or other elevated source) and background regions*

Element	Industrial/ urban ($\mu g\,g^{-1}$ dry wt)	Source (Reference)	Background ($\mu g\,g^{-1}$ dry wt)	Reference
Class A metals				
Ce	2.2–7.2	Power plant (Olmez *et al.*, 1985)	0.18–0.89	Gough *et al.* (1988)
Na	1000–6000	Sea aerosols (Nieboer *et al.*, 1978)	50–1000	Nieboer *et al.*, (1978)
U	3.0–151	Mining (Boileau *et al.*, 1982)	0.5–1.0	Beckett *et al.* (1982)
Borderline metals				
As	128–11 400	Gold smelter (Hocking *et al.*, 1978)	0.06–2.21	Puckett (1978)
Ni	8–312	Nickel smelting (Tomassini *et al.*, 1976)	1.7–5.5	Puckett (1978)
V	150–578	Wood pulp mill (Laaksovirta and Olkkonen, 1979)	0.17–9.7	Puckett (1978)
Zn	1000–25 000	Smelter (Nash, 1975)	10–30	Nash (1975)
Class B metals				
Cu	15–250	Nickel smelting (Tomassini *et al.*, 1976)	0.7–5	Puckett and Burton (1981)
	1000–4900	Cuprous rock (Alstrup and Hansen, 1977)		
Hg	0.48–0.87	Chlor-alkali plant (Lodenius and Laaksovirta, 1979)	0.009–0.101	Pakarinen and Häsänen (1983)
Pb	111–270	Helsinki (Laaksovirta *et al.*, 1976)	0.4–9.2	Puckett (1978)
Nonmetals				
F	260–940	Fertilizer factory (Takala *et al.*, 1978)	2.9–7.8	Takala *et al.* (1978)
S	470–4800	Sulfate deposition gradient (Takala *et al.*, 1985)	101–961	Puckett (1978)
			170–320	Tomassini *et al.* (1976)

12.5 Compartmentalization of elements within lichens

Ions may exist in different compartments within lichens, and hence are not evenly distributed throughout lichen thalli. By analyzing sequential extract fractions from lichens, Brown and Beckett (1984) estimated distribution of cations into (1) an intercellular and surface fraction, (2) ion exchange site fraction, (3) intracellular fraction, and (4) residual fraction. They made (1) two half-hour extractions in deionized water to determine the extracellular and surface fraction, (2) two one-hour elutions in 20 mM $NiCl_2$ at pH 5.4 to determine the ions bound to the extracellular anionic binding sites (Ni, a borderline element, will competitively displace the more common cations, such as Ca, K, Mg, and Na), (3) a 30-minute treatment with boiling deionized water to rupture cell membranes and release soluble intracellular ions, and (4) boiling residual fraction in concentrated nitric acid to dryness and redissolving in molar nitric acid and thereby solubilizing elements in the residual fraction. The degree to which these extraction procedures actually do yield data reflecting these compartments needs further critical investigation. For example, the assumption that the residual fraction is primarily part of the intracellular fraction may not be correct. External particulate trapping (see above) is probably a common occurrence both in unpolluted (e.g. atmospheric aerosols of pedogenic origin) and polluted environments. It is unlikely that the recommended washing procedures would remove all or even a large fraction of the particulates trapped within the lichen thalli. Nevertheless, it does provide an extremely useful conceptual framework for discussion. It is, of course, essential to know where elements are located to understand their physiological importance.

12.6   Deposition studies

Lichens are useful as biomonitors in atmospheric deposition studies for several reasons (Puckett 1988; Garty 2001). Many species have broad geographic distributions and this allows documentation of spatial patterns. In addition, as perennial, slow-growing organisms lichens maintain a fairly uniform morphology with time and do not shed their parts as readily as vascular plants (e.g. leaves). Thus, morphological changes in the lichens themselves with time affect deposition characteristics minimally. In a few cases, such as with the reindeer lichen *Cladonia stellaris*, and the epiphyte *Evernia prunastri*, annual branching patterns are formed, and therefore documentation of temporal trends is possible in cases where contaminants are not mobile once deposited. Furthermore, the lack of a waxy cuticle and associated stomates, as occurs on vascular plant leaves, means that many contaminants can be absorbed over the entire lichen surface. Perhaps most importantly lichens are capable of accumulating many elements to concentrations that vastly exceed their physiological requirements and therefore deposition patterns are distinguishable from normal elemental loadings.

12.6.1   *Deposition of smelted metals to lichens*

Studies on elemental accumulation in lichens around pollution sources are numerous (Nash 1989). A few examples are cited in Table 12.1, particularly for more recent studies. In most cases distinctly more elevated concentrations are found near pollution sources than in areas farther away. In the case of isolated pollution sources, concentrations decrease logarithmically with distance and background concentrations defined across distances where concentrations are relatively constant (Nieboer and Richardson 1981). The definition of what are appropriate background concentrations may actually be quite complex, due to such problems as heterogeneity of substrates, influence of non-pollution sources on accumulation, and constraints imposed by analytical detection limits. In addition, the widespread occurrence of industrialization within temperate latitudes has led to long-distance transport of pollutants into remote areas.

Another challenge is assessing the variable effect of Earth crustal contributions on elemental loading within the lichens. One solution is to use the procedure of calculating enrichment factors (EF) for any element (X) in the lichen relative to crustal rock or soil (Puckett and Finegan 1980) according to the following formula:

$$EF = \frac{(X/\text{reference element}) \text{ in lichen}}{(X/\text{reference element}) \text{ in crustal rock}}$$

Scandium and titanium are particularly useful indicators of crustal material because they have no known biological function, and aluminum may have some utility as well. Puckett and Finegan used scandium in conjunction with average rock composition applicable for a number of localities throughout northern Canada. By focusing on ratios it was possible to obtain understanding of heterogeneous data with respect to absolute values. For example, the elements Al, Cr, Co, Fe, Na, Ti, and V all had mean EF ratios less than 5 (for example, although the absolute values of Al varied markedly, the ratio of Al/Sc was relatively constant), and this was interpreted as meaning that the origin was primarily crustal material. In contrast, EF values for other elements were higher and ranged up to 1958 for Cl in *Thamnolia subuliformis*. Elements with EF values generally above 100 for the 14 most common lichen species included Sb, Cl, Pb, and S, and these high EF ratios were interpreted as indicating noncrustal origin.

12.6.2 *Deposition of mercury to lichens*

Due to contamination of fish and other vertebrates Hg has become of major concern in recent years (Nriagu and Pacyna 1988). It is released from coal combustion, and hence is associated with widely distributed electrical grids, and more locally around point sources, such as chloralkali plants. Mercury is often emitted roughly 50% in vapor form (Hg^0) and 50% in various water soluble forms, such as HgCl or $CHHgCl_2$. Lichens have been used effectively to monitor point source emissions (Steinnes and Krog 1977; Sensen and Richardson 2002) and also to monitor long-range Hg transport (Evans and Hutchinson 1996). In addition, they have been used to monitor emissions associated with geothermal areas (Bennett and Wetmore 1999) and old mining areas (Bargagli *et al.* 1987), and from volcanoes (Barghigiani *et al.* 1990).

12.6.3   *Deposition of inorganic ions to lichens*

Deposition studies are not limited to metals, but also include anion deposition associated with acid rain and persistent organics associated with agricultural and industrial activities. For example, both Bruteig (1993) in Norway and Takala *et al.* (1985) in Finland have documented a sulfate deposition gradient along south to north transects within their countries. In more southern regions nitrogen pollutants are common in regions with high vehicular traffic. In Los Angeles, Boonpragob *et al.* (1989) documented nitrate deposition, a major portion of which was probably HNO_3, to a transplanted lichen of over 170 μmol g^{-1} over a 10-week period during the summer dry period.

12.6.4   *Deposition of organics to lichens*

Persistent organic pollutants have the major advantage that most do not occur naturally and hence ambiguities concerning background levels are not a concern. South to north gradients of several chlorinated hydrocarbons, polychlorinated biphenyls (PCBs) congeners and hexachlorocyclohexanes (HCHs) isomers were recently established by Muir *et al.* (1993) in Canada, and similarly Carlberg *et al.* (1983) found south to north gradients in Norway for chlorinated hydrocarbons, phthalates and polyaromatic hydrocarbons (PAHs). Polychlorinated biphenyls (PCBs) have been monitored by Garty *et al.* (1982) in Israel. Chlorinated hydrocarbons are widely used as pesticides, and have entered the Earth's biogeochemical pathways, with accumulation being documented in lichens in such remote locations as Antarctica (Bacci *et al.* 1986) and northern Sweden (Villeneuve and Holm 1984). These studies are testament to the efficiency with which these pollutants are subject to long-distance dispersal. In addition, they have recently been used to monitor dioxins and furans (Augusto *et al.* 2004).

12.7   Metal toxicity

12.7.1   *Zinc and cadmium toxicity*

Deposition of trace metals as accumulated in lichens has frequently been investigated (Section 12.6), but the potential toxicity of these metals (e.g.

Puckett 1976) is infrequently considered, in part because other air pollutants, such as SO_2, may be more important at many sites. However, Nash (1975) has argued that zinc may well be the most lichen-toxic component of the pollutants around a zinc smelter at Palmerton, PA, USA. The smelter, situated just north of the first ridge of the Appalachian Mountains, began operation in the late 1800s and by the early 1970s only 7 lichens remained in comparison with 77 found in a control area over 50 km east-northeast (Nash 1972). Zinc was found in surface soils to be as high as 135 000 ppm and Cd as high as 1750 ppm (Buchauer 1973). The zone where eleveated levels of SO_2 were found extended only $c.$ 8 km east and 6 km west of the smelter, but the corresponding zones of elevated Zn and Cd extended 22–25 km east and $c.$ 17 km west. Because Zn and Cd caused significant reduction in net photosynthesis at concentrations of 300–500 ppm Zn or Cd, and because Zn was present in 10 times higher concentrations, it was inferred that Zn was the most important pollutant (Nash 1975).

Subsequent work in England with 1 mM zinc solutions by Brown and Beckett (1984) has shown a reduction in gross photosynthesis to 5–40% of controls in 10 species of cyanobacterial-containing lichens in the genera *Collema*, *Lichina*, *Lobaria*, *Nephroma*, *Peltigera*, and *Sticta*. In addition, for *Peltigera horizontalis*, reduction in photosynthesis to less than 50% of controls was demonstrated for 100 μM solutions of Cd, Cu, or Zn. These results may help explain why *Peltigera* species were found in the control area, but not near the Pennsylvanian zinc smelter (Nash 1972). In other work Beckett and Brown (1983) investigated *Peltigera* populations on a variety of natural and old abandoned mine sites. By using the following formula, zinc tolerance index (ZTI) can be calculated as:

$$ZTI = 100 \times \frac{\text{Rate of photosynthesis after metal treatment}}{\text{Rate of photosynthesis after water treatment}}$$

where metal treatment was submersion in 100 μM Zn solutions for 30 minutes. Responses ranged from 10% to 88%, with the more tolerant populations coming from the mine sites.

12.7.2   Nickel toxicity

Two huge metal smelting complexes have become infamous: one in the vicinity of Sudbury, Ontario, and the second on the Kola Peninsula on the Russian side of the border. Vegetation in huge areas around these complexes has been largely destroyed (Richardson *et al.* 1980; Koptsik *et al.* 2004). With the cessation of operations at the former site, partial recovery has occurred (Gunn *et al.* 1995; Bačkor and Fahselt 2004), but even in the absence of SO_2, apparently toxic effects of metals are persistent. Earlier investigations around Sudbury, including toxicological aspects, when the smelters were in full swing are well summarized by Richardson *et al.* (1980). In the Kola area, *Cladonia stellaris* has proved a useful surrogate receptor for Ni deposition (Grodzinska *et al.* 1999), even though Ni exposure at Kola concentrations is known to have a severe effect on membrane permeability (Hyvärinen *et al.* 2000).

12.7.3   *Copper toxicity*

Copper smelters are well known to adversely affect lichen communities (LeBlanc *et al.* 1974; Mikhailova and Scheidegger 2001). As with other smelter situations, SO_2 emissions may well be a major toxic factor, although copper may also be toxic. Branquinho *et al.* (1997) have demonstrated membrane leakage and reduction in chlorophyll fluorescence parameters in relation to Cu uptake. Similarly Baćkor and Dzubaj (2004) documented the development of chlorosis in relation to Cu exposure. Also Hauck and Zöller (2003) demonstrated substantial soredial growth reduction for *Hypogymnia physoides* in response to Cu treatments. Also Monnet *et al.* (2005) have found that the aquatic lichen *Dermatocarpon luridum* is a good bioindicator of Cu in solution.

12.7.4   *Magnesium toxicity*

Initially, Hauck *et al.* (2001, 2002a) established a strong correlation between epiphytic lichen abundance and Mn bark concentrations and Mn/Ca ratios in spruce forest dieback-affected regions of the Harz Mountains in northern Germany. Subsequently, similar field evidence was documented in northern New York (Schmull and Hauck 2003), western Montana (Hauck and Spribille 2005), and British Columbia (Hauck and Paul 2005), and his group has now built a strong case based on experimental as well as field evidence for Mn being an important site factor (Hauck *et al.* 2006). The widely distributed *Hypogymnia physodes* was shown to be more sensitive to Mn than the sulfur-tolerant *Lecanora conizaeoides*, in part because the latter species has a greater capacity to sequester Mn in P granules (Hauck *et al.* 2002b, c). In addition, growth of the common soredia of the former species is quite sensitive to Mn concentrations (Hauck *et al.* 2002b). Cyanolichens, many of which are sensitive to polluted areas, exhibited considerable toxicity to Mn (Hauck *et al.* 2006).

12.8   Metal tolerance

Lichens occurring on metaliferous substrates accumulate significant concentrations of the metals and form distinct species assemblages (e.g. the alliance Acarosporion sinopicae with two associations called Acarosporetum sinopicae Hil. 1923, Lecanoretum epanorae Wirth 1972, and Lecideion inopis Purvis), that reflect variation in metal content of the substrate. The most common genera containing some species tolerant of high metals include *Acarospora*, *Aspcilia*, *Cladonia*, *Lecanora*, *Lecidea*, *Porpidia*, *Rhizocarpon*, *Stereocaulon*, and *Tremolecia*. Many species exhibit a high degree of specificity for these substrates and thus may have disjunct distributions reflecting the availability of these substrates.

Naturally occurring elevated concentrations can be quite extraordinary (e.g. Table 12.1). In addition, in the Harz Mountains of Germany, Lange and Ziegler (1963) report iron contents within the range of 0.6 to 5.5% dry weight and Noeske *et al.* (1970) reported the following concentrations ($\mu g\ g^{-1}$) of 60 to 410 Cu; 125 to 2750 Pb; 150 to 436 Zn and 170 to 270 Sb in five lichens. In Norway and Sweden around abandoned copper mines, Purvis (1984) reported copper concentrations of 5.3 to 5.9% in two species. Similarly in California, Czehura (1977) reported copper values of 0.25 to 2.3% in four species. Finally, in the vineyards of central Europe *Lecanora vinetorum*, with Cu concentrations up to 5000 $\mu g\ g^{-1}$, is famous for its tolerance of antifungal, copper-containing sprays.

Lange and Ziegler (1963) proposed that tolerance mechanisms might involve (1) inherent cytoplasmic tolerance, (2) cytoplasmic immobilization and detox-ification of ions by chemical combination, and (3) transport (or retention) of ions to regions external to the plasmalemma and even the cell wall. Today there is good evidence for several different mechanisms involving the latter possibility, but we have relatively little evidence for or against the first two possibilities. In addition, Beckett and Brown (1984*b*) have demonstrated that competition among ions may lead to reduced toxic ion uptake into the cytoplasm.

External sequestering of metals as part of crystals formed externally on the hyphae is one mechanism for avoiding toxicity. For example, Purvis (1984) showed by a combination of X-ray diffraction, infrared spectroscopy, and elec-tron probe microanalysis that the lichens *Acarospora rugulosa* and *Lecidea theiodes* with *c.* 5% Cu by dry weight had sequestered most of the copper as copper oxalate crystals. Subsequently Purvis *et al.* (1987) have shown that two other copper-tolerant lichens, *Acarospora smaragdula* and *Lecidea lactea*, are able to form Cu–norstictic acid complexes in the cortices of distinctively green specimens. Norstictic acid is a very common lichen secondary product that crystallizes externally on lichen hyphae.

In the case of polluted situations, it is clear that major portions of the metals are present in insoluble form (such as metallic oxides) and hence may be physiologically inactive unless solubilized. In such situations the concentra-tions of cations at cell wall exchange sites may be 2–4 orders of magnitude lower than total metallic concentrations determined for the lichen as a whole.

Competitive effects of different ions in different concentrations may also affect uptake. For example, Beckett and Brown (1984*a*) showed that equimolar concentrations of the cations Co, Mg, Mn, and Ni inhibit 40–50% and Zn inhibits

32% of the intracellular uptake of cadmium. Because metals such as zinc and cadmium frequently co-occur, competitive effects may well result in low internal concentrations. In particular Beckett and Brown (1984b) observed reduced intracellular cadmium uptake (V_{max} of 2.39 μmol g^{-1} h^{-1} versus control V_{max} of 3.93 μmol g^{-1} h^{-1}) in Peltigera populations collected respectively from an old mine site where enhanced zinc and cadmium concentrations occurred in the soil and a control area. Thus, competitive exclusion of ions from the cytoplasm is an additional tolerance mechanism.

The relation between the extracellular cation exchange pool in the cell wall and intracellular uptake needs further investigation. Initially, Brown and Beckett (1984) suggested that the extracellular pool might act as a buffer to the influx of cations. However, more recently they have recognized that it may instead be a source of cations, because binding to external exchange sites is reversible. In particular for both Cladonia and Peltigera species, Brown and Beckett (1985) demonstrated mobilization of cadmium from extracellular sites and transfer into the cytoplasm during incubation in deionized water following initial saturation of external exchange sites with 300 μM Cd solutions. Further information in these areas will be critical to our understanding of cytoplasmic concentrations of metals.

Once metals are within the cytoplasm it is known that binding centers in proteins and enzymes satisfy the reactivity requirements of Class A, Class B, or borderline ions. Specificity of metals occupying these binding centers is related to their size and geometry as well as ligand type. Consequently, Ochiai (1977) divided mechanisms of metal-ion toxicity into three categories: (1) blocking of the essential biological function groups of proteins and enzymes, (2) displacing the essential metal ions in proteins and enzymes, or (3) modifying the active conformation of proteins and enzymes. There is good evidence that at least some lichens are tolerant of higher internal concentrations of metal ions than other lichens, but particular cellular mechanisms that allow such tolerance are not currently known. In future studies it may thus be useful to determine if the tolerant species (ecotypes, etc.) possess some specific cellular mechanism for preventing toxicity by one or more of these mechanisms.

12.9   Hydrologic cycle alteration

The poikilohydric nature of lichens enables them to influence the hydrologic cycle, even in some systems where their biomass is small. In arctic and subarctic regions where lichen mats cover the ground continuously, lichens impede evaporation from the soil to the point that soils may be continually saturated when not frozen (Kershaw and Rouse 1971). Even epiphytic lichens may alter incoming hydrological inputs to ecosystems. For example, in an oak woodland in California, epiphytic lichens intercepted 7.5% of the incoming precipitation over a three-year period (Knops *et al.* 1996). Although it has yet to be quantified, the interception of fog and dew water by lichens in coastal deserts may well represent a significant fraction of hydrological inputs to those systems, particularly in regions where precipitation rarely occurs, such as the Atacama Desert of South America. In interior arid and semi-arid regions lichens occur commonly as part of the cryptogamic crusts that are particularly extensive on undisturbed soils. These crusts not only intercept precipitation, but also facilitate infiltration of water into the soil, which, in the absence of the cryptogams, frequently has an impervious $CaCO_3$ layer at or near the soil surface (Harper and Marble 1988).

12.10 Mineral cycling at the ecosystem level: interception of dry and wet deposition

Mineral cycling can be divided into intrasystem and intersystem cycling (Waring and Schlesinger 1985). The intrasystem cycle involves the movement of nutrients within an ecosystem, and the intersystem cycle involves the flux of nutrients between ecosystems and their interconnections with global cycles (Fig. 12.2). Within a vascular plant dominated ecosystem the majority of nutrients are found within the internal nutrient cycle flowing from the soil nutrient pool into the plants and by leaching or decomposition back to the soil pool. Intersystem inputs occur as wet and dry deposition to canopies from atmospheric sources and as weathering of underlying bedrock. Intersystem output occurs primarily with hydrologic movement of ions and particles through soils,

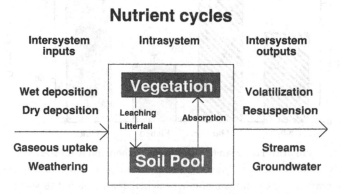

Fig. 12.2 Conceptual diagram of the intra- and intersystem nutrient cycles of an ecosystem.

although release of gases is also important in the nitrogen and sulfur cycles. Evidence is now mounting that lichens may influence daily, monthly, and seasonal patterns in mineral cycling involving both the intra- and intersystem cycles. In part this is related to their relatively large surface areas across which nutrient exchange occurs.

The interception of atmospheric nutrients by lichens has implications for many forest ecosystems because these nutrients originate primarily from outside the ecosystem. For example, in a California oak woodland Knops *et al.* (1996) compared 20 trees stripped of their dominant epiphytic lichen, *Ramalina menziesii*, with 20 trees where the lichen remained intact, and calculated that respectively 2.85 and 0.15 kg ha^{-1} y^{-1} of N and P was captured by the lichen from sources outside the ecosystem. Although these numbers are small relative to deposition of these elements in some agricultural areas, they represent a highly significant input to nutrient-poor ecosystems. In this case lichen capture of nutrients exceeded bulk deposition (a measure of sedimentation) by a factor of approximately three. As a result, productivity of the whole ecosystem may be enhanced.

The capture of these allogenic nutrients by epiphytic lichens affects the immediate return of nutrients in throughfall from forest canopies to the soil surface. For example, in the same California woodland Knops *et al.* (1996) found that the epiphytic lichens enhanced deposition of total N, NO$_3$, organic N, Ca, Mg, Na, and Cl, they reduced deposition of SO$_4$, and did not affect the deposition of NH$_4$, K, and total P. In addition, lichens, through their litterfall and

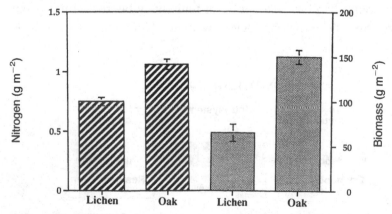

Fig. 12.3 Comparison of mean ($n = 20$) annual lichen and leaf litterfall nitrogen (bars with diagonal lines) and biomass for a Californian oak woodland, in which *Ramalina menziesii* is the dominant epiphytic lichen, for the period 16 May 1991 to 15 May 1992 (Knops *et al.* 1996). The vertical lines centered through the means represent two standard errors.

subsequent decomposition, affect longer-term nutrient deposition to soils. In the case of the California ecosystem, much of the nitrogen in leaves is remobilized into the tree stem before abscission, but this is, of course, impossible in lichens that have no vascular connection to their host trees. As a result lichen litter was almost twice as rich in nitrogen as leaves and contributed substantially more nitrogen deposition than would be inferred from the mass of the lichen litter alone (Fig. 12.3). In addition, lichens may have significantly higher quantities of micronutrients, such as Fe, than vascular plant leaves, although the reverse is true for some macronutrients (Knops *et al.* 1991). Once lichen litter is deposited on the soil surface, it may affect leaf decomposition. Knops *et al.* (1996) found that phosphorus was initially sequestered into the lichen and that concomitantly leaf decomposition was slowed.

Nutrients present in epiphytic lichens may originate from both intra- and interecosystem sources and it is relatively difficult to distinguish the two sources. For example, nutrients leached out of leaves may be subsequently absorbed by lichens hanging below and hence originate from the intraecosystem nutrient cycle. In contrast, both wet and dry depositions contribute nutrients from outside the ecosystem (Knops *et al.* 1991). Wet deposition includes not only rain, but also snow, dew, and fog. Lichens are particularly effective at capturing both dew and fog and these sources are frequently richer in nutrients than rain. Dry deposition includes sedimentation of large aerosols (greater than 2–10 μm in diameter), impaction of smaller aerosols, and gaseous uptake. Sedimentation may result in the incorporation of particles within the lichen thalli. In contrast, impaction involves the impingement of aerosols onto the lichen's surface. Although deposition rates of aerosols to lichens in wind tunnel studies have not specifically been investigated, the small, dissected, and pendulous nature of many common lichen epiphytes leads to the inference that impaction rates onto such lichens may well exceed those onto other canopy components. Thus far, nitrogen fixation by lichens has been investigated extensively (Chapter 11), but other nitrogen gases (NO, N_2O, NO_2, etc.) have yet to be investigated systematically. In the case of sulfur gases, the uptake of SO_2 (Winner *et al.* 1988) and carbonyl sulfide (Gries *et al.* 1994) and release of dimethylsulfide and H_2S (Gries *et al.* 1994) are well documented. For SO_2, uptake rates by lichens may substantially exceed those by vascular plants where closed stomata limit absorption.

13

Individuals and populations of lichens

D. FAHSELT

13.1   Individuals?

In lichenology the term "thallus" or "body" is less contentious than "individual," as a lichen thallus is not a single genetic entity. The status of "individual" has sometimes been applied loosely to physically distinct thalli, such as the stalked umbrella-shaped structure of *Umbilicaria*, or to discrete thalli of any lichen species. A more sophisticated approach, and one which is more meaningful biologically, is to consider a lichen "individual" as any thallus material which is genetically uniform with respect to the dominant, or fungal, symbiont. This, of course, requires genetic information which may not always be available.

Even defined according to genetic properties of the mycobiont, lichen "individuals" do not correspond to individuals of most other species. The reason is the high degree of internal thallus complexity. Not only are two or more interdependent primary symbiotic partners closely associated with one another, but there may be more than one strain of each and possibly even an array of other symbionts. In some ways lichens are as much like little communities or ecosystems (Section 1.6) as individuals.

13.2   The primary partners

Lichens are often presented as an association of two symbionts, but because they have evolved several times (Gargas *et al.* 1995; Lutzoni *et al.* 2001) few generalizations are applicable to all. Not only the nature, but also the number of participating symbionts may differ. The bulk of the thallus is almost always formed by the mycobiont, a lichen-forming fungus, which in 99% of lichen species is an ascomycete and in most other cases a basidiomycete. The mycobiont always associates either with a photosynthesizing partner, or

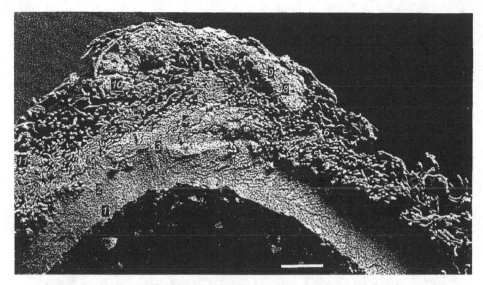

Fig. 13.1 Scattered clusters of photobionts in a podetium of *Cladonia rangiferina* shown by scanning electron microscopy, with locations in this section confirmed by light microscopy. The outer surface is ecorticate. 1, 2, 3 and 4 point to inner cortex; 5, 6 and 7 to loosely arranged medullary hyphae; 8, 9, 10 and 11 to photobiont clusters. (Image by Marcella Trembley.) Scale bar represents 7.5 μm.

photobiont, which is usually either a green alga, termed a chlorobiont (Lange and Wagenitz 2004), or less commonly with a cyanobacterium, or cyanobiont.

Some lichens have two photobionts, one chlorobiont and one cyanobiont, in different parts of the same thallus. There may even be two chlorobiont species, one in the early stages of thallus development and another which predominates later with the two finally coexisting (Friedl 1987). A further variation is a mycobiont, which is capable of lichenization by a photobiont or managing alone as a free-living fungus (Ahmadjian 1967*a*: 45, 55, 73). The sexual stage of such a facultative mycobiont may occur in either form (Wedin *et al.* 2004).

The photobiont in a lichen is typically unicellular and seated in a layer or stratum near the upper or outer surface of the thallus where it is surrounded by mycobiont hyphae. However, in some lichens photobionts are aggregated in clumps (Fig. 13.1) or, alternatively, scattered throughout the medulla. A lichen-forming fungus and its photobiont may differ in the degree of selectivity, and the degree of selectivity may also differ between one lichen and another. Several mycobionts associate preferentially with a single species, or even a single clade, of photobiont (Miao *et al.* 1997; Rambold *et al.* 1998; Kroken and Taylor 2000; Piercey-Normore 2004; Yahr *et al.* 2004), but the same photobiont may be involved in several different lichen species (e.g. Beck 1999; O'Brien *et al.* 2005). Low selectivity of cyanobionts has also been shown in antarctic lichens (Wirtz *et al.* 2003).

13.3   Genetic uniformity of symbionts within a thallus?

Lichens seem to differ from one another in respect to the degree of mycobiont heterogeneity within a thallus. Various kinds of evidence suggest that mycobionts of some lichen species are quite uniform. For example, although several phenolic chemotypes are represented in populations of *Cladonia chlorophaea* in the southern Appalachian Mountains, USA, each individual podetium exhibits lichen products of only one (DePriest 1993*a*). This is consistent with one hyphal strain in each podetium. Most lichens have not been subjected to molecular analysis (Fig. 13.2) and serious concerns have been raised about the stringency of many molecular studies (Parmasto 2004). However, rigorous application of molecular methods provides a direct means of approaching thallus uniformity. Using fungal-specific primers to target regions of the mycobiont genome, as opposed to that of the photobiont, base sequences in introns, which insert themselves into the mycobiont nuclear gene that codes for small subunits of ribosomal RNA (SSU rDNA), have shown uniformity throughout each podetium of *Cladonia subtenuis*. Sequences may in fact be the same throughout an entire, intensely sampled lichen mat and even in detached podetia nearby (Beard and DePriest 1996). "Individuals" in some lichens could thus be rather extensive if defined according to genetic properties of the mycobiont. Thalli of *Letharia* spp. also show mycological simplicity, as each exhibits a single allele at all studied loci (Kroken and Taylor 2001). The mycobiont of *Lobaria pulmonaria*, assessed using random short

Fig. 13.2   A mat of the lichen *Cladonia alpestris* in the ground layer of a *Pinus resinosa* forest. The possibility of basal hyphal connections between the relatively discrete much-branched podetia has not been examined, and as for many other lichens, genetic complexity within and between podetia is as yet undetermined.

repeats of DNA called "microsatellites," likewise shows no variation within thalli (Walser *et al.* 2003).

At the same time there are indications that thalli of other lichens are internally heterogeneous (Bailey 1976). Certain physiological evidence is consistent with the idea of multiple genotypes within thalli of some lichen species. For example, rates of respiration and photosynthesis vary over the surface of *Umbilicaria* spp., and electrophoretic separation of enzyme forms extracted from samples taken throughout a thallus produce different banding patterns (Larson 1983; Larson and Carey 1986). It is possible that differences are related to higher metabolic activity near the umbilicus (Hestmark *et al.* 1997), but the explanation may also involve multiple strains within one thallus (Larson and Carey 1986). Genotypes may in fact coexist, as shown in some podetia of *Cladonia chlorophaea* which varied internally with respect to Group I introns (DePriest 1993c). Because ascospore progeny from different ascomata on the same specimen of *Ochrolechia parella* differed in electrophoretically separated bands of randomly amplified DNA (Murtagh *et al.* 2000), it has been suggested that hyphae of adjacent thalli may have mingled in crowded habitats.

Secondary chemistry may also suggest association of two chemical races of mycobionts within a single thallus (Brodo 1978). In addition, there may be greater differences in growth rates, and in polyol chemistry, among lobes of *Parmelia* than between thalli, another situation that could be explained by involvement of multiple genotypes (Armstrong and Smith 1992). Phenotypic variation within thalli of both *Haematomma ochroleucum* (Laundon 1978) and *Parmelia omphalodes* (Skult 1984) is also consistent with the presence of more than one strain of mycobiont.

Photobionts vary also in the degree of heterogeneity exhibited within one thallus. They may be genetically homogeneous, as in *Pleurosticta acetabulum* where all samples from one thallus proved to be identical to one another in 18 S nrDNA sequences; similarly all were the same within each thallus of *Tremolecia atrata* (Beck and Koop 2001). In addition, only one algal genotype was found per thallus of *Letharia* (Kroken and Taylor 2001).

The first evidence of intrathalline variability in photobionts was the differences between cultures of single algal cells isolated from the same fruiting body of *Cladonia cristatella* (Ahmadjian 1967a). Genetic analysis provides direct indications of inherent differences between photobionts; within thalli of *Evernia mesomorpha*, nearly half of 290 specimens exhibited multiple algal genotypes sometimes even in the same thallus branch (Piercey-Normore 2006).

Genetic diversity could arise from somatic mutation or involvement of multiple propagules in the formation of the thallus. Coalescence of several soredia has been observed in the early stages of development of both *Hypogymnia* and

Physcia (Schuster *et al.* 1985). Fusion of hyphae from several spores was documented in *Xanthoria* (Ott 1987*b*) and numerous fusions among neighboring thalli were detected as well in *Cladonia rangiferina* (Jahns 1987). Hyphal mergers have been seen also in *Umbilicaria* (Hageman 1989) and may not be uncommon in some lichens. Involvement of multiple symbiont genotypes in one thallus may be advantageous in responding to environmental changes over time or in occupying diverse microenvironments within a heterogeneous habitat (Piercey-Normore 2006).

13.4   Secondary fungi

Along with the primary mycobiont and photobiont other symbionts may cohabit in a lichen thallus, and the prevalence of nonmycobiont fungi, sometimes called accessory or secondary fungi, is becoming much more widely appreciated. Although their existence was acknowledged long ago (Zopf 1897; des Abbayes 1953; Ahmadjian 1967*a*), comprehensive studies are relatively recent. Taxonomic literature documenting the diversity of lichenicolous fungi has been expanding rapidly over the last two decades (e.g. Nimis and Poelt 1987; Alstrup and Hawksworth 1990; Hawksworth 2003; Santesson *et al.* 2004), with momentum steadily increasing and the number of described species doubling over this period to approximately 1500 (Lawrey and Diederich 2003). As many as 3000–5000 species have been projected worldwide, or even a larger number (Diederich 2003).

The fruiting bodies of lichen-inhabiting fungal species are sometimes evident on the upper surface of infected thalli, as are discoloration and lesions caused by invasive fungi. Gall-like structures have also been reported (Madelin 1968). Scanning electron microscopy (SEM) may be able to distinguish the hyphae of invasive fungi from those of the mycobiont, on the basis of dimensions, swellings or clamp connections, characteristic surface features, and the tendency to collapse under negative pressure (Fahselt *et al.* 2001; Fig. 13.3). However, some of the ultrastructural features which distinguish secondary inhabitants and mycobionts may be apparent only in axenic culture and not the intact thallus (Sun *et al.* 2002).

Molecular methods have unparalleled potential for analyzing thallus complexity contributed by secondary fungi. For example, more than 500 strains of filamentous fungi were isolated from only 17 thalli of *Cladonia* and *Stereocaulon* in the Black Forest of Germany (Petrini *et al.* 1990), and although some of these were soil pathogens or decomposers, others were strains which associated exclusively with lichens. Fifty-nine lichenicolous fungi were cultured from 16 lichen species in Antarctica (Möller and Dreyfuss 1996). Several secondary

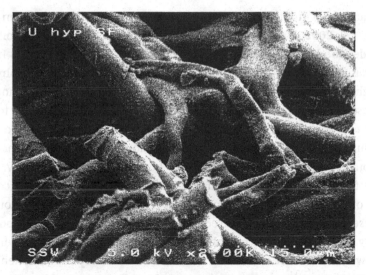

Fig. 13.3 Scanning electron micrograph of medullary hyphae in *Umbilicaria hyper-borea* after burial by ice in Greenland. Using molecular methods a number of sapro-phytes were identified in this material and heterogeneity is demonstrated among hyphae inside the thallus. Large diameter hyphae with thick walls are those of the mycobiont while thinner-walled hyphae, prone to collapsing under negative pressure used to prepare specimens, may be those of an invasive fungus.

fungi typically inhabited thalli of all 12 North American lichens investigated by Sun *et al.* (2002), and more than 1000 genetically different strains of lichen-inhabiting fungi were reported from specimens of Letharietum vulpinae (Peršoh 2004) collected from only six collection sites.

In addition to the better-known lichenicolous fungi evident on the cortex, a rich array of nonobligate microfungi has been revealed in the thallus interior (Suryanarayanan *et al.* 2005). Of most accessory fungi reported inside lichen thalli (Miadlikowska *et al.* 2004a,b), those in *Peltigera* spp., at least, are phylogen-etically distinct from those on the surface and more closely allied with ascomy-cetes found as endophytes in vascular plants. Many endolichenic forms are restricted to the interior (Miadlikowska *et al.* 2004a,b). They may be adapted to particular niches within the thallus as some are limited to certain strata, occur-ring for example in the medulla but not the algal layer (Sun *et al.* 2002).

It appears that secondary fungi may occur throughout a large part of the geographic range of a host lichen. For example, *Pestalotiopsis maculans* shows high constancy in species of *Cladonia*, *Usnea*, and *Parmotrema* over much of North America (Sun *et al.* 2002). High congruity of fungal inhabitants between distant collecting sites of *Peltigera* also supports the idea that secondary fungi in lichens are widespread geographically (Miadlikowska *et al.* 2004a, b).

While accessory fungi may inhabit lichens without apparent harm (Zopf 1897; Sun *et al.* 2002), details of their interactions with primary symbionts are poorly understood. Thalli of *Cladonia* spp. infected by the lichenicolous fungus *Arthrorhaphis aeruginosa* exhibit a number of peculiarities: the chlorophyll *a/b* ratio is higher in infected than in control thalli, quantities of some xanthophylls are increased, and an atypical hydrophobic blue pigment may be formed (Fahselt *et al.* 2000). Thus, it appears that lichen metabolism is altered by an invasive fungus, but the mode of action will undoubtedly depend on the particular secondary inhabitant as well as the host species. Nutritionally, some secondary fungi may be fungicolous and survive by exploiting the cellular resources of other fungi (Hawksworth *et al.* 1995), and *Pestalotiopsis* could be one of these because it is not in direct contact with the photobiont. It possibly utilizes organics leaked from mycobiont hyphae damaged during wet/dry and freeze/thaw cycles (Sun *et al.* 2002). In addition to accessory fungi, lichens host numerous other associates, such as algae, yeasts, protozoans, and bacteria, on or within thalli (Ahmadjian 1967*a*). Mosses establish on lichens as well, as do epilichenic lichens, each with its own photobiont. However, the nature of most of these associations remains largely unexplored.

The multiplicity of inhabitants in a thallus could of course confound many aspects of experimental lichenology, including investigations into genetic diversity of the mycobiont. The possibility of contamination must therefore be anticipated and steps taken to prevent DNA of photobionts or accessory symbionts from being extracted unwittingly and mistaken for that of the primary fungal symbiont. One way of limiting spurious sequences is to selectively dissect thallus material; however, this method is not always exacting. The use of fungal- or algal-specific primers to target particular genomes is beneficial to a point, but would be most effective if discrimination were possible between mycobionts and other fungi. A highly favored approach to genetic analysis of the lichen mycobiont, and the one which most effectively avoids interference by the genomes of minor thallus inhabitants, is to culture the mycobiont axenically and to extract DNA for analysis from pure cultures (Murtagh *et al.* 1999, 2000; Honegger *et al.* 2004).

One example of the confusion which may arise due to contamination was encountered during an attempt to sequence the *Umbilicaria cylindrica* mycobiont in subfossil lichens, frozen for centuries under an ice field in Greenland. Several samples unexpectedly aligned with yeasts rather than Ascomycota (DePriest *et al.* 2000), and it was finally determined that fungus-specific probes had not targeted DNA of the mycobiont, but rather of saprobes that probably invaded after ice melt. In fact one anomalous sequence was closely related to that of a common cold-climate saprophyte found on the frozen remains of the famous

Austrian "Iceman." In this case the DNA of primary interest, that of the myco-biont, may have been damaged during ice burial, leaving saprophyte sequences predominant by the time of collection. Just as the misplacement of subfossil samples in cladistic analysis signaled contamination by extraneous fungi, the wide scattering of replicates in studies of recent lichen material may also indicate involvement of secondary or pathogenic fungi.

13.5   Populations

As reproduction in lichens is usually quite poorly understood, a popula-tion cannot be defined on the basis of interbreeding. Therefore the term "population" is often used to mean any group of conspecific thalli, which are spatially separated from other such groups. Variation seems to be universally present within natural populations, and its full extent can only be appreciated through intense sampling. Polymorphism is probably of value to the lichen for long-term survival, but only heritable variants are of evolutionary significance. Features which depend primarily upon environment, for example plasticity in response to substrate, light availability, or moisture regime (Fahselt and Krol 1989), or that relate to stage of development, for example differences in stage of development, size, age, or sexual expression (Hageman and Fahselt 1986; Golm *et al.* 1993; Ramstad and Hestmark 2001), are less germane.

Although little may be known of the loci and alleles responsible, striking differences between thalli growing together in uniform environments, particu-larly if of comparable size and age, probably do have a genetic basis. One example is the familiar lichen *Cladonia cristatella* (Ahmadjian 1967a; Fig. 13.4) with adjacent podetia in natural populations contrasting markedly in features such as form and color. Experimentally, common gardens and reciprocal trans-plants also provide evidence of heritability, but precise detail on the genetic

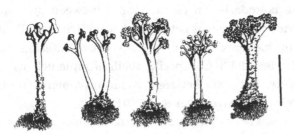

Fig. 13.4   Highly polymorphic podetia typically found within populations of *Cladonia cristatella*. Apothecia differ in form, but also in color, ranging from light orange to bright red to maroon. Spore cultures from different podetia show marked physiological differences as well. Scale bar represents 1 cm.

structure of populations is best obtained through stringent molecular analysis. Sequencing of bases in DNA/RNA, or occasionally amino acids in encoded proteins, has now been accomplished in a number of lichens.

13.6   Innate differences between thalli within populations

The degree of inherent variability appears to be partly linked with breeding systems, with populations of apotheciate species of umbilicate lichens showing more enzyme variation among thalli than those reproducing asexually (Fahselt 1989, 1995). Genetic analysis of *Letharia* species likewise reveals fewer genotypes in populations lacking fruiting bodies than in sexual populations (Grube and Kroken 2000). Nevertheless, sterile lichens also exhibit polymorphism, possibly because diverse genotypes are acquired through occasional immigration and variants with slightly differing ecological preferences are favored in particular microsites within a population (DePriest 1993a). Thalli might also differ from one another due to development from genetically distinctive propagules, singly or in combination, and parasexuality could also contribute to intrathalline differences within populations, as could somatic mutation.

As it is useful to describe the extent of genetic variation within populations, several statistics have been adopted for this purpose. Commonly employed for diploids have been the proportion of polymorphic loci, mean number of segregating alleles and average heterozygosity (Gillespie 1991), but expressions used in lichenology include the number of different sequences (strains) and the number of restriction fragment patterns in a population. The number of alleles/number of samples (Walser *et al.* 2005), and percent divergence among strains (e.g. Dyer *et al.* 2001), have likewise been used as measures of diversity. The number of haplotypes and haplotype frequencies are other expressions that characterize variation within populations (Lindblom and Ekman 2006).

In free-living fungi, genetically different thalli may interact antagonistically and produce distinctive boundaries in contact zones between incompatible genotypes. Similarly genetic differences between adjacent lichen thalli may be reflected in pigmented or thickened margins, perhaps due to differing alleles at incompatibility loci (Clayden 1997). Conspecific thalli of some lichens exhibit vegetative incompatability, but tests between axenic mycobiont cultures of other species have been inconclusive (Dyer *et al.* 2001).

Genetic differentiation has been directly demonstrated within populations of several lichens. One is *Cladonia chlorophaea*, with podetia displaying varied banding patterns after electrophoresis of restriction fragments of nuclear SSU rDNA (DePriest 1993b). Most lichen mats include several mycobiont genotypes with podetia differing from each other on the basis of insertions and deletions of highly mobile introns (DePriest 1993c). Although the untranscribed introns have no known impact on phenotype of the thallus, variation in length, base sequences, and position of insertion are extremely useful in assessing dynamics of these populations.

Analysis of randomly amplified polymorphic DNA (RAPD) of mycobiont cultures revealed distinct genetic markers discriminating among thalli in woodland populations of *Graphis scripta* and *Phaeographis dendritica* (Murtagh et al. 1999). Variation in DNA fingerprints based on sequences amplified using fungal-specific primers was evident as well in both *Graphis scripta* and *Xanthoria elegans* (Dyer et al. 2001). Because the targeted loci were randomly chosen and their functional importance unknown, an unusually large number of primers was used to increase coverage of the genome, and the highest genetic diversity values observed were comparable to those in free-living fungi. In all sampled populations of *Lobaria pulmonaria* random short repeats of DNA at six loci likewise show variation among thalli (Walser et al. 2003, 2005), and considerable polymorphism was also reported within local populations of *Stereocaulon* (Döring and Wedin 2004). Variation was suggested as well in randomly amplified sequences from even a small number of samples per population of *Usnea filipendula* (Heibel et al. 1999).

Seven haplotypes of sequences coding for cytochrome oxidase (*cox 1* gene) were sampled in populations of *Cladonia subcervicornis*, with three to five haplotypes found in each population (Printzen and Ekman 2003). Based on two spacer regions of the nuclear ribosomal DNA several haplotypes were found in *Xanthoria parietina* (Lindblom and Ekman 2006), and high intraspecific variation was observed in all populations. One haplotype was found in nearly half of all samples, in contrast to the majority of haplotypes which only occurred with low frequencies. No clones extended further than a few meters.

In fact two assessments have been made of variation in *Xanthoria parietina*. The Norwegian study using total internal transcribed spacer (ITS) and partial intergenic spacer (IGS) sequences in 30 sample thalli per population on rocks or bark suggested levels of genetic diversity similar to those in other lichen-forming fungi (Lindblom and Ekman 2006). A study of populations in other parts of Europe using nuclear SSU rDNA and the H1 gene region coding for the protein hydrophobin were amplified and sequenced from single ascospore cultures (Honegger et al. 2004). In this case DNA populations were also variable, but *X. parietina* showed far less genetic diversity than five other species and its generally homogeneous character was confirmed by lesser variability in pigmentation, secondary chemistry, and growth rates of cultures.

Several considerations affect estimates of variability. Some genetic regions are obviously more polymorphic than others: for example, in ribosomal clusters internal transcribed spacers are more variable than the gene for 5.8S rDNA (Scherrer and Honegger 2003). Of loci examined in populations of *Lobaria pulmonaria*, two were variable while four others exhibited no polymorphism whatsoever (Zoller *et al.* 1999). Two data sets from *Letharia*, involving sequences from one versus several loci, were evidently at variance with one another because the two generated quite different geneologies (Kroken and Taylor 2001). Sampling protocols may affect the observed variation, as may the geographic range over which samples are gathered, and variability could also be overestimated if fungal primers engage other fungi besides the mycobiont. Additionally, levels of genetic diversity of one lichen could presumably differ from one population to the next, because selective forces, genetic drift, or sexuality could vary in each. After all, even mats within the same sampling site have been known to differ with respect to the degree of genetic variability (DePriest 1993*b*).

For photobionts there is less information concerning intrapopulational variation. In scrub vegetation in southern Florida, USA, several ITS genotypes of *Trebouxia* were represented in each study site, but most *Cladonia* mycobionts associated with only a single clade (e.g. Yahr *et al.* 2004). Provisional indications of intrapopulational genetic diversity of *Trebouxia* photobionts in randomly amplified rDNA from *Xanthoria parietina* were astonishingly high (Nyati *et al.* 2004).

13.7   Genetic variation inferred from enzyme polymorphism

While analysis of enzyme polymorphism is more labor intensive than DNA, inherent variability in a population can be inferred from these direct gene products. Protein yield from lichen thalli is generally low, but genetic regions coding for protein are probably of more biological significance than randomly chosen regions with no apparent function. Proteins are less variable than DNA because of redundancy in the genetic code: for example, substitutions in the third base of a codon do not always alter the corresponding amino acids. The relative heterogeneities of DNA and protein are well illustrated in *Xanthoria parietina* where variation in amino acid sequences of hydrophobin is less than that in the DNA coding for it (Scherrer and Honegger 2003).

Sequencing of lichen proteins has not been a common practice, however, and protein variation at the population level has so far been most commonly approached through electrophoresis of isozymes (Fahselt 2001) or multiple enzyme forms. A minimal mass of hyphae is required to permit detection of proteins, so enzymes extracted from whole thalli are mainly products of the mycobiont (Fahselt 1985; Fahselt and Hageman 1994) rather than photobionts. For the same reason it might be expected that extracts would not include appreciable titre of enzymes from a secondary thallus inhabitant, although

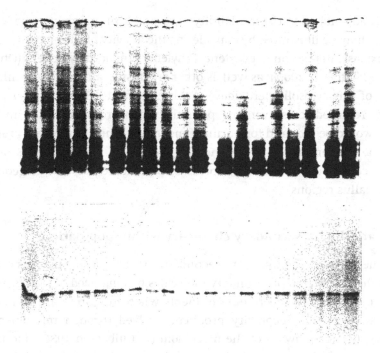

Fig. 13.5   Electrophoretic separation of multiple enzyme forms in extracts of
thalli from one stand of *Umbilicaria mammulata*. (Upper panel) Several bands of
esterase are polymorphic. (Lower panel) Mannitol dehydrogenase is invariant.

heavy infection might contaminate samples. Assuming that interference from
secondary symbionts is insignificant, the varying electrophoretic enzyme patterns are indicative of mycobiont diversity in corresponding coding regions.
Several enzymes analyzed to reflect overall variation of the genome show
polymorphism in populations of more than two dozen North American lichens,
including species of *Stereocaulon*, *Cladonia*, *Xanthoria*, and *Umbilicaria* (Hageman
and Fahselt 1990; Fig. 13.5). Differences in multiple enzyme forms have also
been reported within populations of *Cetraria* spp. in both Europe and North
America (Mattson 1991) and in a Middle Eastern desert lichen, *Caloplaca aurantia*
(Nevo *et al.* 1997).

Polymorphism of enzymes can be illustrated graphically by ordination of
presence/absence band data, with axial scores computed in multidimensional
space and variation expressed in terms of standard deviations from the mean
tendency of each population on the first few axes (Jancey 1966; Hageman and
Fahselt 1990). Variability may also be illustrated using Renyi's generalized
entropy (Fahselt *et al.* 1997), or statistics such as the Simpson–Gini diversity
index, the Shannon–Weaver diversity index, or the number of different banding
patterns exhibited within a given population (Hageman and Fahselt 1990).

Possible advantages to the availability of several forms of an enzyme in one population may be illustrated by considering the activity of the enzyme, ascorbate peroxidase, in relation to ethylene. Ethylene, produced by lichens (Ott and Zwoch 1992; Ott *et al.* 2000*a*) as well as other organisms, almost totally inhibits one form of this cytosolic peroxidase (Caldwell *et al.* 1998) and, because this enzyme is required for scavenging potentially dangerous peroxide, another form of it would be required for normal functioning when ethylene is present. As ethylene concentrations are higher in growing or marginal parts of the thallus, it could be that different forms of peroxidase are advantageous in different thallus regions.

13.8   Variation of secondary chemistry within populations

The lichen commitment to secondary chemistry (Chapter 7) is consistent with the production of secondary molecules by long-lived or slow-growing species (Coley 1988) in severe environments where resources are in limited supply (Lawrey 1986). Secondary products, evolved through refinement of basic biosynthetic pathways of the mycobiont (C. Culberson 1986), and those in a thallus or population are often structurally related derivatives of phenolic carboxylic acid derivatives. However, compounds may also be synthesized along diverse pathways, as in some *Cladonia* spp. which, although morphologically uniform, produce secondary substances belonging to the usnic acid complex, the psoromic acid complex, as well as the rangiformic acid complex (Ruoss 1987). Variation in secondary products within a population growing under heterogeneous conditions (e.g. Rundel 1969) may reflect a physiological or an evolutionary response to environment; however, coexistence of diverse chemistries under uniform conditions, for example of light, substrate, and moisture, is consistent with the status of lichen substances as genetically determined products of the mycobiont (C. Culberson *et al.* 1988). Chemically distinct ecotypes are in some cases characterized by different geographic ranges and in areas where these overlap a variety of chemotypes may intermingle (C. Culberson *et al.* 1985).

Secondary products are numerous and may be present in high concentrations, often 5% of the thallus by mass, or sometimes as much as 10–20%. These substances must therefore be of considerable importance to a lichen. Nevertheless, quantitative differences in secondary products have consistently been found among thalli in all populations studied. Phenolic concentrations in *Cladonia alpestris* (Fahselt 1984; Fig. 13.6) illustrate the idea of Hestmark (2000) that few thalli in a population exhibit the mean state of a given character and most show instead a condition deviating to some degree. Densitometry of thin layer chromatography plates reveals significant differences in the concentrations of

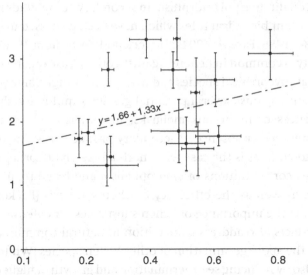

Fig. 13.6 Variation in the quantities of usnic acid and perlatolic acid in *Cladonia alpestris* within one population (mg/ml). Some thalli have double or quadruple the concentrations of others and seem to represent a range of approaches to resource allocation.

gyrophoric acid between thalli of *Umbilicaria muhlenbergii* growing only a few meters apart. Concentrations of polyphenolics may also differ significantly among nearby colonies of *Xanthoparmelia cumberlandia*, among clumps of *Cladonia turgida* in one small collecting area, and in thalli of *Lasallia papulosa* sometimes just centimeters apart. The amounts of secondary products in populations of *Tephromela atra* also vary, as is evident from the large standard errors around mean amounts of secondary products (Hesbacher *et al.* 1996). Intrapopulational differences in levels of polyphenolic acids are apparently long-standing, as a wide range of concentrations was documented by high performance liquid chromatography in both subfossil and recent populations of Umbilicariaceae (Fahselt and Alstrup 1997).

The total absence of a substance that is typical for a population or species, or at least the inability to detect it, may be viewed as an extreme expression of quantitative variation, and usnic-acid-deficient thalli occurring rarely in lichen populations provide one example. Such strains have also been found in *Cladonia* spp. and in *Arctoparmelia centrifuga* (Clayden 1992). Similarly, one chemotype of *Ramalina siliquosa* is unable to produce depsidones, although these compounds are characteristic of related taxa (W. Culberson *et al.* 1993). Molecular explanations to account for such deficiencies could include down-regulation of required synthetic enzymes or debilitating base alterations in corresponding coding regions. Usnic acid deficiency, for example, might reflect activity of the enzyme required to cyclize a 9-carbon polyketide into the usnic acid precursor.

Quantitative and qualitative polymorphism in secondary compounds may be related to their important biological roles which have been discussed in several reviews (Lawrey 1984, 1986; Fahselt 1994*a*; Huneck and Yoshimura 1996). First, antimicrobial activity is common in lichens, against both disease organisms and soil microbes, such as mycorrhizal species and wood decay fungi. Some provide broad-spectrum defense against several potential grazers simultaneously, with the feeding preferences of numerous mollusks, mammals, and arthropods influenced by the presence or absence of secondary products. Feeding may be focused on the photobiont, as is the case with moth larvae on *Cladonia pocillum*. Such activity reduces concentrations of chlorophylls *a* and *b* and total carotenoids in the thallus as well as the efficiency of photosynthesis (Baćkor *et al.* 2003), thus illustrating the importance of lichen substances for defense.

Polyphenolic products also address competition in natural communities by severely decreasing the spore germination of other lichen species (Whiton and Lawrey 1984), and also by reducing seed germination and growth of higher plant seedlings. These substances are thus potentially capable of securing a lichen advantage in vegetational communities, and on mine tailings it appeared they may in fact even preclude establishment of successional tree species (Goldner *et al.* 1986). Such allelopathic effects may be indirect, mediated through interactions with required mycorrhizae, and would be most effective at times of seedling recruitment. Another contribution of lichen products is that they help to seal the outer surface of mycobiont hyphae and to restrict movement of water and solutes in the thallus to the apoplastic continuum inside hyphal walls (Chapter 3). This permits intercellular spaces to be maintained dry for free exchange of metabolically important gases (Honegger 1991*a*; Armaleo 1993).

In addition polyphenolics, located on or near the upper surface of the lichen thallus, are well positioned to intercept both visible and ultraviolet light and many strongly absorb solar radiation. Although high UV irradiance degrades these compounds (Swanson and Fahselt 1997; Begora and Fahselt 2001) little protection may be required by some lichens against UV because the cortex itself permits little transmittance even after polyols are removed with acetone (Solhaug and Gauslaa 2004*b*). Parietin, however, protects *Xanthoria parietina* against intense photosynthetically active radiation (Solhaug and Gauslaa 2004*b*).

Because carbon directed into secondary chemistry becomes unavailable for growth, reproduction, or cell maintenance, a variety of carbon deployment alternatives perhaps helps to ensure continuity of a population under a variety of circumstances. The utility of any particular chemotype probably depends upon the challenge of the moment. Thalli with copious defense compounds may outperform under some kinds of duress, while lower concentrations, or different chemistries, may have an advantage under other circumstances.

13.9   Physiological polymorphism within lichen populations

Cyanide generation is one polymorphic process assessed in lichen populations. While most populations of *Dermatocarpon* are uniformly cyanogenic, one third of those investigated included acyanogenic thalli (Bergman and Eginger 1990). A proportion of the population thus remains viable although thalli may lack the ability to produce, or release, hydrogen cyanide.

Growth rates differ as well among thalli in the same population. Mycobionts of *Cladonia cristatella* isolated by Ahmadjian (1967*a*) show different capacities for growth. Physiologically contrasting variants in populations of *Lecanora muralis* (Seaward 1976, 1982*a*) also have differing growth parameters. Two strains are conscripted into the population at different times of the year, with fast growers establishing in winter and those growing more slowly becoming incorporated in spring. The two strains are recruited also under different levels of atmospheric pollution. Clearly two growth strategies would provide more scope for the population than one.

In the past, the variation in natural populations which tended to confound physiological field investigations was regarded as random (Kershaw 1977). However, the rates of photosynthetic gas exchange in *Parmelia praesignis*, which differed greatly among thalli, exhibited a normal distribution such as is characteristic of polygenic traits (Link and Nash 1984*a*). This finding is not surprising because of the many enzymes, and gene loci, involved in carbon fixation. Maintenance of photosynthetic variants in a population probably increases the availability of an advantageous strain for performing under a range of circumstances.

Physiological performance of lichen thalli varies within natural populations, but whether there is a genetic basis for these differences is usually not well understood. Several lichenological studies demonstrate the significance of size and age to water relations and photosynthesis, for example, but the relationship to genotype remains ill-defined (Hestmark 2000). It has been shown directly, however, that lichen populations growing in different environments may be characterized by genetic differences, as the cultured ascomycete of *Xanthoria parietina* growing on bark may be distinguished from that on rock substrates on the basis of spacer sequences in nuclear ribosomal DNA (Lindblom and Ekman 2006).

13.10   Genetic exchange

Morphological intermediates between distinctive extremes suggest genetic exchange, for example between two common morphological species

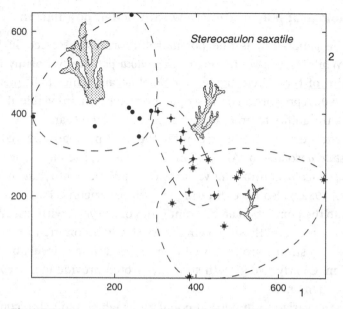

Fig. 13.7 Two common morphological forms of *Stereocaulon saxatile*, with an intermediate which intermingles in some populations. The possibility of genetic exchange between the two extreme forms is suggested by intermediate morphology as well as by ordination based on electrophoretic banding patterns of enzymes.

of *Aspicilia* (Ekman and Fröberg 1988), or as do intermediates between enzymatically distinctive morphotypes of *Stereocaulon saxatile* (e.g. Fahselt 1991; Fig. 13.7). The first direct demonstration of sexuality in lichens, however, was provided by C. Culberson *et al.* (1988), who analyzed sporelings from the same apothecia in the *Cladonia chlorophaea* complex in geographic areas where ranges of chemotypes overlap. Secondary chemistry of sister sporelings, after lichenization by a photobiont, in some cases differed from that of the parent podetium and instead resembled other chemotypes nearby. This provided proof of genetic exchange between members of the species complex, and allowed the inference that spermatia of one chemotype must have accessed trichogynes of another. Likewise, two chemotypes of *Ramalina cuspidata* have also been shown to interbreed because some ascospores differ from the haploid maternal parent and exhibit instead chemistry of another strain (W. Culberson and Culberson 1994). Marked differences in chemistry among sporelings derived from a single apothecium of *Lecanora dispersa* (Leuckert *et al.* 1990) likewise suggest the involvement of genetically different nucleii in spore production.

Mycobiont cultures of single spores from the same ascus of *Cladonia cristatella* differ physiologically as well, and vary in growth rates, shape, and aggression toward potential photobionts (Ahmadjian 1967*a*); the same single-spore cultures exhibit different electrophoretic patterns of enzyme forms (Fahselt 1987).

DNA from three other species of *Cladonia* shown by RAPD polymerase chain reaction to differ genetically in spore cultures from the same podetium is evidence of heterothallism and probable outbreeding (Seymour *et al.* 2005a). Similarities in DNA between different clades in the genus *Letharia* also provide evidence of interbreeding, because base sequences at several independent loci could only be explained by genetic exchange (Kroken and Taylor 2001).

In contrast, most members of the *Ramalina siliquosa* complex breed true in quadrats where several chemical races are mixed (W. Culberson and Culberson 1994). Genetic uniformity of RAPD markers in sibling spores in both *Graphis scripta* and *Ochrolechia parella* is strong evidence of self-fertilization (Murtagh *et al.* 2000). The extremely low polymorphism within sets of sibling spores of *Xanthoria parietina* indicates that this lichen may also be selfing (Honegger *et al.* 2004). While limits to gene flow in lichens are poorly understood, barriers could be chromosomal or genic and even the timing of gamete production could prevent genetic exchange. Another possibility is that, although spores are subject to long-distance wind dispersal, occasional disseminules from outlying populations are simply overwhelmed by an abundance of those produced locally (W. Culberson and Culberson 1994).

Barriers to gene exchange foster even greater differences between populations than within and differences between populations are consistent with impediments to interbreeding. For example, sequences of ITS regions and partial IGS regions in DNA of *Xanthoria parietina* (Lindblom and Ekman 2006) reveal more variation between than within collection sites, and the same is shown by DNA fingerprints of randomly amplified sequences of both *Xanthoria elegans* and *Graphis scripta* (Dyer *et al.* 2001). Multiple enzyme forms of *Umbilicaria muhlenbergii* (Hageman 1989; Fig. 13.8) are also more similar among thalli from each of the sampling sites than between sites. Restrictions to interbreeding may help to preserve locally adapted gene complexes, because introduction of genotypes from widely different populations may produce outbreeding depression, especially in later generation hybrids (e.g. Edmands 1999; Montalvo and Ellstrand 2001).

In some lichens most gene exchange occurs over surprisingly short distances, sometimes even less than 1 m (DePriest 1993a; W. Culberson and Culberson 1994), indicating that a fairly local concept of breeding population may be appropriate for at least some lichens. Gene migration may be restricted even within one collection site, as there are striking differences in secondary chemistry among mats of *Cladonia chlorophaea* in the southern Appalachians, USA (DePriest 1993a). On a wider scale, concentrations of secondary products differ significantly among some sampling sites (e.g. Hesbacher *et al.* 1996). Similarly, although four phylogenetically distinctive clades of *Letharia* are sympatric, one

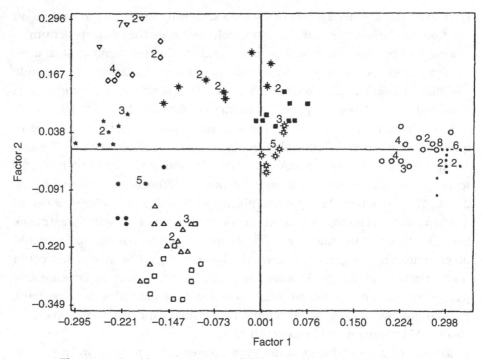

Fig. 13.8   Graphic representation of differences in electrophoretically produced banding patterns of eight enzymes in samples from ten *Umbilicaria muhlenbergii* sites in central Ontario, Canada. Enzyme variation within sampling sites is illustrated on a plot of the first two axes produced by principal components analysis. Closer affinities are evident between thalli within populations than between. Each symbol represents samples from a different collecting site or population.

clade, *L. rugosa*, has alleles at several loci which are unique and not shared with the others (Kroken and Taylor 2001). Also, the prevailing strains of the *Asterochloris* group of the photobiont *Trebouxia* in *Cladonia subtenuis* vary depending upon the geographic region (Yahr *et al.* 2004).

Are there physical barriers to gene flow in lichens? Alleles restricted to either coastal or inland populations of *Lobaria pulmonaria* in British Columbia, Canada, suggest barriers to gene exchange (Walser *et al.* 2005), which in this case appear to be geographical. However, neither the degree of physical separation nor orientation with prevailing winds explains all patterns of interpopulation variation; for example, multiple enzyme forms found in different populations of *Umbilicaria* (Hageman and Fahselt 1992) cannot be strictly explained by either. There is significantly reduced gene flow among the northernmost of four populations studied in *Cladonia subcervicornis* and the other three. While ascospores are often viewed as long-distance propagules it may not be the case for this lichen (Printzen and Ekman 2003). Distance, however, is not always an effective barrier, because two clades of *Letharia*, *L. rugosa* and *L. vulpina*, interbreed in spite of widely different geographic ranges (Kroken and Taylor 2001).

13.11   Recent evolution

Cladograms bespeak of past evolutionary diversification among lichen groups, but evolutionary changes appear to be ongoing, with taxonomically difficult species complexes possibly reflecting flux or transition. Sequencing of several loci has revealed genetically distinct entities in *Letharia*, which, although not yet morphologically distinct, may be on the road to attaining species status (Kroken and Taylor 2001). The extreme mobility of introns in populations of both *Cladonia chlorophaea* (DePriest 1993a) and *Cladonia subtenuis* (Beard and DePriest 1996) also suggests continuing evolution at the molecular level.

Some recent evolutionary changes have been documented specifically in relation to human activity. For example, certain types of disturbance may be associated with decreased polymorphism. In *Lobaria pulmonaria*, the level of variation did not change due to logging for charcoal production, but polymorphism was less in logged and burned forests than in relatively undisturbed sites (Werth et al. 2005) possibly due to reduced frequency of deciduous host trees. Likewise, the variability of enzyme forms in *Stereocaulon saxatile* was lower in close proximity to mines emitting particulates containing metals and radionucleides (Fahselt et al. 1997; Fig. 13.9). Effects on genotypic diversity will probably depend on the type of perturbation and its severity, as well as on the challenge it presents to particular lichen species.

Because exposure to unstable radioactive nuclei increases mutation rates, the net reduction in enzyme diversity in *Stereocaulon saxatile* near uranium mines (Fahselt et al. 1997) was unexpected. However, such a decrease in enzyme diversity, as compared with uncontaminated sites, would be understandable if genotypes intolerant to mine dust were eliminated by selection. *Parmelia sulcata* was similar in that the number of amplification products of the rRNA gene cluster was lower in sites where it had previously encountered sulfur dioxide than in populations with no history of exposure (Crespo et al. 1999); the genotypes that recolonized successfully evidently represented only a subset of those in the original population. However, enzyme polymorphism in *Umbilicaria mammulata* sampled along mine-site transects showed no net change in polymorphism under exposure to mine dust, so this lichen may have been less susceptible and selection less stringent.

Lecanora muralis is an example of an urban lichen which has evolved resistance to elevated levels of pollutants such as sulfur dioxide (Seaward 1976, 1982a). In this species a resistant superrace, with modified morphological,

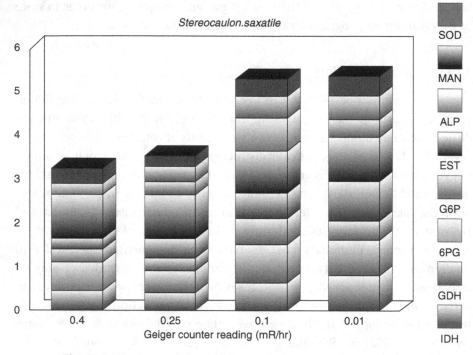

Fig. 13.9 Enzyme variation within stands of *Stereocaulon saxatile* at differing distances from a mine producing dust involving uranium and other heavy metals. Variability is expressed as number of different electrophoretic patterns and is partitioned according to the contributions of eight enzyme systems stacked in the order shown.

physiological and ecological traits, seems to have developed from more rural stock and finally achieved the equivalent of ecotypic status. Distinctive enzyme banding patterns in *Cladonia mitis* growing near geothermal vents (Fahselt 1992) perhaps reflect adjustments to sulfurous conditions.

Photobionts are particularly sensitive to pollution, but *Trebouxia jamesii* appears to be adapted to substrates containing metals, as it occurs in all nine species of lichens established on rocks with high metal concentrations (Beck 1999). These samples have physiological characteristics not seen elsewhere. One isolated algal strain was even more tolerant than the others and could survive not only field concentrations of metals, but also much higher levels tested in liquid culture. Similarly, *T. erici*, the photobiont of *Cetraria islandica*, evolved tolerance for copper and cobalt on mine-spoil heaps in Slovakia. Unlike wild populations with no history of metal contamination, populations in contaminated sites developed the ability to maintain the integrity of chlorophyll *a* in spite of metal exposure (Baćkor and Zetikova 2003). To test whether selection could produce tolerance in normal populations, wild type *T. erici* was cultured axenically and subjected to gradually increasing concentrations of copper over a period of three years. After this time a tolerant strain was in fact produced, and tolerance was maintained through several further generations (Baćkor and Váczi 2002). The capacity for a lichen to reconcile with high metal contents must be harbored in the wild type gene pool and then entrenched through selection during the course of prolonged exposure.

In natural communities of free-living fungi, the species composition shifts under exposure to experimentally enhanced UV-B, according to the sensitivities of individual species (Newsham *et al.* 1997; Duguay and Klironomos 2000). Correspondingly, the relative importance of lichen genotypes probably changes also in response to physically and chemically altered environments. Over time directional selection presumably produces resistance in more and more evolutionary lines. For example, only a few species have so far been able to establish on 70-year-old mine tailings near Sudbury, Canada (Baćkor and Fahselt 2003), while many more have become accommodated to 400-year-old mine heaps in Europe (Nash 1989; Huneck *et al.* 1990).

Although not yet reported in lichens, DNA may also evolve in relation to its own molecular stability. For example, in extreme thermophiles a high percentage of C–G pairs, with three rather than two hydrogen bonds across the DNA double helix, is considered an adaptation to a stress such as high temperature (Brock 1978).

It might be assumed that important functional links between photobionts and mycobionts, and possibly even other associates, would mean that these partners evolve simultaneously. However, this is certainly not always the case because sequencing in the Cladoniaceae has shown that the fungus and alga have evolved quite independently of each other (Piercey-Normore and DePriest 2001). This can be understood in view of low specificity between some partners (Kroken and Taylor 2000; Piercey-Normore and DePriest 2001).

14

Environmental role of lichens

M. R. D. SEAWARD

Although earlier literatures provided an insight into the uniqueness of lichens, it gave little hint of the major role these apparently insignificant organisms play in the shaping of the physical and biological environment of our planet and their importance in maintaining its equilibrium. Their role as biological weathering agents in the development of soils, for example, was formerly considered in a geological context only, but recent research has shown that these organisms are capable of biodeteriorating stone substrates within a relatively short timescale. Information is now available to demonstrate that lichens can often contribute substantial biomass and support a high biodiversity of micro- and macroorganisms, creating complex food webs and adding significantly to energy flow (Chapter 10) and mineral cycling (Chapter 12).

The disappearance of lichens, due to many aspects of human interference in the natural world, has therefore led inexorably to environmental impoverishment. Lichens are natural sensors of our changing environment: the sensitivity of particular lichen species and assemblages to a very broad spectrum of environmental conditions, both natural and unnatural, is widely appreciated. Lichens are therefore used increasingly in evaluating threatened habitats, in environmental impact assessments, and in monitoring environmental perturbations, particularly those resulting from a disturbingly large and growing number of chemical pollutants (Chapter 15). Nevertheless, lichens undoubtedly represent one of the most successful forms of symbiosis in nature. They are to be found worldwide, exploiting not only all manner of natural, usually stable, micro- and macroenvironments, but in many cases adapting to extreme conditions, including some brought about by human disturbance. This chapter deals with ecological aspects of lichen biology with an emphasis on dispersal,

Lichen Biology, ed. Thomas H. Nash III. Published by Cambridge University Press.
© Cambridge University Press.

establishment, pedogenesis, biodeterioration, succession, community structure, animal interactions, conservation, and environmental monitoring.

14.1   Dispersal

Although the discharge, dispersal and establishment of reproductive propagules in fungi have been extensively studied, these processes in lichenized fungi are still underresearched (Bailey 1976). The development, form, and role of the various reproductive structures found in lichens are described and reviewed in Chapter 4. These are basically ascomata and conidiomata which disperse only the mycobiont of a lichen, and vegetative propagules which are capable of dispersing both mycobiont hyphae and photobiont cells simultaneously.

Ascus discharge is probably similar to that observed in other ascomycetes. In most mature lichens, ascospores are discharged perennially, but wet weather often favors their liberation. However, in some species discharge is affected by drying and consequent rupturing of asci. There are conflicting views on seasonality of discharge, but lichens most probably release ascospores whenever asci are under stress due to a shift in water balance. Although light has no apparent effect on ascospore discharge in most lichens, diurnal periodicity has been observed, with some species tending to increase discharge under light conditions. Temperature extremes and sulfur dioxide tend to depress discharge, but higher humidity levels probably increase it, as does precipitation. many species releasing ascospores within one hour of moistening; this is not the case with certain types of vegetative propagule liberation.

Transportation of discharged ascospores is more easily facilitated by structures such as stalked ascomata and erect podetia which raise them above the thallus surface. Generally, however, ascospores, varying in size from c. 1 to 500 μm, are forcibly ejected, usually to a distance of 3 to 50 mm from the fruiting body, depending upon size and number of ascospores clustered in a projectile; this is sufficient for them to be dispersed by one or more agents (wind, water, and/or animals). The wind speeds necessary for effective ascospore liberation depend on the form of the ascomata and the size and sculpturing of the ascospores. Those in mature perithecia (e.g. *Cyphelium inquinans*) require speeds in excess of $5\,\mathrm{m\ s^{-1}}$; such speeds would be equally effective for soredia dislodgement, as determined by experimentation (Bailey 1976). Wind speeds of 30 to $35\,\mathrm{m\ s^{-1}}$, commonly experienced for example in polar landscapes supporting lichens, would transport them long distances. Small and lightweight ascospores are obviously best suited to long-range transport by wind, but farinose soredia and small thallus fragments are also capable of being dispersed over large distances (Bailey 1976).

Water can both liberate and transport ascospores via splash droplets, surface trickles, streams, and rivers in much the same way as for vegetative propagules, but, to date, this area is understudied. It is unlikely that flowing water is important in carrying ascospores to a suitable habitat except in the case of aquatic species, but lichen fragments have been found in arctic drift ice and meltwater (Thomson 1972).

Vegetative propagation, involving whole thalli or fragments thereof or specialized structures (e.g. soralia or isidia), is a common method of reproduction in lichens. Transport is facilitated by invertebrates and to a lesser extent by vertebrates (mainly via mammalian fur and bird feathers). Isidia form discrete outgrowths from the cortex and are usually constricted at their bases to facilitate easy detatchment from the thallus by mechanical means. A variety of other vegetative propagules containing one or more photobiont cells and attendant fungi are also found, mainly as scale- or leaf-like segments or chain-like filaments or buds.

Erratic and vagrant (or vagant) species (Weber 1977; Rosentreter 1993; Pérez 1994) possess the remarkable ability to re-establish themselves independently elsewhere by the detachment of major portions or even whole thalli. This is brought about by frost, insolation, or animals acting on a parent body often established on an unstable substrate, as found in arid areas. They also have the potential for scattering ascospores (and indeed whole ascomata), and other propagules and thallus fragments whilst mobile. Alternatively, vegetative reproduction "by growth" occurs in some foliose, squamulose, and lobate crustose lichens where radially developing thalli die out at the center, leaving only an outer circle of outwardly-growing lobes. Separate thalli occasionally regain more or less their radial form some distance from the original thallus. Circles up to 1 m in diameter have been recorded for saxicolous species; this may also occur in terricolous species, but it is less easily detected. Certainly, many species inhabit large areas for a considerably longer time than could be expected from the life-span of a single thallus; this fact may be responsible for some exaggerated ideas about the age of individual thalli.

Soredia, commonly found in nature independent of the parent thallus, have proved highly amenable to in situ field transplant and in vitro investigations of the thalli producing them. They have been recorded from all media and a wide variety of substrates. Such soredia have been detected not only in the air spora worldwide, but also in snowfalls, glacial ice, etc. in remote regions (Bailey 1976). Laboratory studies have shown that soredia can be detached from dry thallus surfaces by wind speeds as low as $2\,\mathrm{m\,s^{-1}}$, the effectiveness of propagule liberation being dependent upon the water status of the lichen (Armstrong 1992). Splash droplets are also effective in the local dispersal of soredia, particularly in the case of those contained in cup-shaped podetia, where they can be splashed to distances of 75 to 90 cm; droplets falling onto soredia borne in soralia on flat thallus surfaces are less effective (Bailey 1966). Soredia (despite their water repellence), isidia and other propagules, including thallus fragments, are liberated by rainwater and form downward streaks on vertical and inclined surfaces supporting lichens, such as tree trunks, rocks, and churchyard memorials. Most of these processes are only locally effective. However, Tapper (1976) demonstrated that soredia were capable of being dispersed from tree to tree up to 30 m from their source, presumably by wind action.

A wide variety of invertebrates, including springtails (Collembola), barklice (Psocoptera), lacewings (Neuroptera), and moths (Lepidoptera), feed off and shelter among lichen thalli. In so doing, they accidentally coat their bodies with propagules, particularly soredia. This is advantageous to the animal in terms of protective camouflage, but also aids propagule dispersal. Fragmentation of thalli by invertebrate grazing, which can be highly selective, will greatly assist dispersal. Although dispersal in most cases is local, wider propagation is potentially provided by flying insects. Some insect larvae have cases partially constructed out of lichen fragments; propagation occurs not only by animal movement, but also by ecdysis when the old, lichen-covered skins are shed. A few weevils (Coleoptera) actually have carapaces which facilitate the growth of lichens on them for protective crypsis. Mites (Acari), particularly oribatids (Cryptostigmata), are commonly to be found in association with lichens. One oribatid species in Papua New Guinea actually inhabits lichens and fungi growing on the backs of large, flightless weevils.

Slugs and snails (Mollusca) are known to disperse propagules, not only by fragmentation as a result of selective grazing, but also by transporting them via body slime or passing them out in feces (Peake and James 1967; McCarthy and Healy 1978). Maritime pyrenocarpous lichens are capable of establishing themselves, mainly as immersed perithecia, on shells of both living (and dead) mollusks, such as limpets, whose movements, albeit very slow, provide the possibility for limited ascospore dispersal and by their grazing action effect thallus fragmentation, enabling propagules to be transported to new sites by wave action.

Many vertebrates are also involved in lichen dispersal, such as the fragmenting of thalli by mammals trampling fruticose lichens or the transport of long-lived lichens capable of generating propagules on the carapaces of the Galapagos giant land tortoises. Birds and mammals aid dispersal when lichen propagules adhere to feathers, feet, skin or hair, or by collecting material for use as food or construction of nests. Long-distance lichen dispersal by birds may account for lichen biogeographical distributions (Chapter 16). Human activities can play a significant role in accidental introductions (e.g. lichens may be unintentionally transported to new habitats on unbarked timber, firewood, peat, or stone), leading in some cases to an increase in the number of cosmopolitan taxa.

There is great advantage in adopting asexual reproductive strategies: the production of large numbers of propagules with appropriate viable bionts can readily exploit new micro- and macroenvironments, but they lack the potential for genetic recombination and therefore all capability for further evolutionary development. Clearly both dispersal capacity and habitat/substrate suitability need to be considered (Section 14.2). Studies of old-growth forests in Sweden (Ockinger et al. 2005) and North America (Sillett et al. 2000) have demonstrated the relative importance of these two variables; in the case of the Swedish study, time–space analyses revealed that dispersal capacity, as determined, for example, by distances between similar trees newly and formerly colonized by Lobaria pulmonaria (mean being 35 m, maximum 75 m), was probably the most important factor in limiting the local distribution of this lichen, but habitat-quality factors may be important on a smaller scale.

14.2   Establishment

Efficient scattering of reproductive propagules in no way guarantees their successful establishment and development: suitable substrates and environmental conditions are necessary when they come to rest following one or more deposition processes involving sedimentation, gravitation, and washout by rain. Propagule impact on various surfaces leading to successful development of thalli is accounted for by efficient trapping, and indeed electrostatic charges generated by lichen ascospores have been suggested for this phenomenon in some species (Garrett 1972).

Undoubtedly, lichens are capable of colonizing a wide range of substrates (Brodo 1973), not only natural surfaces provided by rocks, soil, bark, decorticated wood, leaves, and animal carapaces, but also industrial materials (Brightman and Seaward 1977) such as plastic, rubber (Fig. 14.1), metals, and glass. Field experiments have shown the importance of surface texture and microtopography for the successful attachment and survival of propagules (Armstrong 1988). Although surface topography may provide a favorable environment for germination, it is likely that interspecific and intraspecific competition as a result of a build-up of propagules in such niches will be intense, their interactions determining the ultimate composition of the local lichen flora.

Soredial growth on a suitable substrate is quickly followed by thallus development, and in the case of some isidia, tissue differentiation may have already

Fig. 14.1   A diverse nitrophilous lichen flora developed on tractor tyres facilitated in the first instance by the presence of nutrient-enriched soil and dusts. (Photograph by V. John.)

developed prior to establishment. The subsequent processes involved in thallus organization and development are poorly understood, but work on crustose lichen communities that occur on the surfaces of leaves (foliicolous lichens) in humid tropical forests have provided some insight into how successful the various reproductive propagules and strategies are (Sanders 2001*a,b*), as well as details of the microscopic stages of lichen ontogeny and the patterns of subsequent development (Sanders and Lücking 2002).

Successful germination and establishment of the various types of lichen propagule are dependent not only on characteristics of the available substrate (e.g. texture, pH, nutrient status, microtopography), but also numerous other environmental factors, such as humidity and temperature (Barkman 1958; Brodo 1973; Topham 1977). In addition, competition with established lichens and other organisms for space, light, and nutrients by the newly developing propagule can be considerable; such competition may arise as a consequence of mechanical action, inhibition of gaseous exchange, reduction in light intensity, or allelopathic or toxic chemicals.

Lichens may allelopathically interact with other organisms; for example, inhibition of germination of moss spores and angiosperm seeds is demonstrated (Lawrey 1984). The halo effect created by lichens on algal colonization (Fig. 14.2), frequently found on horizontal stone surfaces, such as table tombstones, may be due to allelopathic inhibition; other microorganisms are similarly affected, extracts from certain lichens exerting antibacterial and antifungal action (Section 7.8). The possibility of intraspecific allelopathic effects determining the growth and spatial distribution of lichens has been postulated by Seaward

Fig. 14.2 Allelopathy in lichens as demonstrated here by *Hypogymnia tubulosa* on a table tombstone, its thallial chemistry creating an adjacent zone devoid of algae. (Photograph by M. R. D. Seaward.)

(1982*a*). All the above examples refer to mature thalli, but it is reasonable to suppose that actively germinating ascospores and other propagules, thallus initials, and propagated lobes and fragments are all capable of producing chemicals with similar inhibitory properties.

The factors which control the processes of development and growth (morphogenesis, Chapter 5) are poorly understood; for example, little is known of the growth rate of thallus initials, a phase crucial to both the establishment and competitive ability of the new thalli, or how long a period must pass before lichens establish themselves on a particular substrate, or how long before this can actually be detected, or how successful are the thallus initials. Recent work on lichen recolonization has shown that careful microscopic examination of substrates revealed widespread, abundant early stages of thallus development, but their survival is often tenuous, particularly in stressful environments (Seaward 1997).

14.3   Pedogenesis and biodeterioration

It is highly probable that lichens were early colonizers of terrestrial habitats on our planet, coping with harsh environments and contributing to the evolution of atmospheres more suited to a much wider variety of life forms. However, their major long-term role has been as biological weathering agents, with a pedogenic action which is both physical and chemical in nature (Syers and Iskander 1973; Jones 1988). Despite early controversy concerning the pedogenic significance of lichens, their effectiveness in the biodeterioration of rocks has been clearly demonstrated by recent research, which has revealed that substantial quantities of the substrate can be degraded even over relatively short periods of time. Furthermore, lichens have the capacity to accumulate elements such as nitrogen, phosphorus, and sulfur, thereby increasing the potential bioavailability of these elements to successive life forms which may replace lichens during soil development. Organic material derived from lichen decomposition, together with detached particles of the substrate and atmospherically derived dusts trapped by thalli, all contribute to the development of soils. However, in a few cases, lichens have been considered as being bioprotective, the thallus acting as an umbrella to protect the underlying rock from rain, pollutants, salt deposition, and insolation (Carter and Viles 2003, 2005).

The weathering action of saxicolous lichens can be physical, due to penetration by rhizinae and expansion and contraction of thalli, and/or chemical, due to carbon dioxide, oxalic acid, and the complexing action of lichen substances. The latter have a low but significant solubility in water, forming soluble metal complexes under laboratory conditions when they react with minerals and rocks, particularly limestone (Syers and Iskander 1973). Oxalic acid, formerly considered of minor importance in the biodeterioration process, has been proved otherwise. The widespread occurrence of metal oxalates, particularly calcium oxalate, in lichens (and in nature generally), the nature of the thallus–substrate interface, and the chemical disruption of the substrate are significant components of the weathering process.

Of considerable interest, as exemplified by the burgeoning literature on the subject, is the effect of lichens on stone surfaces, particularly ancient monuments (St. Clair and Seaward 2004). Detailed spectroscopic studies have demonstrated the highly destructive properties of calcium oxalate produced by lichen thalli (Fig. 14.3), particular attention being directed towards the dramatic effects caused by the action of certain aggressive lichen species on historic monuments and other works of art, where biodeterioration processes have been shown to be devastatingly destructive within a surprisingly short timescale (Seaward and Edwards 1997). Such action on natural substrates is clearly of significance in a pedogenetic context, since lichens are usually regarded as weathering agents on a geological timescale. Biodeterioration studies have revealed that calcium oxalate encrustations of some aggressive lichens can be produced at the thallus–substrate interface up to depths of almost 2 mm within a period of less than 12 years, and in one case, for a typical thallus with a diameter of 1 cm, it has been calculated that 135 mg of calcium carbonate is converted into calcium oxalate monohydrate at the interface. On Italian Renaissance frescoes, 1 m^2 of fresco and underlying plaster has probably been converted into more than 1 kg of calcium oxalate; furthermore, with the incorporation of calcite and gypsum

Fig. 14.3 *Dirina massiliensis* forma *sorediata* on calcareous church wall to show extensive development of calcium oxalate at the thallus–substrate interface. (Photograph by M. R. D. Seaward.)

into the thallus encrustation, it is likely that more than four times this amount of the underlying substrate has been chemically and physically disturbed (Seaward and Edwards 1997).

The impact of lichen weathering of rocks on a global scale has been, and continues to be, important in terms of climatic consequences and the habitability of our planet: their disappearance from particular ecosystems would be critical over major areas. It has been noted that if today's weathering were to take place under completely abiotic conditions, dramatic increases in global temperature would result (Schwartzman and Volk 1989).

14.4   Community structure

The fundamental ecological needs of particular species are reflected in broad classifications based on their substrates (saxicolous, corticolous, lignicolous, muscicolous, terricolous, foliicolous), although many species transcend these boundaries, and on their pH requirements (acidophilous, neutrophilous, calciphilous) and on ambient nutrient status (oligotrophic, eutrophic, hypertrophic) (Wirth 2001; Seaward and Coppins 2004). Lichens display a remarkable resilience, being able to establish themselves over much of the world's terrestrial surface, often in extreme environments inhospitable to most other forms of life, but can also be highly susceptible to disturbance, particularly that generated by a very wide range of human activities. Although many communities where lichens play a significant or indeed dominant role have been studied in some detail, many remote and currently inaccessible parts (polar, arid, and tropical regions) of the world undoubtedly harbor undiscovered lichenological riches, as exemplified, for example, by biodiversity counts (usually tentative) of tropical countries and their rain forests; in one case, a single tree was found to support no fewer than 175 species (Aptroot 2001).

Lichen species and assemblages are faithful to particular habitats and as such make ideal environmental indicators. Their spatial and temporal patterns are ecologically delimited by one or more factors. Lichen communities are often species rich and densely crowded with complex biotic interactions. In terms of determining the ecological processes in a community, spatial patterns need to be determined for habitats and ecosystems ranging, for example, from an individual rock to scree covering an entire mountainside. Each scale gives a different perspective: at a larger scale lichen distributions are correlated with the physical features, such as slope and aspect, while at a much smaller scale a rock face is composed of many microhabitats, as reflected in the elaborate pattern of the different lichen species (John and Dale 1991).

Zonal distributions of marine (littoral) and maritime (supralittoral) lichens on rocky shores in temperate and subtemperate regions have been widely studied and documented, their eye-catching color delimitations clearly manifesting the major factors involved (Fletcher 1980). Less is known about the zonation of lichens on seashores in both colder and warmer waters, but in the latter case they are usually replaced near lower tidal lines by cyanobacteria and, with increasing distance from the sea, dictated more by water and nutrients derived from land sources. The composition of the flora is naturally related to aspect, and thereby exposure to wind and wave action, and to the geology of the underlying rocks, which may affect the lichen pattern, but the overriding zonation depends on the capacity of the different species to withstand longer or shorter periods of immersion in seawater, or greater or lesser amounts of seawater spray (Feige 1973; Nash and Lange 1988). Local differences in species composition are usually due to bird activity (ornithocoprophily) and agricultural practices in adjacent land.

Lichens of freshwater habitats have until recently been poorly studied. Since many are submerged for only part of the year, many should be considered as amphibious (Santesson 1939); they belong to relatively few taxonomic groups (mainly Verrucariaceae and Collemataceae), are mostly pyrenocarps, and their phycobionts are normally cyanobacteria or coccoid green algae which naturally occur in a free-living state in freshwater (Aptroot and Seaward 2003); but a few, such as *Peltigera hydrothyria* (= *Hydrothyria venosa*), are apparently submerged throughout the year. Freshwater lichens do not occur in standing water, only in habitats subjected to running water or wave action (lake shores, rivers, streams, springs, waterfalls, weirs, and rock run-off). More than 100 species have been described mainly from these habitats in temperate regions, but new species are continually being found in subtropical and tropical regions. Usually only a few species are present in a particular habitat. Freshwater lichens have been used as indicators of water quality (Davis *et al.* 2000), their presence emphasizing the need for conserving particular water courses and habitats. They also have pronounced vertical zonations (upper terrestrial, fluvial, fluvial mesic, and submerged zones). These zones on rock surfaces or lichen demarcations on trees have monitoring value in terms of determining natural hydrological disturbances and those as a result of human activities (Section 13.11) (Gregory 1975; Hale 1984; Beckelhimer and Weaks 1986; Timoney and Marsh 2004), and also most probably global warming.

The phytosociological approach for analysing plant communities, with its own elaborate nomenclature to define alliances, noda, etc., was developed in continental Europe in the 1920s. Although not practiced worldwide, it has been effectively employed to demonstrate the wide variety and complexity of lichen communities (Barkman 1958; Wirth 1972; James *et al.* 1977; Roux 1981); specific lichen groupings can be used to define (qualitatively, and often quantitatively) their contribution to a habitat (e.g. from a pebble to a rocky mountain outcrop) or to a major landscape feature (e.g. from a small woodland to a tropical rain forest). Such sampling analyses have proved important in both small- and large-scale environmental surveys, particularly for evaluating a wide range of natural perturbations and those as a result of human activities, and for conservation purposes. The spectrum of lichen communities ranges from those almost or wholly composed of a single lichen species (monospecific communities, e.g. in polluted environments) to those of such complexity that individual thalli or species are exploiting niches influenced by a variety of microenvironmental factors. Naturally, the size of the sample must be representative of the ecological unit under investigation, but all too often microtopographical features, which may or may not be influential on lichen pattern and distribution, are either overlooked or overemphasized.

An alternative to phytosociological investigations are multivariate analyses of community data (Gaugh 1982; Kershaw and Looney 1985; McCune and Grace 2002), from which graphed ordinations, in two or three dimensions, are derived; these often explain 50–60% or more of the overall variation due to multiple species sampled in multiple stands (different locations). Substantial ecological insight can be gained by relating environmental variables to the ordination axes. Many of the earlier techniques required statistical assumptions, such as multivariate normality, that are not often available from vegetation data, but the development of nonmetric multidimensional scaling (NMS), a nonparametric technique, does not make these assumptions (McCune and Grace 2002).

Vectors depicting relative importance of environmental variables can readily be superimposed on NMS ordinations (Jovan and McCune 2004).

The wide range of lichen communities delimited through phytosociological analyses and other means has determined not only the diversity, frequency, cover value, and biomass of the different species, but also the relative importance of the lichens relative to the other components of the community or ecosystem; lichen-rich systems are self-evident, but in many cases where the lichens contribute a relatively low biomass, they may nonetheless assume greater lichenological importance in terms of, for example, invertebrate associations (e.g. Wessels and Wessels 1991) and nutrient turnover/mineral cycling (Chapter 12).

Of increasing interest to lichenologists, as well as ecologists and environmental scientists, is the urban ecosystem. For almost two centuries, urban areas, mostly in the developed world, have been of lichenological interest purely in terms of their lack or paucity of lichens, air quality being identified as the limiting factor (Chapter 15), but the contributions of the many factors operating in a changing urban climate and complex landscape, although appreciated, were not fully evaluated. With the implementation of clean air policies in recent years, the urban environment is lichenologically unrecognizable: without the major limiting factor, a diverse landscape, in terms of habitats, substrates, and chemistry (mainly nutrient enrichment), is there to be exploited (Seaward 1997). Some cities could soon be on a par with lichen-dominated ecosystems, since an urban landscape of structures constructed from a variety of natural and manufactured building materials can provide suitable substrates for the establishment of a diverse lichen mosaic.

14.5   Succession

With the passage of time, different lichen species and assemblages are subsumed by other lichen cover as a consequence of a particular succession, in many cases eventually giving way to bryophytes and vascular plants unless interrupted by a controlling factor, more than likely induced by humans, that prevents the sequence. Although the ecology of the later phases of such successions have been widely studied and extensively cited in classic textbooks, changes in the development of the different lichen phases, spatially and temporally, have received less attention. However, sequential data are gradually being assembled to highlight the nature and role of lichens (and bryophytes), more particularly for epiphytic communities on the twigs and branches of specific host trees, showing how they adjust to the growth and changes in the bark characteristics of the host, to changes in microclimatic factors, such as moisture and light, and to competition for space (Stone 1989; Hilmo 1994; Ruchty *et al.* 2001). At any given time, usually at a mature phase in the development of a tree, a woodland, or a forest, consideration will be given to the vertical distribution of epiphytes from ground to uppermost canopy level (Eversman *et al.* 1987), including their diversity and composition on the bole, both vertically and according to aspect (Harris 1971). In view of many environmental variables affecting forest epiphytes, experiments to examine the effect of height in the canopy on the growth rates of transplanted lichens treated the vertical gradient as a complex suite of factors, simplified to one variable, namely canopy position (Antoine and McCune 2004). Much of the above refers to work undertaken on representative trees or in reasonably uniform tree stands; however, woodlands vary laterally as well as vertically and "hot spots" of lichen diversity are to be expected (Peterson and McCune 2003).

14.6   Ecosystem dynamics

Ecological work in recent years has been directed mainly towards the study of growth rates and colonization, ecophysiology, and air pollution monitoring. Much less is known of the part played by lichens in ecological dynamics and nutrient and energy budgets. In any study of ecological dynamics, it is necessary to determine the key components in order to reveal the relative importance of lichens for a particular ecosystem. Boreal coniferous forests, cold deserts, dune systems, hot arid and semi-arid lands, maritime rocks, high altitudes, and tropical rain forests contain examples of ecosystems where lichens often contribute a significant proportion of the biomass and/or biodiversity. It is estimated that 8–10% of terrestrial ecosystems are dominated by lichens (Larson 1987); furthermore, lichens are often abundant in particular strata of other ecosystems, such as terricolous mats in boreal forests or arboreal epiphytes in temperate rain forests.

Although it has proved difficult to estimate lichen biomass of tall forests, it is possible to do this by sampling litterfall, which, according to studies in the Cascade Range, USA (McCune 1994), has shown that the biomass of the lichen litter was about one-hundredth of that remaining on the trees. Lichen biomass on subalpine firs on Mount Baker, Washington, contributed *c.* 1750 kg ha^{-1} (Rhoades 1981). Other North American forest systems showed a wide variation in lichen biomass, and consequently mineral capital, which seldom accounted for more than 10% of the annual above-ground turnover of a particular element (Pike 1978). Detailed studies have shown, for instance, that the epiphyte *Ramalina menziesii* plays an important role in the annual turnover of biomass and macronutrients (Boucher and Nash 1990*b*); using destructive sampling techniques, they calculated the standing biomass of this lichen to be 706 kg ha^{-1}, representing 94% of the total epiphytic lichens by weight, and that it not only contributed 26.4% of the biomass but also 9.4% of the total litter biomass when compared with the blue oak on which it grew. Canopies supporting such pendulous epiphytes considerably increase their surface area, nutrient input being further enhanced by the scavenging nature of lichens which accumulate aerosols (including particulate trapping) and ions at exchange sites. Photographic, physical, and chemical techniques have been employed to determine the surface area of air-dried samples of R. menziesii collected from a blue oak stand in California where this lichen represents c. 75% (c. 500 kg ha^{-1}) of the total lichen biomass; the total of c. 667 kg ha^{-1} for all pendulous lichen epiphytes at this site is remarkably high when compared with the c. 900 kg ha^{-1} of the oak leaves. Tentative determinations of the surface area of R. menziesii alone for this stand range between 0.15 and 0.7 million m^2 ha^{-1} and for the site where this lichen is more abundant (i.e. 706 kg ha^{-1}), a figure between 0.21 and almost 1 million m^2 ha^{-1} has been calculated (Seaward 1996b). It should be noted that the latter figure represents 100 times the surface area of the ground on which the trees stand. The lower figures for the two sites are undoubtedly underestimates as the higher figures were derived from a more sophisticated technique which determined microdetails of thallus surfaces, including pores. Clearly, loss of such an extensive surface area, with a very high atmospheric scavenging capability, as a result of air pollution and by tree felling is bound to have significant repercussions, not only ecologically but also physiologically since the lichen biomass plays a key role in the water relations of such canopies (Knops et al. 1996; Section 12.9).

Naturally, the biomass and diversity of lichens vary according to a wide range of factors, both climatic and topographical at a micro- and macroenvironmental level, often defined by strata (as in the case of the epiphytic communities described above), zones, or gradients. However, although considerable data are available on these differences in the biodiversity, information on biomass is limited and where available, as in the case of increasing altitude (Wolf 1993), the lichen measurements tend to be combined with bryophytes; however, biomass values for epiphytic lichens have been calculated for various forest types (Wein and Speer 1975; Lang et al. 1980; Pike 1981; Rhoades 1983; Boucher and Nash 1990b). Lichen biomass data have also been assembled for treeless shrublands and bogs (Wein and Speer 1975), hot deserts (Nash et al. 1977; Nash and Moser 1982) and tundra zones, particularly in respect of reindeer and caribou studies; the biomass of nitrogen-fixing lichens and their fixation rates have also been determined from different habitats (Denison 1973; Crittenden 1975; Forman 1975; Becker et al. 1977; Chapter 11).

Despite such detailed studies, our knowledge of the total processes involved is too fragmentary to determine complete energy-flow diagrams and budgets. For an integrated model of the dynamics of an ecosystem containing a significant lichen component, response patterns and typical parameter values for a full range of abiotic and biotic factors involved need to be established. At best, energy-flow and mineral cycling calculations can only be an approximation. Minerals taken up by lichens may subsequently reach other components of the surrounding biosphere via litterfall, leaching, bacterial incorporation, or non-cellular particle formation. Data on lichen decomposition and consequent release of minerals are scarce; it would appear, however, that the rate of lichen decomposition is rapid (half-lives of c. 90–760 days), but generally less than that of vascular plants. Interestingly, common saprophytic agencies rarely appear to be involved in the decomposition process, perhaps due to the inhibiting effect of lichen substances. The abundance and diversity of invertebrates involved in the process can be considerable.

14.7   Animal and lichen interactions

14.7.1   Invertebrates

Lichen associations play an important role in soil fertility and ecological energetics, and their value in energy budgeting, modeling, etc. should therefore not be underestimated. Energetics of lichen faunas are poorly understood due to the wide variety of data from disparate sources: information on ingestion, assimilation, egestion, respiration, growth and death rates, numbers, and biomass needs careful coordination and interpretation (Engelmann 1966; Gerson and Seaward 1977; Seaward 1988). Of special interest to our understanding of microenvironmental processes is the pedogenetic study on the lichenophagous larvae of a South African bagworm by Wessels and Wessels (1991). Clarens sandstone is weathered by endolithic lichens which dissolve the cementing material, thereby loosening quartz crystals which are then used by the larvae to construct the bags in which they live. It is estimated that the larvae contribute 4.4 kg of weathered sandstone per hectare per year to the area; furthermore, the larvae utilize the lichens as a food source, the resulting feces providing 200 g of organic material per hectare per year and contributing to mineral cycling (Fig. 14.4).

Quantitative data on numbers and weight of the various faunal groups per unit area of lichen thallus, thalli, or community on which to base ecological energetic studies have been derived mainly from microhabitat studies, but more attention is now being focused on environments where lichens form a

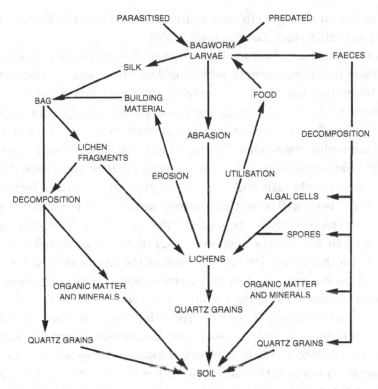

Fig. 14.4 Role of bagworm larvae and lichens in microecological processes associated with Clarens sandstone outcrops in South Africa. (From Wessels and Wessels 1991, with permission.)

major component of the flora, such as littoral zones, arctic and antarctic ecosystems, and certain types of woodland. There is also a lack of factual information relating to food webs and nutrient recycling. The effects of environmental pollutants need more detailed investigation, since air pollution adversely affects lichens, and hence their primary consumers and in turn their predators. A significant uptake of heavy metals, pesticides, and radio-elements by invertebrate grazers via lichens should be expected. The effects of pollutant accumulation on the palatability of lichens, ecosystem dynamics studies to determine residence times of toxins and rate of detoxification, and decomposition are but a few of the many aspects of lichen–invertebrate associations in need of further study. Despite the presence of antiherbivore secondary substances in numerous lichens (Rundel 1978; Section 7.9), the biodiversity and abundance of invertebrates which feed on, shelter in, and are camouflaged by lichens are enormous. Calculations of their biomass and further study of their role in mineral cycling and energy flow will reveal how highly significant invertebrates are to many ecosystems, although a recent study has shown that growth of mature lobes of *Lobaria pulmonaria* is not inhibited by snail grazing (Gauslaa *et al.* 2006*a*).

As well as providing food for invertebrates, lichens provide a milieu that enables them to survive extended periods of desiccation within a thallus micro-habitat, where they can also be protected against predation either by physical concealment or by blending with their background through camouflage or mimicry. Camouflage is effected by these animals in various ways, including covering themselves with lichen fragments and soredia, or in some cases even providing a foundation on which an entire lichen will establish itself. Mimicry has been achieved through evolution of adaptive coloration and body shape. Coloration has been well researched, mainly with respect to industrial melan-ism and more particularly in the study of the ascendancy of melanic morphs of various moth species, the peppered moth (*Biston betularia*) being the best-known exemplar (Majerus 1998). As a result of the rise in air pollution levels in Britain from the early nineteenth century onwards, epiphytic lichens disap-peared over a wide area of the country, and thereby the protective crypsis for light-colored morphs of *B. betularia*. From the 1950s, the rise to dominance of one or two lichen and algal species creating monotonous verdant backgrounds appeared to be almost equally advantageous/disadvantageous to both light- and dark-colored morphs. The ratio between the two morphs is being carefully monitored following the implementation of clean air policies (Cook 2003), since the lichen–algal monoculture, frequently susceptible to fungal infections, e.g. *Athelia arachnoidea*, followed by an initially slow but more recently very significant improvement in the lichen flora, and possible changes to the avian fauna predat-ing the moth, has seen a dramatic reversal in favor of the light-colored morph of *B. betularia*. However, interpretation of the relationship between the two moths and the above-described environmental shifts is complex (Grant *et al.* 1996).

14.7.2   *Vertebrates*

Many bird species use lichens as nesting material, some showing a definite preference for certain types of lichen not only for nest construction (Fig. 14.5) and camouflage, but also, as in the case of bower birds, for purely decorative purposes. The disappearance of epiphytic lichens, mainly due to deforestation and air pollution, has depleted this source of nest-building mate-rial, and, more importantly, the consequent loss of the associated lichenopha-gous insect fauna. The loss of this important food source as well as egg-laying sites for invertebrates affects the complex food webs upon which bird popula-tions depend. Only a few studies have highlighted the significant differences in the invertebrate faunas of unpolluted and polluted environments. Lichens are also used as nesting material by flying squirrels (Hayward and Rosentreter

Fig. 14.5 Nest of chaffinch (*Fringilla coelebs*): the lichen *Parmelia sulcata* has been selectively used despite the availability locally of a wide variety of macrolichens for its construction. (Photograph by S. N. Karabulut.)

1994), which were highly selective in their choice of lichens, as 96% by volume of their nests were constructed of three species of *Bryoria*. Interestingly, these three species contain little or no lichen substances, making them more palatable to these squirrels, which are known to feed off lichens.

A large number of mammal species are known to feed on lichens, although their importance in the animal's diet varies considerably (Richardson and Young 1977). Deer, elk, ibex, gazelle, musk ox, mountain goat, polar bear, lemming, vole, tree mouse, marmot, squirrel, monkeys, and some domestic animals may include lichens in their diets, perhaps fortuitously, but more likely as a means of supplementing their normal diet or as winter feed. However, it would appear that lichens (Usneaceae) form a large component of the diet of the Sichuan snub-nosed monkey throughout the year; from November to April they are its primary food, accounting for more than 43% of its diet, and for much of the remaining part of the year form an important ingredient (Li 2006). This and similar observations on Yunnan snub-nosed monkeys (Kirkpatrick *et al.* 1998) in mixed forests considered the importance of tree species composition and the replenishment rate of lichens.

By far the most important mammalian lichen feeders are reindeer and caribou, the Eurasian and North American subspecies, respectively, of *Rangifer tarandus*. The winter diet of *R. tarandus* contains more than 50% of lichens, consumption being *c.* 3–5 kg day^{-1}, varying with the geographic location of the herd. Animals not only scrape away the snow cover to expose terricolous lichens on which to feed, but also consume arboreal and saxicolous lichens where woodland and rocks prevail. It is clear that the lichens on which R. tarandus feeds have a low nutritive value (Scotter 1965; Richardson and Young 1977), being particularly deficient in protein, calcium, and phosphorus, often leading to overgrazing of this resource and the need for an optimal harvesting policy (Virtala 1992). A level of c. 5% protein, normally regarded as acceptable for the survival of domestic animals, is not attained in the majority of lichens; a decrease in muscle weight over the winter period is therefore inevitable. However, the 7.4% protein content of Ramalina lacera may adequately sustain grazing desert gazelles if regularly available to them (Hawksworth et al. 1984).

Remote imagery has been effectively employed to map changes in lichen habitats (Section 14.10 below), for example in order to better understand population dynamics and define wildlife management in respect of the Canadian caribou (Théau and Duguay 2003; Théau et al. 2005), a demographic explosion of the herd causing severe local degradation of the vegetation cover, particularly the lichen mats (Théau and Duguay 2004).

14.8   Forest management

Clearly detailed inventories from a wide range of woodlands and forests are necessary in order to gauge the associated lichen assemblages in terms of the inherent characteristics of particular tree assemblages, community structure, their age, ecological continuity, and past and present management, as well as determining the spatial (vertical and horizontal) contribution of the different epiphytic species to the arboreal lichen flora. A fundamental challenge of modern forest management is how to extract needed resources, such as timber, without adversely affecting biodiversity and other factors important to long-term ecosystem sustainability. As a precursor to such management, it is necessary to combine design-assisted and model-assisted approaches to interpret systematically collected inventory data on the distribution and ecology of lichens, particularly of rare and vulnerable species (Edwards et al. 2004).

Therefore, for present and future management of woodlands and forests, habitat models to forecast the frequency of occurrence of epiphytic lichen species in a forested landscape under different plans have been developed (McCune et al. 2003) and detailed ecological studies have been undertaken on the effects (detrimental or otherwise) of a variety of forestry practices, such as selective felling (Rolstad et al. 2001), green-tree retention (Sillett and Goslin 1999) and clear-cutting of major areas (Esseen et al. 1997), on the diversity and biomass of epiphytic lichens. One such study on the effects of a fragmented logging pattern on the epiphytic lichens of a boreal spruce forest showed that diasporas were less successful in establishing themselves in logged areas, and since colonization was species-specific, it was recommended that the development of management guidelines should be based on wide scientific knowledge about the life-history characteristics of the species (Hilmo et al. 2005).

Many lichens are host-specific in subtropical to boreal regions, but this is not always the case in tropical rain forests (Sipman and Harris 1989; Sipman 1997; Seaward and Aptroot 2003). Hence, monocultural plantations, woodlands subjected to silvicultural thinning, and rain forests which are selectively logged have reduced tree species richness and as a consequence reduced lichen diversity (Richardson and Cameron 2004); in general, biodiversity increases with forest productivity (Gjerde et al. 2005). Fragmentation of habitat brought about by felling generates a series of interconnected factors, mainly as a result of increased irradiance and exposure, which could increase, reduce, or even bring about extinction of lichen populations, according to the route followed (Gauslaa et al. 2001). The size, shape, and edge effects of forest fragments affect habitat heterogeneity, and thereby lichen diversity (Gignac and Dale 2005). Furthermore, managed forest systems are often devoid of a broad spectrum of niches, such as fallen trees in various states of decortication and decomposition, conducive to the establishment of a wider range of lichen species (Jansova and Soldan 2006). There are only a few studies of lichen vegetation pre- and post-logging (Kantvilas and Jarman 2006).

Little is known about the epiphytic lichens of recovering and restored forests, other than those responding to decreased pollution burdens and usually subjected to liming, fertilizing, seeding, and planting (Anand et al. 2005). In the case of conserving Nephroma occulatum, a rare nitrogen-fixing macrolichen endemic to the Pacific Northwest, according to Rosso et al. (2000), management needs to focus on populations and habitat needs rather than on individuals; from their calculations, cutting with retention of individual trees surrounded by small buffers could result in the eventual loss of this species from the study area. However, perhaps too much attention has been paid to macrolichens, since recent habitat studies in Scottish woodlands have shown that the effect of decreasing woodland extent on epiphytic richness is generally more severe for microlichens (comprising a greater number of rare and specialist species) than the more generalist macrolichens (Ellis and Coppins 2007).

14.9   Conservation

There is a wealth of evidence to show that lichens are greatly affected by environmental disturbance, of both natural and human origin. In consequence, habitats and ecosystems are destroyed not only at a local and regional level, but also on a global scale. The deterioration of lichen floras, reduction in biodiversity and the loss of specific lichen taxa due to natural disasters are beyond our control, but such losses are exacerbated by human mismanagement, particularly as a result of deforestation, agricultural practices, and a wide variety of atmospheric pollutants (see Chapter 15), the latter contributing significantly to global warming, the effects of which are currently being detected by shifts in the ecology and distribution of many lichens. It is therefore increasingly urgent to conserve rare and endemic species of lichens, but in order to address this, it is necessary to have a detailed understanding of the status of the targeted species, a difficult task where ecological and distributional data are incomplete.

It is debatable as to whether habitats or species should be the primary concern: on the one hand, lichen conservation should aim at conserving habitats and regulating pollution and, on the other, red lists of threatened species need to compiled, as, for example, in the case of many European countries; however, it should be noted that some lists are constructed to cater for political/national interests rather than being based on natural geographical/ecological principles (Sérusiaux 1989). Ideally, both approaches are clearly necessary. A review of the principles and priorities of lichen conservation is provided by Seaward (1982b) and more detailed accounts of practices employed in the measurement of biodiversity and the maintenance of lichen habitats are given in Scheidegger et al. (1995b). Undoubtedly, the maintenance of threatened species and lichen biodiversity in general is dependent on good management practices. In assessing lichen loss, one should not overlook human needs, such as the collection of lichens for medicinal use (Richardson 1988), but there is also overzealous collecting of material for economic and cultural purposes (Richardson 1991) and, often more alarmingly, of rarities for reference collections (herbaria and exsiccatae).

Undoubtedly, human beings are currently the paramount agents of lichen destruction, by causing disturbance of ecosystems worldwide, through deforestation, agricultural practices, urbanization, and pollution of air, water and soil, and exploitation of natural resources. It has often proved difficult to make a convincing case to justify the conservation of lichens when weighed against other pressing needs (Seaward 1982b), but numerous measures are currently being explored and practiced worldwide to counteract this (Scheidegger et al. 1995b). However, lichens form such an important component of the complex web of life that their disappearance affects the balance of nature to a surprising degree. This is particularly the case in tundra zones, high altitudes, cold deserts, dune systems, semi-arid lands, deserts, and even urban areas, where they provide vital links in food chains and are important in community development and succession on rocks and soils. Some, but not enough, attention has been paid to global deforestation and the ongoing decline of associated lichen vegetation, more particularly foliose and pendulous forms, but terricolous fruticose species still dominate huge regions of arctic and subarctic landscapes where their future is often uncertain due to commercial exploitation of oil, minerals, timber, and, indeed, lichens themselves, as well as long-range air pollution and global warming.

14.10   Environmental monitoring

There are areas of the globe where the results of lichen denudation are now being detected by means of remote sensing. Such losses may well have climatic repercussions and exert a measurable influence on global warming, as, for example, in the case of the disappearance of epilithic lichens over a very large area of the Canadian shield as a direct consequence of atmospheric pollution formerly emanating from the smelting operations at Sudbury (Seaward 1996b), which has fortunately subsided due to pollution abatement measures, allowing the barren rock surfaces to be recolonized by lichens, thereby restoring their light-absorbing ability (Rollin et al. 1994).

Similar situations prevail in terricolous lichen dominated areas where human disturbance has significantly reduced lichen cover. This is particularly apparent in mountainous regions where skiing, once the pastime of a select few, is now enjoyed by thousands, resulting in the wholesale erosion of lichen-rich vegetational cover; in some cases the entire ecosystem has been buried beneath alien imported materials such as bitumen. Sensitive lichen-dominated ecosystems in hot and cold deserts have been similarly destroyed through human activities. It has been shown, for example, that the microphytic crusts of lichens, mosses, and cyanobacteria in semi-arid regions of eastern Australia contribute up to 27% of the ground cover and decrease reflectance (O'Neill 1994) and that some gypsiferous soils in the intermountain area of the western United States support soil crust communities with a high species diversity, often with a 60–100% ground cover of lichens (e.g. St.Clair et al. 1993; Fig. 14.6). Over a long period of time, lichen crust communities change the physicochemical properties of soil, enhancing their stability and fertility (Belnap and Lange 2005a); their retention and indeed restoration are necessary when human interference affects the natural equilibrium through increasing desertification (Bowker et al. 2005) and land mismanagement (Evans et al. 2001), by, for example, the introduction of domestic animals (Ponzetti and McCune 2001; Warren and Eldridge 2001) and the impact of off-road vehicles, the effects of which are equally profound in the case of mat-lichen ecosystems of boreal forests, tundra, and sand dunes.

Fig. 14.6 Lichens, common components of biological soil crust communities in arid lands, stabilize soil surfaces (e.g. *Psora cerebriformis*, center) and in the case of cyanolichens (*Placidium squamulosum*, left and *Collema tenax*, right) contribute significant amounts of fixed nitrogen to the community. (Photograph by L. St. Clair.)

Changes in reflectance and in ecophysiological responses, such as chlorophyll levels, gaseous exchange, and water absorbance, brought about by anthropogenic disturbances to lichen-dominated communities are being detected in remotely sensed images (Petzold and Goward 1988; O'Neill 1994; Karnieli *et al.* 2001). As a consequence, the environmental significance of variations in these activities is increasingly being recognized. "Hot spot" reflectance signatures of common boreal lichens determined by laboratory studies (Kaasalainen and Rautiainen 2005) provide some insight into how they could be detected by remote sensing. In arctic and alpine areas, lichens possess specialized physiological mechanisms enabling them to photosynthesize and take up water at low temperatures. Lichens may also be effective ice nucleating agents and can therefore initiate freezing of supercooled water at relatively warm temperatures. Such phenomena may be used to interpret data derived from remote sensing (Nordberg and Allard 2002). In the case of global warming, temperature differences in remote areas such as the polar regions could be reflected by the presence or absence of lichen cover.

As well as playing a major role in shaping the natural world, both physically and biologically, lichens are natural sensors of our changing environment (Seaward 2004). Environmental interpretation can be assisted by means of lichens, based on the presence and/or absence of particular species and/or the nature and composition of assemblages indicative of one or more identifiable factors. It is, for example, possible to use the composition of the lichen flora to evaluate habitat and ecosystem stability, often in terms of ecological continuity over time, as in the determination of the age and past management of deciduous woodlands and coniferous forests (Rose 1976).

Information gained from our knowledge of how lichens respond to long-term perturbations and short-term upheavals in nature can be applied to the interpretation and monitoring of environmental changes and disasters brought about through a wide range of human activities. The reaction of lichens to sudden natural events such as fire, volcanic eruptions, and earthquakes on the one hand and to the long-term effects of glaciers, snow, and water on the other can be effectively employed to determine those human impacts which destabilize soil, rock, and water systems. Thus, lichens can often be used as an early warning system for other biota which without remedial action would subsequently suffer stress or indeed extinction through forest and agricultural mismanagement, desertification, urbanization, industrialization, and a whole host of other problems arising from world overpopulation.

Baseline information on lichen assemblages and ecosystems which are ecologically or geographically zoned on the basis of particular natural phenomena has proved invaluable in assessing widespread increases in various pollutants and climatic changes resulting from global warming; for example, distinctive lichen zonations at freshwater and marine waterlines can be affected by acidification, eutrophication (and indeed hypertrophication), and other polluting agencies. Delimitation of such lichen assemblages will change as water levels rise as a consequence of increases in global temperature. From detailed field studies and remote sensing, it should be possible to monitor changes in the lichen flora of terrestrial environments resulting from displaced snow lines, episodic snow-kill, avalanches, seismic landslides and other unstable debris flows, retreating glaciers due to climatic shifts, and flooding (Benedict 1991; Bull et al. 1994; Sonesson et al. 1994; Insarov and Schroeter 2002; Wolken et al. 2005; Wolken 2006). Such time–space investigations can fully exploit lichenometrical techniques (Section 10.7) widely developed over the past few decades (Innes 1983, 1985; McCarroll 1993; Winchester and Harrison 1994; Gob et al. 2003; Solomina and Calkin 2003).

Long-term field investigations involving stringent ecological and phytogeographical criteria through a comprehensive ongoing program of detailed mapping can provide the basis for large-scale monitoring of quantitative and qualitative changes in environmental regimes, ranging from air pollution (Chapter 15) to global warming (e.g. van Herk et al. 2002; Belnap and Lange 2005b); in the case of the latter, poleward movements of specific lichen communities could be readily detected by remote sensing. Intensive lichen monitoring is a necessary component of any program aimed at effective long-term observation of environmental disturbances, both natural and of human origin.

The proper use of lichens as indicators and samplers of ambient conditions is a valuable resource for the environmentalist (Seaward 2004). In theory, techniques which have been developed by ecologists could be employed on a large scale for "low technology" environmental appraisals and impact studies, particularly where on-site instrumentation would be expensive to install and maintain. Unfortunately, most of the methodologies require a fairly detailed understanding of the taxonomy of lichens and the development of protocols for consistency in measurement and interpretation by future researchers; similarly, techniques based on bioassays necessitate depletion of the resource material, rigorous protocols for its collection, preparation and analysis, and sophisticated analytical equipment.

15

Lichen sensitivity to air pollution

T. H. NASH III

Lichens have been recognized as being very sensitive to air pollution for many years (Hawksworth 1971; Nimis *et al.* 2002). In the 1800s independent observations in England, Munich, and Paris documented that lichens were already disappearing from urban areas. By the early 1900s this "city" effect was a widely recognized phenomenon in Europe and was first attributed to coal dust, which was emitted by most homes as well as many industries. Only later did the colorless gas, sulfur dioxide, become recognized as a principal phytotoxic agent. Today the list of air pollutants is much longer and includes oxidants, hydrogen fluoride, some metals (Section 12.7), acid rain, and organics. Certainly the list of potentially toxic substances is not yet fully circumscribed.

The high sensitivity of lichens is related to their biology. Most species live for decades or hundreds of years and a few longer; thus, as perennials, they are subject to the cumulative effect of pollutants. Lichens have no vascular system for conducting water or nutrients; as a consequence, they have developed efficient mechanisms for taking up water and nutrients from atmospheric sources. Fog and dew, major water sources for lichens, often have much higher pollutant concentrations than precipitation, and the lichens' nutrient concentration mechanisms also will concentrate pollutants. Unlike many vascular plants, lichens have no deciduous parts, and hence cannot avoid pollutant exposure by shedding such parts. Furthermore, the lack of stomata and cuticle in lichens means that aerosols may be absorbed over the entire thallus surface; thus, lichens have little biological control over gas exchange, and air pollutant gases are assumed to readily diffuse down to the photobiont layer. Although dehydration allows lichens to survive dry periods, it also concentrates solutions to the point that toxic concentrations may occur. Finally alteration of the

Lichen Biology, ed. Thomas H. Nash III. Published by Cambridge University Press.

symbiotic balance between the photobiont and mycobiont may readily lead to a breakdown of the lichen association.

That lichens are sensitive to air pollution is, of course, a generalization that requires cautious interpretation and limited extrapolation. The distribution patterns of a lichen species may well reflect varying levels of an air pollutant, but variation in its distribution may also be caused by a variety of abiotic or biotic factors as well. Furthermore, all lichens are not equally sensitive to all air pollutants; rather, different lichen species exhibit differential sensitivity to specific air pollutants. Sensitive species may become locally extirpated when a pollutant is present, but at least some tolerant species are likely to persist. This differential sensitivity is, however, very useful when interpreting air pollution effects. Just as canaries are useful in detecting toxic gases in coal mines, lichens can be used to monitor ecosystem health where air pollutants are present (Rosentreter and Eldridge 2002; Will-Wolf 2002). Similarly, lichens can be used to monitor potential human health problems (Cislaghi and Nimis 1997).

15.1   Lichens in relation to sulfur dioxide

The strongest case for using lichens as bioindicators of air pollution involves sulfur dioxide (Grace *et al.* 1985*a*; Seaward 1993; Hawksworth 2002; Nash and Gries 2002). Some forms of coal (and other fuel products) have particularly high levels of sulfur, and its oxidation leads to the formation of sulfur dioxide, one of the major gases associated with acid rain. In fact sulfur dioxide has only an average atmospheric residency time of about 12 hours, because its high solubility in water leads to its trapping in water vapor aerosols and rapid conversion to sulfuric acid, one of the stronger acids.

15.1.1   *Lichen distribution patterns in relation to sulfur dioxide gradients*

Evidence that lichens have been responding to sulfur dioxide (SO_2) has now been obtained from throughout the world (Sugiyama *et al.* 1976; Eversman 1978; Søchting and Johnsen 1978; Türk and Christ 1980; Will-Wolf 1980; Taylor and Bell 1983; Pišút and Lisicka 1985; McCune 1988; Chen *et al.* 1989; Freedman *et al.* 1990; Nimis *et al.* 1990; Hamada *et al.* 1995; Aguiar *et al.* 1998; Aarrestad and Aamlid 1999; Wolseley 2002). For example, Skye (1968) elegantly demonstrated how bark of trees became acidified and lost its buffering capacity with proximity to the city center (in his case a detailed investigation of Stockholm). This caused a shift in lichen epiphytic species composition to those that were more acidophilic.

Probably the most convincing field data came from England (Hawksworth 2002), where a grid of sulfur dioxide monitoring stations was established throughout the country in the mid 1900s. Because SO_2 measurements across the grid covered a wide range of atmospheric SO_2 levels, it was possible to examine patterns in epiphyte community composition on adjacent trees and infer relative SO_2 sensitivities based on those patterns across sites (Hawksworth and Rose 1970). On the basis of differential changes in the epiphytic lichen communities, they established two 10-point scales (one for basic bark and the other for acidic bark) relating long-term SO_2 measurements to changes in the epiphytic lichen communities. Their scales essentially provided elaborate hypotheses about putative differential sensitivity to SO_2 among different lichen species. Even though initial experimental investigations involving SO_2 fumigations did not strongly support the scale (e.g. Baddley et al. 1973), subsequent SO_2 experiments have, in fact, strongly supported the scale (Türk et al. 1974; Nash 1988). Although some of the changes noted in the scale may well have been related to factors other than sulfur dioxide (e.g. Kostner and Lange 1986), the general trends observed by Hawksworth and Rose have proved to be quite robust. The most dramatic evidence in support of this assertion has been the widespread reinvasion of lichens as trophospheric SO_2 levels have recently decreased throughout England (Purvis 2000; Fig. 15.1). Recognition of the detrimental effects of SO_2 had led to legislation limiting SO_2 emissions and implementation of that legislation. The result was a dramatic decrease in measured SO_2 levels, starting in the early 1970s (Laxen and Thompson 1987; Micallef and Colls 1999).

Concomitantly, many of the putatively sensitive species have reinvaded urban areas (Henderson-Sellers and Seaward 1979; Rose and Hawksworth 1981; Hawksworth and McManus 1989, 1992; Bates et al. 1990; Boreham 1992; Gilbert 1992; Seaward 1997). Similar trends are occurring elsewhere, such as Paris (Seaward and Letrouit-Galinou 1991) and Munich (Kandler 1987), although in some cases reinvasions are limited to nitrogen-tolerant species due to continued high levels of nitrate and ammonium deposition levels (e.g. The Netherlands: van Dobben and de Bakker 1996; van Dobben and ter Braak 1998, 1999; van Herk 2002). In parallel with these reinvasions, the abundance of Lecanora conizaeoides has declined dramatically (Bates et al. 2001). That Lecanora was the lichen species judged to be most SO_2 tolerant in Hawksworth and Rose's original scale. It may well have a high sulfur requirement (Bates et al. 2001; Hauck et al. 2001), and it may well have evolved originally around sulfur-emitting fumaroles in Iceland (M. R. D. Seaward, personal communication). Similar trends in the latter species' distribution are occurring in Germany as well (Wirth 1993).

15.1.2   Experimental investigations with sulfur dioxide

The use of lichens as bioindicators of sulfur dioxide has gained much validity from a plethora of experimental investigations involving exposure to

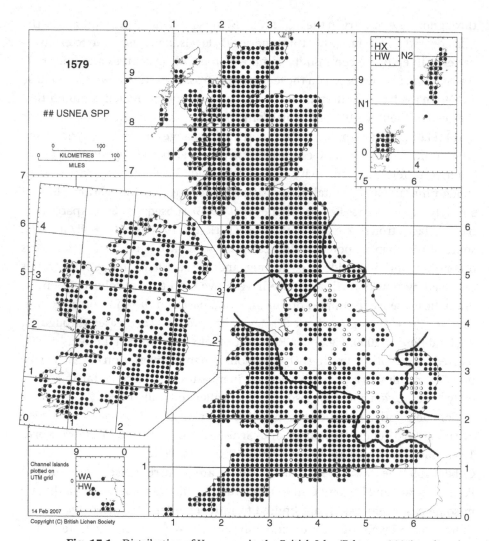

Fig. 15.1 Distribution of *Usnea* spp. in the British Isles (February 2007) to show its disappearance from an area of *c.* 68 000 km² (delimited by solid line) during the period *c.* 1800–1970, and its subsequent re-establishment at numerous sites (in 168 recording units 10 km × 10 km) within that area as a consequence of atmospheric pollutant amelioration: open circles, pre-1960 (usually nineteenth century) records; solid circles, 1960 onwards records. (Update of map published in Seaward 1998.)

controlled levels of SO_2. Because experimental conditions have varied considerably, one needs to evaluate critically the degree to which different investigations are equivalent. For example, initial investigations used closed containers (Pearson and Skye 1965), where only the initial concentration was known, as the subsequent uptake of SO_2 by the lichens and/or the walls of the container was unknown, and thus the exposure treatment remained undefined. A second approach involved exposing the lichens to aqueous solutions of SO_2, which are prepared by bubbling gaseous SO_2 through water and weighing the resultant solution (e.g. Puckett *et al.* 1973). The method is intuitively attractive insofar as physiologically active lichens have aqueous solutions in their protoplasts. Nieboer *et al.* (1977) have proposed a method for calibrating aqueous solutions with gas phase concentrations, and this may be appropriate for defining initial conditions; however, SO_2 in aqueous solutions may decline with time (Richardson and Nieboer 1983). Consequently, the degree to which such experiments can be equivalent to longer-term gaseous experiments, where the lichen is continuously exposed to a constant SO_2 concentration, remains only partially resolved. Probably more important is the fact that almost all lichens are terrestrial organisms, and, as such, typically exist in a gaseous microenvironment; thus, their behavior in an aqueous medium may well be altered. The use of a flow-through gas exchange system, in which SO_2 concentrations are maintained at a constant level, works best if the fumigated lichens are maintained at controlled temperatures and at nearly constant water contents by adjusting the air-flow's dew point to keep the relative humidity high (Türk *et al.* 1974; Gries *et al.* 1997a,b). If one monitors SO_2 concentrations at both the entrance and exit to the exposure chambers, then one can calculate SO_2 uptake by the lichens (e.g. Gries *et al.* 1997b), assuming equilibrium with the chamber walls.

In the field it is possible to conduct small-scale SO_2 fumigations, simply by slowly bleeding SO_2 from a tank through a tube with multiple small orifices, as was done by Moser *et al.* (1980) across a lichen mat in the arctic tundra. After monitoring periodically both the SO_2 concentrations across the top of the lichen mats and the photosynthetic capacity of dominant lichens across the growing season, these authors demonstrated a pattern of decreasing SO_2 injury with increasing distance from the exposure system. Subsequent measurements at the site demonstrated no apparent recovery three years later (Moser *et al.* 1983).

15.1.3   *Lichen sulfur and sulfur dioxide uptake kinetics*

Apparently most sulfur in lichens originates from atmospheric sources (Spiro *et al.* 2002). For example, investigations around sour gas processing plants in Alberta by Case and Krouse (1980) established that the atmospheric SO_2 from the industrial facilities had a unique ^{34}S signature. The same pattern was found in the surrounding lichens; however, the vascular plants of that region had a very different sulfur signature consistent with the inference that the soil was their principal source of sulfur. Furthermore, lichens may differentially take up sulfate from dry and wet deposition to canopies in relatively unpolluted areas (Knops *et al.* 1996). Consequently, low levels of SO_2 may actually enhance the sulfur budget of lichens, and not be detrimental.

Because of the high solubility of SO_2 in water, any moist lichen may be a major sink for SO_2, even when not metabolically active (Gries *et al.* 1997*b*). Initially, absorption of SO_2 leads to higher total sulfur levels in lichen thalli (O'Hare and Williams 1975). A number of field investigations have also demonstrated that total sulfur levels increase with length of exposure of lichen transplants (Richardson and Nieboer 1983). However, the overall relationship between SO_2 exposure and total sulfur (or sulfate) concentrations in lichens is more complex, because severely injured specimens lose their ability to retain sulfur (Galun and Ronen 1988; Häffner *et al.* 2001), a fact probably related to increased membrane permeability (Puckett *et al.* 1977).

It is widely assumed that lichens must be moist to be sensitive to SO_2 (Ferry and Baddeley 1976; Richardson and Nieboer 1983; Grace *et al.* 1985*b*; Wirth 1987). However, Coxson (1988) has experimentally demonstrated that some injury occurs when air-dried lichens are fumigated, a fact that may reflect the residual water present in such lichens and the fact that some SO_2 absorption occurs under such conditions.

Under aqueous conditions, SO_2 rapidly forms the strong acid H_2SO_4, which dissociates into H^+ and HSO_3^- (bisulfite) ions at pHs near neutrality (Nieboer *et al.* 1984). Many physiological effects due to SO_2 exposure are attributed to the resulting acidity (Türk and Wirth 1975; Rennenberg and Polle 1994). In addition, the bisulfite anion is also recognized as being quite toxic, and some experiments have been run with aqueous bisulfite solutions as a surrogate for SO_2 fumigations (Hill 1971; Hällgren and Huss 1975; Silberstein *et al.* 1996*a,b*). Cells gradually oxidize bisulfite to sulfate, which is much less toxic. Across a wide range of SO_2 concentrations and using five different lichen species, approximately 70% of the SO_2 taken up can subsequently be leached as sulfate (Gries *et al.* 1997*b*). Understanding how the lichens deal with the remaining 30% (Section 15.1.5) then becomes the critical issue in understanding known differential sensitivity to SO_2 among lichens.

For experiments of constant duration with moist lichens, SO_2 uptake directly varies with SO_2 concentration (Gries *et al.* 1997*b*). Furthermore, if lichens dry relatively rapidly during experimental SO_2 fumigations, it may appear that responses are only a function of fumigation concentration and are independent of fumigation duration (Huebert *et al.* 1985). However, if lichen water content is maintained at relatively constant levels, then both length of exposure and SO_2 concentration contribute to measured response, as was established from the dose-response investigations of Sanz *et al.* (1992). The idea of SO_2 dose received by a lichen being a function of both duration and concentration is an implicit assumption of the threshold extrapolations proposed by Richardson and Nieboer (1983). Remarkably, calculated thresholds based on laboratory experiments with aqueous SO_2 correspond quite closely to estimates based on field investigations, such as those taken from Moser *et al.* (1980) for *Cladonia rangiferina* (Richardson and Nieboer 1983; Grace *et al.* 1985a,b; Winner *et al.* 1988).

15.1.4   *Lichen physiological responses to SO_2*

A wide range of response variables (Nieboer *et al.* 1976; Richardson and Nieboer 1983; Fields 1988; Galun and Ronen 1988; Geebelen and Hoffmann 2001) have been employed to measure physiological response of lichens, both to field conditions (*in situ* or in transplants) or in laboratory experiments where concentrations of SO_2 are controlled. Initial work focused on chlorophyll degradation (Pearson and Skye 1965; Rao and LeBlanc 1966; Nash 1973; Gries *et al.* 1995; Balaguer and Manrique 1995), which requires moderately high SO_2 fumigation levels. With improvements in techniques to measure carbon dioxide gas exchange, measurements shifted to dark respiration (Baddeley *et al.* 1973; Marsh and Nash 1979; Gordy and Hendrix 1982) or net (or gross) photosynthesis (Puckett *et al.* 1974; Türk *et al.* 1974; Punz 1979; Türk and Christ 1980; Huebert *et al.* 1985; Sanz *et al.* 1992). In general, these investigations have established differential sensitivity among species that closely match inferred sensitivity of the species based on field investigations (Nash 1988), even though field sites are often subjected to a mixture of pollutants (von Arb *et al.* 1990; Garty *et al.* 1997). With aqueous SO_2 experiments, SO_2 response may be modified by the presence of metal ions: decreased in the case of Cu, unaffected by Pb, and increased by Mg, Zn, and Ca (Richardson *et al.* 1979).

Results of gas exchange measurements during SO_2 fumigations have been corroborated by using chlorophyll fluorescence techniques (Gries *et al.* 1995; Calatayud *et al.* 1996; Deltoro *et al.* 1999; Jensen and Kricke 2002), that allow inferences about effects on photosystems I and II (Fig. 15.2). One aspect of the lichen symbiosis is the partial transfer of carbohydrates from the photobiont (alga or cyanobacterium) to the fungus as an immediate result of the dark reactions of photosynthesis (Richardson *et al.* 1968). Fields and St. Clair (1984) and Nieboer *et al.* (1976) have shown that carbohydrate transfer is often decreased following short-term exposure to SO_2.

Inside cells, potassium is the dominant inorganic ion. Puckett *et al.* (1977) and Tomassini *et al.* (1977) demonstrated that a major efflux of K occurred in response to aqueous SO_2 exposure. With increasing SO_2 concentrations, a biphasic loss of K was observed and the results were interpreted as reflecting an increase in permeability of the cells. The magnitude of K release due to SO_2 depended in part on co-occurrence of other ions, increasing in the case of Cu and Pb, unaffected by Mg and Ca, and decreasing in the case of Sr, Ni, or Zn (Nieboer *et al.* 1979). Quantification of K in small concentrations requires accurate

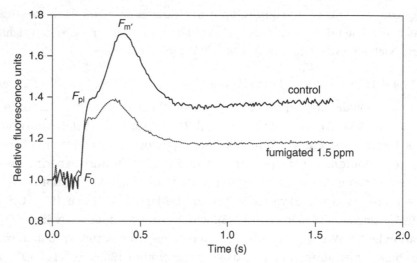

Fig. 15.2 Comparison of rapid fluorescence induction kinetics for *Ramalina menziesii* unfumigated (control) and after 6 hours of fumigation with 1.5 ppm SO_2. F_0, low level of fluorescence; F_{pl}, intermediate level; $F_{m'}$, maximum level.

analytical techniques, and similar results may be obtained by measuring total conductivity with a simple meter in a lichen-rinsed solution (Pearson and Henriksson 1981). Conductivity measurements are often used in field investigations as well (Alebic-Juretic and Arko-Pijevac 1989; Garty *et al.* 1993).

About 10% of lichen species contain cyanobacteria, either as the main photobiont or secondarily in specialized structures called cephalodia (Chapter 11), and many are very sensitive to air pollution (Richardson and Cameron 2004). Many cyanolichens, due to the presence of nitrogenase, are capable of nitrogen fixation. Kytövita and Crittenden (1994) demonstrated that nitrogenase activity was inhibited by acidification. Nitrogen fixation rate, as measured indirectly by acetylene reduction activity (ARA), is often used as a response variable as well. For example, under aqueous conditions, Hällgren and Huss (1975) found that nitrogen fixation reduction occurred at lower SO_2 concentrations than corresponding photosynthetic measurements (^{14}C uptake) in *Stereocaulon paschale*. Reductions in ARA due to SO_2 exposure have also been reported by Sheridan (1979) and Henriksson and Pearson (1981), but their results are of limited value because of the techniques employed (Richardson and Nieboer 1983). Although it is often suggested that nitrogen fixation is more sensitive than the parameters discussed above (Richardson and Nieboer 1983; Fields 1988), additional data are certainly necessary to assess the generality of the assertion.

Reproduction in lichens is also inhibited by SO_2. Reduction in the production of apothecia in the lichenized ascomycetes (98% of the species belong to the ascomycetes) has frequently been observed in polluted areas (LeBlanc and De Sloover 1970; Bedeneau 1982), and ascospore germination may also be inhibited by SO_2 (Belandria *et al.* 1989). Even asexual means of propagation, such as the germination of soredia (Margot 1973), may be inhibited by SO_2.

15.1.5 SO_2 effects on lichen ultrastructure and biochemistry

Eversman and Sigal (1987) and Holopainen and Kärenlampi (1984) have documented ultrastructural changes in two lichens exposed to a range of SO_2 concentrations. In the alga, initial injury included swelling and deformation of the mitochondria, stretching of the chloroplast envelopes, and deformation of pyrenoglobuli. As injury progressed, particularly at the higher concentrations and with longer exposures, the thylakoids stretched and the pyrenoids, chloroplast stroma, nucleus, and mitochondria all degenerated. Injury to the fungus primarily occurred at higher concentrations and included swelling of mitochondria and vesiculation of the mesosome-like organelles. These patterns of injury have been largely confirmed by subsequent investigations (Sharma *et al.* 1982; Eversman and Sigal 1987; Plakunova and Plakunova 1987; Holopainen and Kauppi 1989) in other species. The initial injury symptoms documented by Holopainen and Kärenlampi (1984) were also observed in field-collected lichens from central Finland (Holopainen, 1983).

The pattern of ultrastructural injury corresponds well to biochemical and physiological responses observed in lichens exposed to SO_2. For example, Malhotra and Khan (1983) found reductions in protein and lipid biosynthesis and CO_2 fixation in *Evernia mesomorpha* exposed to 0.34 and 0.1 ppm SO_2, concentrations that correspond well to midrange concentrations employed by Holopainen and Kärenlampi. Because lipids and proteins are essential constituents of cell membranes, these types of changes may well be associated with the observed ultrastructural changes. Similar effects on lichen lipids and fatty acids were found by Bychek-Guschina *et al.* (1999). Changes to the mitochondria would certainly affect respiration and changes to the thylakoids would affect photosynthesis. The most severe ultrastructural effects probably correspond to situations where a major reduction in photosynthesis, increased electrolyte leakage, and destruction of chlorophylls has occurred (see earlier sections).

One of the primary effects of SO_2 absorption is cellular acidification (Pfanz *et al.* 1987; Rennenberg and Polle 1994). At neutral pH (7.0), SO_2 in water dissociates approximately 50% to bisulfite (HSO_3^-) and 50% to sulfite (SO_3^-), but at pH 4.0 sulfite dominates, and this accounts in part for the greater toxicity under lower pHs due to the greater oxidizing power of sulfite (Nieboer *et al.* 1976). To a degree, lichen species may vary in their ability to oxidize sulfite. For example, Miszalski and Niewiadomska (1993) found that a SO_2-tolerant species is a more efficient oxidizer than less tolerant species. Acidification has consequences for many enzyme reactions, as activity is frequently pH dependent. For example, Ziegler (1977) found that sulfite inhibited RubisCO activity in *Pseudevernia furfuracea*. Furthermore, with acidification, carboxylation efficiency is reduced (Price and Long, 1989) and the electron transport system is inhibited (Chen *et al.* 1992). All of these effects would readily reduce photosynthesis. Furthermore, SO_2 inhibited the growth of *Trebouxia*, the most common lichen alga, much more at pH 4.0 than pH 5.0 (Marti 1983). Likewise, lichen net photosynthesis is dramatically reduced at lower pHs (Puckett *et al.* 1973; Türk and Wirth 1975). The ameliorating effect of Ca^{2+} in reducing the effect of SO_2 on photosynthesis (Richardson *et al.* 1979) relates to the ion's ability to reduce the degree of acidification. Likewise, when considering concrete substrates, the occurrence of some otherwise sensitive lichen species in areas with higher SO_2 levels (Gilbert 1970) can be understood in terms of the buffering effect of the substrate. In contrast, poorly buffered tree bark supported few if any lichens (Skye 1968).

The ability to detoxify excessive sulfur may be critical. One pathway in the assimilation of SO_2 leads to the formation of sulfite and its reduction to the gas H_2S, as mediated by the enzyme sulfite reductase (Romagni *et al.* 1997). This process results in dissipation of part of the sulfur loading and is thought to take place in the thylakoids (Rennenberg 1984). Unfumigated lichens are known to release H_2S (Gries *et al.* 1994), but at very low concentrations (0.01–0.04 pmol $g\,dw^{-1}\,s^{-1}$). Fumigations with low levels of SO_2 result in increased H_2S emissions that persisted to a reduced degree following cessation of the fumigation, and differences among species in the amount released was evident (Gries *et al.* 1997a). Furthermore, species varied in their ability to release more H_2S at higher SO_2 levels. However, overall, the magnitude of H_2S release was relatively small (0.4–2.5%) compared with the inferred amount of absorbed SO_2 that was not leached as sulfate. Other trace sulfur gases are known (COS, CH_3SH, DMS, CS_2), and lichens are known to assimilate COS and release DMS and sporadically release small quantities of the other gases (Gries *et al.* 1994), but the magnitude of release was also very small (same order of magnitude as H_2S). Thus, the release of trace sulfur gases in lichens is probably a relatively unimportant mechanism to reduce excess sulfur loading.

Alternatively, sulfur metabolism may be enhanced by increasing the activity of the enzymes, such as ATP sulfurylase or cysteine synthase (Stulen and de Kok 1993), but these enzymes have not yet been investigated in lichens.

Recent work on SO_2 injury mechanisms in plants have focused on the production of free radicals (see also Chapter 8), particularly those occurring in the chloroplast due to sulfite accumulation (Rennenberg and Polle 1994). Under control conditions, photolysis leads to the formation of the superoxide radical, which is normally scavenged by superoxide dismutase (SOD). Hydrogen peroxide is the product of the SOD-mediated reaction, and it can be removed by glutathione, ascorbate, or peroxidases. Sulfite can alter the conformation of enzymes involved in these last two steps (Alscher 1984). At high concentrations, sulfite can be photo-oxidized by the superoxide radical of free oxygen, leading to the formation of hydroxyl radicals that disrupt the lipid fraction of the chloroplast membranes (Peiser and Yang 1985). Such effects correspond well to the types of injury observed by Holopainen (1984) in lichens. For lichens suffering from acute injury, similar effects would likely be manifest in chlorophyll degradation.

Some evidence points to these processes also occurring in lichens as well. Modenesi (1993) provides evidence that free radicals are generated in lichens as a response to SO_2, and Köck et al. (1985) demonstrated that a different form of SOD was formed in Trebouxia in response to sulfite treatments. Using a variety of physiological and biochemical techniques, Silberstein et al. (1996a) carefully documented that Xanthoria parietina was SO_2-tolerant, but that Ramalina duriaei was SO_2-sensitive. Subsequently, they showed that parietin, the orange secondary product present in the upper layer of Xanthoria but absent in Ramalina, could act as a potential antioxidant (Silberstein et al. 1996b). Furthermore, they demonstrated a loss in SOD activity in the sensitive species following treatment with bisulfite, a result that parallels observations following SO_2 fumigations with a different, sensitive lichen, Evernia prunastri (Deltoro et al. 1999). They also found stimulation of SOD in a tolerant species, Ramalina farinacea, as did Kong et al. (1999) for a different species. But Thomas (1999) essentially found no change in SOD activity with shorter-term fumigations of several species. With increasing bisulfite treatment, Silberstein et al. (1996b) found a reduction of peroxidase activity in sensitive species, but an increase in peroxidase activity in tolerant species. Deltoro et al. (1999) found the same pattern (but the differences were not always significant) for peroxidase, ascorbic peroxidase, and catalase following SO_2 fumigation. Thus, there is some evidence that tolerant species can more effectively deal with oxygen radical formation than sensitive species, as supported by differential enzymatic responses.

Furthermore, the SH groups in glutathione (GSH) give it antioxidant properties, and, consequently, GSH may well play a role in protecting tolerant species against excess sulfur (de Kok and Stulen 1993). It is probably the most abundant thiol in plants, and is well known to change its pool size during the initial stages of rewetting dry lichens (Kranner and Grill 1994). Silberstein et al. (1996b) demonstrated some initial increase in the glutathione pool following bisulfite treatment, and Kong et al. (1999) demonstrated a major increase in glutathione with increasing SO_2 fumigations of *Xanthoparmelia mexicana*. Although Romagni *et al.* (1998) did not confirm such a response with SO_2 fumigations, they did demonstrate that a SO_2-tolerant lichen, *Pseudevernia intensa*, had an order of magnitude larger pool size of glutathione than other more sensitive species. The enzyme, glutathione reductase, controls the regeneration of GSH from GSSG (glutathione disulfide). Deltoro *et al.* (1999) reported that fumigation with high SO_2 concentrations resulted in stimulation of growth rate activity in their tolerant species, but Thomas *et al.* (1997) did not find a significant change in glutathione reductase activity with SO_2 fumigations of other lichen species. Thomas did, however, find higher rates of growth rate activity in green-algal lichens than a cyanolichen species. Thus, SO_2-sensitive cyanolichens may not be able to generate as much of the antioxidant GSH as species containing green-algal lichens.

15.2   Oxidants and lichens

Oxidants are a group of strongly oxidizing air pollutants formed photo-chemically by series of chemical reactions involving mixtures of nitrogen oxides and hydrocarbons. Ozone, which was initially identified in the Los Angeles area, is probably the most infamous oxidant, and it is the most important air pollu-tant causing injury to agricultural crops (US EPA 1996). In addition, a number of other oxidants, such as peroxyacetyl-nitrate (PAN), are known to cause injury to plants (Taylor 1969).

Lichen studies have been conducted in the Los Angeles area both in coastal mountain areas (Ross 1982) and in higher mountain forests (Nash and Sigal 1980; Sigal and Nash 1983). Because of negligible home heating requirements in this region, SO_2 emissions are extremely low, and most vascular plant injury patterns are ascribed to oxidants (Miller and McBride 1998). On the basis of historical data and the extensive, recent sampling of lichens, species exhi-biting differential sensitivity to the Los Angeles environs are identified (Sigal and Nash 1983). Very sensitive species, such as *Evernia prunastri*, *Peltigera* spp., *Pseudocyphellaria* spp., and *Ramalina* spp., are no longer found, although herbar-ium records document that many of them were abundant in the late 1800s and early 1900s (Hasse 1913). The sensitive species, such as *Collema nigrescens* and *Usnea* spp., are found in vestigial amounts in protected habitats; moderately tolerant ones, such as *Hypogymnia enteromorpha* (= *H. imshaugii* in this area but called the former in the literature), common, but in reduced abundance com-pared with control areas, and they often exhibited marked morphological anomalies (Sigal and Nash 1983). In contrast, the few tolerant species, such as *Letharia vulpina*, were relatively unaffected. These observations were supported by sampling along oxidant gradients within the highly polluted San Bernardino Mountains of southern California. The gradient reflected not only differences in oxidant dose, but also a range of oxidant injury patterns exhibited by ponderosa pine, as demonstrated by fumigations with ozone (Miller *et al.* 1983). Along this oxidant gradient there was a marked reduction in cover of the dominant two lichens (*Letharia vulpina* and *Hypogymnia imshaugii*) occurring on conifers (Nash and Sigal 1980) and also by species occurring on black oak (Sigal and Nash 1983). Furthermore, experiments demonstrated photosynthetic reduction in *Hypogymnia imshaugii*, *Flavoparmelia caperata* and *Parmelia sulcata* with ozone exposure (Nash and Sigal 1979) and similarly in *Collema nigrescens*, *H. enteromorpha* and *P. sulcata* when treated with PAN (Sigal and Taylor 1979). In contrast, *Ramalina menziesii*, a putatively sensitive species (Sigal and Nash 1983), was unaffected by very short-term ozone fumigations (Ross and Nash 1983). However, overall these results together with the field work provided the first, relatively strong case for asserting that oxidants were having an adverse impact on lichens.

However, in regions where oxidant levels are lower than in southern California, oxidant relationships to lichen community parameters are always recognizable. Herzig and Urech (1991) did show that lichen community patterns in central Switzerland could be related to O_3, as part of multiple regressions with other pollutants; however, McCune (1988) and Lorenzini *et al.* (2003) found no lichen community relationships to ozone gradients, respectively in Indiana (USA) and Italy. The latter observations may reflect the occurrence of less steep oxidant gradients or poorer pollutant estimation at the field sites, as well as lower O_3 levels. More recently it has become apparent that how oxidant pollutant exposure is expressed is critical (Washburn 2006). For example, ozone indices developed for the growing season of agricultural crops are not appropriate, as the lichens are present all year. For the Cincinnati, Ohio, region, an area within 200 km of McCune's study area, Washburn (2005, 2006), established that O_3 indeed had a marked effect on epiphytic lichen communities when exposure over a 5-year period was considered with a nonweighted ozone exposure index (SUM00) (Fig. 15.3). On the basis of this diagram, one can hypothesize that *Physcia americana* (PHYAME), *Physconia detersa* (PHYDET), *P. leucoleiptes* (PHYLEU), *Xanthoria fulva* (XANFUL), and *Phaeophyscia cernohorskyi* (PHACER) are relatively tolerant to O_3; whereas *Rimelia reticulata* (RIMRET), *Parmotrema* sp. 3 (PARMSP), *Phaeophyscia* sp. 1 (PHAES1), and *Phaeophyscia* sp. 2 (PHAES2) are relatively sensitive to O_3.

In addition, a number of recent experimental fumigations with O_3, often with more realistic concentrations and durations, support the assertion that O_3 adversely affects lichens. Many but not all investigations have demonstrated

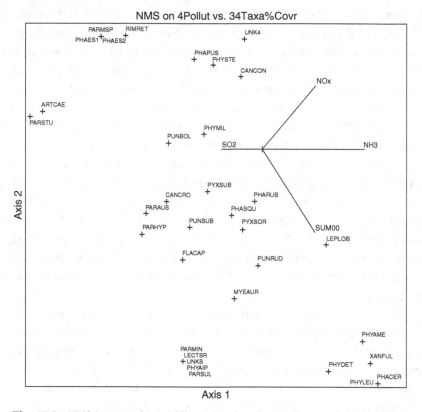

Fig. 15.3 NMS (nonmetric multidimensional scaling) ordination for 34 epiphytic species from 13 stands in the greater Cincinnati, Ohio, area based on Washburn (2005). Approximately 80% of the variation is explained by the ordination. Vectors for four air pollutants (SO_2, NO_x, NH_3 [SUM00 = O_3]) based on air quality measurements over 5 years are superimposed on the ordination. The relative impact of a pollutant is proportional to the length of the vector. Based upon the relative vector lengths, one may conclude that NH_3, O_3, NO_x, and SO_2 explain approximately 25%, 24%, 22%, and 10%, respectively, of the total amount of variation in the lichen community data. Species clusters oriented along the axes of each pollutant reflect putative tolerance (species clumped in the direction of the vector) and sensitivity (species clumped in the opposite direction).

reduced photosynthesis (Sigal and Johnston 1986). Others have demonstrated photosystem II inhibition (Scheidegger and Schroeter 1995; Balaguer *et al.* 1996; Calatayud *et al.* 2000), enhanced K^+ leakage (Tarhanen *et al.* 1997), SOD activity stimulation (inferred by Schlee *et al.* 1995), and ultrastructural changes (Eversman and Sigal 1987; Tarhanen *et al.* 1997) relative to ozone exposure. Differential response among species is evident (within and across investigations) with some species not exhibiting any decline and others being relatively sensitive. Initially most experiments were short term (hours to a couple of days) at high, sometimes unrealistic, concentrations, but more recently longer-term investigations (months) with treatment conditions reflecting environmental conditions have been conducted (e.g. Scheidegger and Schroeter 1995; Balaguer *et al.* 1996). In spite of the expanding experimental evidence demonstrating O_3 effects, some researchers still claim that lichens are not good biomonitors for ozone (Engelbert and Vonarburg 1995).

15.3   Hydrogen fluoride and lichens

In contrast to the ubiquitous occurrence of ozone in terrestrial environments, fluoride distribution is primarily limited to point sources, such as aluminum smelters, fertilizer plants, glassworks, and brick and ceramic manufacturing. Volatile hydrogen fluoride, which is very difficult to retain, is the primary form of fluoride that causes injury, which may occur at extremely low ambient concentrations (ppb or less). Fluoride accumulates in tissues and causes necrosis, literally death of cells that often become strongly discolored. For lichens, critical accumulation apparently occurs at 20–25 ppb (Gilbert 1971; Nash 1971; Roberts and Thompson 1980), above which necrotic patterns develop. Ascospore germination is also inhibited by fluorides (Belandria *et al.* 1989). In addition, ultrastructure is markedly affected by fluorides by granulation and then breakdown of the thylakoids, the appearance of platogobuli among the thylakoids, and the presence of pseudocrystalline structures in the chloroplast and cyctoplasm (Holopainen 1984; Holopainen and Kärenlampi 1985). All of these effects are different from SO_2 effects on ultrastructure. Fluoride accumulation in lichens has also been documented around volcanoes (Notcutt and Davies 1993, 1999). Areas around aluminum smelters have been extensively investigated (LeBlanc *et al.* 1972; Gilbert 1985; Perkins and Millar 1987; Perkins 1992; Bunce 1996), but other sources have been researched to a lesser extent.

15.4   Organopollutants and lichens

Organopollutants exist in a wide variety of forms related to many aspects of oil exploitation and applications of the modern chemical industry, such as the development of pesticides. To a limited extent lichens have been used to monitor deposition patterns (Section 12.6.4). Coastal lichens show little ability to recover after being exposed to crude oil spills (Walker *et al.* 1978). For a marine lichen, *Lichina pygmaea*, Brown (1972) found that the surfactant in the emulsifier used to disperse the soil was more detrimental to the lichen than the oil itself. Once the oil has been refined and combusted in vehicles a wide range of organics are released to the atmosphere, and at least in the case of aldehydes some lichen sensitivity is known (Zambrano *et al.* 2000). Although agricultural impacts on lichens have primarily emphasized ammonia deposition (Brown 1992; Section 11.3), negative effects of herbicides and pesticides are beginning to be investigated (Hallbom and Bergman, 1979; Alstrup 1992; Modenesi 1993; Brown *et al.* 1995; Jensen *et al.* 1999; Newmaster *et al.* 1999). In addition, pesticide usage is known to adversely affect lichen species richness (Bartók 1999), although the effects up the food chain may be of greater concern (Kelly and Gobas 2001).

16

Lichen biogeography

D. J. GALLOWAY

Biogeography is an important, multidisciplinary area of modern science, which is primarily concerned with current and historical distribution patterns of organisms, and operates through observation and detection of pattern at both small and large scales. It attempts meaningful explanations for these observed patterns. Ganderton and Coker (2005) claim that biogeography, because of its multifaceted nature, encourages people to *share* experiences and expertise, and to see the wider value in the approach of a discipline, with which they may be unfamiliar, or to take a sideways look at their own expertise to see how relevant it is for problem solving in another area. Biogeography functions through the development of hypotheses and theories derived from searching for patterns, testing assumptions and predictions, with new observations. It is above all "... a science of synthesis, bringing in the best aspects of many other disciplines, not only of conventional sciences, but also those of history, mathematics, and language. It relates critically to topics of immediate importance, such as global climate change, the ecology of extinctions, environmental management, and world population growth and resource availability" (Ganderton and Coker 2005). There are now many general biogeography texts available: Humphries and Parenti (1999), Crisci *et al.* (2003), MacDonald (2003), Ganderton and Coker (2005), Cox and Moore (2005), and Lomolino *et al.* (2006).

It is now widely accepted that lichens are an ancient group of fungi (Tehler *et al.* 2003; Liu and Hall 2004), with a fossil record stretching back 400 MA to the Early Devonian in the Rhynie chert deposits in Scotland (Taylor *et al.* 1995*a*, 1997, 2004), and to 600 MA in marine phosphorite of the Doushanto Formation at Weng'an in South China (Yuan *et al.* 2005). In the *Protolichenes Hypothesis*, Eriksson (2005:22) postulates that morphological and recent molecular and paleontological studies indicate that the subphylum Pezizomycota very

Lichen Biology, ed. Thomas H. Nash III. Published by Cambridge University Press.
© Cambridge University Press.

probably evolved from a group of lichenized ascomycetes, the hypothetical group *Protolichenes*. Algae and cyanobacteria were available in abundance long before land plants evolved, and the first ascomycetes probably lived in symbiosis with these. Many types of asci occur in lichenized ascomycetes, indicating that they developed along several evolutionary lines over a long time frame. In some groups, there were losses of the ability to live symbiotically with one or more photobionts, resulting in these groups becoming free-living saprophytes or parasites (Lutzoni *et al.* 2001). In the absence of a well-documented fossil record for lichenized fungi, it is possible to infer ancestry through chemical (Galloway 1991a), biogeographical, and molecular data (Printzen and Lumbsch 2000; Myllys *et al.* 2003; DePriest 2004; Miadlikowska and Lutzoni 2004; Palice and Printzen 2004; Thell *et al.* 2004a, b; Wedin *et al.* 2004, 2005b). A stimulating exchange of views on "Lichens: a special case in biogeographical analysis" was reported in Barreno *et al.* (1998) and serves to underline several problem areas in lichen biogeography. A growing series of papers on lichen biogeography in recent years contain much useful information and include: Egea and Torrente (1994), Jørgensen (1996), Goward and Ahti (1997), Hestmark (1997), Ahti (1999, 2000), Louwhoff (2001), Lücking and Kalb (2001), Adler and Calvelo (2002), Liberatore *et al.* (2002), Litterski and Otte (2002), Follmann (2002), Sipman (2002), Lücking (2003), Lücking *et al.* (2003), Otte *et al.* (2002), Thell *et al.* (2004b), Ertz *et al.* (2005), and Sipman (2006a).

16.1   Patterns of distribution

Lichens show distinctive patterns of distribution at both micro and macro levels, and because they are an ancient group of fungi, their mapped distributions show patterns similar to other major groups of organisms. This makes them ideal subjects for biogeographical analysis, though to date, very few such analyses have been attempted. Some lichens appear to have very restricted ranges; others are extremely widely, if not universally, distributed, while others have ranges between these two extremes. It is currently possible to distinguish at least 16 major patterns of distribution in lichens and these are summarized below. As the use of molecular techniques allows increasingly refined concepts of lichen systematics to emerge, it is expected that these biogeographical patterns of distribution will also be refined and emended.

16.1.1   *Cosmopolitan taxa*

Some widespread lichens occur on all land masses and many oceanic islands. For example, the mazediate, calicioid lichens have a number of cosmopolitan taxa including *Calicium*, *Chaenotheca*, *Chaenothecopsis*, *Cyphelium inquinans*, *Microcalicium arenarium*, and *Mycocalicium subtile* (Tibell 1999), and extremely widespread taxa are also common in *Bacidia*, *Caloplaca*, *Candelariella*, *Catillaria*, *Cladonia* (Stenroos 1993; Goward and Ahti 1997; Stenroos *et al.* 2002*a*, *b*), *Collema*, *Collemopsidium* (Mohr *et al.* 2004), *Diploschistes*, *Graphis*, *Lecanora*, *Lecidea*, *Lepraria*, *Lichina*, *Normandina pulchella*, *Peltigera*, *Pertusaria*, *Porina*, *Porpidia*, *Rhizocarpon*, *Teloschistes*, *Thamnolia*, *Usnea*, *Verrucaria*, and *Xanthoria*. The cosmopolitan sterile, alpine species *Thamnolia vermicularis* is widespread in alpine and periglacial environments of the world and, in spite of an extremely conservative morphology, has two well-defined chemodemes within well-defined ranges (Kärnefelt and Thell 1995). Although it is dispersed locally by thallus fragmentation, the mechanism by which its wider dispersal is achieved is not known.

The Parmeliaceae is one of the largest families of lichen-forming fungi, and has a worldwide distribution (Blanco *et al.* 2006), including the taxa *Flavoparmelia caperata*, *F. soredians*, *Hypotrachyna sinuosa*, *Parmelia sulcata*, *Parmotrema perlatum*, *Punctelia subrudecta*, and *Xanthoparmelia pulla*. The Physciaceae also has many cosmopolitan taxa, including: *Amandinea punctata*, *Buellia griseovirens*, *Hyperphyscia adglutinata*, *Phaeophyscia orbicularis*, *Physcia adscendens*, *Physcia caesia*, *Pyxine subcinerea*, and *Rinodina bischoffii*. In New Zealand, 376 taxa (22% of the lichen mycobiota) are cosmopolitan, while in Antarctica, there is a decline in numbers of cosmopolitan species with increasing climatic severity, ranging from 17% of the lichen mycobiota (South Shetland Islands) to only 7% of the lichen mycobiota for continental Antarctica (Øvstedal and Lewis Smith 2001).

16.1.2   Endemic taxa

An endemic element is often present in the lichen mycobiota of a particular geographical region and may consist of genera of very limited distribution. Macaronesia has a number of taxa endemic to the area (Hafellner 1995), including *Usnea geissleriana*, *U. krogiana*, and *U. macaronesica* (Clerc 2006). In western North America endemic taxa include *Ahtiana pallidula* (Thell *et al.* 1995), *Esslingeriana idahoensis*, *Nephroma occultum* (Goward 1995), *Kaernefeltia californica* (Thell and Goward 1996), and *Waynea californica* (Moberg 1990). *Kaernefeltia californica* was one of a number of endemic taxa (including *Heterodermia erinacea*, *Hypogymnia duplicata*, *Niebla ceruchis*, *N. homalea*, *Pilophorus acicularis*, *Platismastia lacunosa*, *Pseudocyphellaria anthraspis*, and *Ramalina menziesii*) from the west coast of North America first collected in the late eighteenth century by the Scottish botanist Archibald Menzies (Galloway 1995*a*). In neighboring Baja California there are characteristic endemic species such as *Dendrographa leucophaea*, *Dirina mexicana*, *Fellhanera nashii*, and *Hubbsia californica*.

In South America there are a variety of endemic taxa in the Roccellaceae (Follmann 2002).

The island of Socotra in the Indian Ocean has a fascinating lichen mycobiota that contains a wealth of relictual and endemic taxa, including *Feigeana*, *Minksia*, *Roccellographa*, and *Simonyella* (Mies and Printzen 1997).

In New Zealand, 23% of the lichen mycobiota are endemic taxa (Galloway 2007), and include: *Argopsis megalospora*, *Austrella brunnea*, *Bunodophoron palmatum*, *Byssoloma adspersum*, *Caloplaca erecta*, *Cladonia pulchra*, *Dirina neozelandica*, *Fuscoderma applanatum*, *Labyrintha implexa*, *Lobaria asperula*, *Megalospora bartlettii*, *Menegazzia dielsii*, *Parmeliella rakiurae*, *Peltularia crassa*, *Phyllisciella aotearoa*, *Placopsis durietziorum*, *Pseudocyphellaria rufoviresens*, *Ramalina inflexa*, *Steinera polymorpha*, *Sticta colinii*, *Umbilicaria murihikuana*, and *Zahlbrucknerella compacta*. Australia has many endemic taxa (McCarthy 2006), including: *Hertelidea eucalypti*, *H. geophila* (Printzen and Kantvilas 2004), *Hueidea* (Kantvilas and McCarthy 2004), *Maronina australiensis* (McCarthy 2004), *Polyblastia australis* (McCarthy 2001), *Siphulella* (Johnston 2001), and *Steinia australis* (Kantvilas 2004). The lichen mycobiota of Antarctica has a high degree of endemism, ranging from 24% (South Georgia) to 50% (continental Antarctica) of the presently known lichen mycobiota (Øvstedal and Lewis Smith 2001).

Endemic taxa are of considerable biogeographical interest; they may either represent the emergence of "new" genera or species from ancestors over a period of isolation, or reflect an "old" or "relict" distribution of a group surviving after widespread extinctions. For example, the rich speciation in usnic-acid-containing species of *Xanthoparmelia* reflects adaptations to increasingly local arid conditions in Africa and Australia (Hale 1990; Elix 1994; Blanco *et al.* 2004), and these may provide an example of the first case. Concomitant with the northward drift of Australia and South Africa into lower, drier latitudes has been the development of a high proportion of rare, endemic species with limited morphological variation but diverse chemistry. The long isolation of New Zealand, Australia, and Antarctica from southern South America has resulted in a high diversity of endemic species in genera such as *Menegazzia* (Bjerke 2005), *Placopsis* (Galloway 2002, 2004*a*, *b*), *Pseudocyphellaria*, and *Sticta* (Galloway 1994*b*, 1997, 2001), with many taxa being vicariants of those occurring in southern South America.

16.1.3   Austral taxa

The biotas of the major southern hemisphere land masses have many similarities with disjunct distributions observed between taxa in New Zealand, South America, and the subantarctic islands (Sanmartín and Ronquist 2004). Such disjunctions are also observed in austral lichen mycobiotas, although the affinities are most noticeable at the generic, rather than the species, level. Two major groupings are found in austral lichen mycobiotas (Galloway 1991*b*, 1996, 2008).

1. Paleoaustral lichens are thought to represent primitive Gondwanan groups poorly adapted for long-distance dispersal. These would date from the Cretaceous (or earlier), when the Protopacific margin of Gondwanaland was available for colonization at times when cool temperate conditions prevailed and a vegetated West Antarctica may have been linked to South America and the South Pacific land areas. Paleoaustral lichens share several features: they occur in cool temperate biomes, often in forests dominated by species of *Nothofagus*, or else in shrublands or grasslands; they have been flexible enough to colonize or recolonize cool temperate habitats after periods of climatic change; and they show disjunct distributions. Examples are: *Bartlettiella fragilis*, *Brigantiaea phaeomma*, *Bryoria austromontana*, *Caloplaca cribrosa*, *Degelia gayana*, *Degeliella versicolor*, *Erioderma leylandii* ssp. *leylandii*, *Haematomma nothofagi*, *Lecidea lygomma*, *Leioderma pycnophorum*, *Mycobilimbia australis*, *Nephroma cellulosum*, *Pannaria sphinctrina*, *Pannoparmelia angustata*, *Placopsis bicolor*, *P. perrugosa*, *P. stenophylla*, *Pseudocyphellaria faveolata*, *Pyrrhospora laeta*, *Umbilicaria zahlbruckneri*, and *Wawea fruticulosa*.

2. Neoaustral lichens are taxa dispersed after the fragmentation of Gondwanaland, mainly between post-Oligocene and the present. They commonly produce copious quantities of vegetative diaspores that allow long-distance transport via birds, ocean currents, or in the West Wind Drift. Presently, dispersal scenarios appear to be increasingly *de rigueur* in explanations of austral disjunctions (Sanmartín and Ronquist 2004; Nathan 2005; de Queiroz 2005; McGlone 2005; Cook and Crisp 2005). Examples of neoaustral lichens include: *Bunodophoron patagonicum*, *Caloplaca cirrochrooides*, *Calycidium polycarpum*, *Fuscoderma amphibolum*, *Leifidium tenerum*, *Menegazzia globulifera*, *Pannaria durietzii*, *P. leproloma*, *Parmelia cunninghamii*, *Parmeliella granulata*, *Placopsis microphylla*, *P. rhodocarpa*, *Pseudocyphellaria glabra*, *P. mallota*, *Psoroma soccatum*, *Steinera sorediata*, *Stereocaulon corticatulum*, and *Trapeliopsis congregans*.

In Valdivian rain forests, 11% of the foliicolous lichen mycobiota recorded by Lücking *et al.* (2003) comprise a distinctive austral element and include: *Badimiella pteridophylla*, *Caprettia setifera*, and *Kantvilasia hians*, formerly known only from rain forest biomes in New Zealand and Tasmania. Austral areas are well covered by lichen checklists and/or floras (Feuerer 2006), and include Argentina (Calvelo and Liberatore 2001), Australia (McCarthy 2006), Chile

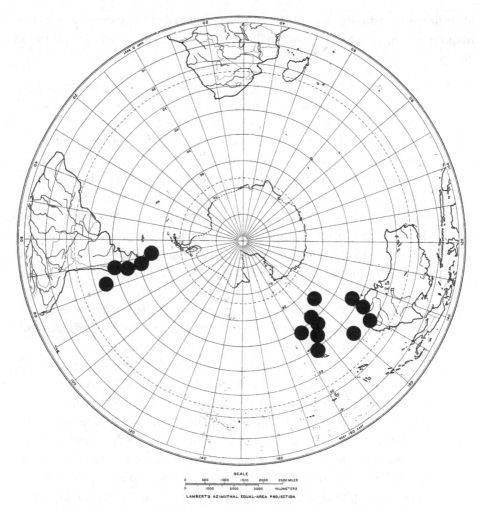

Fig. 16.1 Known distribution of *Pseudocyphellaria glabra*.

(Galloway and Quilhot 1999), New Zealand (Malcolm and Galloway 1997; Pennycook and Galloway 2004; Galloway 2007) and Antarctica (Øvstedal and Lewis Smith 2001).

In view of the importance of the Antarctic ozone hole and its possible effects on lichen communities at high altitudes and latitudes (Galloway 1993; Quilhot *et al.* 1998; Bjerke *et al.* 2002), the neoaustral lichen *Pseudocyphellaria glabra* (Fig. 16.1) has considerable importance as a test organism in monitoring levels of cortical UV-B screening compounds (e.g. usnic acid) throughout its disjunct distribution.

16.1.4   Bipolar taxa

Taxa with bipolar distributions are present in both the northern and southern hemispheres, especially at high latitudes (i.e. in the boreal/arctic and austral zones). They are absent, or very nearly so, from the tropics. Nearly all bipolar lichens occupy similar alpine or tundra-like habitats in both hemispheres, including grasslands, fens, bogs, or fellfield with cushion vegetation and patterned ground. Dwarf shrubs may be present, but trees are largely absent. Bipolar lichens are frequently recorded in periglacial environments, especially in the subantarctic islands. Alpine regions, in which bipolar lichens occur, have long histories of stable, open, unchanged conditions. They are areas of long isolation and are often unusual habitats in areas of high ecological and geographical diversity.

Within the overall pattern of bipolar distribution, six different patterns of bipolar lichen disjunctions are distinguishable: (1) South America–northern hemisphere, (2) South America–Australasia–northern hemisphere, (3) New Zealand–northern hemisphere, (4) Australia–northern hemisphere, (5) circumboreal–circumantarctic, and (6) circumboreal–tropical mountains–circumantarctic. Bipolar lichens account for 10% of the New Zealand lichen mycobiota, some 171 taxa in 81 genera (Galloway 2007), and between 24% (South Sandwich Islands) and 41% of the South Orkney Islands lichen mycobiotas (Øvstedal and Lewis Smith 2001). Examples of bipolar lichens include: *Alectoria nigricans*, *Arthrorhaphis alpina*, and *A. citrinella* (Fig. 16.2), *Bellemera alpina*, *Brigantiaea fuscolutea*, *Bryonora castanea*, *Caloplaca tornoensis*, *Catillaria contristans*, *Cetrariella delisei*, *Cladonia ecmocyna*, *C. mitis* (Myllys et al. 2003), *Frutidella caesioatra*, *Icmadophila ericetorum*, *Lecanora cavicola*, *Lecidella wulfenii*, *Leptogium britannicum*, *Ochrolechia frigida*, *Physconia musicgena*, *P. perisidiosa* (Otte et al. 2002), *Pseudocyphellaria norvegica*, *Psoroma paleaceum* (Timdal and Tønsberg 2006), *Schaereria fabispora*, *Solorina crocea*, *Umbilicaria grisea*, *Usnea sphacelata* (Wirtz et al. 2006), and *Xanthoria elegans*.

Bipolar distributions are found in a large range of systematic groups, both terrestrial and marine, in fish, Mollusca, Foraminifera, Coleoptera, flowering plants, and bryophytes, as well as in lichens. Evidence suggests that bipolar patterns, in some groups at least, are an ancient phenomenon. For additional data on bipolar lichens see Galloway and Aptroot (1995) and Galloway (2003, 2007). The high species richness of the lichen mycobiota of the Skibotn region of northern Norway (709 species) recently recorded by Elvebakk and Bjerke (2006a, b), contains a high number of bipolar taxa.

16.1.5   Paleotropical taxa

The paleotropics include most of Africa, the Arabian Peninsula, most of the Indian subcontinent and the Malesian Archipelago, and the islands of the Pacific Ocean. Many lichen genera have taxa showing such a distribution and include: *Arthonia peraffinis*, *Bactrospora metabola*, *Brigantiaea lobulata*,

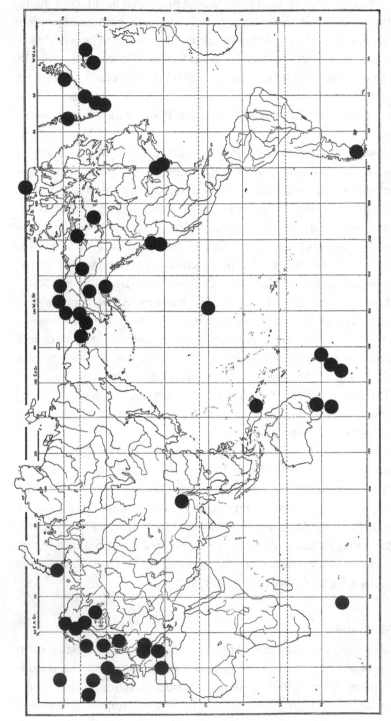

Fig. 16.2 World distribution of *Arthrorhaphis alpina* and *A. citrinella* compiled from literature references and herbarium specimens.

Chroodiscus macrocarpus, Cladia aggregata, Enterographa pallidella, Erioderma sorediatum, Leioderma sorediatum, Loxospora septata, Miltidea ceroplasta, Pannaria fulvescens, Parmelinopsis swinscowii, Phylloporis viridis, Physcia crispa, P. poncinsii, Placopsis fusciduloides, Porina exocha, Pseudocyphellaria argyracea, P. pickeringii, Ramalina celastri, Stereocaulon delisei, Teloschistes velifer, Trichothelium javanicum, Umbilicaria umbilicarioides, Usnea trichodeoides, and *Xanthoparmelia taractica.*

16.1.6   Neotropical taxa

The neotropics comprise tropical South America and include much of Mexico and the Caribbean islands. A number of lichen genera speciate richly in this area, often with high degrees of endemism. Among these may be mentioned *Bactrospora* (Egea and Torrente 1993), Cladoniaceae (Ahti 2000), *Erioderma* (Jørgensen and Arvidsson 2002), hairy species of *Leptogium* (Jørgensen 1997), *Lobaria* (Yoshimura and Arvidsson 1994; Yoshimura 1998), *Parmeliella* (Jørgensen 2000b), *Peltigera* (Vitikainen 1998), and *Sticta* (Galloway 1994b, 1995b; Galloway and Thomas 2004). For additional information on neotropical lichens, see also Aptroot and Sipman (1997) and Sipman (2002, 2006a). A key to neotropical lichens is available on the web (Sipman 2006b).

16.1.7   Pantropical taxa

Pantropical taxa have warm temperate affinities and are found on all major land masses in tropical regions. They are probably derived from equatorial Tethyan environments and habitats and in the Pacific Basin are best developed in lowland habitats. Recent checklists of the foliicolous lichen mycobiota of Mexico (Herrera-Campos *et al.* 2004), and of the lichen mycobiota of Vietnam (Aptroot and Sparrius 2006) for instance, show substantial numbers of taxa from such well-known pantropical genera as: *Aderkomyces, Agonimia, Arthonia, Asterothyrium, Aulaxina, Byssoloma, Chroodiscus, Coccocarpia, Coenogonium, Dirinaria, Fellhanera, Fissurina* (Staiger 2002), *Glyphis* (Staiger 2002), *Graphis, Gyalectidium* (Ferraro *et al.* 2001), *Heterodermia, Hypotrachyna* (Divakar *et al.* 2006), *Malcolmiella, Mazosia, Parmotrema, Phyllopsora, Porina, Pyxine, Sticta, Strigula, Trichothelium, Trypethelium,* and *Usnea. Pseudocyphellaria clathrata* illustrates such a pantropical distribution (Fig. 16.3).

16.1.8   Australasian taxa

There are extremely close links between the floras and vegetation of Australia and New Zealand, which have existed before the isolation of New Zealand through the formation of the Tasman Sea (Pole 1993, 1994; McGlone *et al.* 2001; Lee *et al.* 2001; Winkworth *et al.* 2002, 2005; McDowall 2004; Sanmartín and Ronquist 2004; McGlone 2005; Waters and Craw 2006). At

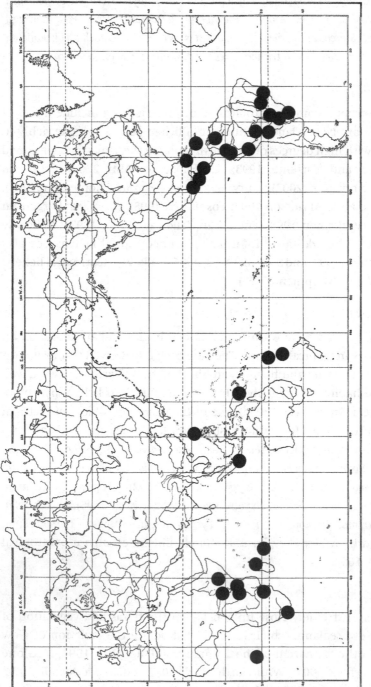

Fig. 16.3 Known distribution of *Pseudocyphellaria clathrata*.

present 304 taxa (18%) in the New Zealand lichen mycobiota also occur in Australia (Galloway 2007). Most shared species are found in southeastern Australia and Tasmania, indicating a close floristic and biogeographical relationship (Galloway 1991*b*, 2007; Rogers 1992). The cool temperate Australasian element is best seen in forest, scrub, and alpine/subalpine grassland vegetation in New Zealand, Tasmania, and southeastern Australia. The climate in these areas is very similar and is marked by oceanic, humid, cloudy conditions with frequent precipitation. Lichens of this element are also commonly associated with species of *Nothofagus*. Macrolichens of Tasmania and southern New Zealand are similar (Kantvilas and Jarman 1999; Kantvilas *et al.* 2002), with *c.* 80% of the Tasmanian species being shared with New Zealand, but with diversity in New Zealand being much greater.

Some of these disjunct Australasian lichens (e.g. *Fuscoderma subimmixta* in Fig. 16.4) are presumed ancient relicts that evolved in Cretaceous times in cool temperate habitats around the protopacific margin of Gondwanaland. They later became isolated in cool temperate habitats on the New Zealand microcontinent (referred to as Neozelandica by geologists), Tasmania, and the uplands of southeastern Australia, after rifting and drift occurred subsequent to the opening of the Tasman Sea and Bass Strait. New Zealand was isolated from eastern Gondwanaland for longer than Tasmania and would have had more opportunities for increased speciation events, as has apparently occurred in *Bunodophoron* (Wedin 1995), *Placopsis*, and *Pseudocyphellaria* (Galloway *et al.* 2001; Galloway 2007).

In addition to relict or paleoaustral elements in the lichen mycobiotas of Tasmania and New Zealand, a distinctive and more recent relationship exists between Australia and New Zealand with many taxa derived from Australia, transported by the prevailing westerly winds (Cook and Crisp 2005; McGlone 2005; Waters and Craw 2006). Additions to the lists of Australasian lichens quoted in Galloway (1991*b*, 1996) include (asterisks refer to lichenicolous fungi): *Arthonia tasmanica*, *Arthrorhaphis citrinella* var. *catolechioides*, *Austropeltum glareosum*, *Bactrospora arthonioides*, *Bunodophoron flacidum*, *Calycidium cuneatum*, *Chroodiscus lamelliferus*, *Cladonia cucullata*, *C. cyanopora*, *C. darwinii*, *C. glebosa*, *C. nudicaulis*, *Degeliella rosulata*, *Enterographa bella*, *Lecanora austrooceanica*, *Leptogium biloculare* (Galloway 1999), *Loxospora solenospora*, *Notocladonia cochleata*, *N. undulata* (Hammer 2003), *Pannaria centrifuga*, *Parasiphula complanata* (Grube and Kantvilas 2006), *Parmeliella subtilis*, *Peltigera tereziana*, *Perigrapha nitida* (Ertz *et al.* 2005), *Placopsis brevilobata*, *Plectocarpon bunodophori*, *P. gallowayi* (Ertz *et al.* 2005), *Polycoccum crespoae*, *Psoroma geminatum*, *Pyrenopsis tasmanica*, *Ramboldia stuartii*, *Santessoniella pulchella*, *Stirtoniella kelica* (Galloway *et al.* 2005), *Topeliopsis decorticans*, *Xanthoparmelia alexandrensis*, *X. amplexula*, *X. cheelii*, *X. exillima*, *X. notata*, *X. reptans*, *X. semiviridis*, *X. thamnoides*, *X. verdonii*, and *X. waiporiensis*.

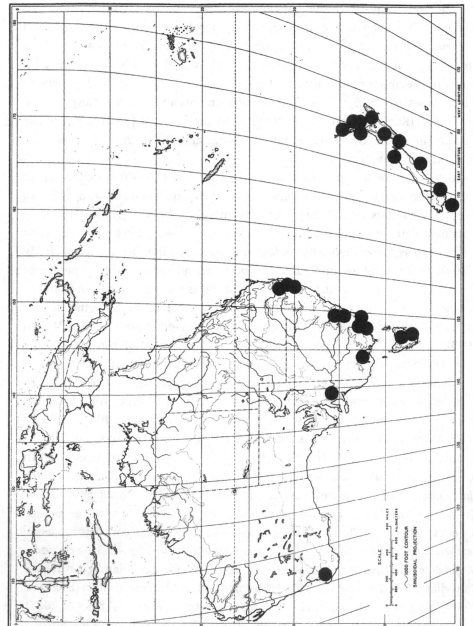

Fig. 16.4 Known distribution of *Fuscoderma subimmixta*.

16.1.9   Western Pacific taxa

Taxa in this element, often called the Indo-Malayan element, are most strongly represented in southeast Asian lichen mycobiotas. They also occur to some extent northwards to Japan, westwards to India and even Africa, occasionally south and east to northeastern Australia, and even to New Zealand. Paleotropical taxa also have this distribution, but they generally range more widely eastwards into the Pacific, often as far east as the Galapagos or Juan Fernandez archipelagos. Taxa with such a distribution include: *Agonimia pacifica*, *Amandinea decedens*, *Calopadia subcoerulescens*, *Coccotrema porinopsis*, *Dictyonema moorei*, *Diploschistes sticticus*, *Heterodermia microphylla*, *Lecanora galactinica*, *Pannaria globuligera*, *Pannoparmelia wilsonii*, *Parmelia erumpens*, *Porina corrugata*, *Rinodina reagens*, *Thysanothecium scutellatum* (Fig. 16.5), and the genera *Cetrelia*, *Nephromopsis* (Randlane and Saag 1998; Thell *et al.* 2005), *Parmelaria*, and *Parmelinella* (Randlane *et al.* 1997; Blanco *et al.* 2006).

16.1.10   Circum-Pacific taxa

Several lichens show a wide distribution around the Pacific Ocean, essentially following the margins of the Pacific plate. These include: *Caloplaca rosei*, *Chaenothecopsis sanguinea*, *Hypogymnia pulverata*, *Leioderma sorediatum* (Fig. 16.6), *Mastodia tessellata*, *Parmelia kerguelensis*, *Placopsis cribellans*, *Spilonema dendroides*, and *Xanthoparmelia digitiformis*.

16.1.11   Atlantic taxa

A number of lichens are restricted to Atlantic islands from the Faeroes to Macaronesia (see map in Galloway 1994a: 392). Other species occurring in these areas are also found in oceanic habitats of the British Isles, western Europe or Scandinavia (see Hafellner 1995; Jørgensen 1996; Berger and Aptroot 2003; Jørgensen 2005). These include: *Bactrospora homalotropa*, *Byssoloma croceum*, *Degelia atlantica*, *Fuscopannaria atlantica*, *Gomphillus calycioides*, *Herteliana taylorii*, *Lepraria umbricola*, *Leptogium burgessii*, *L. hibernicum*, *Micarea coppinsii*, *Opegrapha multipuncta*, *Porina atlantica*, *Pseudocyphellaria lacerata*, *Pyrenula hibernicum*, *Sticta canariensis*, and *Thelotrema isidioides*.

16.1.12   Eastern North America – western European (amphi-Atlantic) taxa

The present-day distribution of biota in the North Atlantic is largely a product of Quaternary climate and environmental change. Two contrasting hypotheses (see Brochmann *et al.* 2003) are currently invoked to explain observed distributions: (1) the *tabula rasa* (clean slate) hypothesis argues that the whole biota has immigrated after the last glacial period, and (2) the *glacial*

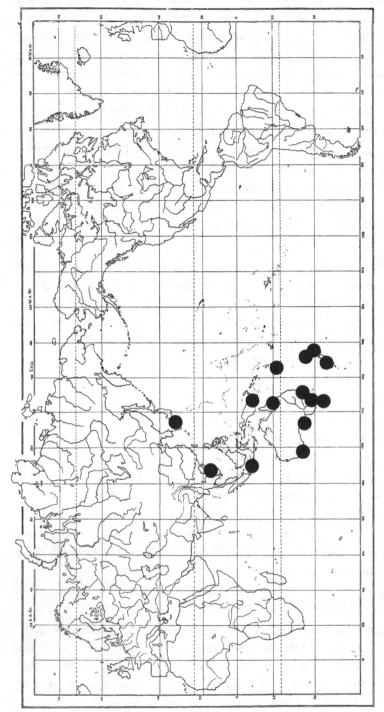

Fig. 16.5 Known distribution of *Thysanothecium scutellatum*.

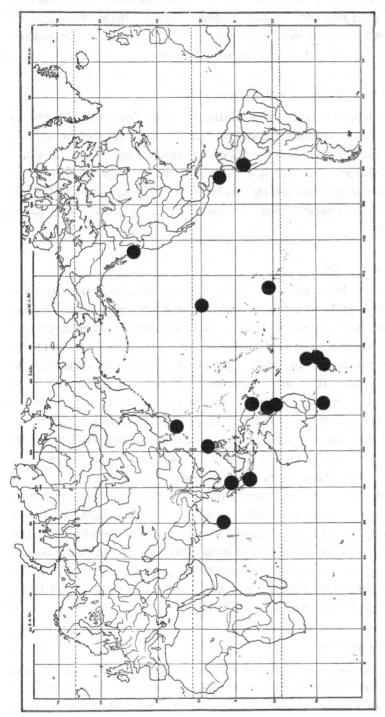

Fig. 16.6 Known distribution of *Leioderma sorediatum*.

survival hypothesis invokes persistence of relics in ice-free refugia during the Pleistocene glaciations. A small number of lichens show an amphi-Atlantic distribution (Dahl 1998) and include: *Cladonia strepsilis*, *Erioderma pedicellatum* (Scheidegger 1998), *Lasallia pustulata*, *Pycnothelia papillaria*, *Rhizocarpon timdalii* (Ihlen and Fryday 2002), and *Stereocaulon dactylophyllum*.

16.1.13   *Western North American – western European taxa*

Several lichen taxa are confined either to mountains or to highly humid climates of western North America and western Europe. They include *Agyrium rufum*, *Anzina carneonivea*, *Buellia crystallifera*, *Cavernularia hultenii* (Printzen and Ekman 2002; Printzen *et al.* 2003), *Cliostomum leprosum* (Goward *et al.* 1996; Ekman 1997), *Gyalideopsis alnicola*, *Japewia subaurifera*, *Kaernefeltia merrillii* (Thell and Goward 1996), *Lecidella laureri*, *Mycoblastus caesius*, and *Rinodina disjuncta* (Tønsberg 1992).

16.1.14   *Mediterranean taxa*

Regions of the world with Mediterranean climates frequently have physiognomically similar vegetation, which is highly adapted to fire. A good discussion of phytogeographic patterns in the Mediterranean lichen mycobiota is given in Nimis (1996), who recognizes four, broadly defined phytoclimatic elements: (1) a northern element; (2) a temperate element; (3) a humid, sub-tropical element; and (4) an arid, subtropical element. *Physconia venusta*, an example of a "Mediterranean" lichen, is mapped by Litterski and Otte (2002). Several lichens are known from southern Europe, northern Africa, Macaronesia, and southern California/northern Baja California (Barreno 1991; Nash *et al.* 2002, 2004) and include taxa best known from arid or semi-arid areas, for example *Anema progidulum*, *Aspicilia desertorum*, *Chromatochlamys muscorum*, *Culbersonia nubila*, *Digitothyrea*, *Diploschistes diacapsis*, *Fulgensia desertorum*, *Fuscopannaria mediterranea*, *Heppia* spp., *Koerberia sonomensis*, *Peltula psammophila*, *Placidium fringens*, *Porocyphus coccodes*, *Toninia tristis*, and *Tornabea scutellifera*. Several lichen checklists from the Mediterranean region are available and include the Iberian Peninsula and the Balearic Islands (Llimona and Hladun 2001); Italy (Nimis and Martellos 2003); Morocco (Egea 1996); Tunisia (Seaward 1996*a*); Israel (Galun and Mukhtar 1996), and Mediterranean Turkey (John 1996).

16.1.15   *American–Asian taxa*

This disjunction, especially the eastern North American – eastern Asian pattern, is extremely ancient and is also known in seed plants and bryophytes. Lichens also showing this disjunction include species in *Allocetraria*, *Cetrelia*,

Collema, *Erioderma* (Jørgensen 2001), *Fuscopannaria* (Jørgensen 2000a), and *Oropogon* (Esslinger 1989).

16.1.16   South American – African taxa

Kärnefelt (1990) discusses and maps several lichen taxa that are disjunct between South America and southern Africa. These include: *Caloplaca elegantissima*, *C. isidiosa*, *C. ochraceofulva*, *C. subunicolor*, *Coelopogon epiphorellus* (Fig. 16.7), *Leprocaulon gracilescens*, *Peltula clavata*, *Placidium acarosporoides*, *Umbilicaria haplocarpus* (Hestmark 1997), and *Xanthomendoza mendozae* (Kondratyuk and Kärnefelt 1997).

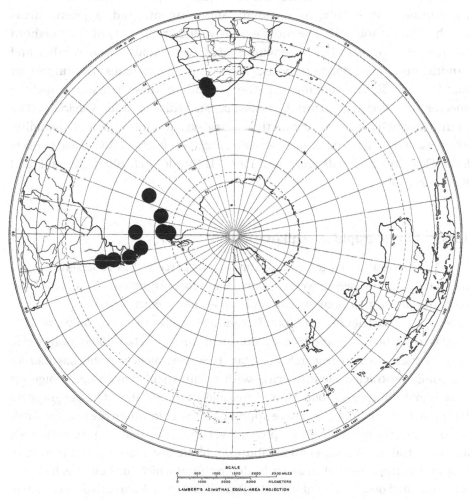

Fig. 16.7   Known distribution of *Coelopogon epiphorellus* (after Kärnefelt 1990).

16.1.17   *Southern xeric taxa*

Several lichens share a distribution pattern of southern Africa, Western Australia, South Australia and the dry, inland lake basins of southern New Zealand, with habitats in these regions characterized by winter-rainfall, summer-drought climates. This is a trans-Indian Ocean element which is well-documented in the distinctive Cape flora (Galley and Linder 2006). Examples are: *Digitothyrea rotundata, Diploschistes hensseniae, Xanthoparmelia subimitatrix*, and *X. tegeta*.

16.1.18   *Boreal arctic–alpine taxa*

This element is developed in the holarctic region of the northern hemisphere, including most of North America, Europe, and Asia but excluding the Arabian Peninsula, the Indian subcontinent, and adjacent areas south of the Himalayas. The most extensive lichen biomass of the northern hemisphere occurs in lichen-dominated woodlands, heathlands, and tundra throughout northern boreal forest and tundra zones. The important mat-forming lichens dominating the ground in these situations include species of *Alectoria, Cetraria, Cladonia, Stereocaulon*, and *Vulpicida*. The family Parmeliaceae has distinctive boreal–arctic–alpine genera including *Allantoparmelia, Allocetraria, Arctoparmelia, Asahinea, Brodoa, Coelocaulon, Cornicularia, Flavocetraria, Pseudevernia*, and *Vulpicida* (Brodo *et al.* 2001; Elvebakk and Bjerke 2006*b*).

16.2   Biogeographical explanations

Distribution patterns traditionally are perceived as representing affinities or elements (see above) in a particular biota. Arguments or models have been developed that help explain and sometimes predict observed patterns of geographical distribution of biota. Recent textbooks of biogeography, such as MacDonald (2003), Cox and Moore (2005), and Lomolino *et al.* (2006), summarize current methods and research agendas. However, biogeography has had an unusually protracted identity crisis with its aims still remaining ambiguous (Ebach and Humphries 2003; Humphries and Ebach 2004), and biogeographers failing to develop fully integrative approaches to determining the roles of Earth history and ecology in the geography of diversification (Riddle 2005). Vicariance and dispersal are fundamental attributes of biotic distributions, and phylogeography has the potential to assist in determining which of these mechanisms has generated observable patterns (McDowall 2004; Cook and Crisp 2005; Cowie and Holland 2006; Crisci *et al.* 2006).

16.2.1   Migrationist biogeography

The migrationist or dispersalist tradition has wide currency among biogeographers. It has its roots in the views of Darwin and Wallace, who invoked a theory of chance migration from single centers of origin. The distributions of taxa are mapped and interpretations drawn. Although there is no generally agreed method of analysis, centers of origin are frequently proposed, as are routes and modes of migration. These are assumed to be the major cause of presently observed distributions. Ecology is also regarded as a major factor in determining distribution. For instance, climate (especially water availability) as a limiting factor in the distribution of the biota, especially lichens, in the Antarctic terrestrial environment is discussed in Kappen (2000, 2004) and Øvstedal and Lewis Smith (2001). No special use of taxonomy is made in analysis involving migrationist models. Taxa occurring in presumed centers of origin (diversity) are sometimes assumed to be primitive, sometimes advanced. Many examples of migrationist explanations of distribution patterns are found in the lichenological literature, e.g. Cladoniaceae (Goward and Ahti 1997), Parmeliaceae (Louwhoff 2001), foliicolous lichens (Lücking and Kalb 2001; Lücking 2003), *Siphula* (Kantvilas 2002), North American lichens (Weber 2003), *Physconia* (Otte *et al.* 2002), *Cladonia* (Litterski and Ahti 2004), and *Melanelia* (Otte *et al.* 2005).

16.2.2   Cladistic biogeography

Cladistic or vicariance biogeographers use cladistics to generate a taxonomic cladogram for each group of taxa studied. They then replace each taxon with the area in which it occurs, to produce an area cladogram. Area cladograms from different groups are compared to determine congruence. Distribution is mapped as presence/absence in area units. It is assumed that an ancestral species occurred over the whole area and that parts were isolated by major geological events, such as plate tectonics, mountain building, and various geomorphological processes, giving rise to allopatric speciation. Recent developments and applications of cladistic biogeography are given in Humphries and Parenti (1999), Brooks (2004), Humphries and Ebach (2004), and Morrone (2005). There is an increasing number of studies utilizing cladistic approaches in the lichenological literature, of which the following are recent examples: Ekman (2001), Stenroos *et al.* (2002b), Gaya *et al.* (2003), Tibell (2003), Tehler *et al.* (2004), and Wirtz *et al.* (2006).

16.2.3   Panbiogeography

The panbiogeographic method (Craw *et al.* 1999; Grehan 2001; Heads 2004, 2005) was developed by Leon Croizat, whose important insight into biotic distribution patterns was that they are highly repetitive. These repetitive patterns he called "generalized tracks" – lines on a map that summarize distributions of many diverse individual taxa. He rejected chance dispersal as the cause of generalized tracks, because the pattern is independent of the individual taxon's dispersal abilities. The "main massing" is emphasized rather than the limits of distribution. Tracks applied to global patterns of distribution have no correlation whatever with modern geography but extend across all of the world's oceans and continents.

Distributions are generalized by identifying major tracks (tracks correlate with former zones of disturbance, such as geological folds or fracture zones, or coastlines and their environs), nodes (track junctions, and centers of endemism), and gates (major nodes). One such node occurs around the Tasman and Coral Sea, where major crustal deformation occurred throughout the Paleozoic, Mesozoic, and Cenozoic eras (Heads 2001, 2004). The method avoids assumptions of any particular mode of dispersal, although adherents emphasize geological history and de-emphasize migration for some taxa. However, it is accepted that certain "weedy" taxa are actively migrating. Terrane tectonics is often of special interest and were used to explain lichen distributions of species of *Anzia, Icmadophila splachnirima, Pannaria, Peltigera*, and *Pseudocyphellaria* along the alpine fault in New Zealand (Heads 1997, 1998; Heads and Craw 2004). But panbiogeography as a research agenda has been well and truly deconstructed in recent times (McDowall 2004), and McGlone (2005) feels that "perhaps it can now be decently laid to rest."

16.2.4   *Phylogeography*

Phylogeography is a recent and fast-developing field of enquiry that studies processes controlling the geographic distributions of lineages by analyzing the genealogies of populations and genes. It takes a population genetic and phylogenetic perspective on biogeography; and historical population movements such as range expansions, vicariant fragmentation, and various dispersal events are invoked to explain current geographic distributions patterns of populations and of genetic lineages (Avise 1998; Riddle and Hafner 2004). Current development of phylogeography as a discipline has concentrated on the historical *phylo-* component through the utilization of phylogenetic analysis. In contrast, the spatial-*geographic* component is not a prominent feature of many existing phylogenetic approaches and has often been dealt with in a relatively naive fashion (Kidd and Ritchie 2006). Phylogeography of the lichen *Cavernualria hultenii*, which is widely disjunct across North America and Europe, is discussed by Printzen *et al.* (2003). Increasing numbers of phylogeographic investigations of lichen populations and communities are to be expected, consequent upon the current, rapid uptake of phylogenetic methods in systematic lichenology. See, for example, recent work on parmelioid lichens (Blanco *et al.* 2006; Divakar *et al.* 2006) and the *Neuropogon* group of *Usnea* (Wirtz *et al.* 2006).

16.3   Lichen biogeography and conservation issues

Over the past three decades nearly all arguments about the conservation of nature on this planet have involved the issue of biological diversity and ways to preserve it (Groombridge 1992; Heywood 1995; Humphries *et al.* 1995; Margules and Pressey 2000). Lichens have not remained immune from such enquiries, and several studies testify to the contemporary importance of conservation of lichens and their habitats (Brown *et al.* 1994; Goward 1994, 1995, 1996; Wolseley 1995; Kantvilas 2000; Johnson and Galloway 2002). In their recent book, Ganderton and Coker (2005: chapter 15) state that "in terms of biogeography, conservation can be seen as one more element in the dynamic interactions between species and their natural environment" an interaction that seeks to maintain a desired range of taxa and habitats often against prevailing ecological and environmental changes. Many aspects of our environment are changing locally or globally as a result of human activities. These include temperature, carbon dioxide, rainfall, UV radiation, ozone, acidification, and nitrification, with some of these changes having direct effects on lichen communities (Galloway 1993). Conservation strategies, such as selection of protected areas, now need to take into account changes in environmental factors and human-induced events, such as rapid climate change that may shift the environmental conditions of a protected area in a way that may no longer support the taxa or ecosystems the protected area attempted to conserve (Crisci *et al.* 2006). How current views on phylogenetic information might be applied in conservation management is discussed by Purvis *et al.* (2005). In this compilation, four main areas are identified: (1) whether the "species" is the most appropriate unit of conservation, (2) how knowledge of the origin of biodiversity patterns might help to conserve them, (3) how phylogeny can assist in understanding what factors underpin extinction risk, and (4) what the evolutionary past can tell us about the evolutionary future.

Lichens, over their long evolutionary history, have seen many catastrophic changes to the planet's terrestrial environments and are probably much better placed, with their unique microbial symbiotic systems, than we are to survive some future anthropogenic mass extinction episode.

17

Systematics of lichenized fungi

A. TEHLER AND M. WEDIN

Historically, lichen fungi have through most of the nineteenth and twentieth centuries been arranged in its own class, Lichenes, based on their symbiotic life form as expressed by their composite thalli. By convenience, the name of the lichen fungus is usually but inaccurately applied to the feature referred to as a lichen, as if that was an organism. Actually, lichens are small ecosystems (Section 1.6), comprising associations with two or more components, an algal producer and a fungal consumer. The components are individual organisms; lichens are not. Consequently, lichens per se cannot be classified into natural systems because they have no phylogeny. Today lichen fungi are classified together with other chitinous fungi and incorporated into a common fungal system.

17.1   Systematics

Systematics – the science of studying the diversity and hierarchy of nature – is not only the oldest natural science, but is also a science where the modern development is progressing at a dramatic pace. Systematics is built up by four major parts: *taxonomy* (the delimitation and description of taxa), *nomenclature* (the formal naming of taxa), *phylogeny* (the natural relationships among taxa), and *classification* (the organization of taxa into a hierarchical system). In the present treatment, we focus on the phylogeny and classification of lichen fungi. Taxonomic methods and nomenclatural rules and principles are beyond the scope of the present book.

Humans have always had a need to classify organisms according to useful principles. The current principle, that a classification should reflect their natural relationships (the phylogeny) resulting in higher taxa made up by

Lichen Biology, ed. Thomas H. Nash III. Published by Cambridge University Press.

genealogically closely related lower taxa, is comparatively recent. This principle goes back to ideas from Haeckel, Darwin, and Hennig. The now generally accepted principle that taxa should be monophyletic (i.e. include all descendants from a common ancestor, and only these) was introduced by Haeckel, and strongly advocated by Hennig, who claimed that the monophyly criterion is what makes taxonomic groups natural. Monophyletic groups are characterized by uniquely derived traits – synapomorphies – that make them possible to distinguish through phylogenetic analysis. Nonmonophyletic, paraphyletic groups are usually recognized by the lack of the same feature that was used to characterize the corresponding monophyletic group, for example the phanerogams (presence of seed) and the cryptogams (lack of seed). The paraphyletic nature of the cryptogams becomes evident when realizing that all other organisms outside the phanerogams lack seeds, not just the cryptogams. Polyphyletic groups are recognized by having one or more character that has evolved independently in different unrelated groups.

17.2   History

Before Schwendener's revolutionary discovery (Schwendener 1869), it was not understood that lichens were actually "double-organisms" composed of one or several algal components, the photobiont or photobionts, and a fungal component, the mycobiont. Before that date lichens were indeed thought of as single organisms and accordingly were treated as a systematic unit, *Lichenes*, separate from fungi, mosses, and algae.

From a nomenclatural point of view the starting date for lichen fungi is the same as that for phanerogams, namely Linnaeus "Species Plantarum" (Linnaeus 1753), and thus different from that of other fungi. Despite the excellence of Linnaeus' work in classifying the flowering plants, it contained only a few lichen fungi, "lichens." Furthermore, Linnaeus, who lumped only some 80 different species into the single genus *Lichen*, treated these and other lichens very inadequately. The Swedish scientist Erik Acharius (1757–1819), also considered as the father of lichenology, erected the first proper classification of lichen fungi, "lichens." He accomplished this in a series of publications beginning with *Lichenographiae Svecicae Prodromus* (Acharius 1798), continuing with *Methodus qua omnus detectos Lichenes* (Acharius 1803), *Lichenographia universalis* (Acharius 1810), and finally *Synopsis methodica lichenum* (Acharius 1814).

The systematic consequences of Schwendener's discovery of the true nature of lichens being dual organisms was first utilized by the Finnish scientist Vainio (1890), who endeavored to classify at least some higher ranks of the lichen fungi together with the other fungi. In the beginning of the twentieth century,

Zahlbruckner, one of the most prominent and important persons in the study of lichen fungi, "lichens," published a system that has had a huge impact on the classification of lichen fungi. Although his system was very similar to that of Vainio's, Zahlbruckner unfortunately chose the conservative view of his colleague Reinke (1894–1896) that the "lichens" were actually to be regarded as a systematic unit. Zahlbruckner (1907, 1926) treated lichens as the subclass Lichenes, a level equal to the ascomycetes and the basidiomycetes. This, perhaps practical but artificial, view has prevailed during most of the twentieth century. Despite the forceful argument and attempt by Santesson (1952) and others to integrate lichen fungi into a natural fungal system, lichen fungi have, nevertheless, been considered a matter for lichenologists, quite distinct from nonlichen fungi, which similarly have been considered a matter for mycologists. Consequently, euascomycete fungi were divided into two separate and about equally large systems, one with lichens and one with fungi. Hawksworth (1985) estimated the number of lichens to be 13 500 species, of which nearly all are euascomycetes. Thus, approximately half of some 28 000 euascomycete species are lichenized. Only in the last decade of the twentieth century are efforts becoming successful in bridging the phylogenetic inconsistency of the artificial division of "lichens" and "fungi."

In the recent history of lichen fungal classification, several morphological character complexes have been claimed to show the true natural relationships, and subsequent classifications have mainly been constructed using these characters. The pattern and characteristics in the development (ontogeny) of the ascomata was hugely influential in the lichen classification presented by Henssen and Jahns (1974), and ontogenetic characteristics continue to provide valuable information (Döring and Lumbsch 1998; Lumbsch et al. 2001a). The character complex that otherwise has played the most important role in recent lichen fungal classification is the structure of the apical apparatus of the ascus. Ascus apex characters were hardly in use for classification purposes in lichen fungi until the work of Hafellner (1984), probably the single most influential publication in lichen systematics in the latter half of the twentieth century. Hafellner postulated that the ascus type would be invariable within taxa corresponding to genera and families. The ascus type rapidly became a popular a-priori criterion for lichen systematists to distinguish higher taxa, particularly within the Lecanorales, and for assigning numerous newly described genera to families. Rambold and Triebel (1992) and Hafellner et al. (1993) proposed a revised subordinal classification of the Lecanorales, based explicitly on the assumption of low evolutionary rates in ascus apex morphology. The assumption here is that ascus characters are "conservative" and only rarely transformed in evolution, and hence can be presumed to be useful tools to characterize natural groups. This is a still largely untested assumption, which has been criticized by Tibell (1998). Although some ascus traits may well be synapomorphies for monophyletic groups, as suggested by Peršoh et al. (2004), several ascus-based groupings have been recently shown not to be monophyletic (Wedin et al. 2005b). We still lack a major comprehensive assessment of ascus evolution in light of phylogenetic data. A general conclusion may be that it is more important to identify monophyletic groups by testing character congruence through phylogenetic analyses, rather than "defining" taxa by a-priori judgment of character importance.

Eriksson and his Outline of Ascomycota (most recently 2006b) is by far the most important achievement in the continuing effort to reach a general classification of all ascomycetes, whether lichenized or nonlichenized. Most importantly, the integration is leading to an actual and practical exchange of information between lichenologists and mycologists at meetings, workshops, and excursions. For a detailed historical survey of various ascomycete systems, the reader is referred to Hawksworth (1985).

17.3   Phylogeny

The polyphyletic origin of major groups of fungal-like taxa is generally acknowledged with Fungi (-mycota). They include the Oomycota (included in the Straminopila), Plasmodiophoromycota, Myxomycota, Dictyosteliomycota, and Acrasiomycota, evolving in different lineages as related to major or well-known eukaryotes (Alexopoulos et al. 1996). Fungal-like groups, such as "myxomycetes," oomycetes, and plasmodiophoromycetes, are not included in the Fungi and are believed to be more closely related to other eukaryotes (Tehler 1988; Hawksworth 1991; Lipscomb et al. 1998; Tehler et al. 2000).

Evidence clearly indicates that the chitinous fungi, "Fungi," constitute a monophyletic group including Chytridiomycota, Zygomycota, Ascomycota, and Basidiomycota (for references see Hawksworth et al. 1995). Their monophyly is supported by both morphological, chemical, and other characters, such as chitin in the cell walls, hyphal organization, heterotrophism by absorption of predigested food, α-aminoadafic (AAA) lysine synthetic pathway, and the flagellar apparatus, if present (Tehler 1988). Monophyly is also supported by DNA sequence data (Bruns et al. 1991; Wainright et al. 1993; Lipscomb et al. 1998; Tehler et al. 2000; 2003; Lutzoni et al. 2004).

Alexopoulos et al. (1996) and Hawksworth et al. (1995) use the kingdom name Fungi for this group, equaling to Eumycota, as earlier used by Tehler (1996) and Tehler et al. (2000). For this book we will use Fungi or chitinous fungi for that particular group of organisms since nearly all of its members share cell walls with chitin (Prillinger 1982; Tehler 1988, 1995).

Various suggestions on the evolution of the Fungi have been put forward. On the basis of the lysine biosynthesis pathway, Vogel (1964) postulated that the nonflagellate fungi, Amastigomycota, together with the nonlysine-producing protozoa and metazoa were derived from the euglenids. Cavalier-Smith (1987) suggested a choanoflagellate as a common ancestor of fungi and animals. None of these suggestions received any statistical group support from parsimony analysis of SSU rDNA sequences (Lipscomb *et al.* 1998). More recently, it was suggested that the Microsporidia constituted the sister group to the Fungi (Hirt *et al.* 1999), but that hypothesis has been contradicted by data from both ribosomal sequences (Lipscomb *et al.* 1998) and protein sequences (Tanabe *et al.* 2002). We conclude that the hypothesis describing a sister group relationship between chitinous fungi and animals is today the most well established and most widely adopted among mycologists and phylogeneticists (Wainright *et al.* 1993; Lipscomb *et al.* 1998; Tehler *et al.* 2000, 2003; Lutzoni *et al.* 2004).

The evolution of fungal symbiosis, today advanced into mycorrhizae, lichens, and pathogens, is probably very old. It might date back to a common ancestor that evolved a strategy with factors for recognizing autotrophic organisms and structures for infecting them (Tehler *et al.* 2000). Such an event would make a synapomorphic feature for a group of chitinous fungi, including all Dikaryomycota and the Glomeromycota, informally called *Symbiomycota* by Tehler *et al.* (2003). This would suggest that the symbiosis between fungi and photosynthetic plants evolved prior to the colonization of land by plants and not as a result of the colonization process. Otherwise, and less likely, the factors and structures to acquire a symbiotic lifestyle would be polyphyletic and evolving as separate events in the Glomeromycota, Ascomycota, and Basidiomycota (Tehler *et al.* 2003). The hypothesis that fungi developed symbiotic partnerships with autotrophic organisms in an aquatic environment long before they emerged on land (Tehler *et al.* 2003) was recently supported by paleontological evidence (Yuan *et al.* 2005).

Even if fungal symbiosis is uniquely derived for a group, *Symbiomycota*, it does not seem likely that the evolution of lichenization is uniquely derived for a monophyletic group of lichenized fungi, judging from its presence in various distantly related ascomycete groups as well as among the basidiomycetes. On the basis of various primitive symbiotic life stages, certain fungi have evidently evolved independently into larger lichenized groups (Gargas *et al.* 1995; Tehler *et al.* 2003; Lutzoni *et al.* 2004). It is important to realize that lichen-forming fungi are no more a systematic unit than are parasites, trees, or water plants.

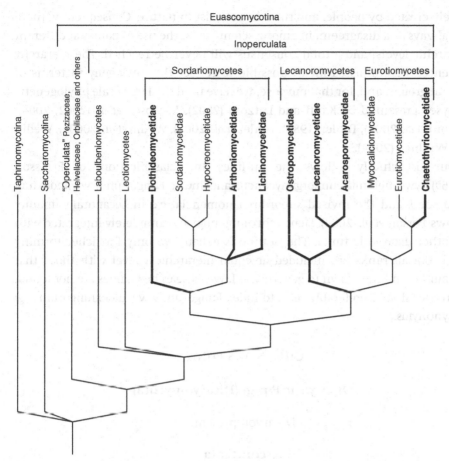

Fig. 17.1 Relationships in the phylum Ascomycota based on a summary of phylogenies mainly from Wedin *et al.* (2005*b*), Lutzoni *et al.* (2004) and Tehler *et al.* (2003). Lichenized taxa are marked with thick lines and with names in bold.

Nearly all lichen fungi are ascomycetes; only some 20 species are basidiomycetes. Furthermore, within the ascomycetes, all lichen fungi belong to any of the three classes: the Sordariomycetes, the Lecanoromycetes, or the Eurotiomycetes (Fig. 17.1). Of these classes, the Lecanoromycetes is nearly exclusively lichenized and also contains the overwhelming majority of all lichen-forming species. Natural relationships within the Euascomycotina are beginning to be resolved, and major and minor groups are emerging, even though much work remains to be done.

17.4   Classification

Classification is the ordering of groups. The use and definition of genera and other supraspecific taxa is and must be subjective, since such units are merely created by people, and they do not exist in nature. Consequently, there will always be a disagreement among scientists on the use of names at different hierarchic levels, and a total consensus will never be reached. The hierarchy presented here aims as far as possible to reflect the phylogeny in terms of natural groups; and, for that purpose, we have used the large-scale phylogenetic analyses presented by Kauff and Lutzoni (2002), Lumbsch *et al.* (2002, 2004), Lutzoni *et al.* (2004), Tehler (1988), Tehler *et al.* (2003), Wedin *et al.* (2000*a*), Wedin and Wiklund (2004).

Our list chiefly follows the outlines and classifications of Eriksson (2006*b*) www.fieldmuseum.org/myconet/, and Index Fungorum www.indexfungorum.org/, and the revised subordinal nomenclature in Lecanorales mainly follows Peršoh *et al.* (2004). Lichen-forming fungi are completely integrated with the other dikaryotic fungi. The outline is exhaustive only for lichen-forming fungi, but all ranks are included at equal hierarchical level with those that contain lichen fungi down to generic level. When searched names are not found, the reader should preferably refer to Index Fungorum www.indexfungorum.org/ for synonyms.

CHITINOUS FUNGI

Dikaryotic Fungi (Dicaryomycota)

Dikaryon present.

Ascomycota

Ascomata monokaryotic with the dikaryon feeding on the monokaryon. The Ascomycota was highly supported as monophyletic by both Tehler *et al.* (2003) and Lutzoni *et al.* (2004).

Subphylum Taphrinomycotina

No lichen fungi present. The Taphrinomycotina were resolved in a polytomy together with the Pneumocystidiomycetes, Schizosaccharomycetes, and the Neolectales in Tehler *et al.* (2003).

Subphylum Saccharomycotina

No lichen fungi present. The Saccharomycotina were resolved in an unnamed sister pair group together with the Euascomycotina with a statistical group support of 88% (Tehler *et al.* 2003).

Subphylum Euascomycotina (=Pezizomycotina)

Lichen fungi numerously present. Hamathecial tissues developed; yeast phase absent; production of melanin from pentaketide biosynthesis. The Euascomycotina were significantly supported as monophyletic by both Tehler *et al.* (2003) and Lutzoni *et al.* (2004). The Euascomycotina is the sister group to the Saccharomycotina (see above).

"Supraclass" Operculata

Ascus opening by a lid. No lichen fungi present. The operculates are probably paraphyletic. They contain the Pezizaceae and the Helvellaceae, Morchellaceae, Tuberaceae, and others, including the inoperculate Orbiliaceae in an unsupported grade group (Tehler *et al.* 2003), or in an insignificantly supported group that corresponds to the Class Pezizomycetes by Eriksson (2006*b*) and Lutzoni *et al.* (2004).

"Supraclass" Inoperculata

Ascus regularly opening by a pore or sometimes evanescent. Within the Euascomycotina, the inoperculate euascomycetes, Inoperculata, are supported as monophyletic by the exclusion of the Orbiliomycetes (Tehler *et al.* 2003).

Class Leotiomycetes

No lichen fungi present.

Class Laboulbeniomycetes

No lichen fungi present.

Class Sordariomycetes

Subclass Arthoniomycetidae

This subclass was resolved as monophyletic by Lutzoni *et al.* (2004).

ORDER ARTHONIALES   (Grube 1998; Tehler 1990.)

Arthoniaceae.   *Amazonomyces, Arthonia, Arthothelium, Coniarthonia, Cryptothecia, Helminthocarpon, Paradoxomyces, Sporostigma, Stirtonia.*

Chrysotrichaceae.   *Byssocaulon, Chrysothrix.*

Roccellaceae.   (Tehler *et al.* 2004; Tehler and Irestedt 2007.) *Ancistrospora, Bactrospora, Camanchaca, Chiodecton, Combea, Cresponea, Dendrographa, Dichosporidium, Diplogramma, Dirina, Dolichocarpus, Enterodictyon, Enterographa, Erythrodecton, Feigeana, Follmanniella, Gorgadesia, Graphidastra, Haplodina, Hubbsia, Ingaderia, Lecanactis, Lecanographa, Llimonaea, Mazosia, Melampylidium, Minksia, Opegrapha, Pentagenella, Pseudolecanactis, Pulvinodecton, Roccella, Roccellaria, Roccellina, Roccellographa, Sagenidiopsis, Sagenidium, Schismatomma, Schizopelte, Sclerophyton, Sigridea, Simonyella, Streimannia, Syncesia, Tania.*

Arthoniales genera of unsettled family position: *Arthophacopsis, Catarraphia, Sipmania, Synarthonia, Tylophorella, Wegea.*

Subclass Dothideomycetidae

This is a very large, heterogeneous and probably paraphyletic subclass with nearly 700 genera out of which only a few are lichen-forming distributed in the order Pleosporales. None of the orders **Capnodiales**, **Dothideales**, **Hysteriales**, **Jahnulales**, **Myriangiales** or **Patellariales** contains any lichen fungi. *Arthrorhaphis*, formerly placed in Patellariales, Arthrorhaphidaceae, has moved to the class Lecanoromycetes, subclass Ostropomycetidae (Lumbsch *et al.* 2007).

ORDER PLEOSPORALES.   Most species in this order are nonlichenized. There is no consensus between authors exactly which genera are lichenized and which are nonlichenized. At least five out of numerous families not mentioned here contain lichen fungi. Pleosporales was supported as monophyletic in (Tehler *et al.* 2003.)

Arthopyreniaceae.   *Arthopyrenia.*

Dacampiaceae.   *Dacampia.*

Microtheliopsidaceae.   *Microtheliopsis.*

Naetrocymbaceae.   *Naetrocymbe.*

Pyrenothricaceae.   *Pyrenothrix.*

Dothideomycetes genera of uncertain position: *Mastodia, Mycoporopsis, Thelenidia.*

Subclass Hypocreomycetidae

No lichen fungi present.

Subclass Lichinomycetidae

Exclusively lichen fungi.

ORDER LICHINALES

Gloeoheppiaceae.   *Gloeoheppia, Gudelia, Pseudopeltula*.

Heppiaceae.   *Corynecystis, Epiphloea, Heppia, Pseudoheppia*.

Lichinaceae.   *Anema, Calotrichopsis, Cryptothele, Digitothyrea, Edwardiella, Ephebe, Euopsis, Finkia, Gonohymenia, Gyrocollema, Harpidium, Jenmania, Lecidopyrenopsis, Lemmopsis, Lempholemma, Leprocollema, Lichina, Lichinella, Lichinodium, Mawsonia, Metamelanea, Paulia, Peccania, Phloeopeccania, Phylliscidiopsis, Phylliscidium, Phyllisciella, Phylliscum, Porocyphus, Pseudopaulia, Psorotichia, Pterygiopsis, Pyrenocarpon, Pyrenopsis, Stromatella, Synalissa, Thelignya, Thermutis, Thermutopsis, Thyrea, Zahlbrucknerella*.

Peltulaceae.   *Neoheppia, Peltula, Phyllopeltula*.

Subclass Sordariomycetidae

No lichen fungi present.

Class Lecanoromycetes

(Lumbsch *et al.* 2004; Peršoh *et al.* 2004; Reeb *et al.* 2004; Wedin *et al.* 2005*b*.)
This class contains nearly exclusively lichenized fungi.

Subclass Acarosporomycetidae

ORDER ACAROSPORALES.   (Reeb *et al.* 2004; Wedin *et al.* 2005*b*.)

Acarosporaceae.   *Acarospora, Glypholecia, Lithoglypha, Pleopsidium, Polysporina, Sarcogyne, Thelocarpella, Timdalia*.

Candelariaceae.   *Candelaria, Candelariella, Candelina, Placomaronea*.

Acarosporales genus of uncertain family position: *Pycnora*.

Subclass Ostropomycetidae

(Lumbsch *et al.* 2007.)

ORDER AGYRIALES.   (Lumbsch *et al.* 2001*a, b*; Schmitt *et al.* 2003.) This order is probably paraphyletic in the following circumscription.

Agyriaceae.   *Amylora, Anzina, Aspiciliopsis, Coppinsia, Lithographa, Orceolina, Placopsis, Placynthiella, Ptychographa, Rimularia, Trapelia, Trapeliopsis, Xylographa*.

Anamylopsoraceae.   *Anamylopsora.*

Schaereriaceae.   *Schaereria.*

ORDER GYALECTALES.   (Kauff and Lutzoni 2002; Kauff and Büdel 2005.)
This order is probably paraphyletic in this circumscription.

Coenogoniaceae.   *Coenogonium, Dimerella.*

Gyalectaceae.   *Belonia, Bryophagus, Cryptolechia, Gyalecta, Pachyphiale, Ramonia, Semigyalecta.*

ORDER GRAPHIDALES

Gomphillaceae.   (Grube *et al.* 2004; Lücking *et al.* 2005.) *Actinoplaca, Aderkomyces, Aplanocalenia, Arthotheliopsis, Asterothyrium, Aulaxina, Calenia, Caleniopsis, Diploschistella, Echinoplaca, Ferraroa, Gomphillus, Gyalectidium, Gyalideopsis, Hippocrepidea, Jamesiella, Lithogyalideopsis, Paratricharia, Psorotheciopsis, Rubrotricha, Sagiolechia, Tricharia.*

Graphidaceae.   (Staiger 2002; Staiger *et al.* 2006.) *Acanthothecis, Anomalographis, Carbacanthographis, Diorygma, Dyplolabia, Fissurina, Glaucinaria, Glyphis, Graphis, Gymnographa, Gymnographopsis, Hemithecium, Leiorreuma, Phaeographina, Phaeographis, Platygramme, Platythecium, Sarcographa, Solenographa, Thalloloma, Thecaria.*

Solorinellaceae.   *Gyalidea, Solorinella.*

Thelotremataceae.   Thelotremataceae forms an unresolved, paraphyletic group with Graphidaceae according to Staiger *et al.* (2006), who suggest that it should be treated within the family Graphidaceae. *Ampliotrema, Chroodiscus, Diploschistes, Ingvariella, Myriotrema, Nadvornikia, Ocellularia, Phaeotrema, Platygrapha, Polistroma, Pseudoramonia, Reimnitza, Thelotrema, Topeliopsis, Tremotylium.*

ORDER TRICHOTHELIALES.   (Grube *et al.* 2004.)

Myeloconidaceae.   *Myeloconis.*

Porinaceae.   *Clathroporina, Polycornum, Porina, Segestria, Trichothelium.*

Trichotheliales genus of uncertain family position: *Amphorothecium.*

ORDER PERTUSARIALES.   (Schmitt *et al.* 2006.)

Coccotremataceae.   (Schmitt *et al.* 2001.) *Coccotrema, Parasiphula.*

Ochrolechiaceae.   Schmitt *et al.* (2006) suggest resurrecting this family for *Ochrolechia, Varicellaria* and parts of *Pertusaria* that currently have no generic names. *Ochrolechia, Varicellaria.*

Pertusariaceae.   (Schmitt and Lumbsch 2004.) *Loxosporopsis, Monoblastia, Pertusaria, Thamnochrolechia.*

Hymeneliaceae.   *Eiglera, Hymenelia, Ionaspis, Tremolecia.*

Megasporaceae.   *Aspicilia, Lobothallia, Megaspora.*

Icmadophilaceae. (Platt and Spatafora 2000; Grube and Kantvilas 2006.) *Dibaeis, Icmadophila, Pseudobaeomyces, Siphula, Siphulella, Thamnolia.*

ORDER OSTROPALES

Stictidaceae. (Wedin *et al.* 2005*a*.) *Absconditella, Petractis, Schizoxylon, Stictis* (*incl. Conotrema*) *Thelopsis, Topelia.*

Ostropomycetidae families of uncertain order position:

Baeomycetaceae. (Lumbsch *et al.* 2004, 2007.) *Ainoa, Baeomyces, Phyllobaeis.*
Arctomiaceae. (Lumbsch *et al.* 2005.) *Arctomia, Gregorella, Wawea.*
Protothelenellaceae. (Schmitt *et al.* 2005*b*.) *Protothelenella, Thrombium.*
Arthrorhaphidaceae. (Lumbsch *et al.* 2007.) *Arthrorhaphis.*
Thelenellaceae. (Schmitt *et al.* 2005*b*.) *Chromatochlamys, Thelenella.*
Phlyctidaceae. (Wedin *et al.* 2005*b*.) *Phlyctis.*

Subclass Lecanoromycetidae

ORDER LECANORALES

SUBORDER PELTIGERINEAE. (Wiklund and Wedin 2003; Miadlikowska and Lutzoni 2004.)

Peltigeraceae. *Peltigera, Solorina.*
Lobariaceae. *Lobaria, Pseudocyphellaria, Sticta.*
Nephromataceae. *Nephroma.*
Massalongiaceae. (Wedin *et al.* 2007.) *Leptochidium, Massalongia, Polychidium.*
Coccocarpiaceae. (Wedin and Wiklund 2004.) *Coccocarpia, Peltularia, Spilonema, Spilonemella, Steinera.*
Collemataceae. *Collema, Homothecium, Leciophysma, Leightoniella, Leptogium, Physma, Ramalodium, Staurolemma.*
Pannariaceae. (Ekman and Jørgensen 2002.) *Austrella, Degelia, Degeliella, Erioderma, Fuscoderma, Fuscopannaria, Kroswia, Leioderma, Lepidocollema, Moelleropsis, Pannaria, Parmeliella, Protopannaria, Psoroma, Psoromidium, Santessoniella, Siphulastrum.*
Placynthiaceae. *Hertella, Hueella, Koerberia, Placynthiopsis, Placynthium, Vestergrenopsis.*

SUBORDER LECANORINEAE

Lecanoraceae. (Ekman and Wedin 2000; Wedin *et al.* 2000*a*; Arup *et al.* 2007.) *Arctopeltis, Bryodina, Bryonora, Carbonea, Cladidium, Claurouxia, Clauzadeana, Diomedella, Edrudia, Japewia, Lecanora, Lecidella, Maronina, Miriquidica, Myrionora, Protoparmeliopsis, Psorinia, Punctonora, Pyrrhospora, Ramalinora, Ramboldia, Rhizoplaca, Sagema, Traponora, Tylothallia, Vainionora.*

Cladoniaceae.   (Stenroos and DePriest 1998; Wedin *et al.* 2000*a*; Stenroos *et al.* 2002*b*.) *Calathaspis, Cladia, Cladonia, Gymnoderma, Heterodea, Heteromyces, Metus, Myelorrhiza, Notocladonia, Pilophorus, Pycnothelia, Ramalea, Sphaerophoropsis, Squamella, Thysanothecium.*

Stereocaulaceae.   (Ekman and Tønsberg 2002; Myllys *et al.* 2005; Högnabba 2006.) *Hertelidea, Lepraria, Stereocaulon.*

Squamarinaceae.   *Squamarina.*

Parmeliaceae.   (Blanco *et al.* 2004, 2005; Divakar *et al.* 2006; Crespo *et al.* 2007.) *Ahtia, Ahtiana, Alectoria, Allantoparmelia, Allocetraria, Almbornia, Anzia, Arctocetraria, Arctoparmelia, Asahinea, Brodoa, Bryocaulon, Bryoria, Bulborrhizina, Bulbothrix, Canoparmelia, Cavernularia, Cetraria, Cetrariastrum, Cetrariella, Cetrariopsis, Cetrelia, Coelopogon, Cornicularia, Coronoplectrum, Dactylina, Esslingeriana, Evernia, Everniopsis, Flavocetraria, Flavoparmelia, Flavopunctelia, Himantormia, Hypogymnia, Hypotrachyna, Imshaugia, Kaernefeltia, Karoowia, Letharia, Lethariella, Masonhalea, Melanelia, Melanelixia, Melanohalea, Menegazzia, Myelochroa, Namakwa, Neopsoromopsis, Nephromopsis, Nesolechia, Nimisia, Nodobryoria, Omphalodiella, Omphalodium, Omphalora, Oropogon, Pannoparmelia, Parmelia, Parmelina, Parmeliopsis, Parmotrema, Parmotremopsis, Phacopsis, Placoparmelia, Platismatia, Pleurosticta, Protoparmelia, Protousnea, Pseudephebe, Pseudevernia, Pseudoparmelia, Psiloparmelia, Psoromella, Punctelia, Relicina, Relicinopsis, Sulcaria, Tuckermanella, Tuckermanopsis, Usnea, Vulpicida, Xanthomaculina, Xanthoparmelia.*

Cetradoniaceae.   *Cetradonia.*

Gypsoplacaceae.   (Arup *et al.* 2007.) These two genera, which together may form a monophyletic group, are the closest known relatives to Parmeliaceae, but the classification of them is still uncertain and tentative. *Gypsoplaca.*

Haematommaceae.   *Haematomma.*

Mycoblastaceae.   (Arup *et al.* 2007.) *Calvitimela, Mycoblastus, Tephromela.*

SUBORDER PSORINEAE

Pilocarpaceae.   Andersen and Ekman (2005) found Micareaceae and Ectolechiaceae to be better treated as synonyms to Pilocarpaceae. *Badimia, Badimiella, Bapalmuia, Barubria, Bryogomphus, Byssolecania, Byssoloma, Calopadia, Calopadiopsis, Fellhanera, Fellhaneropsis, Kantvilasia, Lasioloma, Lobaca, Loflammia, Logilvia, Micarea, Psilolechia, Roccellinastrum, Sporopodiopsis, Sporopodium, Szczawinskia, Tapellaria.*

Psoraceae.   (Andersen and Ekman 2005.) *Eremastrella, Glyphopeltis, Protoblastenia, Protomicarea, Psora, Psorula.*

Ramalinaceae.   (Ekman 2001; Andersen and Ekman 2005.) *Adelolecia, Arthrosporum, Bacidia, Bacidina, Bacidiopsora, Biatora, Catinaria, Cenozosia,*

Cliostomum, Compsocladium, Crocynia, Crustospathula, Echidnocymbium, Frutidella, Heppsora, Herteliana, Jarmania, Krogia, Lecania, Megalaria, Mycobilimbia, Myxobilimbia, Niebla, Phyllopsora, Physcidia, Ramalina, Ramalinopsis, Rolfidium, Schadonia, Scutula, Speerschneidera, Squamacidia, Stirtoniella, Thamnolecania, Tibellia, Toninia, Toniniopsis, Trichoramalina, Vermilacina, Waynea.

SUBORDER PHYSCIINEAE

Catillariaceae. The subordinal placement of this family is tentative. *Austrolecia, Catillaria, Halecania, Placolecis, Solenopsora, Sporastatia, Xanthopsorella.*

Physciaceae. (Wedin *et al.* 2000b.) Contrary to the classification by Eriksson (2006b), the mazaediate representatives of Physciaceae do not form a natural group. If Physciaceae is divided into two families (Physciaceae and Caliciaceae), Caliciaceae will include *Buellia* and segregates (*Amandinea, Diplotomma, Hafellia,* etc.). *Acolium, Acroscyphus, Amandinea, Anaptychia, Australiena, Buellia, Calicium, Coscinocladium, Cratiria, Culbersonia, Cyphelium, Dermatiscum, Dermiscellum, Dimelaena, Diploicia, Diplotomma, Dirinaria, Gassicurtia, Hafellia, Heterodermia, Hyperphyscia, Hypoflavia, Mischoblastia, Mobergia, Monerolechia, Phaeophyscia, Phaeorrhiza, Physcia, Physciella, Physconia, Pyxine, Redonia, Rinodina, Rinodinella, Santessonia, Stigmatochroma, Tetramelas, Texosporium, Thelomma, Tholurna, Tornabea, Tylophoropsis.*

SUBORDER SPHAEROPHORINEAE.   (Wedin and Döring 1999.)

Calycidiaceae. The subordinal placement of this genus is tentative. *Calycidium.*

Sphaerophoraceae. *Austropeltum, Bunodophoron, Leifidium, Neophyllis, Sphaerophorus.*

SUBORDER RHIZOCARPINEAE

Rhizocarpaceae. (Ihlen and Ekman 2002.) *Catolechia, Epilichen, Poeltinula, Rhizocarpon.*

SUBORDER TELOSCHISTINEAE

Brigantiaeaceae. The subordinal placement of this family is tentative. *Argopsis, Brigantiaea.*

Letrouitiaceae. *Letrouitia.*

Megalosporaceae. *Austroblastenia, Megaloblastenia, Megalospora.*

Teloschistaceae. *Caloplaca, Cephalophysis, Fulgensia, Huea, Ioplaca, Josefpoeltia, Seirophora, Teloschistes, Xanthodactylon, Xanthomendoza, Xanthopeltis, Xanthoria.*

SUBORDER LECIDEINEAE

Lecideaceae. Lecideaceae and Porpidiaceae are most likely paraphyletic relative to each other (Buschbom and Mueller 2004; Buschbom and Barker 2006.) as are the large genera *Lecidea* and *Porpidia,* which also have several small genera nested inside. Here, we follow a traditional view of the classification, awaiting further progress. *Bahianora, Cecidonia, Cryptodictyon, Lecidea, Lecidoma, Lopacidia, Pseudopannaria, Rhizolecia.*

Porpidiaceae. *Amygdalaria, Bellemerea, Catarrhospora, Clauzadea, Farnoldia, Immersaria, Koerberiella, Labyrintha, Paraporpidia, Poeltiaria, Poeltidea, Porpidia, Schizodiscus, Stephanocyclos, Xenolecia.*

Lecanorales taxa of unsettled position:
Families:
Aphanopsidaceae. *Aphanopsis, Steinia.*
Biatorellaceae. *Biatorella, Maronella, Piccolia.*
Miltideaceae. *Miltidea.*
Genera: *Auriculora, Bartlettiella, Botryolepraria, Buelliastrum, Helocarpon, Leprocaulon, Lopadium, Lopezaria, Nimisiostella, Notolecidea, Psorotichiella, Ravenelula, Scoliciosporum, Sporacestra, Stenhammarella, Strangospora, Wadeana.*

ORDER UMBILICARIALES. (Wedin *et al.* 2005*b*.) The circumscription of this group is very tentative and it is not clear if it constitutes a monophyletic group. In Tehler *et al.* (2003), the sister pair *Umbilicaria* and *Lasallia* resolve as sister group to the Eurotiomycetes and perhaps should be included in that class. In Lutzoni *et al.* (2004), only *Lasallia* is nested with the Eurotiomycetes clade; whereas *Umbilicaria* goes with the Ostropomycetidae clade. Here we follow Wedin *et al.* (2005*b*), which is also the more traditional view, but there is no statistical significance to support either of the alternative placements.

Umbilicariaceae. *Lasallia, Umbilicaria.*
Elixiaceae. *Elixia.*
Fuscideaceae. *Fuscidea, Hueidea, Loxospora, Maronea, Orphniospora, Ropalospora, Sarrameana.*

Umbilicariales genera of unsettled position: *Boreoplaca, Hypocenomyce, Ophioparma.*

Lecanoromycetidae taxa of unsettled position:
Families:
Thelocarpaceae. *Melanophloea, Sarcosagium, Thelocarpon.*
Dactylosporaceae. *Dactylospora.*
Pachyascaceae. *Pachyascus.*
Vezdaeaceae. *Vezdaea.*
Genera. *Biatoridium, Bouvetiella, Corticifraga, Corticiruptor, Eschatogonia, Korfiomyces, Podotara.*

Class Eurotiomycetes

Subclass Chaetothyriomycetidae

ORDER CHAETOTHYRIALES.   No lichen fungi present.

ORDER PYRENULALES
Monoblastiaceae.   *Acrocordia, Anisomeridium, Monoblastia.*
Pyrenulaceae.   *Anthracothecium, Lithothelium, Pyrenula, Pyrgillus, Sulcopyrenula.*
Requienellaceae.   No lichen fungi present.
Strigulaceae.   *Oletheriostrigula, Phylloblastia, Strigula.*
Trypetheliaceae.   *Astrothelium, Campylothelium, Exiliseptum, Laurera,*
    Polymeridim, Pseudopyrenula, Trypetheliopsis, Trypethelium.
Xanthopyreniaceae.   *Collemopsidium, Pyrenocollema.*

ORDER VERRUCARIALES.   Mainly containing lichen fungi.
Adelococcaceae.   No lichen fungi present.
Verrucariaceae.   *Agonimia, Anthracocarpon, Awasthiella, Bagliettoa, Bellemerella,*
    Bogoriella, Catapyrenium, Clauzadella, Clavascidium, Dermatocarpon, Endocarpon,
    Haleomyces, Henrica, Heterocarpon, Heteroplacidium, Involucropyrenium, Kalbiana,
    Leucocarpia, Macentina, Neocatapyrenium, Normandina, Placidiopsis, Placidium,
    Placocarpus, Placopyrenium, Placothelium, Polyblastia, Psoroglaena, Rhabdopsora,
    Scleropyrenium, Spheconisca, Staurothele, Thelidiopsis, Thelidium, Trimmatothele,
    Trimmatothelopsis, Verrucaria.

Subclass Eurotiomycetidae

No lichen fungi present.

Subclass Mycocaliciomycetidae

No lichen fungi present.

Euascomycotina (= Pezizomycotina), taxa of unsettled position:
Families:
Coniocybaceae.   This is a mazaediate group, which is not related to the
    mazaedia-producing pin-lichens in the Lecanoromycetes, but the true
    relationships are currently unclear (Tibell 1984). *Chaenotheca, Sclerophora.*
Microcaliciaceae.   This enigmatic mazaediate group is very isolated and its
    relationships are unknown (Tibell 1984). *Microcalicium.*
Genus: *Schistophoron.* The genus contains only one, mazaediate species.
Basidiomycota.   Metabasidium present.

Class Homobasidiomycetes

Subclass Agaricomycetidae

ORDER POLYPORALES.   With the exception of the following fungi, this order is exclusively nonlichen forming.

Atheliaceae.   The cyanobacterial photobiont is located in the basidiocarp; hence a lichen thallus in the strict sense is not formed. *Dictyonema*.

ORDER AGARICALES.   With the exception of the following fungi, this order is exclusively nonlichen forming.

Clavariaceae.   The mycelium is only associated with the photobiont and an actual thallus is not formed. *Multiclavula*.

Tricholomataceae.   Some species of this genus are lichenized. In some species the mycelium is only associated with the photobiont, forming a green layer that covers the substrate (*Botrydina*). In other species a well-developed squamulose thallus is developed (*Coriscium*). *Omphalina*.

"IMPERFECT LICHEN FUNGI"

Many populations of lichen fungi are invariably sterile. Fruiting bodies never develop, or have never been found, and phylogenetic affinities remain enigmatic. Some species have developed taxa with either sexual or asexual means of dispersal; and, in many cases, it is possible to refer the sterile taxa to fertile counterparts on the basis of similarity in thallus morphology, anatomy, and chemistry. The genetic mechanisms and functions that regulate the two dispersal strategies are not understood. In lichenology, the phenomenon has been referred to as the species pair concept (Poelt 1970, 1972). The concept was challenged by Tehler (1982) and has since then been quite lively debated, most recently by Buschbom and Mueller (2005, see further references therein). Most sterile taxa have been systematically determined with the help of molecular data, but some still remain without systematic position: *Cystocoleus* (Ascomycota), *Racodium* (Ascomycota), *Phyllobathelium* (Phyllobatheliaceae, Ascomycota), *Sarcopyrenia* (Ascomycota).

Appendix: Culture methods for lichens and lichen symbionts

E. STOCKER-WÖRGÖTTER AND A. HAGER

Many topics of lichen biology deal with questions about how the different symbiotic partners (mycobionts, photobionts, cyanobionts) interact within the lichen thallus. The separation, isolation, and culture of the lichen symbionts or components offer researchers insights into functional aspects of the lichen symbiosis, such as identifying parameters essential for their growth in the aposymbiotic state or triggers for producing secondary metabolites (polyketides, shikimic acid derivatives, etc.) in culture. Furthermore, culturing provides a means for investigating how lichen symbionts respond to each other, how they recognize each other through chemical signals, and how a functional symbiosis is established. Many of these fundamental problems in lichenology have been investigated (Chapter 5), but not fully resolved. Apart from the questions that arise from investigating only the typical lichen bionts, one can utilize more advanced molecular methods to study other associated partners of the lichen symbiosis, including molds, yeasts, lichenicolous fungi, lichenicolous lichens, and parasitic bacteria located on the surface of or within the thalli.

Laboratory culture

Over the past 20 years many culture experiments have been undertaken to improve culture methods for lichen symbionts, in general, and also to re-establish lichen symbioses (di- or tripartite partnerships of ascomycetous and basidiomycetous lichens) under artificial conditions (Fig. App. 1). Such experiments help to answer basic questions, like how the lichen fungus transforms from a relatively unstructured mycelium into a highly organized thallus.

Lichen Biology, ed. Thomas H. Nash III. Published by Cambridge University Press.
© Cambridge University Press.

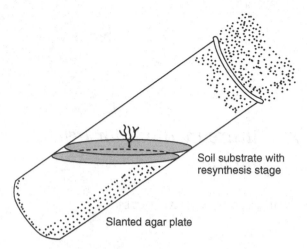

Fig. App.1   Resynthesis tube for dipartite lichen symbioses (e.g. Stocker-Wörgötter and Elix 2006).

Such resynthesis experiments can significantly extend our knowledge about symbiont coordination and steps in thallus ontogeny. A thallus represents a "super structure" that is formed by autonomous fungal (hyphae) and algal and/ or cyanobacterial cells. Such a developmental process is different from multicellular growth forms (e.g. higher plants), which develop "real" tissues by starting with one single cell. In the strict sense, lichens do not form tissues, so the term "tissue culture," introduced by Yamamoto (1990) should be avoided and replaced by "thallus fragment culture" or "Yamamoto method."

One major objective is to establish conditions under which particular polyketides are expressed in cultured mycobionts, and to understand the nature and "chemical history" of the "lichen substances" by analyses of polyketide synthase (PKS) genes. This requires using several model lichens to scale up the output of cultured mycobiont (mycelia). To this end, we have established a culture collection that contains 250 axenically grown mycobionts and 50 pure cultures of photobionts and cyanobionts.

In general, most lichen-forming fungi cannot be grown under the same stable culture conditions that have been successfully used with higher plant tissue cultures or with cultures of molds and actinomycetous bacteria. Lichen fungi require more complex nutrients and need varying culture conditions involving physiological and environmental stresses. Recent advances in culturing techniques have destroyed the myth that lichen fungi or lichens are difficult to culture.

In nature, lichen fungi have adapted to various extreme environments, and such harsh conditions play an essential role in culturing. Consequently, to be successful, laboratory experiments with lichens and lichenized fungi have to be combined with extensive field studies to document those environmental conditions. To control culture conditions, electronically adjustable culture chambers can be used to simulate such variables as light–dark regimes, temperature fluctuations, and moisture fluctuations, necessary to increase or slow down the growth rates of mycobionts, and to provide a combination of factors which are necessary to induce secondary metabolite production. Several culture chambers have been adapted to simulate a selection of overall climatic conditions, including those dominating in the tropics, in temperate regions, or in polar regions.

Culture methods

DNA analyses as a means of establishing the identity of the mycobionts

Natural lichen thalli often grow together with other microorganisms, including bacteria, epiphytic and parasitic fungi, mosses, etc. To evaluate whether culture contamination is occurring by lichen-associated organisms (especially the fungi), the cultured mycobionts should be routinely identified by DNA analyses on the basis of ITS sequences. Only mycobiont isolates that have been confirmed as the "right" mycobiont should be subcultured and prepared for large-scale culturing. Total DNA is extracted from cultured mycobionts for comparison with that obtained from the voucher specimens. Lyophylized pieces of individual mycobiont and parent thalli are ground, using liquid nitrogen, in 1.5 ml Eppendorf tubes. Total genomic DNA is extracted after the DTAB/CTAB protocol, introduced by Armaleo and Clerc (1995) and Arup and Grube (1998).

DNA amplification and purification

The ITS-regions, the 5.8 region and the flanking parts of the small and large subunits (SSU 18S and LSU 28S) of the rDNA, may be amplified using a Gene Amp PCR System thermal cycler. Primers for the PCR are ITS1, ITS2, ITS3, and ITS4 designed for lichen fungi. The PCR mix contains 1.25 units of Dynazyme Taq polymerase (Finnzymes), 0.2 mM of each of the four dNTPs, 0.5 μM of each primer, and 10–50 ng genomic DNA. The PCR products are cleaned by polyethylene glycol (PEG) precipitation, and the complementary strands of the DNA are sequenced using a Dye Terminator Ready Reaction Kit (Perkin Elmer) following the instructions of the manufacturer. The second purification is performed using a Quiaquick PCR product purification kit.

Isolation of the lichen fungus

Depending on the objective of the research, two mycobiont isolation methods may be performed: spore isolation and "thallus fragment culture"

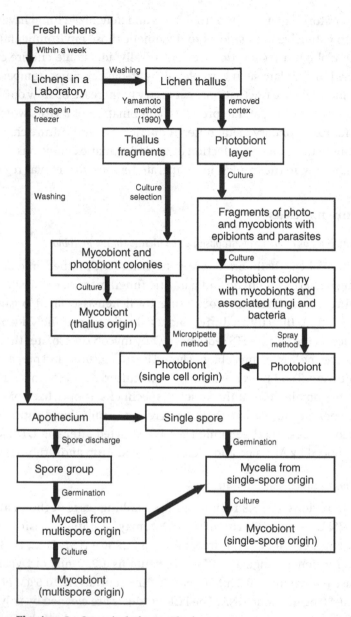

Fig. App.2 Spore isolation method.

(Yamamoto method; modified from Yamamoto 1990). Both methods are extensively described by Yoshimura *et al.* (2002), Yamamoto *et al.* (2002), and Stocker-
Wörgötter (2002*a*, *b*) and are illustrated in Fig. App. 2 and Fig. App. 3. Single
spore isolates are very useful for genetic experiments, and secondary lichen
metabolites can provide specific genetic markers (W. Culberson and Culberson

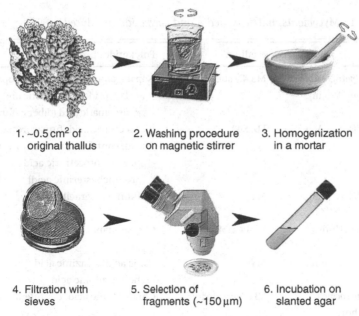

1. ~0.5 cm² of   2. Washing procedure   3. Homogenization
   original thallus   on magnetic stirrer   in a mortar

4. Filtration with   5. Selection of   6. Incubation on
   sieves   fragments (~150 µm)   slanted agar

Fig. App.3   Overview – steps of thallus fragment culture.

1994; Stocker-Wörgötter *et al.* 2004; Stocker-Wörgötter and Elix 2004). Further discussion below highlights resynthesis experiments of a basidiomycetous lichen (*Omphalina hudsoniana*).

Large-scale culturing

Pharmaceutical and biotechnological uses of lichen fungi (Stocker-Wörgötter 2005) requires large quantities of identified fungal materials for extraction. Typically, large-scale culturing uses liquid media in culture bottles and fermenters. Most lichen fungi can be cultured in liquid and semiliquid media. Some lichen pigments, anthraquinones (e.g. chrysophanol, naphthazarines) and soluble compounds are produced in liquid culture and these can be readily accumulated (e.g. in shake flask cultures). Most of the polyketides that are formed by medullary hyphae are not formed in liquid media. For this reason, the final step to induce polyketide production involves using solid media in large-sized (diameter of 15–20 cm) petri dishes, from which approximately 10 mg to 10 g of substance can be collected. Most of the polyketides formed in the cultures are mixtures that have to be purified by further chemical separation methods (preparative chromatography, selective extraction, etc.).

Table App. 1 shows a selection of nutrient media that have been used to induce polyketide biosynthesis ("lichen substances").

Table App. 1. *Mycobionts, nutrient media and polyketides produced.*

Lichen fungi	Medium[a]	Polyketides produced
Bunodophoron patagonicum (C.W. Dodge) Wedin	MS 4% sucrose	Isopatagonic and 2-O-methylisopatagonic acid (depsides), ascomatic and norascomatic acid (dibenzofurans)
Cetraria islandica (L.) Ach.	LB, S4%	Protocetraric acid, fumarprotocetraric acid, confumarprotocetraric acids, succin-protocetraric acid (protolichesterinic acid)
Cladonia bellidiflora (Ach.) Shaer.	MY	Bellidiflorin, graciliformin
Cladonia furcata (Huds.) Schrad.	LB 4% ribitol	Chrysophanol
Cladonia salmonea Stenroos	MY	Usnic acid, salazinic acid, rhodocladonic acid
Cryptothecia rubrocincta (Ehrenb.) Thor.	LB 4% erythriol	Chiodectonic acid, confluentic acid
Evernia divaricata (L.) Ach.	LB + B	Divaricatic acid
Haematomma persoonii (Fée) Massal.	LB 4% ribitol 4% sorbitol	Isosphaeric acid, chloroatranorin, sphaerophorin, russulone
Haematomma stevensiae Rogers	S4%	Haematommone, russulone
Heterodea muelleri (Hampe) Nyl.	LB + S, MS, S2%	Diffractaic acid, barbatic acid
Lecanora rupicola (L.) Zahlbr.	LB	Lecanoric acid, sordidone, eugenitol, atranorin (haematommic acid)
Lobaria fendleri (Tuck. Ex Mont.) Lindau	MS	Gyrophoric and 4-O-methylgyrophoric acid
Lobaria spathulata (Inum.) Yoshim.		Methylorsellinate, lecanoric and gyrophoric acids, telephoric acid (shikimic acid pathway)
Neuropogon sphacelatus (R.Br.) D.J. Galloway	LB	Usnic acid
Peltigera aphthosa Willd	MIX, then 1.8 % BBM Agar	Tenuiorin
Protousnea magellanica (Mont.) Krog	LB, S4%	Usnic acid, sekikaic and subsekikaic acid, 4'-O-demethylsekikaic and 4'-O-demethylsub-sekikaic acids
Ramalina peruviana Ach.	Liquid LB	Atranorin, sekikaic acid
Solorina crocea (L.) Ach.	MIX	Solorinic and disolorinic acids, hybocarpone
Stereocaulon ramulosum (Sw.) Rausch	MY, S4%	Perlatolic acid, stenosporic acid, divaricatic acid
Umbilicaria mammulata (Ach.) Tuck.	PDA	Gyrophoric acid

Table App. 1. *(cont.)*

Lichen fungi	Medium[a]	Polyketides produced
Xanthoparmelia flavescentireagens (Gyeln.) D.J. Galloway	MY, MS, PDA, LB	Usnic acid, norlobaridone, loxodin, divaricatic acid
Xanthoria elegans (Link) Th. Fr.	LB	Parietin, 1-*O*-methylparietin, emodin, 1-*O*-methylemodin, teloschistin, teloschistin monoacetate, 1-*O*-methylphyscionbisanthrone, physcion-bisanthrone

[a] For explanation of abbreviations, see Table App. 3.

Culture and resynthesis of basidiolichens as exemplified by *Omphalina hudsoniana* (Jenn.) Bigelow

Basidiolichens, like *Omphalina* spp., are distributed in arctic and alpine habitats. The majority of *Omphalina* species are not lichenized and do not form any thallus-like structures. But two of the most prominent species, *O. ericetorum* and *O. hudsoniana*, are lichen-forming and live symbiotically with green algae (e.g. *Coccomyxa* sp.). *Omphalina hudsoniana* is an exciting subject to study (Stocker-Wörgötter 2001), as this fungus forms various associations with green algae and cyanobacteria, in addition to other bacteria that have yet to be identified. Recent results show that algal colonies of both *Coccomyxa* sp. and *Hyalococcus* sp. (a common photobiont of the lichen genus *Dermatocarpon*) can be present in the *Omphalina* thalli (Figs. App. 4–9). All isolated *Omphalina* mycobionts were significantly cyanotrophic (depending on a loose association with cyanobacteria, such as *Nostoc*, *Scytonema*, or *Stigonema*) for growth and are successfully cultured on nutrient agar (Murashige–Skoog medium) enriched with nitrogenous compounds and sugars, such as sucrose, and polyols (ribitol, sorbitol, mannitol). Extended experiments revealed that the type of nitrogen source (a preference for a particular genus or species of cyanobacterium) and also the kind of carbon source did not play an essential role. Apparently any kind of bound nitrogen and sugars can be utilized by these fungi. Additionally, the mycobionts need temporary cold temperature treatments (e.g. 6 weeks in a refrigerator, freezer at 4 °C and/or −10 to −20 °C) to thrive. As was demonstrated in various resynthesis experiments with lichens, functional thalli with a cortex (with *Omphalina hudsoniana*) developed only after desiccation of several weeks. Omphalinoid fruiting bodies ("mushrooms") were formed by the mycobionts growing on a layer of cyanobacteria (*Nostoc*).

Tables App. 2 and App. 3 show the culture media for photobionts and mycobionts, respectively.

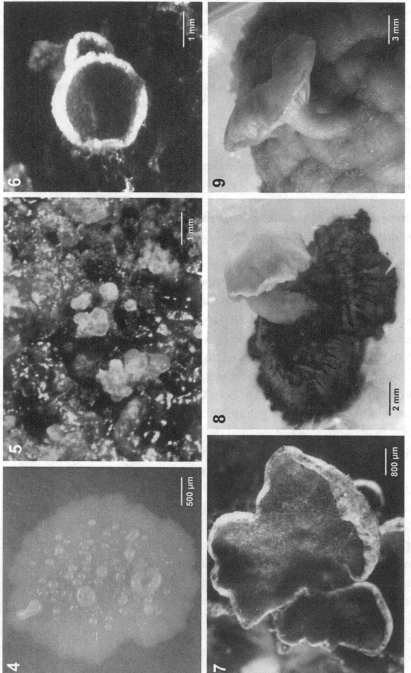

Figs. App.4–9 Fig. App.4. Mycobiont of *Omphalina hudsoniana*, isolated from the green lobules with initiation stages of fruiting bodies, on modified MS-medium. Fig. App.5. Tiny lobules formed in a resynthesis experiment on *Nostoc* colonies (cyanotrophyl). Fig. App.6. Cup-shaped thalli, grown after 6 months. Fig. App.7. Fully developed lobules on soil, 18 months old. Fig. App.8. *Omphalina hudsoniana* mycobiont forming a basidiocarp, grown on a layer of *Nostoc*, on Sabouraud 4% sucrose medium, 36 months. Fig. App.9. Mycobiont, grown on synthetic medium, Sabouraud 6% sucrose 3% sorbitol medium, 42 months.

Table App. 2. *Culture media for photobionts*

Medium	Stock solutions	Amount
Bold's Basal Medium	$NaNO_3$	10.0 g / 400 ml Aqua dest.
(BBM) (Deason and	K_2HPO_4	3.0 g / 400 ml Aqua dest.
Bold 1960)	$CaCl_2 \cdot 2H_2O$	1.0 g / 400 ml Aqua dest.
	KH_2PO_4	7.0 g / 400 ml Aqua dest.
	$MgSO_4 \cdot 7H_2O$	3.0 g / 400 ml Aqua dest.
	NaCl	1.0 g / 400 ml Aqua dest.

Mix 10 ml from each stock solution with 940 ml Aqua dest. and add 1 ml of the four trace element solutions written below.

Trace element solutions and amount

50.0 g EDTA and 31.0 g KOH soluble in 1000 ml Aqua dest.

4.98 g $FeSO_4 \cdot 7H_2O$ soluble in 1000 ml acidulated Aqua dest. (1.0 ml H_2SO_4 conc. in 1000 ml Aqua dest.)

11.42 g H_3BO_3 soluble in 1000 ml Aqua dest.

Following salts soluble in 1000 ml Aqua dest.:

$ZnSO_4 \cdot 7H_2O$	8.82 g
MoO_3	0.71 g
$Co(NO_3) \cdot 6H_2O$	0.49 g
$MnCl_2 \cdot 4H_2O$	1.44 g
$CuSO_4 \cdot 5H_2O$	1.57 g

Add 15 g agar for solid medium.

Medium	Stock solutions	Amount
3 × N BBM (Brown	As above except:	30.0 g / 400 ml Aqua dest.
and Bold 1964)	$NaNO_3$	

Medium	Ingredients	Amount
Organic Nutrient	$1 \times N$ BBM	970 ml
Medium for	Proteose peptone	10 g
Trebouxia	Glucose	20 g
(Ahmadjian 1967b)	Agar	15 g

Make up to 1000 ml with double distilled water.

Medium	Ingredients	Amount
MDM Medium for	KNO_3	1 g
Cyanobacteria	$MgSO_4 \cdot 7H_2O$	250 mg
(Watanabe 1960)	K_2HPO_4	250 mg
	NaCl	100 mg
	$CaCl_2 \cdot 2H_2O$	10 mg
	Fe solution	1 ml
	A5 solution	1 ml
	Agar	15 g

Make up to 1000 ml with double distilled water.

Table App. 2. *(cont.)*

Medium	Stock solutions	Amount
Fe solution	$FeSO_4$	1 g
	Distilled water	500 ml
	Concentrated H_2SO_4	2 drops
A5 solution	H_3BO_3	286 mg
	$MnSO_4 \cdot 7H_2O$	250 mg
	$ZnSO_4 \cdot 7H_2O$	22.2 mg
	$CuSO_4 \cdot 5H_2O$	7.9 mg
	Na_2MoO_4	2,1 mg
	Distilled water	100 ml

Table App. 3. *Culture media for mycobionts*

Medium	Ingredients	Amount
Water Agar Medium (WA)	Agar	20 g
	Make up to 1000 ml with double distilled water.	
Lilly & Barnett Medium (LB)	D-Glucose	10.0 g
(Lilly and Barnett 1951)	L-Asparagine	2.0 g
	$MgSO_4 \cdot 7H_2O$	0.5 g
	KH_2PO_4	1.0 g
	Agar	15.0 g
	$Fe(NO_3)_3 \cdot 9H_2O$	0.2 mg
	$ZnSO_4 \cdot 7H_2O$	0.2 mg
	$MnSO_4 \cdot H_2O$	0.1 mg
	Thiamine-HCl (Vit. B1)	100 µg
	Biotin	5 µg
	Make up to 1000 ml with double distilled water.	
LB + S (Esser 1976)	Add 20 ml soil extract.	
LB + B	Add 20 ml bark extract.	
Malt Yeast Extract Medium (MY)	Malt extract	20 g
(Yamamoto 1990)	Yeast extract	2 g
	Agar	20 g
	Make up to 1000 ml with double distilled water.	
Murashige & Skoog Medium (MS)	Malt extract	2 g
(Murashige and Skoog 1962)	Caseine hydrolysate	2 g
	Mannitol	20 g
	Saccharose	40 g
	Agar	18 g
	Murashige Mineral Salts	1 mg
	Make up to 1000 ml with double distilled water.	

Table App. 3. *(cont.)*

Medium	Ingredients	Amount
Potato Dextrose Agar (PDA)	PDA	39 g
	Make up to 1000 ml with double distilled water.	
Sabouraud 2% glucose agar (S2%)	Polypeptone	10 g
	Glucose	20 g
	Agar	18 g
	Make up to 1000 ml with double distilled water.	
Sabouraud 4% glucose agar (S4%)	Polypeptone	10 g
	Glucose	40 g
	Agar	18 g
	Make up to 1000 ml with double distilled water.	
MIX Medium	Peptone from meat	8 g
	Peptone from caseine	8 g
	Malt extract	20 g
	Yeast extract	3 g
	NaCl	5 g
	Glucose	40 g
	Agar	15 g
	Make up to 1000 ml with double distilled water.	

References

Aarrestad, P. A. and Aamlid, D. (1999). Vegetation monitoring in South-Varanger, Norway – species composition of ground vegetation and its relation to environmental variables and pollution impact. *Environmental Monitoring and Assessment*, **58**, 1–21.

Abba', S., Ghignone, S. and Bonfante, P. (2006). A dehydration-inducible gene in the truffle *Tuber borchii* identifies a novel group of dehydrins. *BMC Genomics*, **7**, 39.

Acharius, E. (1798). *Lichenographiae Svecicae Prodromus*. Linköping: D.G. Björn.

Acharius, E. (1803). *Methodus qua omnus detectos Lichenes*. Stockholm: F.D.D. Ulrich.

Acharius, E. (1810). *Lichenographia universalis*. Göttingen: F. Dandewerts.

Acharius, E. (1814). *Synopsis methodica lichenum*. Lund.

Adams, G. C. and Kropp, B. R. (1996). *Athelia arachnoidea*, the sexual state of *Rhizoctonia carotae*, a pathogen of carrot in cold storage. *Mycologia*, **88**, 459–472.

Adler, M. T. and Calvelo, S. (2002). Parmeliaceae s. str. (lichenized Ascomycetes) from Tierra del Fuego (southern South America) and their world distribution patterns. *Mitteilungen aus dem Institut für Allgemeine Botanik Hamburg*, **30–32**, 9–24.

Adler, M. T., Fazio, A., Bertoni, M. D., *et al.* (2004). Culture experiments and DNA verification of a mycobiont isolated from *Punctelia praesignis* (Parmeliaceae, lichenized Ascomycotina). *Bibliotheca Lichenologica*, **88**, 1–8.

Aguiar, L. W., Martau, L., de Oliveira, M. L. A. A. and Martins-Mazzitelli, S. M. A. (1998). Efeitos do dióxido de enxofre (SO_2) em liquens, Rio Grande do Sul, Brasil. *Iheringia Botanica*, **50**, 67–73.

Ahmadjian, V. (1967a). *The Lichen Symbiosis*. Toronto: Blaisdell Publishing.

Ahmadjian, V. (1967b). A guide to the algae occurring as lichen symbionts: isolation, culture, cultural physiology and identification. *Phycologia*, **6**, 127–160.

Ahmadjian, V. (1973). Methods of isolating and culturing lichen symbionts and thalli. In *The Lichens*, ed. V. Ahmadjian and M. E. Hale, pp. 653–659. London: Academic Press.

Ahmadjian, V. (1988). The lichen alga *Trebouxia*: does it occur free-living? *Plant Systematics and Evolution*, **158**, 243–247.

Ahmadjian, V. (1993). *The Lichen Symbiosis*. New York: John Wiley.

Ahmadjian, V. (1995). Lichens are more important than you think. *BioScience*, **45**, 123–124.

Ahmadjian, V. (2001). *Trebouxia*: reflections on a perplexing and controversial lichen photobiont. In *Symbiosis*, ed. J. Seckbach, pp. 373–383. Dordrecht: Kluwer Academic.

Ahmadjian, V. and Heikkilä, H. (1970). The culture and synthesis of *Endocarpon pusillum* and *Staurothelse clopima*. *Lichenologist*, **4**, 259–267.

Ahmadjian, V. and Jacobs, J. B. (1981). Relationship between fungus and alga in the lichen *Cladonia cristatella* Tuck. *Nature (London)*, **389**, 169–172.

Ahti, T. (1999). Biogeography. *Nordic Lichen Flora*, **1**, 7–8.

Ahti, T. (2000). Cladoniaceae. *Flora Neotropica Monograph*, **78**, 1–362.

Alebic-Juretic, A. and Arko-Pijevac, M. (1989). Air pollution damage to cell membranes in lichens – results of simple biological test applied in Rijeka, Yugoslavia. *Water, Air, and Soil Pollution*, **47**, 25–33.

Alexopoulos, C. J., Mims, C. W. and Blackwell, M. (1996). *Introductory Mycology*. 4th edn. New York: John Wiley.

Allan, A. and Fluhr, R. (1997). Two distinct sources of elicited reactive oxygen species in tobacco epidermal cells. *Plant Cell*, **9**, 1559–1572.

Allen, J. F., Mullineaux, C. W., Sanders, C. E. and Melis, A. (1989). State transitions, photosystem stoichiometry adjustment and non-photochemical quenching in cyanobacterial cells acclimated to light absorbed by photosystem I or photosystem II. *Photosynthesis Research*, **22**, 157–166.

Alpert, P. (1988). Survival of a desiccation tolerant moss, *Grimmia laevigata*, beyond its observed microdistributional limits. *Journal of Bryology*, **15**, 219–227.

Alscher, R. (1984). Effects of SO_2 on light-modulated enzyme reactions. In *Gaseous Air Pollutants and Plant Metabolism*, ed. M. J. Koziol and F. R. Whatley, pp. 181–200. London: Butterworths.

Alstrup, V. (1992). Effects of pesticides on lichens. *Bryonora*, **9**, 2–4.

Alstrup, V. and Hansen, E. S. (1977). Three species of lichens tolerant of high concentrations of copper. *Oikos*, **29**, 290–293.

Alstrup, V. and Hawksworth, D. L. (1990). The lichenicolous fungi from Greenland. *Meddelelser om Grønland, Bioscience*, **31**, 1–90.

Amthor, J. S. (1995). Higher plant respiration and its relationships to photosynthesis. In *Ecophysiology of Photosynthesis*, ed. E. D. Schulze and M. M. Caldwell, pp. 71–101. Berlin: Springer.

Anagnostidis, K. and Komárek, J. (1985). Modern approach to the classification system of cyanophytes. 1 – Introduction. *Archiv für Hydrobiologie, Supplementband*, **71** (*Algological Studies*, **38/39**), 291–302.

Anagnostidis, K. and Komárek, J. (1988). Modern approach to the classification system of cyanophytes. 3 – Oscillatoriales. *Archiv für Hydrobiologie, Supplementband*, **80** (*Algological Studies*, **50/53**), 327–472.

Anagnostidis, K. and Komárek, J. (1990). Modern approach to the classification system of cyanophytes. 5 – Stigonematales. *Algological Studies*, **59**, 1–73.

Anand, M., Laurence, S. and Rayfield, B. (2005). Diversity relationships among taxonomic groups in recovering and restored forests. *Conservation Biology*, **19**, 955–962.

Andersen, H. L. and Ekman, S. (2005). Disintegration of the Micareaceae (lichenized Ascomycota): a molecular phylogeny based on mitochondrial rDNA sequences. *Mycological Research*, **109**, 21–30.

Anonymous (2004). Vagrant lichen charged in elk deaths. *Castilleja*, **23**, 1.

Antoine, M. E. (2004). An ecophysiological approach to quantifying nitrogen fixation by *Lobaria oregana*. *Bryologist*, **107**, 82–87.

Antoine, M. E. and McCune, B. (2004). Contrasting fundamental and realized ecological niches with epiphytic lichen transplants in an old-growth *Pseudotsuga* forest. *Bryologist*, **107**, 163–173.

Aptroot, A. (1987). Terpenoids in tropical Pyxinaceae (lichenized fungi). In *XIV International Botanical Congress, Abstracts*, Berlin, ed. W. Greuter, B. Zimmer and H.-D. Behnke, p. 5-04-7.

Aptroot, A. (2001). Lichenized and saprobic fungal biodiversity of a single *Elaeocarpus* tree in Papua New Guinea, with the report of 200 species of ascomycetes associated with one tree. *Fungal Diversity*, **6**, 1–11.

Aptroot, A. and Seaward, M. R. D. (2003). Freshwater lichens. In *Freshwater Mycology*, ed. C. K. Tsui and K. D. Hyde, pp. 101–110. Hong Kong: Fungal Diversity Press.

Aptroot, A. and Sipman, H. J. M. (1997). Diversity of lichenized fungi in the tropics. In *Biodiversity of Tropical Microfungi*, ed. K. D. Hyde, pp. 93–106. Hong Kong: Hong Kong University Press.

Aptroot, A. and Sparrius, L. B. (2006). Additions to the lichen flora of Vietnam, with an annotated checklist and bibliography. *Bryologist*, **109**, 358–371.

Archer, A. W. and Elix, J. A. (1993). Additional new taxa and a new report of *Pertusaria* (lichenised Ascomycotina) from Australia. *Mycotaxon*, **49**, 143–150.

Archer, D., Eggink, G., Schweizer, M. Stymne, S. and Rathledge, G. (1999). *Manipulation of Lipid Metabolism Aimed at the Production of Fatty Acids and Polyketides*. Final report, Contr. No. Air 2-CT94-967. Internet Publication www.biomatnet.org.secure/Air/S1045.htm.

Archibald, P. A. (1975). *Trebouxia* DePuymaly (Chlorophyceae, Chlorococcales) and *Pseudotrebouxia* (Chlorophyceae, Chlorosarcinales). *Phycologia*, **14**, 125–137.

Armaleo, D. (1993). Why do lichens make secondary products? In *XV International Botanical Congress Abstracts*, ed. M. Furuya, p. 11. Yokohama, Japan: International Union of Biological Sciences.

Armaleo, D. and Clerc. P. (1990). Lichen chimeras: DNA analysis suggests that one fungus forms two morphotypes. *Experimental Mycology*, **15**, 1–10.

Armaleo, D. and Clerc, P. (1995). A rapid and inexpensive method for the purification of DNA from lichens and their symbionts. *Lichenologist*, **27**, 207–213.

Armstrong, R. A. (1974). A comparison of the growth-curves of the foliose lichen *Parmelia conspersa* determined by a cross-sectional study and by direct measurement. *Environmental and Experimental Botany*, **32**, 221–227.

Armstrong, R. A. (1988). Substrate colonization, growth, and competition. In *CRC Handbook of Lichenology*, Vol. 2, ed. M. Galun, pp. 3–16. Boca Raton: CRC Press.

Armstrong, R. A. (1992). Soredial dispersal from individual soralia in the lichen *Hypogymnia physodes* (L.) Nyl. *Environmental and Experimental Botany*, **32**, 55–63.

Armstrong, R. A. (1993). Factors determining lobe growth in foliose lichen thalli. *New Phytologist*, **124**, 675–679.

Armstrong, R. A. and Smith, S. N. (1992). Lobe growth variation and the maintenance of symmetry in foliose lichen thalli. *Symbiosis*, **12**, 145–158.

Aronson, J. M. (1977). Cell walls and intracellular polysaccharides of Leptomitales. *Abstracts Second International Mycological Congress A–L*, ed. H. E. Bigelow and E. G. Simmons, p. 19. Tampa: IMC-2, Inc.

Arup, U. and Grube, M. (1998). Molecular systematics of *Lecanora* subgenus *Placodium*. *Lichenologist*, **30**, 415–425.

Arup, U., Ekman, S., Grube, M., Mattsson, J. E. and Wedin, M. (2007). The sister group relation of Parmeliaceae (Lecanorales, Ascomycota). *Mycologia*, **99**, 42–49.

Arup, U., Ekman, S., Lindblom, L. and Mattsson, J. E. (1993). High performance thin layer chromatography (HPTLC), an improved technique for screening lichen substances. *Lichenologist*, **25**, 61–71.

Asahina, Y. and Shibata, S. (1954). *Chemistry of Lichen Substances*. Tokyo: Japan Society for the Promotion of Science.

Ascaso, C., Wierzchos, J. and de los Rios, A. (1995). Cytological investigations of lithobiontic microorganisms in granitic rocks. *Botanica Acta*, **108**, 474–481.

Aspray, T., Jones, E., Whipps, J. and Bending, G. (2006). Importance of mycorrhization helper bacteria cell density and metabolite localization for the *Pinus sylvestris / Lactarius rufus* symbiosis. *FEMS Microbiology Ecology*, **56**, 25–33.

Augusto, S., Pinho, P., Branquinho, C., et al. (2004). Atmospheric dioxin and furan deposition in relation to land-use and other pollutants: a survey with lichens. *Journal of Atmospheric Chemistry*, **49**, 53–65.

Avise, J. C. (1998). The history and purview of phylogeography: a personal reflection. *Molecular Ecology*, **7**, 371–379.

Bacci, E., Calamari, D., Gaggi, C., et al. (1986). Chlorinated hydrocarbons in lichen and moss samples from the Antarctic Peninsula. *Chemosphere*, **15**, 747–754.

Baćkor, M. and Dzubaj, A. (2004). Short-term and chronic effects of copper, zinc and mercury on the chlorophyll content of four lichen photobionts and related alga. *Journal of the Hattori Botanical Laboratory*, **95**, 271–284.

Baćkor, M. and Fahselt, D. (2004). Physiological attributes of the lichen *Cladonia pleurota* in heavy metal-rich and control sites near Sudbury (Ont., Canada). *Environmental and Experimental Botany*, **52**, 149–159.

Baćkor, M. and Váczi, P. (2002). Copper tolerance in the lichen photobiont *Trebouxia erici* (Chlorophyta). *Environmental and Experimental Botany*, **48**, 11–20.

Baćkor, M. and Zetikova, J. (2003). Effects of copper, cobalt and mercury on the chlorophyll content of lichens *Cetraria islandica* and *Flavocetraria cucullata*. *Journal of the Hattori Botanical Laboratory*, **93**, 175–187.

Baćkor, M., Dvorsky, K. and Fahselt, D. (2003). Influence of invertebrate feeding on the lichen *Cladonia pocillum*. *Symbiosis*, **34**, 281–291.

Baddeley, M. S., Ferry, B. W. and Finegan, E. J. (1973). Sulphur dioxide and respiration in lichens. In *Air Pollution and Lichens*, ed. B. W. Ferry, M. S. Baddeley and D. L. Hawksworth, pp. 299–313. Toronto: University of Toronto Press.

Badger, M. R., Pfanz, H., Büdel, B., Heber, U. and Lange, O. L. (1993). Evidence for the functioning of photosynthetic CO_2 concentration mechanisms in lichens containing green algal and cyanobacterial photobionts. *Planta*, **191**, 57–70.

Bailey, R. H. (1966). Studies on the dispersal of lichen soredia. *Journal of the Linnean Society, Botany*, **59**, 479–490.

Bailey, R. H. (1976). Ecological aspects of dispersal and establishment in lichens. In *Lichenology: Progress and Problems*, ed. D. H. Brown, D. L. Hawksworth and R. H. Bailey, pp. 215–247. New York: Academic Press.

Balaguer, L. and Manrique, E. (1995). Factors which determine lichen response to chronic fumigations with sulphur dioxide. *Cryptogamic Botany*, **5**, 215–219.

Balaguer, L., Manrique, E., de los Rios, A., *et al.* (1999). Long-term responses of the green algal lichen *Parmelia caperata* to natural CO_2 enrichment. *Oecologia*, **119**, 166–174.

Balaguer, L., Valladares, F., Ascaso, C., *et al.* (1996). Potential effects of rising tropospheric concentrations of CO_2 and O_3 on green-algal lichens. *New Phytologist*, **132**, 641–652.

Bargagli, R., Iosco, F. P. and Barghigiani, C. (1987). Assessment of mercury dispersal in an abandoned mining area by soil and lichen analysis. *Water, Air, and Soil Pollution*, **36**, 219–225.

Barghigiani, C., Bargagli, R., Siegel, B. Z. and Siegel, S. M. (1990). A comparative study of mercury distribution on the Aeolian volcanoes, Vulcano and Stromboli. *Water, Air, and Soil Pollution*, **53**, 179–188.

Barkman, J. J. (1958). *Phytosociology and Ecology of Cryptogamic Epiphytes*. Assen: Van Gorcum.

Barreno, E. (1991). Phytogeography of terricolous lichens in the Iberian Peninsula and the Canary Islands. *Botanika Chronika*, **10**, 199–210.

Barreno, E., Grube, M., Le Bois, L., *et al.* (1998). Forum discussion. Lichens: a special case in biogeographical analysis. *International Lichenologial Newsletter*, **31**, 18–24.

Barták, M., Solhaug, K. A., Vráblíková, H. and Gaulaa, Y. (2006). Curling during desiccation protects the foliose lichen *Lobaria pulmonaria* against photoinhibition. *Oecologia*, **149**, 553–560.

Bartók, K. (1999). Pesticide usage and epiphytic lichen diversity in Romanian orchards. *Lichenologist*, **31**, 21–25.

Bates, J. W., Bell, J. N. B. and Farmer, A. M. (1990). Epiphyte recolonisation of oaks along a gradient of air pollution in south-east England, 1979–1990. *Environmental Pollution*, **68**, 81–99.

Bates, J. W., Bell, J. N. B. and Massara, A. C. (2001). Loss of *Lecanora conizaeoides* and other fluctuations of epiphytes on oak in S. E. England over 21 years with declining SO_2 concentrations. *Atmospheric Environment*, **35**, 2557–2568.

Bauer, H. (1984). Net photosynthetic CO_2 compensation concentrations of some lichens. *Zeitschrift für Pflanzenphysiologie*, **114**, 45–50.

Beard, K. H. and DePriest, P. T. (1996). Genetic variation within and among mats of the reindeer lichen, *Cladina subtenuis*. *Lichenologist*, **28**, 171–182.

Beck, A. (1999). Photobiont inventory of a lichen community growing on heavy-metal-rich rock. *Lichenologist*, **31**, 501–510.

Beck, A. (2002). Photobionts: diversity and selectivity in lichen symbioses. *International Lichenological Newsletter*, **35**, 18–24.

Beck, A. and Koop, H. U. (2001). Analysis of the photobiont population in lichens using a single-cell manipulator. *Symbiosis*, **31**, 57–67.

Beck, A., Friedl, T. and Rambold, G. (1998). Selectivity of photobiont choice in a defined lichen community: inferences from cultural and molecular studies. *New Phytologist*, **139**, 709–720.

Beck, A., Kasalicky, T. and Rambold, G. (2002). Myco-photobiontal selection in a Mediterranean cryptogam community with *Fulgensia fulgida*. *New Phytologist*, **153**, 317–326.

Beckelhimer, S. L. and Weaks, T. E. (1986). Effects of water transported sediment on corticolous lichen communities. *Lichenologist*, **18**, 339–347.

Becker, V. E. (1980). Nitrogen fixing lichens in forests of the southern Appalachian Mountains of North Carolina. *Bryologist*, **83**, 29–39.

Becker, V. E., Reeder, J. and Stetler, R. (1977). Biomass and habitat of nitrogen fixing lichens in an oak forest in the North Carolina Piedmont. *Bryologist*, **80**, 93–99.

Beckett, A. (1981). Ascospore formation. In *The Fungal Spore: Morphogenetic Controls*, ed. G. Turian and H. R. Hohl, pp. 107–129. London: Academic Press.

Beckett, P. J., Boileau, L. J. R., Padovan, D., Richardson, D. H. S. and Nieboer, E. (1982). Lichens and mosses as monitors of industrial activity associated with uranium and lead accumulation patterns. *Environmental Pollution (Series B)*, **4**, 91–107.

Beckett, R. P. (1995). Some aspects of the water relations of lichens from habitats of contrasting water status studied using thermocouple psychrometry. *Annals of Botany*, **76**, 211–217.

Beckett, R. P. (1997). Pressure-volume analysis of a range of poikilohydric plants implies the existence of negative turgor in vegetative cells. *Annals of Botany*, **79**, 145–152.

Beckett, R. P. and Brown, D. H. (1983). Natural and experimentally-induced zinc and copper resistance in the lichen genus *Peltigera*. *Annals of Botany*, **52**, 43–50.

Beckett, R. P. and Brown, D. H. (1984a). The control of cadmium uptake in the lichen genus *Peltigera*. *Journal of Experimental Botany*, **35**, 1071–1082.

Beckett, R. P. and Brown, D. H. (1984b). The relationship between cadmium uptake and heavy metal uptake tolerance in the lichen genus *Peltigera*. *New Phytologist*, **97**, 301–311.

Beckett, R. P. and Minibayeva, F. V. (2007). Rapid breakdown of exogenous extracellular hydrogen peroxide by lichens. *Physiologia Plantarum*, **129**, 588–596.

Beckett, R. P., Marschall, M. and Laufer, Z. (2005a). Hardening enhances photoprotection in the moss *Atrichum androgynum* during rehydration by increasing fast rather than slow-relaxing quenching. *Journal of Bryology*, **27**, 7–12.

Beckett, R. P., Mayaba, N., Minibayeva, F. V. and Alyabyev, A. J. (2005b). Hardening by partial dehydration and ABA increase desiccation tolerance in the cyanobacterial lichen *Peltigera polydactylon*. *Annals of Botany*, **96**, 109–115.

Beckett, R. P., Minibayeva, F. V., Vylegzhanina, N. V. and Tolpysheva, T. (2003). High rates of extracellular superoxide reduction by lichens in the Suborder Peltigerineae correlate with indices of high metabolic activity. *Plant, Cell and Environment*, **26**, 1827–1837.

Bedeneau, M. (1982). Reproduction in vitro des effets de la pollution par le dioxyde de soufre sur quelques lichens. *Annales des Sciences Forestieres*, **39**, 165–178.

Bedford, D. J., Schweizer, E., Hopwood, D. A. and Khosla, C. (1995). Expression of a functional fungal polyketide synthase in the bacterium *Streptomyces coelicolor* A3(2). *Journal of Bacteriology*, **177**, 4544–4548.

Begora, M. and Fahselt, D. (2001). Photolability of secondary compounds in some lichen species. *Symbiosis*, **31**, 3–22.

Belandria, G., Asta, J. and Nurit, F. (1989). Effects of sulphur dioxide and fluoride on ascospore germination of several lichens. *Lichenologist*, **21**, 79–86.

Belnap, J. (2001). Factors influencing nitrogen fixation and nitrogen release in biological soil crusts. In *Biological Soil Crusts: Structure, Function, and Management*, ed. J. Belnap and O. L. Lange, pp. 241–261. Berlin: Springer.

Belnap, J. (2002). Nitrogen fixation in biological soil crusts from southeast Utah, USA. *Biological Fertility of Soils*, **35**, 128–135.

Belnap, J. and Lange, O. L. (eds.) (2003). *Biological Soil Crusts: Structure, Function, and Management*: Ecological Studies 150. Berlin: Springer.

Belnap, J. and Lange, O. L. (2005a). Lichens and microfungi in biological soil crusts: community structure, physiology, and ecological functions. In *The Fungal Community: Its Organization and Role in the Ecosystem*, Vol. 3, ed. J. Dighton, J. F. White and P. Oudemans, pp. 117–138. Boca Raton: CRC Press.

Belnap, J. and Lange, O. L. (2005b). Biological soil crusts and global changes: what does the future hold? In *The Fungal Community: Its Organization and Role in the Ecosystem*, Vol. 3, ed. J. Dighton, J. F. White and P. Oudemans, pp. 697–712. Boca Raton: CRC Press.

Benedict, J. B. (1991). Experiments on lichen growth. II. Effects of a seasonal snow cover. *Arctic and Alpine Research*, **23**, 189–199.

Benner, J. W. and Vitousek, P. M. (2007). Development of a diverse epiphyte community in response to phosphorous fertilization. *Ecological Letters*, **10**, 628–636.

Benner, J. W., Conroy, S., Lunch, C. K. Toyoda, N. and Vitousek, P. M. (2007). Phosphorus fertilization increases the abundance and nitrogenase activity of the cyanolichen *Pseudocyphellaria crocata* in Hawaiian montane forests. *Biotropica*, **39**, 400–405.

Bennett, J. P. and Wetmore, C. M. (1999). Geothermal elements in lichens of Yellowstone National Park, USA. *Environmental and Experimental Botany*, **42**, 191–200.

Berger, F. and Aptroot, A. (2003). Further contributions to the flora of lichens and lichenicolous fungi of the Azores. *Arquipélago*, **19A**, 1–12.

Bergman, D. E. and Ebinger, J. E. (1990). Cyanogenesis in the lichen genus *Dermatocarpon*. *Castanea*, **55**, 207–210.

Beschel, R. E. (1961). Dating rock surfaces by lichen growth and its application to glaciology and physiography (lichenometry). In *Geology of the Arctic*, Vol. 2, ed. G. O. Raasch, pp. 1044–1062. Toronto: University of Toronto Press.

Bewley, J. D. (1979). Physiological aspects of desiccation tolerance. *Annual Review of Plant Physiology*, **30**, 195–238.

Bewley, J. D. and Krochko, J. E. (1982). Desiccation tolerance. In *Physiological Plant Ecology*. Vol. II: *Water Relations and Carbon Assimilation*, ed. O. L. Lange, P. S. Nobel, C. B. Osmond and H. Ziegler, pp. 325–378. Encyclopedia of Plant Physiology 12B. Berlin: Springer.

Bilger, W., Rimke, S., Schreiber, U. and Lange, O. L. (1989). Inhibition of energy-transfer to photosystem II in lichens by dehydration: different properties of reversibility with green and blue-green phycobionts. *Journal of Plant Physiology*, **134**, 261–268.

Bingle, L. E., Simpson, T. J. and Lazarus, C. M. (1999). Ketosynthase domain probes identify two subclasses of fungal polyketide synthase genes. *Fungal Genetics and Biology*, **26**, 209–223.

Bischoff, H. W. and Bold, H. C. (1963). Phycological studies. IV. Some soil algae from Enchanted Rock and related algal species. *University of Texas Publication*, **6318**, 1–95.

Bjelland, T. and Ekman, S. (2005). Fungal diversity in rock beneath a crustose lichen as revealed by molecular markers. *Microbial Ecology*, **49**, 598–603.

Bjerke, J. W. (2005). Synopsis of the lichen genus *Menegazzia* (*Parmeliaceae*, lichenized Ascomycotina) in South America. *Mycotaxon*, **91**, 423–454.

Bjerke, J. W., Elvebakk, A., Dominguez, B. and Dahlbäck, A. (2005). Seasonal trends in usnic acid concentrations of Arctic, alpine and Patagonian populations of the lichen *Flavocetraria nivalis*. *Phytochemistry*, **66**, 337–344.

Bjerke, J. W., Lerfall, K. and Elvebakk, A. (2002). Effects of ultraviolet radiation and PAR on the content of usnic and divaricatic acids in two arctic-alpine lichens. *Photochemical and Photobiological Sciences*, **1**, 678–685.

Bjerke, J. W., Zielke, M. and Solheim, B. (2003). Long-term impacts of simulated climatic change on secondary metabolism, thallus structure and nitrogen fixation activity in two cyanolichens from the Arctic. *New Phytologist*, **159**, 361–367.

Björkman, O. (1981). Responses to different quantum flux densities. In *Physiological Plant Ecology*. Vol. I: *Responses to the Physical Environment*, ed. O. L. Lange, P. S. Nobel, C. B. Osmond and H. Ziegler, pp. 57–108. Encyclopedia of Plant Physiology 12A. Berlin: Springer.

Björkman, O., Boardman, N. K., Anderson, J. A., *et al.* (1972). Effect of light intensity during growth of *Atriplex patula* on the capacity of photosynthetic reactions, chloroplast components and structure. *Carnegie Institution Washington Yearbook*, **71**, 115–135.

Black, M. and Pritchard, H. W. (2002). *Desiccation and Survival in Plants: Drying Without Dying*. Oxon: CABI Publishing.

Blaha, J., Baloch, E. and Grube, M. (2006). High photobiont diversity associated with the euryoecious lichen-forming ascomycete *Lecanora rupicola* (Lecanoraceae, Ascomycota). *Biological Journal of the Linnean Society*, **88**, 283–293.

Blanco, O., Crespo, A., Divakar, P. K., Elix, J. A. and Lumbsch, H. T. (2005). Molecular phylogeny of parmotremoid lichens (Ascomycota, Parmeliaceae). *Mycologia*, **97**, 150–159.

Blanco, O., Crespo, A., Elix, J. A., Hawksworth, D. L. and Lumbsch, H. T. (2004). A molecular phylogeny and a new classification of parmelioid lichens containing *Xanthoparmelia*-type lichenan (Ascomycota: Lecanorales). *Taxon*, **53**, 959–975.

Blanco, O., Crespo, A., Ree, R. H. and Lumbsch, H. T. (2006). Major clades of parmelioid lichens (Parmeliaceae, Ascomycota) and the evolution of their morphological and chemical diversity. *Molecular Phylogenetics and Evolution*, **39**, 52–69.

Blum, O. B. (1973). Water relations. In *The Lichens*, ed. V. Ahmadjian and M. E. Hale, pp. 381–400. New York: Academic Press.

Boardman, N. K. (1977). Comparative photosynthesis of sun and shade plants. *Annual Review of Plant Physiology*, **28**, 355–377.

Boardman, N. K., Anderson, J. M., Thorne, S. E. and Björkman, O. (1972). Photochemical reactions of chloroplasts and components of the photosynthetic electron transport chain in two rainforest species. *Carnegie Institution Washington Yearbook*, **71**, 107–114.

Boileau, L. J. R., Beckett, P. J., Lavoie, P., Richardson, D. H. S. and Nieboer, E. (1982). Lichens and mosses as monitors of industrial activity associated with uranium mining in northern Ontario, Canada. I. Field procedures, chemical analyses and interspecies comparisons. *Experimental Pollution (Series B)*, **4**, 69–84.

Boison, G., Mergel, A., Jolkver, H. and Bothe, H. (2004). Bacterial life and dinitrogen fixation at a gypsum rock. *Applied and Environmental Microbiology*, **70**, 7070–7077.

Boissière, M.-C. (1982). Cytochemical ultrastructure of *Peltigera canina*: some factors relating to its symbiosis. *Lichenologist*, **14**, 1–28.

Boissière, M.-C. (1987). Ultrastructural relationship between the composition and the structure of the cell wall of the mycobiont of two lichens. *Bibliotheca Lichenologica*, **25**, 117–123.

Bold, H. C. and Wynne, M. J. (1985). *Introduction to the Algae. Stucture and Reproduction*. 2nd edn. Englewood Cliffs: Prentice Hall.

Boonpragob, K., Nash, T. H., III, and Fox, C. A. (1989). Seasonal deposition patterns of acidic ions and ammonium to the lichen *Ramalina menziesii* Tayl. in southern California. *Environmental and Experimental Botany*, **29**, 187–197.

Borecký, J. and Vercesi, A. (2005). Plant uncoupling mitochondrial protein and alternative oxidase: energy metabolism and stress. *Bioscience Reports*, **25**, 271–286.

Boreham, S. (1992). A study of corticolous lichens on London plane *Platanus* × *hybrida* trees in West Ham Park, London. *London Naturalist*, **71**, 61–71.

Bothe, H. and Loos, E. (1972). Effect of far red light and inhibitors on nitrogen fixation and photosynthesis in the blue-green alga *Anabaena cylindrica*. *Archiv für Microbiologie*, **86**, 241–254.

Bothe, H., Distler, E. and Eisbrenner, G. (1978). Hydrogen metabolism in blue-green algae. *Biochimie*, **60**, 277–289.

Bottomley, P. J. and Stewart, W. D. P. (1977). ATP and nitrogenase activity in nitrogen-fixing heterocystous blue-green algae. *New Phytologist*, **79**, 625–638.

Boucher, V. L. and Nash, T. H., III (1990a). Growth pattern in *Ramalina menziesii* in California: coastal vs. inland populations. *Bryologist*, **93**, 295–302.

Boucher, V. L. and Nash, T. H., III (1990b). The role of the fruticose lichen *Ramalina menziesii* in the annual turnover of biomass and macronutrients in a blue oak woodland. *Botanical Gazette*, **151**, 114–118.

Boustie, J. and Grube, M. (2005). Lichens – a promising source of bioactive secondary metabolites. *Plant Genetic Resources*, **3**, 273–287.

Bowker, M. A., Belnap, J., Davidson, D. W. and Phillips, S. L. (2005). Evidence for micronutrient limitation of biological soil crusts: importance to arid-lands restoration. *Ecological Applications*, **15**, 1941–1951.

Branquinho, C., Brown, D. H. and Catarino, F. (1997). The cellular location of Cu in lichens and its effects on membrane integrity and chlorophyll fluorescence. *Environmental and Experimental Botany*, **38**, 165–179.

Brightman, F. H. and Seaward, M. R. D. (1977). Lichens of man-made substrates. In *Lichen Ecology*, ed. M. R. D. Seaward, pp. 253–293. London: Academic Press.

Broady, P. A. and Ingerfeld, M. (1993). Three new species and a new record of chaetophoracean (Chlorophyta) algae from terrestrial habitats in Antarctica. *European Journal of Phycology*, **28**, 25–31.

Brochmann, C., Gabrielsen, T. M., Nordal, I., Landvik, J. Y. and Elven, R. (2003). Glacial survival or *tabula rasa*? The history of North Atlantic biota revisited. *Taxon*, **52**, 417–450.

Brock, T. D. (1978). *Thermophilic Microorganisms and Life at High Temperatures*. New York: Springer.

Brodo, I. M. (1973). Substrate ecology. In *The Lichens*, ed. V. Ahmadjian and M. E. Hale, pp. 401–441. New York: Academic Press.

Brodo, I. M. (1978). Changing concepts regarding chemical diversity in lichens. *Lichenologist*, **10**, 1–11.

Brodo, I. M. and Richardson. D. H. S. (1978). Chimeroid associations in the genus *Peltigera*. *Lichenologist*, **10**, 157–170.

Brodo, I. M., Sharnoff, S. D. and Sharnoff, S. (2001). *Lichens of North America*. New Haven: Yale University Press.

Brooks, D. R. (2004). Reticulations in historical biogeography: the triumph of time over space in evolution. In *Frontiers of Biogeography: New Directions in the Geography of Nature*, ed. M. V. Lomolino and L. R. Heaney, pp. 125–144. Sunderland: Sinauer Associates.

Brouwer, R. (1962). Distribution of dry matter in the plant. *Netherland Journal of Agricultural Sciences*, **10**, 399–408.

Brown, D. H. (1972). The effect of Kuwait crude oil and a solvent emulsifier on the metabolism of the marine lichen, *Lichina pygmaea*. *Marine Biology*, **12**, 309–315.

Brown, D. H. (1992). Impact of agriculture on bryophytes and lichens. In *Bryophytes and Lichens in a Changing Environment*, ed. J. W. Bates and A. M. Farmer, pp. 259–283. Oxford: Clarendon Press.

Brown, D. H. and Beckett, R. P. (1984). Uptake and effect of cations on lichen metabolism. *Lichenologist*, **16**, 173–188.

Brown, D. H. and Beckett, R. P. (1985). The role of the cell wall in the intracellular uptake of cations by lichens. In *Lichen Physiology and Cell Biology*, ed. D. H. Brown, pp. 247–258. New York: Plenum Press.

Brown, D. H., Ascaso, C. and Rapsch, S. (1987). Ultrastructural changes in the pyrenoid of the lichen *Parmelia sulcata* stored under controlled conditions. *Protoplasma*, **136**, 136–144.

Brown, D. H., MacFarlane, J. D. and Kershaw, K. A. (1983). Physiological-environmental interactions in lichens. XVI. A re-examination of resaturation respiration phenomena. *New Phytologist*, **93**, 237–246.

Brown, D. H., Standell, C. J. and Miller, J. E. (1995). Effects of agricultural chemicals on lichens. *Cryptogamic Botany*, **5**, 220–223.

Brown, M. J., Jarman, S. K. and Kantvilas, G. (1994). Conservation and reservation of non-vascular plants in Tasmania, with special reference to lichens. *Biodiversity and Conservation*, **3**, 263–278.

Brown, R. M., Jr. and Bold, H. C. (1964). Comparative studies of the algal genera *Tetracystis* and *Chlorococcum*. Phycological Studies V. *University of Texas Publications*, **6417**, 1–213.

Brunauer, G. and Stocker-Wörgötter, E. (2005). Culture of lichen fungi for future production of biologically active compounds. *Symbiosis*, **38**, 187–201.

Brunauer, G., Grube, M., Muggia, L. and Stocker-Wörgötter, E. (2006). Gene bank of PKS from the mycobiont of *Xanthoria elegans*. Published in NCBI Database.

Brunner, U. and Honegger, R. (1985). Chemical and ultrastructural studies on the distribution of sporopollenin-like biopolymers in 6 genera of lichen phycobionts. *Canadian Journal of Botany*, **63**, 2221–2230.

Bruns, T. D., White, T. J. and Taylor, J. W. (1991). Fungal molecular systematics. *Annual Review of Ecology and Systematics*, **22**, 525–564.

Bruteig, I. E. (1993). The epiphytic lichen *Hypogymnia physodes* as biomonitor of atmospheric nitrogen and sulphur deposition in Norway. *Environmental Monitoring and Assessment*, **26**, 27–47.

Bryant, J. P., Chapin, F. S., III and Klein, D. R. (1983). Carbon nutrient balance of boreal plants in relation to vertebrate herbivory. *Oikos*, **40**, 357–368.

Bubrick, P. and Galun, M. (1980a). Proteins from the lichen *Xanthoria parietina* which bind to phycobiont cell walls. Correlation between binding patterns and cell wall cytochemistry. *Protoplasma*, **104**, 167–173.

Bubrick, P. and Galun, M. (1980b). Symbiosis in lichens: differences in cell wall properties of freshly isolated and cultured phycobionts. *FEMS Microbiology Letters*, **7**, 311–313.

Bubrick, P., Galun, M. and Frensdorff, A. (1981). Proteins from the lichen *Xanthoria parietina* which bind to phycobiont cell walls. Localization in the intact lichen and cultured mycobiont. *Protoplasma*, **105**, 207–211.

Bubrick, P., Galun, M. and Frensdorff, A. (1984). Observations on free-living *Trebouxia* de Puymaly and *Pseudotrebouxia* Archibald, and evidence that both symbionts from *Xanthoria parietina* (L.) Th. Fr. can be found free-living in nature. *New Phytologist*, **97**, 455–462.

Buchauer, M. J. (1973). Contamination of soil and vegetation near a zinc smelter by zinc, cadmium, copper, and lead. *Environmental Science and Technology*, **7**, 131–135.

Büdel, B. (1982). Phycobionten der Lichinaceen. Diplom-Thesis. Marburg: Universität Marburg.

Büdel, B. (1987). Zur Biologie und Systematik der Flechtengattungen *Heppia* und *Peltula* im südlichen Afrika. *Bibliotheca Lichenologica*, **23**, 1–105.

Büdel, B. (1990). Anatomical adaptations to the semiarid/arid environment in the lichen genus *Peltula*. *Bibliotheca Lichenologica*, **38**, 47–61.

Büdel, B. (1992). Taxonomy of lichenized procaryotic blue-green algae. In *Algae and Symbioses*, ed. W. Reisser, pp. 301–324. Bristol: Biopress Limited.

Büdel, B. and Henssen, A. (1983). *Chroococcidiopsis* (Cyanophyceae), a phycobiont in the lichen family Lichinaceae. *Phycologia*, **22**, 367–375.

Büdel, B. and Henssen, A. (1988). Zwei neue *Peltula*-Arten von Südafrika. *International Journal of Mycology and Lichenology*, **2**, 235–249.

Büdel, B. and Lange, O. L. (1991). Water status of green and blue-green phycobionts in lichen thalli after hydration by water vapor uptake: do they become turgid? *Botanica Acta*, **104**, 361–366.

Büdel, B. and Lange, O. L. (1994). The role of cortical and epicortical layers in the lichen genus *Peltula*. *Cryptogamic Botany*, **4**, 262–269.

Büdel, B. and Wessels, D. C. (1986). *Parmelia hueana* Gyeln., a vagrant lichen from the Namib Desert, SWA/Namibia. I. Anatomical and reproductive adaptations. *Dinteria*, **18**, 3–15.

Bull, W. B. (1996). Dating San Andreas fault earthquakes with lichenometry. *Geology*, **24**, 111–114.

Bull, W. B. and Brandon, M. T. (1998). Lichen dating of earthquake-generated regional rockfall events, Southern Alps, New Zealand. *Geological Society of America Bulletin*, **110**, 60–84.

Bull, W. B., King, J., Kong, F., Moutoux, T. and Phillips, W. M. (1994). Lichen dating of coseismic landslide hazards in alpine mountains. *Geomorphology*, **10**, 253–264.

Bunce, H. W. F. (1996). Methods of monitoring smelter emission effects on a temperate rain forest. *Fluoride*, **29**, 241–251.

Bungartz, F., Garvie, L. A. J. and Nash, T. H., III (2004). Anatomy of the endolithic Sonoran Desert lichen *Verrucaria rubrocincta* Breuss: implications for biodeterioration and biomineralization. *Lichenologist*, **36**, 55–73.

Burkholder, P. R. and Evans, A. W. (1945). Further studies on the antibiotic activity of Lichens. *Bulletin of the Torrey Botanical Club*, **72**, 157–164.

Burkholder, P. R., Evans, A. W., McVeigh, I. and Thornton, H. K. (1944). Antibiotic activity of Lichens. *Proceedings of the National Academy of Sciences, USA*, **30**, 250–255.

Buschbom, J. and Barker, D. (2006). Evolutionary history of vegetative reproduction in *Porpidia* s.l. (lichen-forming Ascomycota). *Systematic Biology*, **55**, 417–484.

Buschbom, J. and Mueller, G. M. (2004). Resolving evolutionary relationships in the lichen-forming genus *Porpidia* and related allies (Porpidiaceae, Ascomycota). *Molecular Phylogenetics and Evolution*, **32**, 66–82.

Buschbom, J. and Mueller, G. M. (2005). Testing "species pair" hypotheses: evolutionary processes in the lichen-forming species complex *Porpidia flavocoerulescens* and *Porpidia melinodes*. *Molecular Biology and Evolution*, **23**, 574–586.

Butin, H. (1954). Physiologisch-ökologische Untersuchungen über den Wasserhaushalt und die Photosynthese bei Flechten. *Biologisches Zentralblatt*, **73**, 459–502.

Butler, M. J. and Day, A. W. (1998). Fungal melanins: a review. *Canadian Journal of Microbiology*, **44**, 1115–1136.

Bychek-Guschina, I. A., Kotlova, E. R. and Heipieper, H. (1999). Effects of sulfur dioxide on lichen lipids and fatty acids. *Biochemistry (Moscow)*, **64**, 61–65.

Calatayud, A., Sanz, M.-J., Calvo, E., Barreno, E. and del Valle-Tascon, S. (1996). Chlorophyll *a* fluorescence and chlorophyll content in *Parmelia quercina* thalli from a polluted region of northern Castellón (Spain). *Lichenologist*, **28**, 49–65.

Calatayud, A., Tempe, P. J. and Barreno, E. (2000). Chlorophyll *a* fluorescence emission, xanthophylls cycle activity, and net photosynthetic responses to ozone in some foliose and fruticose lichen species. *Photosynthetica*, **38**, 281–286.

Caldwell, C. F., Turano, F. J. and McMahon, M. B. (1998). Identification of two cytosolic ascorbate peroxidase cDNAs from soybean leaves and characterization of their products by functional expression in *E. coli*. *Planta*, **204**, 120–126.

Calvelo, S. and Liberatore, S. (2001). Checklist of Argentinian lichens (version 2). Online: www.biologie.uni-hamburg.de/checklists.argen_12.htm.

Campbell, D. (1996). Complementary chromatic adaptation alters photosynthetic strategies in the cyanobacterium *Calothrix*. *Microbiology*, **142**, 1255–1263.

Campbell, D., Hurry, V., Clarke, A. K., Gustafsson, P. and Öquist, G. (1998). Chlorophyll fluorescence analysis of cyanobacterial photosynthesis and acclimation. *Microbiology and Molecular Biology Reviews*, **62**, 667–683.

Cane, D. E., Walsh, C. T. and Khosla, C. (1998). Harnessing the biosynthetic code: combinations, permutations, and mutations. *Science*, **282**, 63–68.

Cardinale, M., Puglia, A. M. and Grube, M. (2006). Molecular analysis of lichen-associated bacterial communities. *FEMS Microbiology Ecology*, **57**, 484–495.

Carlberg, G. E., Ofstad, E. B., Drangsholt, H. and Steinnes, E. (1983). Atmospheric deposition of organic micropollutants in Norway studied by means of moss and lichen analysis. *Chemosphere*, **12**, 341–356.

Carter, N. E. A. and Viles, H. A. (2003). Experimental investigations into the interactions between moisture, rock surface temperatures and an epilithic lichen cover in the bioprotection of limestone. *Building and Environment*, **38**, 1225–1234.

Carter, N. E. A. and Viles, H. A. (2005). Bioprotection explored: the story of a little known earth surface process. *Geomorphology*, **67**, 273–281.

Case, J. W. and Krouse, H. R. (1980). Variations in sulphur content and stable sulphur isotope composition of vegetation near a SO_2 source at Fox Creek, Alberta, Canada. *Oecologia*, **44**, 248–257.

Casely, A. F. and Dugmore, A. J. (2004). Climate change and 'anomalous' glacier fluctuations: the southwest outlets of Myrdalsjokull, Iceland. *Boreas*, **33**, 108–122.

Casselman, K. D. (2001). *Lichen Dyes: The New Source Book*. Mineola, NY: Dover Publications.

Castenholz, R. W. and Waterbury, J. B. (1989). Group I. Cyanobacteria. In *Bergey's Manual of Systematic Bacteriology*, Vol. 3, ed. J. T. Staley, P. Bryant, N. Pfennig and J. G. Holt, pp. 1710–1806. Baltimore: Williams and Wilkins.

Cavalier-Smith, T. (1987). The origin of fungi and pseudofungi. In *Evolutionary Biology of the Fungi*, ed. A. D. M. Rayner, C. N. Brasier and D. Moore, pp. 339–353. Cambridge: Cambridge University Press.

Chamberlain, A. C. (1970). Interception and retention of radioactive aerosols by vegetation. *Atmospheric Environment*, **4**, 57–78.

Chapin, F. S., III (1991). Integrated responses of plants to stress. *BioScience*, **41**, 29–36.

Chapin, F. S., III, Bloom, A. J., Field, C. B. and Waring, R. H. (1987). Plant responses to multiple environmental factors. *BioScience*, **37**, 49–57.

Chapman, R. L. (1984). An assessment of the current state of our knowledge of the Trentepohliaceae. In *Systematics of the Green Algae*. ed. D. E. G. Irvine and D. M. John, pp. 233–250. London: Academic Press.

Chen, G.-X., Kazmir, J. and Cheniae, G. M. (1992). Photoinhibition of hydroxylamine-extracted photosystem II membranes: studies of the mechanism. *Biochemistry*, **31**, 11 072–11 083.

Chen, S., Wu, D. and Wu, J. (1989). Using lichen communities as SO_2 pollution monitors. *Journal of Nanjing Normal University (Natural Science)*, **12**, 77–82.

Chooi, Y. H., Stalker, D., Louwhoff, S. and Lawrie, A. (2006). The search for a polyketide synthase gene producing beta-orsellinic acid and methylphloroacetophenone as precursors of depsidones and usnic acid in the lichen *Chondropsis semiviridis*. Poster presentation, International Mycological Congress, IMC 8, Cairns, Australia.

Cislaghi, C. and Nimis, P. L. (1997). Lichens, air pollution and lung cancer. *Nature*, **387**, 463–464.

Clarke, A. K., Campbell, D., Gustafsson, P. and Öquist, G. (1995). Dynamic responses of photosystem II and phycobilisomes to changing light in the cyanobacterium *Synechococcus* sp. PCC 7942. *Planta*, **197**, 553–562.

Clauzade, G. and Roux, C. (1976). *Les Champignons Lichénicoles non Lichénisés*. Montpellier: Université des Sciences et Techniques du Languedoc.

Clayden, S. R. (1992). Chemical divergence of eastern North American and European populations of *Arctoparmelia centrifuga* and their sympatric usnic acid-deficient chemotypes. *Bryologist*, **95**, 1–4.

Clayden, S. R. (1997). Intraspecific interactions and parasitism in an association of *Rhizocarpon lecanorinum* and *R. geographicum*. *Lichenologist*, **29**, 533–545.

Clerc, P. (2006). Synopsis of *Usnea* (lichenized Ascomycetes) from the Azores with additional information on the species in Macaronesia. *Lichenologist* **38**, 191–212.

Codogno, M., Poelt, J. and Puntillo, D. (1989). *Umbilicaria freyi* spec. nova und der Formenkreis von *Umbilicaria hirsuta* in Europa. *Plant Systematics and Evolution*, **165**, 55–69.

Cohn, F. (1853). Untersuchungen über die Entwicklungsgeschichte microskopischer Algen und Pilze. *Novorum actorum academiae caesareae leopoldinae-carolinae naturae curiosorum*, **24**, 101–256.

Coley, P. D. (1988). Effects of plant growth rate and leaf lifetime on the amount and type of anti-herbivore defense. *Oecologia*, **74**, 531–536.

Collins, C. R. and Farrar, J. F. (1978). Structural resistances to mass transfer in the lichen *Xanthoria parietina*. *New Phytologist*, **31**, 71–78.

Common, R. S. (1991). The distribution and taxonomic significance of lichenan and isolichenan in the Parmeliaceae (lichenized Ascomycotina), as determined by iodine reactions. I. Introduction and methods. II. The genus *Alectoria* and associated taxa. *Mycotaxon*, **41**, 67–112.

Cook, L. G. and Crisp, M. D. (2005). Directional asymmetry of long-distance dispersal and colonization could mislead reconstructions of biogeography. *Journal of Biogeography*, **32**, 741–754.

Cook, L. M. (2003). The rise and fall of the *carbonaria* form of the peppered moth. *Quarterly Review of Biology*, **78**, 399–417.

Cordeiro, L. M. C., Reis, R. A., Cruz, L. M., Stocker-Wörgötter, E., Grube, M. and M. I. (2005). Molecular studies of photobionts of selected lichens from the costal vegetation of Brazil. *FEMS Microbiology Ecolology*, **54**, 381–390.

Cowan, D. A., Green, T. G. A. and Wilson, A. T. (1979*a*). Lichen metabolism. 1. The use of tritium labelled water in studies of anhydrobiotic metabolism in *Ramalina celastri* and *Peltigera polydactyla*. *New Phytologist*, **82**, 489–503.

Cowan, D. A., Green, T. G. A. and Wilson, A. T. (1979*b*). Lichen metabolism. 2. Aspects of light and dark physiology. *New Phytologist*, **83**, 761–769.

Cowan, I. R., Lange, O. L. and Green, T. G. A. (1992). Carbon-dioxide exchange in lichens: determination of transport and carboxylation characteristics. *Planta*, **187**, 282–294.

Cowie, R. H. and Holland, B. S. (2006). Dispersal is fundamental to biogeography and the evolution of biodiversity on oceanic islands. *Journal of Biogeography*, **33**, 193–198.

Cox, C. B. and Moore, P. D. (2005). *Biogeography: An Ecological and Evolutionary Approach*. Oxford: Blackwell Publishing.

Cox, R. J., Hitchman, T. S., Byron, K. J., *et al.* (1997). Post-translational modification of heterologously expressed Streptomyces Type II polyketide synthase acyl carrier proteins. *FEBS Letters*, **405**, 267–272.

Coxson, D. S. (1988). Recovery of net photosynthesis and dark respiration on rehydration of the lichen, *Cladina mitis*, and the influence of prior exposure to sulphur dioxide while desiccated. *New Phytologist*, **108**, 483–487.

Coxson, D. S. (1991). Impedance measurement of thallus moisture content in lichens. *Lichenologist*, **23**, 77–84.

Coxson, D. S. and Curteanu, M. (2002). Decomposition of hair lichens (*Alectoria sarmentosa* and *Bryoria* spp.) under snowpack in montane forest, Cariboo Mountains, British Columbia. *Lichenologist*, **34**, 395–402.

Coxson, D. S. and Nadkarni, N. M. (1995). Ecological roles of epiphytes in nutrient cycles of forest ecosystems. In *Forest Canopies*, ed. M. D. Lowman and N. M. Nadkarni, pp. 495–543. London: Academic Press.

Coxson, D. S., Webber, M. R. and Kershaw, K. A. (1984). The thermal operating environment of corticolous and pendulous tree lichens. *Bryologist*, **87**, 197–202.

Craw, R. C., Grehan, J. R. and Heads, M. J. (1999). *Panbiogeography: Tracking the History of Life*. Oxford Biogeography Series. Oxford: Oxford University Press, **11**, 1–238.

Crespo, A., Bridge, P. D., Hawksworth, D. L., Grube, M and Cubero, O. F. (1999). Comparison of rRNA genotypic variability in the lichen-forming fungus *Parmelia sulcata* from long established and recolonizing sites following sulfur dioxide amelioration. *Plant Systematics and Evolution*, **217**, 177–183.

Crespo, A., Lumbsch, H. T., Matsson, J.-E., *et al.* (2007). Testing morphology-based hypotheses of phylogenetic relationships in Parmeliaceae (Ascomycota) using three ribosomal markers and the nuclear RPB-1 gene. *Molecular Phylogenetics and Evolution*, **44**, 812–824.

Crews, T. E., Kurina, L. M. and Vitousek, P. M. (2001). Organic matter and nitrogen accumulation and nitrogen fixation during early ecosystem development in Hawaii. *Biogeochemistry*, **52**, 259–279.

Crisci, J. V., Katinas, L. and Posadas, P. (2003). *Historical Biogeography: An Introduction*. Cambridge: Harvard University Press.

Crisci, J. V., Sala, O. E., Katinas, L. and Posadas, P. (2006). Bridging historical and ecological approaches to biogeography. *Australian Systematic Botany*, **19**, 1–10.

Crittenden, P. D. (1975). Nitrogen fixation on the glacial drift of Iceland. *New Phytologist*, **74**, 41–49.

Crittenden, P. D. (1983). The role of lichens in the nitrogen economy of subarctic woodlands: nitrogen loss from the nitrogen-fixing lichen *Stereocaulon paschale* during rainfall. In *Nitrogen as an Ecological Factor*, ed. L. Boddy, R. Marchant and D. J. Read, pp. 43–68. Oxford: Blackwell Scientific Publications.

Crittenden, P. D. (1989). Nitrogen relations of mat-forming lichens. In *Nitrogen, Phosphorus and Sulphur Utilization by Fungi*, ed. L. Boddy, R. Marchant and D. J. Read, pp. 243–268. Cambridge: Cambridge University Press.

Crittenden, P. D. (1996). The effect of oxygen deprivation on inorganic nitrogen uptake in an Antarctic macrolichen. *Lichenologist*, **28**, 347–354.

Crittenden, P. D. (1998). Nutrient exchange in an Antarctic macrolichen during summer snowfall snow melt events. *New Phytologist*, **139**, 697–707.

Crittenden, P. D. and Kershaw, K. A. (1978). A procedure for the simultaneous measurement of net CO_2-exchange and nitrogenase activity in lichens. *New Phytologist*, **80**, 393–401.

Crittenden, P. D. and Kershaw, K. A. (1979). Studies on lichen-dominated systems. XXII. The environmental control of nitrogenase activity in *Stereocaulon paschale* in spruce-lichen woodland. *Canadian Journal of Botany*, **53**, 236–254.

Crittenden, P. D., Kalucka, I. and Oliver, E. (1994). Does nitrogen supply limit the growth of lichens? *Cryptogamic Botany*, **4**, 143–155.

Crittenden, P. D., Llimona, X. and Sancho, L. (2004). Nitrogenase activity in *Thyrea* spp. - preliminary results. In *Book of Abstracts of the 5th IAL Symposium, Lichens in Focus*, ed. T. Randlane and A. Saag, p. 44. Tartu: Tartu University Press.

Crowe, J. H., Crowe, L. M. and Chapman, D. (1984). Preservation of membranes in anhydrobiotic organisms: the role of trehalose. *Science*, **223**, 701–703.

Culberson, C. F. (1972). Improved conditions and new data for the identification of lichen products by a standardized thin-layer chromatographic method. *Journal of Chromatography*, **72**, 113–125.

Culberson, C. F. (1986). Biogenetic relationships of the lichen substances in the framework of systematics. *Bryologist*, **89**, 91–98.

Culberson, C. F. and Ammann, K. (1979). Standardmethode zur Dünnschichtchromatographie von Flechtensubstanzen. *Herzogia*, **5**, 1–24.

Culberson, C. F. and Culberson, W. L. (1976). Chemosyndromic variation in lichens. *Systematic Botany*, **1**, 325–339.

Culberson, C. F. and Elix, J. A. (1989). Lichen substances. In *Methods in Plant Biochemistry*, Vol. 1: *Plant Phenolics*, ed. P. M. Dey and J. B. Harborne, pp. 509–535. London: Academic Press.

Culberson, C. F. and Johnson, A. (1976). A standardized two-dimensional thin-layer chromatographic method for lichen products. *Journal of Chromatography*, **128**, 253–259.

Culberson, C. F., Culberson, W. L. and Johnson, A. (1981). A standardized TLC analysis of β-orcinol depsidones. *Bryologist*, **84**, 16–29.

Culberson, C. F., Culberson, W. L. and Johnson, A. (1985). Orcinol-type depsides and depsidones in the lichens of the *Cladonia chlorophaea* group (Ascomycotina, Cladoniaceae). *Bryologist*, **88**, 380–387.

Culberson, C. F., Culberson, W. L. and Johnson, A. (1988). Gene flow in lichens. *American Journal of Botany*, **75**, 1135–1139.

Culberson, C. F., Hale, Jr., M. E., Tønsberg, T. and Johnson, A. (1984). New depsides from the lichens *Dimelaena oreina* and *Fuscidea viridis*. *Mycologia*, **76**, 148–160.

Culberson, C. F., Nash, T. H., III and Johnson, A. (1979). 3-α-Hydroxybarbatic acid, a new depside in chemosyndromes of some *Xanthoparmeliae* with β-orcinol depsides. *Bryologist*, **82**, 154–161.

Culberson, W. L. (1967). Analysis of chemical and morphological variation in the *Ramalina siliquosa* species complex. *Brittonia*, **19**, 333–52.

Culberson, W. L. (1986). Chemistry and sibling speciation in the lichen-forming fungi: ecological and biological considerations. *Bryologist*, **89**, 123–131.

Culberson, W. L. and Culberson, C. F. (1967). Habitat selection by chemically differentiated races of lichens. *Science*, **158**, 1195–1197.

Culberson, W. L. and Culberson, C. F. (1968). The lichen genera *Cetrelia* and *Platismatia* (Parmeliaceae). *Contributions from the United States National Herbarium*, **34**, 449–558.

Culberson, W. L. and Culberson, C. F. (1994). Secondary metabolites as a tool in ascomycete systematics: lichenized fungi. In *Ascomycetes Systematics: Problems and Perspectives in the Nineties*, ed. D. L. Hawksworth. pp. 155–163. New York: Plenum Press.

Culberson, W. L., Culberson C. F. and Johnson, A. (1977). *Pseudevernia furfuracea* – *olivetorina* relationships and ecology. *Mycologia*, **69**, 604–614.

Culberson, W. L., Culberson, C. F. and Johnson, A. (1993). Speciation in the lichens of the *Ramalina siliquosa* complex (Ascomycotina, Ramalinaceae): gene flow and reproductive isolation. *American Journal of Botany*, **80**, 1472–1481.

Curtis, C. J., Emmett, B. A., Grant, H., et al. (2005). Nitrogen saturation in UK moorlands: the critical role of bryophytes and lichens in determining retention of atmospheric N deposition. Journal of Applied Ecology, 42, 507–517.

Czehura, S. J. (1977). A lichen indicator of copper mineralization, Lights Creek District, Plumas County, California. Economic Geology, 72, 796–803.

Dahl, E. (1998). The Phytogeography of Northern Europe (British Isles, Fennoscandia and Adjacent Areas). Cambridge: Cambridge University Press.

Dahl, E. and Krog, H. (1973). Macrolichens of Denmark, Finland, Norway and Sweden. Oslo: Universitetsforlaget.

Dahlkild, A., Kallersjo, M., Lohtander, K. and Tehler, A. (2001). Photobiont diversity in the Physciaceae (Lecanorales). Bryologist, 104, 527–536.

Dahlman, L. and Palmqvist, K. (2003). Growth in two foliose tripartite lichens Nephroma arcticum and Peltigera aphthosa – empirical modeling of external versus internal factors. Functional Ecology, 17, 821–831.

Dahlman L., Näsholm, T. and Palmqvist, K. (2002). Growth, nitrogen uptake, and resource allocation in the two tripartite lichens Nephroma arcticum and Peltigera aphthosa during nitrogen stress. New Phytologist, 153, 307–315.

Dahlman, L., Persson, J., Näsholm, T. and Palmqvist, K. (2003). Carbon and nitrogen distribution in the green algal lichens Hypogymnia physodes and Platismatia glauca in relation to nutrient supply. Planta, 217, 41–48.

David, J. C. (1987). Studies on the genus Epigloea. Systema Ascomycetum, 6, 217–221.

David, K. A. and Fay, P. (1977). Effects of long term treatment with C_2H_2 on N_2-fixing microorganisms. Applied Environmental Microbiology, 34, 640–646.

Davis, W. C., Gries, C. and Nash, T. H., III (2000). The ecophysiological response of the aquatic lichen Hydrothyria venosa to nitrate in terms of weight and photosynthesis over long periods of time. Bibliotheca Lichenologica, 75, 201–208.

Davison, J. (1988). Plant beneficial bacteria. Nature Bio/Technology, 6, 282–286.

Deason, T. R. and Bold, H. C. (1960). Phycological studies. I. Exploratory studies of Texas soil algae. University of Texas Publications, 6022, 1–70.

Debuchy, R. and Turgeon, B. (2006). Mating-type structure, evolution and function in Euascomycetes. In Growth, Differentiation and Sexuality, The Mycota, Vol. 1, ed. U. Kües and R. Fischer, pp. 293–323. Heidelberg: Springer.

De Kok, L. J. and Stulen, I. (1993). Role of glutathione in plants under oxidative stress. In Sulfur Nutrition and Assimilation in Higher Plants, ed. L. J. De Kok, I. Stulen, H. Rennenberg, C. Brunold and W. E. Rauser, pp. 295–313. The Hague: SPB Academic Publishing.

de los Ríos, A. and Grube, M. (2000). Host-parasite interfaces of some lichenicolous fungi in the Dacampiaceae (Dothideales, Ascomycota). Mycological Research, 104, 1348–1353.

de los Ríos, A., Wierzchos, J., Sancho, L. G., Green, T. G. A. and Ascaso, C. (2005). Ecology of endolithic lichens colonizing granite in continental Antarctica. Lichenologist, 37, 383–395.

Deltoro, V. I., Gimeno, C., Calatayud, A. and Barreno, E. (1999). Effects of SO_2 fumigations on photosynthetic CO_2 gas exchange, chlorophyll a fluorescence emission and antioxidant enzymes in the lichens *Evernia prunastri* and *Ramalina farinacea*. *Physiologia Plantarum*, **105**, 648–654.

Dembitsky, V. M. and Tolstikov, G. A. (2005). *Organic Metabolites of Lichens*. Novosibirsk: Publishing House of SB RAS.

Demmig-Adams, B. (2006). Linking the xanthophyll cycle with thermal energy dissipation. *Photosynthesis Research*, **76**, 73–80.

Demmig-Adams, B., Máguas, C., Adams W. W., III, *et al.* (1990). Effect of high light on the efficiency of photochemical energy conversion in a variety of lichen species with green and blue-green phycobionts. *Planta*, **180**, 400–409.

Denison, W. C. (1973). Life in tall trees. *Scientific American*, **228**, 74–80.

Denison, W. C. (1979). *Lobaria oregana*, a nitrogen-fixing lichen in old-growth Douglas fir forests. In *Symbiotic Nitrogen Fixation in the Management of Temperate Forests*, ed. J. C. Gordon, C. T. Wheeler and D. A. Perry, pp. 266–275. Corvallis: Oregon State University.

DePriest, P. T. (1993a). Variation in the *Cladonia chlorophaea* complex. I. Morphological and chemical variation in Southern Appalachian populations. *Bryologist*, **96**, 555–563.

DePriest, P. T. (1993b). Variation in southern Appalachian populations of the *Cladonia chlorophaea* complex. II. ribosomal DNA variation. *Bryologist*, **97**, 117–126.

DePriest, P. T. (1993c). Small subunit rDNA variation in a population of lichen fungi due to optional group-I introns. *Gene*, **134**, 314–325.

DePriest, P. T. (2004). Early molecular investigations of lichen-forming symbionts: 1986–2001. *Annual Review of Microbiology*, **58**, 273–301.

DePriest, P. T., Ivanova, N. V., Fahselt, D., Alstrup, V. and Gargas, A. (2000). Sequences of psychrophilic fungi amplified from glacier-preserved ascolichens. *Canadian Journal of Botany*, **78**, 1450–1459.

de Queiroz, A. (2005). The resurrection of oceanic dispersal in historical biogeography. *Trends in Ecology and Evolution*, **20**, 68–73.

des Abbayes, H. (1953). Travaux sur les lichens parus de 1939 à 1952. *Bulletin Siciétié Botanique de France*, **100**, 83–123.

de Vera, J.-P., Horneck, G., Rettberg, P. and Ott, S. (2003). The potential of the lichen symbiosis to cope with extreme conditions of outer space. I. Influence of UV radiation and space vacuum on the vitality of lichen symbiosis and germination capacity. *International Journal of Astrobiology*, **1**, 285–293.

de Vera, J.-P., Horneck, G., Rettberg, P. and Ott, S. (2004). The potential of the lichen symbiosis to cope with extreme conditions of outer space. II. Germination capacity of lichen ascospores in response to simulated space conditions. *Advances in Space Research*, **33**, 1236–1243.

Dibben, M. J. (1971). Whole-lichen culture in a phytotron. *Lichenologist*, **5**, 1–10.

Diederich, P. (1996). The lichenicolous heterobasidiomycetes. *Bibliotheca Lichenologica*, **61**, 1–198.

Diederich, P. (2003). Review. *Bryologist*, **106**, 629–630.

Dietz, S., Büdel, B., Lange, O. L. and Bilger, W. (2000). Transmittance of light through the cortex of lichens from contrasting habitats. In *Aspects in Cryptogamic Research. Contributions in Honour of Ludger Kappen*, ed. B. Schroeter, M. Schlensog and T. G. A. Green, pp. 171–182. Berlin-Stuttgart: Gebrüder Borntraeger Verlagsbuchhandlung.

Divakar, P. K., Crespo, A., Blanco, O. and Lumbsch, H. T. (2006). Phylogenetic significance of morphological characters in the tropical *Hypotrachyna* clade of parmelioid lichens (Parmeliaceae. Ascomycota). *Molecular Phylogenetics and Evolution*, **40**, 448–458.

Döbbeler, P. (1984). Symbiosen zwischen Gallertalgen und Gallertpilzen der Gattung *Epigloea* (Ascomycetes). *Beihefte zur Nova Hedwigia*, **79**, 203–239.

Döbbeler, P. and Poelt, J. (1981). *Arthropyrenia endobrya* spec nov., eine hepaticole Flechte mit intrazellulärem Thallus aus Brasilien. *Plant Systematics and Evolution*, **138**, 275–281.

Döring, H. and Lumbsch, H. T. (1998). Ascoma ontogeny: is this character set of any use in the systematics of lichenized ascomycetes? *Lichenologist*, **30**, 489–500.

Döring, H. and Wedin, M. (2004). Infraspecific variation within *Stereocaulon* species complexes – genetic markers, individuals, populations and species. In *Lichens in Focus*, ed. T. Randlane and A. Saag, p. 26. Tartu: Tartu University Press.

Drew, E. A. (1966). Some aspects of the carbohydrate metabolism of lichens. Ph.D. thesis. Oxford: University of Oxford.

Duchesne, L. C. and Larson, W. (1989). Cellulose and the evolution of plant life. *BioScience*, **4**, 238–241.

Duguay, K. J. and Klironomos, J. M. (2000). Direct and indirect effects of enhanced UV-B radiation on the decomposing and competitive abilities of saprobic fungi. *Applied Soil Ecology*, **14**, 157–164.

Durrell, L. W. and Newsom, I. E. (1939). *Colorado's Poisonous and Injurious Plants*. Fort Collins: Colorado Experiment Station.

Dyer, P. S., Murtagh, G. J. and Crittenden, P. D. (2001). Use of RAPD-PCR DNA fingerprinting and vegetative incompatibility tests to investigate genetic variation within lichen-forming fungi. *Symbiosis*, **31**, 213–229.

Easton, R. M. (1994). Lichens and rocks: a review. *Geoscience Canada*, **21**, 59–76.

Ebach, M. C. and Humphries, C. J. (2003). Ontology of biogeography. *Journal of Biogeography*, **30**, 959–962.

Edmands, S. (1999). Heterosis and outbreeding depression in interpopulation crosses spanning a wide range of divergence. *Evolution*, **53**, 1757–1768.

Edwards, T. C., Jr., Cutler, D. R., Geiser, L., Alegria, J. and McKenzie, D. (2004). Assessing rarity of species with low detectability: lichens in Pacific Northwest forests. *Ecological Applications*, **14**, 414–424.

Egea, J. M. (1996). Catalogue of lichenized and lichenicolous fungi of Morocco. *Bocconea*, **6**, 19–114.

Egea, J. M. and Torrente, P. (1993). The lichen genus *Bactrospora*. *Lichenologist*, **25**, 211–255.

Egea, J. M. and Torrente, P. (1994). El género de hongos liquenizados *Lecanactis* (Ascomycotina). *Bibliotheca Lichenologica* **54**, 1–205.

Ekman, S. (1997). The genus *Cliostomum* revisited. *Symbolae Botanicae Upsalienses*, **32**, 17–28.

Ekman, S. (2001). Molecular phylogeny of the *Bacidiaceae* (*Lecanorales*, lichenized Ascomycota). *Mycological Research*, **105**, 783–797.

Ekman, S. and Fröberg, S. (1988). Taxonomical problems in *Aspicilia contorta* and *A. hoffmannii*: an effect of hybridization? *International Journal of Mycology and Lichenology*, **3**, 215–225.

Ekman, S. and Jørgensen, P. M. (2002). Towards a molecular phylogeny for the lichen family Pannariaceae (Lecanorales, Ascomycota). *Canadian Journal of Botany*, **80**, 625–634.

Ekman, S. and Tønsberg, T. (2002). Most species of *Lepraria* and *Leproloma* form a monophyletic group closely related to *Stereocaulon*. *Mycological Research*, **106**, 1262–1276.

Ekman, S. and Wedin, M. (2000). The phylogeny of the families Lecanoraceae and Bacidiaceae (lichenized Ascomycota) inferred from nuclear SSU rDNA sequences. *Plant Biology*, **2**, 350–360.

Elix, J. A. (1991). The lichen genus *Relicina* in Australasia. In *Tropical Lichens: Their Systematics, Conservation and Ecology*, ed. D. J. Galloway. Systematics Association Special Volume 43, pp. 17–34. Oxford: Clarendon Press.

Elix, J. A. (1993). Progress in the generic delimitation of *Parmelia sensu lato* lichens (Ascomycotina: Parmeliaceae) and a synoptic key to the Parmeliaceae. *Bryologist*, **96**, 359–383.

Elix, J. A. (1994). *Xanthoparmelia*. *Flora of Australia*, **55**, 201–308.

Elix, J. A. (1999). Detection and identification of secondary lichen substances: contributions by the Uppsala school of lichen chemistry. *Symbolae Botanicae Upsalienses*, **32**, 103–121.

Elix, J. A. and Ernst-Russell, K. D. (1993). *A Catalogue of Standardized Thin Layer Chromatographic Data and Biosynthetic Relationships for Lichen Substances*, 2nd edn. Canberra: Australian National University.

Elix, J. A, Giralt, M. and Wardlaw, J. H. (2003). New chloro-depsides from the lichen *Dimelaena radiata*. *Bibliotheca Lichenologica*, **86**, 1–7.

Elix, J. A., Johnston, J. and Parker, J. L. (1988). A computer program for the rapid identification of lichen substances. *Mycotaxon*, **31**, 89–99.

Ellis, C. J. and Coppins, B. J. (2007). 19th century woodland structure controls stand-scale epiphyte diversity in present-day Scotland. *Diversity and Distributions*, **13**, 84–91.

Ellis, C. J., Crittenden, P. D., Scrimgeour, C. M. and Ashcroft, C. J. (2005). Translocation of ^{15}N indicates nitrogen recycling in the mat-forming lichen *Cladonia portentosa*. *New Phytologist*, **168**, 423–434.

Elstner, E. F. and Oßwald, W. (1994). Mechanisms of oxygen activation during plant stress. *Proceeding of the Royal Society of Edinburgh*, **102**, 131–154.

Elvebakk, A. and Bjerke, J. W. (2006a). The Skibotn area in North Norway – an example of very high lichen species richness far to the north. *Mycotaxon*, **96**, 141–146.

Elvebakk, A. and Bjerke, J. W. (2006b). The Skibotn area in North Norway - an example of very high lichen species richness far to the north: a supplement with an annotated list of species. Online: www.mycotaxon.com/resources/weblists.html.

Engelbert, R. and Vonarburg, C. (1995). Lichen diversity and ozone impact in rural areas of central Switzerland. *Cryptogamic Botany*, **5**, 252-263.

Engelmann, M. D. (1966). Energetics, terrestrial field studies, and animal productivity. *Advances in Ecological Research*, **3**, 73-115.

Englund, B. (1977). The physiology of the lichen *Peltigera aphthosa*, with special reference to the blue-green phycobiont (*Nostoc* sp.). *Physiologia Plantarum*, **41**, 298-304.

Englund, B. (1978). Effects of environmental factors on acetylene reduction by intact thallus and excised cephalodia of *Peltigera aphthosa* Willd. *Ecological Bulletin (Stockholm)*, **26**, 234-246.

Englund, B. and Myerson, H. (1974). *In situ* measurement of nitrogen fixation at low temperatures. *Oikos*, **25**, 183-187.

Enriquez, S., Duarte, C. M., Sand-Jensen, K. and Nielsen, S. L. (1996). Broad-scale comparison of photosynthetic rates across phototrophic organisms. *Oecologia*, **108**, 197-206.

Eriksson, O. E. (2005). Ascomyceternas ursprung och evolution - Protolichenes-hypotesen. *Svensk Mykologisk Tidskrift*, **26**, 22-29.

Eriksson, O. E. (ed.) (2006a). Notes on ascomycete systematics Nr. 4324. *Myconet*, **12**, 83-101. Online: www.fieldmuseum.org/myconet/

Eriksson, O. E. (ed.) (2006b). Outline of Ascomycota, *Myconet*, **12**, 1-82. Online: www.fieldmuseum.org/myconet/.

Ernst, A., Kirschenlohr, H., Diez, J. and Boger, P. (1984). Glycogen content and nitrogenase activity in *Anabaena variabilis*. *Archiv für Microbiologie*, **140**, 120-125.

Ertl, L. (1951). Über die Lichtverhältnisse in Laubflecten. *Planta*, **39**, 245-270.

Ertz, D., Christnach, C., Wedin, M. and Diederich, P. (2005). A world monograph of the genus *Plectocarpon* (Roccellaceae, Arthoniales). *Bibliotheca Lichenologica*, **91**, 1-155.

Esseen, P.-A. and Renhorn, K. E. (1998). Mass loss of epiphytic lichen litter in a boreal forest. *Annals Botanica Fennica*, **35**, 211-217.

Esseen, P.-A., Ehnström, B., Ericson, L. and Sjöberg, K. (1997). Boreal forests. *Ecological Bulletins*, **45**, 16-47.

Esser, K. (1976). *Kryptogamen: Blaualgen, Algen, Pilze, Flechten*. Berlin: Springer.

Esslinger, T. L. (1977). A chemosystematic revision of the brown Parmeliae. *Journal of the Hattori Botanical Laboratory*, **42**, 1-211.

Esslinger, T. L. (1989). Systematics of *Oropogon* (Alectoriaceae) in the New World. *Systematic Botany Monographs*, **28**, 1-111.

Ettl, H. and Gärtner, G. (1984). Über die Bedeutung der Cytologie für die Algentaxonomie, dargestellt an *Trebouxia* (Chlorellales, Chlorophyceae). *Plant Systematics and Evolution*, **148**, 135-147.

Ettl, H. and Gärtner, G. (1995). *Syllabus der Boden-, Luft- und Flechtenalgen*. Stuttgart: Gustav Fischer.

Evans, C. A. and Hutchinson, T. C. (1996). Mercury accumulation in transplanted moss and lichens at high elevation sites in Quebec. *Water, Air, and Soil Pollution*, **90**, 475–488.

Evans, D. J. A., Archer, S. and Wilson, D. J. H. (1999). A comparison of the lichenometric and Schmidt hammer dating techniques based on data from the proglacial areas of some Icelandic glaciers. *Quaternary Science Reviews*, **18**, 13–41.

Evans, J. R. (1983). Nitrogen and photosynthesis in the flag leaf of wheat (*Triticum aestivum* L.). *Plant Physiology*, **72**, 297–302.

Evans, J. R. (1989). Photosynthesis and nitrogen relationships of C-3 plants. *Oecologia*, **78**, 9–19.

Evans, R. D. and Johansen, J. R. (1999). Microbiotic crusts and ecosystem processes. *Critical Reviews in Plant Sciences*, **18**, 183–225.

Evans, R. D., Belnap, J. Garcia-Pichel, F. and Phillips, S. L. (2001). Global change and the future of biological soil crusts. In *Biological Soil Crusts: Structure, Function and Management*, ed. J. Belnap and O. L. Lange, pp. 417–429. Berlin: Springer.

Eversman, S. (1978). Effects of low-level SO_2 on *Usnea hirta* and *Parmelia chlorochroa*. *Bryologist*, **81**, 368–377.

Eversman, S. and Sigal, L. L. (1987). Effects of SO_2, O_3, and SO_2 and O_3 in combination on photosynthesis and ultrastructure of two lichen species. *Canadian Journal of Botany*, **65**, 1806–1818.

Eversman, S., Johnson, C. and Gustafson, D. (1987). Vertical distribution of epiphytic lichens on three tree species in Yellowstone National Park. *Bryologist*, **90**, 212–216.

Fahselt, D. (1984). Interthalline variability in levels of lichen products within stands of *Cladina stellaris*. *Bryologist*, **87**, 50–56.

Fahselt, D. (1985). Multiple enzyme forms in lichens. In *Lichen Physiology and Cell Biology*, ed. D. H. Brown, pp. 129–143. New York: Plenum Publishing.

Fahselt, D. (1987). Electrophoretic analysis of esterase and alkaline phosphatase enzyme forms in single spore cultures of *Cladonia cristatella*. *Lichenologist*, **19**, 71–75.

Fahselt, D. (1989). Enzyme polymorphism in sexual and asexual umbilicate lichens from Sverdrup Pass, Ellesmere Island, Canada. *Lichenologist*, **21**, 279–285.

Fahselt, D. (1991). Enzyme similarity as an indicator of evolutionary divergence: *Stereocaulon saxatile* H. Magn. *Symbiosis*, **11**, 119–130.

Fahselt, D. (1992). Geothermal effects on multiple enzyme forms in the lichen *Cladonia mitis*. *Lichenologist*, **24**, 181–192.

Fahselt, D. (1994a). Secondary biochemistry of lichens. *Symbiosis*, **16**, 117–165.

Fahselt, D. (1994b). Carbon metabolism in lichens. *Symbiosis*, **17**, 127–182.

Fahselt, D. (1995). Lichen sexuality from the perspective of multiple enzyme forms. *Cryptogamic Botany*, **5**, 137–143.

Fahselt, D. (2001). Analysing lichen enzymes by isoelectricfocussing. In *Protocols in Lichenology*, ed. I. Kranner, A. Varma and P. Beckett, pp. 307–331. Berlin: Springer.

Fahselt, D. and Alstrup, V. (1997). High performance liquid chromatography of phenolics in recent and subfossil lichens. *Canadian Journal Botany*, **75**, 1148–1154.

Fahselt, D. and Hageman, C. (1994). Rhizine and upper thallus isozymes in umbilicate lichens. *Symbiosis*, **16**, 95–103.

Fahselt, D. and Krol, M. (1989). Biochemical comparison of two ecologically distinctive forms of *Xanthoria elegans* in the Canadian High Arctic. *Lichenologist*, **21**, 135–145.

Fahselt, D., Krol, M., Hüner, N. and Tønsberg, T. (2000). Pigmentation of *Cladonia* infected by the lichenicolous fungus *Arthrorhaphis aeroguinosa*. *Lichenologist*, **32**, 300–303.

Fahselt, D., Madzia, S. and Alstrup, V. (2001). Scanning electron microscopy of invasive fungi in lichens. *Bryologist*, **104**, 24–39.

Fahselt, D., Tavares, S. and Mazdia, S. (1997). Isozyme variation in lichens in relation to mine dust exposure. In *Progress and Problems in Lichenology in the Nineties*, ed. R. Türk and R. Zorer. *Bibliotheca Lichenologica*, **68**, 111–127.

Farkas, E. E. and Sipman, H. J. M. (1993). Bibliography and checklist of foliicolous lichenized fungi up to 1992. *Tropical Bryology*, **7**, 93–148.

Farkas, E. and Sipman, H. J. M. (1997). Checklist of foliicolous lichenized fungi – after Farkas and Sipman (1993), with additions to 1996. *Abstracta Botanica*, **21**, 173–206.

Farrar, J. F. (1976a) Ecological physiology of the lichen *Hypogymnia physodes*. I. Some effects of constant water saturation. *New Phytologist*, **77**, 93–103.

Farrar, J. F. (1976b). Ecological physiology of the lichen *Hypogymnia physodes*. II. Effects of wetting and drying cycles and the concept of physiological buffering. *New Phytologist*, **77**, 105–113.

Farrar, J. F. (1976c). The lichen as an ecosystem: observation and experiment. In *Lichenology: Progress and Problems*, ed. D. H. Brown, D. L. Hawksworth and R. H. Bailey, pp. 385–406. London: Academic Press.

Farrar, J. F. (1976d). The uptake and metabolism of phosphate by the lichen *Hypogymnia physodes*. *New Phytologist*, **77**, 127–134.

Farrar, J. F. (1988). Physiological buffering. In *CRC Handbook of Lichenology*, Vol. 2, ed. M. Galun, pp. 101–105. Boca Raton: CRC Press.

Farrar, J. F. and Smith, D. C. (1976). Ecological physiology of the lichen *Hypogymnia physodes*. III. The importance of the rewetting phase. *New Phytologist*, **77**, 115–125.

Feige, G. B. (1973). Untersuchungen zur Ökologie und Physiologie der marinen Blaualgenflechte *Lichina pygmaea* Ag. II. Die Reversibilität der Osmoregulation. *Zeitschrift für Pflanzenphyiologie*, **68**, 415–421.

Feige, G. B. and Jensen, M. (1992). Basic carbon and nitrogen metabolism of lichens. In *Algae and Symbioses*, ed. W. Reisser, pp. 277–299. Bristol: Biopress Limited.

Feige, G. B., Lumbsch, H. T., Huneck, S. and Elix, J. A. (1993). The identification of lichen substances by a standardized high-performance liquid chromatographic method. *Journal of Chromatography*, **646**, 417–427.

Fenn, M. E., Baron, J. S., Allen, E. B., *et al.* (2003). Ecological effects of nitrogen deposition in the western United States. *BioScience*, **53**, 404–420.

Ferber, T. (2002). The age and origin of talus cones in the light of lichenometric research. The Skalnisty and Zielony talus cones, High Tatra Mountains, Poland. *Studia Geomorphologica Carpatho-Balcanica*, **36**, 77–90.

Ferraro, L. I., Lücking, R. and Sérusiaux, E. (2001). A world monograph of the lichen genus *Gyalectidium* (Gomphillaceae). *Botanical Journal of the Linnean Society*, **137**, 311–345.

Ferry, B. W. and Baddeley, M. S. (1976). Sulphur dioxide uptake in lichens. In *Lichenology: Progress and Problems*, ed. D. H. Brown, D. L. Hawksworth and R. H. Bailey, pp. 407–418. London: Academic Press.

Feuerer, T. (ed.) (2006). Checklists of lichens and lichenicolous fungi. Version 1 June 2006. Online: www.checklists.de.

Fewer, D., Friedl, T. and Büdel, B. (2002). *Chroococcidiopsis* and heterocyst-differentiating cyanobacteria are each other's closest living relatives. *Molecular Phylogenetics and Evolution*, **41**, 498–506.

Fiechter, E. (1990). Thallusdifferenzierung und intrathalline Sekundärstoffverteilung bei Parmeliaceae (Lecanorales, lichenisierte Ascomyceten). Inauguraldissertation. Zürich: Universität Zürich.

Fields, R. D. (1988). Physiological responses of lichens to air pollutant fumigations. In *Lichens, Bryophytes and Air Quality*, ed. T. H. Nash, III and V. Wirth, pp. 175–200. Bibliotheca Lichenologica 30. Berlin-Stuttgart: J. Cramer.

Fields, R. D. and St. Clair, L. L. (1984). The effects of SO_2 on photosynthesis and carbohydrate transfer in the two lichens: *Collema polycarpon* and *Parmelia chlorochroa*. *American Journal of Botany*, **71**, 986–998.

Fletcher, A. (1980). Marine and maritime lichens of rocky shores: their ecology, physiology and biological interactions. In *The Shore Environment*, Vol. 2: *Ecosystems*, ed. J. H. Price, D. E. G. Irvine and W. F. Farnham, pp. 789–842. London: Academic Press.

Fletcher, J. (2002). Coordination of cell proliferation and cell fate decisions in the angiosperm shoot apical meristem. *BioEssays*, **24**, 27–37.

Fogg, G. E., Fay, P. and Walsby, A. E. (1973). *The Bluegreen Algae*. London: Academic Press.

Follmann, G. (2002). South America as diversity centre of the lichen family Roccellaceae (Arthoniales). *Mitteilungen aus dem Institut für Allgemeine Botanik Hamburg*, **30–32**, 61–77.

Forman, R. T. T. (1975). Canopy lichens with blue-green algae: a nitrogen source in a Columbian rain forest. *Ecology*, **56**, 1176–1184.

Forman, R. T. T. and Dowden, D. L. (1977). Nitrogen fixing lichen roles from desert to alpine in the Sangre de Cristo Mountains, New Mexico. *Bryologist*, **80**, 561–70.

Foyer, C. and Halliwell, B. (1976). The presence of glutathione and glutathione reductase in chloroplasts: a proposed role in ascorbic acid metabolism. *Planta*, **133**, 21–25.

Frank, H. A., Young, A., Britton, G. and Cogdell, R. J. (1999). *The Photochemistry of Carotenoids*. Berlin: Springer.

Freedman, B., Zobens, V., Hutchinson, T. C. and Gizyn, W. I. (1990). Intense, natural pollution affects Arctic tundra vegetation at the Smoking Hills, Canada. *Ecology*, **71**, 492–503.

Fremstad, E., Paal, J. and Möls, T. (2005). Impacts of increased nitrogen supply on Norwegian lichen-rich alpine communities: a 10-year experiment. *Journal of Ecology*, **93**, 471–481.

Friedl, T. (1987). Thallus development and phycobionts of the parasitic lichen *Diploschistes muscorum*. *Lichenologist*, **19**, 183–191.

Friedl, T. (1989a). Comparative ultrastructure of pyrenoids in *Trebouxia* (Microthamniales, Chlorophyta). *Plant Systematics and Evolution*, **164**, 145–159.

Friedl, T. (1989b). Systematik und Biologie von Trebouxia (Microthamniales, Chlorophyta) als Phycobiont der Parmeliaceae (lichenisierte Ascomyceten). Ph.D. thesis. Bayreuth: Universtät Bayreuth.

Friedl, T. (1993). New aspects of the reproduction by autospores in the lichen alga *Trebouxia* (Microthamniales, Chlorophyta). *Archiv für Protistenkunde*, **143**, 153–161.

Friedl, T. (1997). The evolution of the green algae. In Bhattacharaya, D. (Ed.), *Origins of Algae and Their Plastids*, ed. D. Bhattacharaya, pp. 87–101. Wien: Springer.

Friedl, T. and Gärtner, G. (1988). *Trebouxia* (Pleurastrales, Chlorophyta) as a phycobiont in the lichen genus *Diploschistes*. *Archiv für Protistenkunde*, **135**, 147–158.

Friedl, T. and Zeltner, C. (1994). Assessing the relationships of some coccoid green lichen algae and the Microthamniales (Chlorophyta) with 18S rRNA gene sequence comparisons. *Journal of Phycology*, **30**, 500–506.

Friedmann, E. I. (1982). Endolithic microorganisms in the Antarctic cold desert. *Science*, **215**, 1045–1053.

Friedmann, E. I. and Sun, H. J. (2005). Communities adjust their temperature optima by shifting producer-to-consumer ratio, shown in lichens as models. I. Hypothesis. *Microbial Ecology*, **49**, 523–527.

Fritz-Sheridan, R. P. (1985). Impact of simulated acid rain on nitrogenase activity in *Peltigera aphthosa* and *P. polydactyla*. *Lichenologist*, **17**, 27–31.

Fritz-Sheridan, R. P. and Coxson, D. S. (1988a). Nitrogen fixation on the tropical volcano, La Soufrière (Guadeloupe): nitrogen fixation, photosynthesis and respiration during the prevailing cloud/shroud climate by *Stereocaulon virgatum*. *Lichenologist*, **20**, 41–61.

Fritz-Sheridan, R. P. and Coxson, D. S. (1988b). Nitrogen fixation on the tropical volcano, La Soufrière (Guadeloupe): the interaction of temperature, moisture, and light with net photosynthesis and nitrogenase activity in *Stereocaulon virgatum* and response to periods of insolation shock. *Lichenologist*, **20**, 63–81.

Fröberg, L., Berg, C. O., Baur, A. and Baur, B. (2001). Viability of lichen photobionts after passing through the digestive tract of a land snail. *Lichenologist*, **33**, 543–545.

Fujii, I., Watanabe, A., Sankawa, U. and Ebizuka Y. (2001). Identification of *Claisen* cyclase domain in fungal polyketide synthase WA, a naphthapyrone synthase of *Aspergillus nidulans*. *Chemical Biology*, **8**, 187–197.

Gagunashvili, A. and Andrésson, O. (2006). Heterologous expression of a lichen polyketide synthase in filamentous fungi. EUKETIDES Meeting, Turku, Finland.

Gaio-Oliveira, G., Dahlman, L., Máguas, C. and Palmqvist, K. (2004a). Growth in relation to microclimatic conditions and physiological characteristics of four *Lobaria pulmonaria* populations in two contrasting habitats. *Ecography*, **27**, 13–28.

Gaio-Oliveira, G., Dahlman, L., Martins-Loução, M. A., Máguas, C. and Palmqvist, K. (2005a). Nitrogen uptake in relation to excess supply and its effects on the lichens *Evernia prunastri* (L.) Ach and *Xanthoria parietina* (L.) Th. Fr. *Planta*, **220**, 794–803.

Gaio-Oliveira, G., Dahlman, L., Palmqvist, K. and Máguas, C. (2004b). Ammonium uptake in the nitrophytic lichen *Xanthoria parietina* and its effects on vitality and balance between symbionts. *Lichenologist*, **36**, 75–86.

Gaio-Oliveira, G., Dahlman, L., Palmqvist, K. and Máguas, C. (2005b). Responses of the lichen *Xanthoria parietina* (L.) Th. Fr. to varying thallus nitrogen concentrations. *Lichenologist*, **37**, 171–179.

Gaio-Oliveira, G., Moen, J., Danell, O. and Palmqvist, K. (2006). Effect of simulated reindeer grazing on the re-growth capacity of mat-forming lichens. *Basic and Applied Ecology*, **7**, 109–121.

Galley, C. and Linder, H. P. (2006). Geographical affinities of the Cape flora, South Africa. *Journal of Biogeography*, **33**, 236–250.

Galloway, D. J. (1991a). Chemical evolution in the order Peltigerales: triterpenoids. *Symbiosis*, **11**, 327–344.

Galloway, D. J. (1991b). Phytogeography of Southern Hemisphere lichens. In *Quantitative Approaches to Phytogeography*, ed. P. L. Nimis and T. J. Crovello, pp. 233–262. Dordrecht: Kluwer Academic.

Galloway, D. J. (1993). Global environmental change: lichens and chemistry. *Bibliotheca Lichenologica*, **53**, 87–95.

Galloway, D. J. (1994a). *Pseudocyphellaria lacerata* new to the Faeroe Islands. *Lichenologist*, **26**, 391–393.

Galloway, D. J. (1994b). Studies on the lichen genus *Sticta* (Schreber) Ach. I. Southern South American species. *Lichenologist*, **26**, 223–282.

Galloway, D. J. (1995a). The extra-European lichen collections of Archibald Menzies MD, FLS (1754–1842). *Edinburgh Journal of Botany*, **52**, 95–139.

Galloway, D. J. (1995b). Studies on the lichen genus *Sticta* (Schreber) Ach. III. Notes on species described by Bory de St-Vincent, William Hooker, and Delise, between 1804 and 1825. *Nova Hedwigia*, **61**, 147–188.

Galloway, D. J. (1996). Lichen biogeography. In *Lichen Biology*, ed. T. H. Nash III, pp. 199–216. Cambridge: Cambridge University Press.

Galloway, D. J. (1997). Studies on the lichen genus *Sticta* (Schreber) Ach. IV. New Zealand species. *Lichenologist*, **29**, 105–168.

Galloway, D. J. (1999). Notes on the lichen genus *Leptogium* (Collemataceae, Ascomycota) in New Zealand. *Nova Hedwigia*, **69**, 317–355.

Galloway, D. J. (2001). *Sticta. Flora of Australia*, **58A**, 78–97.

Galloway, D. J. (2002). Taxonomic notes on the lichen genus *Placopsis* (Agyriaceae: Ascomycotina) on southern South America, with a key to species. *Mitteilungen aus dem Institut für Allgemeine Botanik Hamburg* **30–32**, 70–107.

Galloway, D. J. (2003). Additional lichen records from New Zealand 40. *Buellia aethalea* (Ach.) Th. Fr., *Catillaria contristans* (Nyl.) Zahlbr., *Frutidella caesioatra* (Schaer.) Kalb, *Placynthium rosulans* (Th.Fr.) Zahlbr. and *Pseudocyphellaria mallota*. *Australasian Lichenology*, **53**, 20–29.

Galloway, D. J. (2004a). *Placopsis hertelii* (*Agyriaceae, Ascomycota*) endemic to New Zealand, with descriptions of four additional new species of *Placopsis* (Nyl.) Linds, from New Zealand. *Bibliotheca Lichenologica*, **88**, 147–161.

Galloway, D. J. (2004b). New lichen taxa and names in the New Zealand mycobiota. *New Zealand Journal of Botany*, **42**, 105–120.

Galloway, D. J. (2007). *Flora of New Zealand Lichens, Including Lichen-forming and Lichenicolous Fungi*, 2nd edn. Lincoln: Manaaki Whenua.

Galloway, D. J. (2008). Southern Hemisphere lichens. *New Zealand Journal of Botany* (In review.)

Galloway, D. J. and Aptroot, A. (1995). Bipolar lichens: a review. *Cryptogamic Botany*, **5**, 184–191.

Galloway, D. J. and Quilhot, W. (1998) ["1999"]. Checklist of Chilean lichen-forming and lichenicolous fungi. *Gayana (Botanica)*, **55**, 111–185.

Galloway, D. J. and Thomas, M. A. (2004). *Sticta*. In *Lichen Flora of the Greater Sonoran Desert Region*, Vol. 2, ed. T. H. Nash III, B. D. Ryan, P. Diederich, C. Gries and F. Bungartz, pp. 513–524. Tempe: Lichens Unlimited.

Galloway, D. J., Hafellner, J. and Elix, J. A. (2005). *Stirtoniella*, a new genus for *Catillaria kelica* (*Lecanorales: Ramalinaceae*). *Lichenologist*, **37**, 261–271.

Galloway, D. J., Kantvilas, G. and Elix, J. A. (2001). *Pseudocyphellaria*. *Flora of Australia*, **58A**, 47–77.

Galun, M. (1988a). *Handbook of Lichenology*, Vols. 1, 2 and 3, Boca Raton: CRC Press.

Galun, M. (1988b). Carbon metabolism. In *Handbook of Lichenology*, Vol. I, ed. M. Galun, pp. 147–156. Boca Raton: CRC Press.

Galun, M. and Bubrick, P. (1984). Physiological interactions between the partners of the lichen symbiosis. In *Cellular Interactions: Encyclopedia of Plant Physiology*, ed. H. F. Linskens and J. Heslop-Harrison, pp. 362–401. Berlin: Springer.

Galun, M. and Mukhtar, A. (1996). Checklist of the lichens of Israel. *Bocconea*, **6**, 149–171.

Galun, M. and Ronen, R. (1988). Interactions of lichens and pollutants. In *CRC Handbook of Lichenology*, Vol. III, ed. M. Galun, pp. 55–72. Boca Raton: CRC Press.

Galun, M. and Shomer-Ilan, A. (1988). Secondary metabolic products. In *CRC Handbook of Lichenology*, Vol. III, ed. M. Galun, pp. 3–8. Boca Raton: CRC Press.

Ganderton, P. and Coker, P. (2005). *Environmental Biogeography*. Harlow, Essex: Pearson Education Ltd.

Garbaye, J. (1994). Helper bacteria: a new dimension to the mycorrhizal symbiosis. *New Phytologist*, **128**, 197–210.

Garcia-Molina, F., Hiner, A. N., Fenoll, L. G., *et al.* (2005). Mushroom tyrosinase: catalase activity, inhibition, and suicide inactivation. *Journal of Agricultural and Food Chemistry*, **53**, 3702–3709.

Gargas, A., DePriest, P. T., Grube, M. and Tehler, A. (1995). Multiple origins of lichen symbiosis in fungi suggested by SSU rDNA phylogeny. *Science*, **268**, 1492–1495.

Garrett, R. M. (1972). Electrostatic charges on freshly discharged lichen ascospores. *Lichenologist*, **5**, 311–313.

Gärtner, G. (1985). Die Gattung *Trebouxia* PUYMALY (Chlorellales, Chlorophyceae). *Archiv für Hydrobiologie, Supplementband*, **74**, (*Algological Studies*, **41**), 495–548.

Gärtner, G. (1992). Taxonomy of symbiotic eukaryotic algae. In *Algae and Symbioses*, ed. W. Reisser, pp. 325–338. Bristol: Biopress Limited.

Gärtner, G. and Ingolic, E. (2003). Further studies on *Desmococcus* brand emend. Vischer (Chlorophyta, Trebouxiophyceae) and a new species *Desmococcus spinocystis* sp. nov. from soil. *Biologia, Bratislava*, **58**, 517–523.

Garty, J. (2001). Biomonitoring atmospheric heavy metals with lichens: theory and application. *Critical Review in Plant Sciences*, **20**, 309–371.

Garty, J., Cohen, Y., Kloog, N. and Karnieli, A. (1997). Effect of air pollution on cell membrane integrity, spectral reflectance and metal and sulfur concentrations in lichens. *Environmental Toxicology and Chemistry*, **16**, 1396–1402.

Garty, J., Galun, M. and Kessel, M. (1979). Localization of heavy metals and other elements accumulated in the lichen thallus. *New Phytologist*, **82**, 159–168.

Garty, J. Karary, Y., Harel, J. and Lurie, S. (1993). The impact of air pollution on the integrity of cell membranes and chlorophyll in the lichen *Ramalina duriaei* (De Not.) Bagl. transplanted to industrial sites in Israel. *Archives of Environmental Contamination and Toxicology*, **24**, 455–460.

Garty, J., Perry, A. S. and Mozel, J. (1982). Accumulation of polychlorinated biphenyls (PCBs) in the transplanted lichen *Ramalina duriaei* in air quality biomonitoring experiments. *Nordic Journal of Botany*, **2**, 583–586.

Gaugh, H. G., Jr. (1982). *Multivariate Analysis in Community Ecology*. Cambridge: Cambridge University Press.

Gauslaa, Y. (2006). Trade-off between reproduction and growth in the foliose old forest lichen *Lobaria pulmonaria*. *Basic and Applied Ecology*, **7**, 455–460.

Gauslaa, Y. and McEvoy, M. (2005). Seasonal changes in solar radiation drive acclimation of the sun-screening compound parietin in the lichen *Xanthoria parietina*. *Basic and Applied Ecology*, **6**, 75–82.

Gauslaa, Y. and Solhaug, K. A. (1998). The significance of thallus size for the water economy of the cyanobacterial old-forest lichen *Degelia plumbea*. *Oecologia*, **116**, 76–84.

Gauslaa, Y. and Solhaug K. A. (1999). High-light damage in air-dry thalli of the old forest lichen *Lobaria pulmonaria* – interactions of irradiance, exposure duration and high temperature. *Journal of Experimental Botany*, **50**, 697–705.

Gauslaa, Y. and Solhaug, K. A. (2001). Fungal melanins as a sun screen for symbiotic green algae in the lichen *Lobaria pulmonaria*. *Oecologia*, **126**, 462–471.

Gauslaa, Y. and Ustvedt, E. M. (2003). Is parietin a UV-B or a blue-light screening pigment in the lichen *Xanthoria parietina*? *Photochemical and Photobiological Sciences*, **2**, 424–432.

Gauslaa, Y., Holien, H., Ohlson, M. and Solhøy, T. (2006a). Does snail grazing affect growth of the old forest lichen *Lobaria pulmonaria*? *Lichenologist*, **38**, 587–593.

Gauslaa, Y., Lie, M., Solhaug, K. and Ohlson, M. (2006b). Growth and ecophysiological acclimation of the foliose lichen *Lobaria pulmonaria* in forests with contrasting light climates. *Oecologia*, **147**, 406–416.

Gauslaa, Y., Ohlson, M., Solhaug, K. A., Bilger, W. and Nybakken, L. (2001). Aspect dependent high-irradiance damage to two transplanted foliose forest lichens *Lobaria pulmonaria* and *Parmelia sulcata*. *Canadian Journal of Forest Research*, **31**, 1639–1649.

Gaya, E., Lutzoni, F., Zoller, S. and Navarro-Rosinés, P. (2003). Phylogenetic study of *Fulgensia* and allied *Caloplaca* and *Xanthoria* species (Teloschistaceae, lichen-forming Ascomycota). *American Journal of Botany*, **90**, 1095–1103.

Geebelen, W. and Hoffmann, M. (2001). Evaluation of bio-indication methods using epiphytes by correlating with SO_2-pollution parameters. *Lichenologist*, **33**, 249–260.

Gehrig, H., Schüssler, A. and Kluge, M. (1996). *Geosiphon pyriforme*, a fungus forming endocytobiosis with *Nostoc* (cyanobacteria), is an ancestral member of the Glomales: evidence by SSU rRNA analysis. *Journal of Molecular Evolution*, **43**, 71–81.

Geitler, L. (1932). Cyanophyceae von Europa unter Berücksichtigung der anderen Kontinente. In *Rabenhorst's Kryptogamenflora von Deutschland, Österreich und der Schweiz*, 2nd edn., Vol. 14, ed. R. Kolkwitz, pp. 1–1196. Leipzig: Akademische Verlagsgesellschaft.

Geitler, L. (1934). Beiträge zur Kenntnis der Flechtensymbiose. IV, V. *Archiv für Protistenkunde* **82**, 51–85.

Geitler, L. (1937). Beiträge zur Kenntnis der Flechtensymbiose. VI. Die Verbindung von Pilz und Alge bei den Pyrenopsidaceen *Synalissa, Thyrea, Peccania* und *Psorotichia*. *Archiv für Protistenkunde*, **88**, 161–179.

Gerson, U. and Seaward, M. R. D. (1977). Lichen-invertebrate associations. In *Lichen Ecology*, ed. M. R. D. Seaward, pp. 69–119. London: Academic Press.

Geyer, M., Feuerer, T. and Feige, G. B. (1984). Chemie und Systematik in der Flechtengattung *Rhizocarpon*: Hochdruckflüssigkeitschromatographie (HPLC) der Flechten-Sekundärstoffe der *Rhizocarpon superficiale*-Gruppe. *Plant Systematics and Evolution*, **145**, 41–54.

Gignac, D. and Dale, M. R. T. (2005). Effects of fragment size and habitat heterogeneity on cryptogam diversity in the low-boreal forest of western Canada. *Bryologist*, **108**, 50–66.

Gilbert, O. L. (1970). Further studies on the effect of sulphur dioxide on lichens and bryophytes. *New Phytologist*, **69**, 605–627.

Gilbert, O. L. (1971). The effect of airborne fluorides on lichens. *Lichenologist*, **5**, 26–32.

Gilbert, O. L. (1985). Environmental effects of airborne fluorides from aluminium smelting at Invergordon, Scotland 1971–1983. *Environmental Pollution, Series A*, **39**, 293–302.

Gilbert, O. L. (1992). Lichen reinvasion with declining air pollution. In *Bryophytes and Lichens in a Changing Environment*, ed. J. W. Bates and A. M. Farmer, pp. 159–177. Oxford: Clarendon Press.

Gillespie, J. H. (1991). *The Causes of Molecular Evolution*. Oxford: Oxford University Press.

Gjerde, I., Sætersdal, M., Rolstad, J., et al. (2005). Productivity–diversity relationships for plants, bryophytes, lichens and polypore fungi in six northern forest landscapes. *Ecography*, **28**, 705–720.

Gob, F., Oetit, F., Bravard, J. P., Ozer, A. and Gob, A. (2003). Lichenometric application to historical and subrecent dynamics and sediment transport of a Corsican stream (Figarella River, France). *Quaternary Science Reviews*, **22**, 2111–2124.

Goebel, K. (1926). Die Wasseraufnahme der Flechten. *Berichte der deutschen botanischen Gesellschaft*, **44**, 158–161.

Goffinet, B. and Bayer, R. J. (1997). Characterization of mycobionts of photomorph pairs in the Peltigerineae (lichenized ascomycetes) based on internal transcribed spacer sequences of the nuclear ribosomal DNA. *Fungal Genetics and Biology*, **21**, 228–237.

Goldner, W. R., Hoffman, F. M. and Medve, R. J. (1986). Allelopathic effects of *Cladonia cristatella* on ectomycorrhizal fungi common to bituminous strip-mine spoils. *Canadian Journal of Botany*, **64**, 1586–1590.

Golm, G. T., Hill, P. S. and Wells, H. (1993). Life expectancy in a Tulsa cemetery: growth and population structure of the lichen *Xanthoparmelia cumberlandia*. *American Midland Naturalist*, **129**, 373–383.

Gombert, S., Asta, J. and Seaward, M. R. D. (2003). Correlation between the nitrogen concentration of two epiphytic lichens and the traffic density in an urban area. *Environmental Pollution*, **123**, 281–290.

Gordy, V. R. and Hendrix, D. L. (1982). Respiratory response of the lichens *Ramalina stenospora* Mull. Arg. and *Ramalina complanata* (Sw.) Ach. to azide, cyanide, salicylhydroxamic acid and bisulfate during thallus hydration. *Bryologist*, **85**, 361–374.

Gorin, P. A. J., Baron, M. and Iacomini, M. (1988). Storage products of lichens. In *CRC Handbook of Lichenology*, Vol. 3, ed. M. Galun, pp. 9–24. Boca Raton: CRC Press.

Gough, L. P., Severson, R. C. and Jackson, L. L. (1988). Determining baseline element composition of lichens. I. *Parmelia sulcata* at Theodore Roosevelt National Park, North Dakota. *Water, Air, and Soil Pollution*, **38**, 157–167.

Goward, T. (1994). Notes on old growth-dependent epiphytic macrolichens in inland British Columbia, Canada. *Acta Botanica Fennica*, **150**, 31–38.

Goward, T. (1995). *Nephroma occultum* and the maintenance of lichen diversity in British Columbia. *Mitteilungen der Eidgenössischen Forschungsanstalt für Wald, Schnee und Landschaft*, **70**, 93–101.

Goward, T. (1996). *Lichens of British Columbia: Rare Species and Priorities for Inventory*. Victoria: Province of British Columbia, Ministry of Forests Research Program. Working Paper 08/1996, pp. i–viii + 1–34.

Goward, T. and Ahti, T. (1997). Notes on the distributional ecology of the Cladoniaceae (lichenized ascomycetes) in temperate and boreal western North America. *Journal of the Hattori Botanical Laboratory*, **82**, 143–155.

Goyal, R. and Seaward, M. R. D. (1982). Metal uptake in terricolous lichens. III. Translocation in the thallus of *Peltigera canina*. *New Phytologist*, **90**, 85–98.

Grace, B., Gillespie, T. J. and Puckett, K. J. (1985a). Sulphur dioxide threshold concentration values for *Cladina rangiferina* in the Mackenzie Valley, N. W. T. *Canadian Journal of Botany*, **63**, 806–812.

Grace, B., Gillespie, T. J. and Puckett, K. J. (1985b). Uptake of gaseous sulphur dioxide by the lichen *Cladina rangiferina*. *Canadian Journal of Botany*, **63**, 797–805.

Grace, J. (1997). Toward models of resource allocation by plants. In *Plant Resource Allocation*, ed. F. A. Bazzaz and J. Grace, pp. 279–291. San Diego: Academic Press.

Gradstein, S. and Lücking, R. (1997). Synthesis of the Symposium (on Foliicolous Cryptogams) and priorities for future research. *Abstracta Botanica*, **21**, 207–214.

Grant, B. S., Owen, D. F. and Clarke, C. A. (1996). Parallel rise and fall of melanic peppered moths in America and Britain. *Journal of Heredity*, **87**, 351–357.

Green, T. G. A. and Lange, O. L. (1995). Photosynthesis in poikilohydric plants: a comparison of lichens and bryophytes. In *Ecophysiology of Photosynthesis*, ed. E. D. Schulze and M. M. Caldwell, pp. 71–101. Berlin: Springer.

Green, T. G. A., Büdel, B., Heber, U., *et al.* (1993). Differences in photosynthetic performance between cyanobacterial and green algal components of lichen photosymbiodemes measured in the field. *New Phytologist*, **125**, 723–731.

Green, T. G. A., Büdel, B., Meyer, A., Zellner, H. and Lange, O. L. (1997). Temperate rainforest lichens in New Zealand: light response of photosynthesis. *New Zealand Journal of Botany*, **35**, 493–504.

Green, T. G. A., Horstmann, J., Bonnett, H., Wilkins, A. and Silvester, W. B. (1980). Nitrogen fixation by members of the Stictaceae (Lichens) of New Zealand. *New Phytologist*, **84**, 339–348.

Green, T. G. A., Meyer, A., Büdel, B., Zellner, H. and Lange, O. L. (1995). Diel patterns of CO_2-exchange for six lichens from a temperate rain forest in New Zealand. *Symbiosis*, **18**, 251–273.

Green, T. G. A., Schlensog, M., Sancho, L. G., *et al.* (2002). The photobiont (cyanobacterial or green algal) determines the pattern of photosynthetic activity within a lichen photosymbiodeme: evidence obtained from *in situ* measurements of chlorophyll *a* fluorescence. *Oecologia*, **130**, 191–198.

Green, T. G. A., Schroeter, B., Kappen, L., Seppelt, R. D. and Maseyk, K. (1998). An assessment of the relationship between chlorophyll *a* fluorescence and CO_2 gas exchange from field measurements on a moss and lichen. *Planta*, **206**, 611–618.

Green, T. G. A., Schroeter, B. and Sancho, L. G. (1999). Plant life in Antarctica. In *Handbook of Functional Plant Ecology*, ed. F. I. Pugnaire and F. Valladares, pp. 495–543. New York: Marcel Dekker, Inc.

Green, T. G. A., Schroeter, B. and Sancho, L. G. (2007). Plant life in Antarctica. In *Functional Plant Ecology*, 2nd edn., ed. F. I. Pugnaire and F. Valladares, pp. 389–433. New York: Marcel Dekker.

Green, T. G. A., Snelgar, W. P. and Wilkins, A. L. (1985). Photosynthesis, water relations and thallus structure of Stictaceae lichens. In *Lichen Physiology and Cell Biology*, ed. D. H. Brown, pp. 57–75. New York: Plenum Press.

Gregory, K. J. (1975). Lichens and the determination of river channel capacity. *Earth Surface Processes*, **1**, 273–285.

Grehan, J. R. (2001). Panbiogeography from tracks to ocean basins: evolving perspectives. *Journal of Biogeography*, **28**, 413–429.

Gries, C., Nash, T. H., III and Kesselmeier, J. (1994). Exchange of reduced sulfur gases between lichens and the atmosphere. *Biogeochemistry*, **23**, 25–39.

Gries, C., Romagni, J. G., Nash, T. H., III, Kuhn, U. and Kesselmeier, J. (1997a). The relation of H_2S release to SO_2. *New Phytologist*, **136**, 703–711.

Gries, C., Sanz, M.-J. and Nash, T. H., III (1995). The effect of SO_2 fumigation on CO_2 gas exchange, chlorophyll fluorescence and chlorophyll degradation in different lichen species from western North America. *Cryptogamic Botany*, **5**, 239–246.

Gries, C., Sanz, M.-J., Romagni, J. G., *et al.* (1997b). The uptake of gaseous sulphur dioxide by non-gelatinous lichens. *New Phytologist*, **135**, 595–602.

Griffith, M. and Yaish, M. W. F. (2004). Antifreeze proteins in overwintering plants: a tale of two activities. *Trends in Plant Science*, **9**, 399–405.

Grime, J. P. (1979). *Plant Strategies and Vegetation Processes*. Chichester: Wiley.

Grime, J. P., Hodgson, J. P. and Hunt, R. (1988). *Comparative Plant Ecology: a Functional Approach to Common British Species*. London: Unwin Hyman.

Grodzinska, K., Godzik, B. and Bienkowski, P. (1999). *Cladina stellaris* (Opiz) Brodo as a bioindicator of atmospheric deposition on the Kola Peninsula, Russia. *Polar Research*, **18**, 105–110.

Groombridge, B., ed. (1992). *Global Biodiversity: Status of the Earth's Living Resources*. London: Chapman and Hall.

Grube, M. (1998). Classification and phylogeny in the Arthoniales (lichenized Ascomycetes). *Bryologist*, **101**, 377–391.

Grube, M. and Blaha, J. (2003). On the pylogeny of some polyketide synthase genes in the lichenized genus *Lecanora*. *Mycological Research*, **107**, 1419–1426.

Grube, M. and de Los Ríos, A. (2001). Observations on *Biatoropsis usnearum*, a lichenicolous heterobasidiomycete, and other gall-forming fungi, using different microscopical techniques. *Mycological Research*, **105**, 1116–1122.

Grube, M. and Hafellner, F. (1990). Studien an flechtenbewohnenden Pilzen der Sammelgattung *Didymella* (Ascomycetes, Dothideales). *Nova Hedwigia*, **51**, 283–360.

Grube, M. and Kantvilas, G. (2006). *Siphula* represents a remarkable case of morphological congruence in sterile lichens. *Lichenologist*, **38**, 241–249.

Grube, M. and Kroken, S. (2000). Molecular approaches and the concept of species and species complexes in lichenized fungi. *Mycological Research*, **104**, 1284–1294.

Grube, M., Baloch, E. and Lumbsch, H. T. (2004). The phylogeny of Porinaceae (Ostropomycetidae) suggests a neotenic origin of perithecia in Lecanoromycetes. *Mycological Research*, **108**, 1111–1118.

Guenther, J. E. and Melis, A. (1990). Dynamics of photosystem II heterogeneity in *Dunaliella salina* (green alga). *Photosynthesis Research*, **23**, 195–203.

Gugger, M. F. and Hoffmann, L. (2004). Polyphyly of true branching cyanobacteria (Stigonematales). *International Journal of Systematic and Evolutionary Microbiology*, **54**, 349–357.

Gunn, J., Keller, W., Negusanti, J., *et al.* (1995). Ecosystem recovery after emission reductions: Sudbury, Canada. *Water, Air, and Soil Pollution*, **85**, 1783–1788.

Gunther, A. J. (1988). Effects of simulated acid rain on nitrogenase activity in the lichen genus *Peltigera* under field and laboratory conditions. *Water, Air, and Soil Pollution*, **38**, 379–385.

Gunther, A. J. (1989). Nitrogen fixation by lichens in a subarctic Alaskan watershed. *Bryologist*, **92**, 202–208.

Gupta, V. (2005). Application of lichenometry to slided materials in the Higher Himalayan landslide zone. *Current Science*, **89**, 1032–1036.

Guzow-Krzeminska, B. (2006). Photobiont flexibility in the lichen *Protoparmeliopsis muralis* as revealed by ITS rDNA analyses. *The Lichenologist*, **38**, 469–476.

Haas, D. and Keel, C. (2003). Regulation of antibiotic production in root-colonizing *Pseudomonas* spp. and relevance for biological control of plant disease. *Annual Review of Phytopathology*, **41**, 117–153.

Hafellner, J. (1984). Studien in Richtung einer naturlicheren Gliederung der Sammelfamilien Lecanoraceae und Lecideaceae. Beitrage zur Lichenologie. Festschrift J. Poelt. *Beihefte zur Nova Hedwigia*, **79**, 241–371.

Hafellner, J. (1995) A new checklist of lichens and lichenicolous fungi of insular Laurimacaronesia including a lichenological bibliography for the area. *Fristchiana* **5**, 3–132.

Hafellner, J., Hertel, H., Rambold, G. and Timdal, E. (1993). *A New Outline of the Lecanorales*. Graz: Privately published by the authors.

Häffner, E., Lomský, B., Hynek, V., *et al.* (2001). Air pollution and lichen physiology. Physiological responses of different lichens in a transplant experiment following an SO_2-gradient. *Water, Air, and Soil Pollution*, **131**, 185–201.

Hageman, C. M. (1989). Enzyme electromorph variation in the lichen family Umbilicariaceae. Ph.D. thesis. London, Ontario: University of Western Ontario.

Hageman, C. M. and Fahselt, D. (1986). A comparison of isozyme patterns of morphological variants in the lichen *Umbilicaria muhlenbergii* (Ach.) Tuck. *Bryologist*, **89**, 285–290.

Hageman, C. M. and Fahselt, D. (1990). Enzyme electromorph variation in the lichen family Umbilicariaceae: within-stand polymorphism in umbilicate lichens of eastern Canada. *Canadian Journal of Botany*, **68**, 2636–2643.

Hageman, C. M. and Fahselt, D. (1992). Geographical distance and enzyme polymorphisms in the lichen *Umbilicaria mammulata*. *Bryologist*, **93**, 316–323.

Hahn, S. C., Tenhunen, J. D., Popp, P. W., Meyer, A. and Lange, O. L. (1993). Upland tundra in the foothills of the Brooks Range, Alaska: diurnal CO_2 exchange patterns of characteristic lichen species. *Flora*, **188**, 125–143.

Hale, M. E. (1973). Growth. In *The Lichens*, ed. V. Ahmadjian and M. E. Hale, pp. 473–492. New York: Academic Press.

Hale, M. E. (1974). *The Biology of Lichens*. 2nd edn. London: Edward Arnold.

Hale, M. E. (1983). *The Biology of Lichens*. 3rd edn. London: Edward Arnold.

Hale, M. E. (1984). The lichen line and high water levels in a freshwater stream in Florida. *Bryologist*, **87**, 261–265.

Hale, M. E. (1990). A synopsis of the lichen genus *Xanthoparmelia* (Vainio) Hale (Ascomycotina, Parmeliaceae). *Smithsonian Contributions to Botany*, **74**, 1–250.

Hallbom, L. and Bergman, B. (1979). Influence of certain herbicides and a forest fertilizer on the nitrogen fixation by the lichen *Peltigera praetextata*. *Oecologia*, **40**, 19–27.

Hällgren, J.-E. and Huss, K. (1975). Effects of SO_2 on photosynthesis and nitrogen fixation. *Physiologia Plantarum*, **34**, 171–176.

Hallingbäck, T. (1991). Blue-green algae and cyanophilic lichens are threatened by air pollution and fertilization. *Svensk Botanisk Tidskrift*, **85**, 87–104.

Hallingbäck, T. and Kellner, O. (1992). Effects of simulated nitrogen rich and acid rain on the nitrogen-fixing lichen *Peltigera aphthosa* (L.) Willd. *New Phytologist*, **120**, 99–103.

Halliwell, B. (2006). Reactive species and antioxidants. Redox biology is a fundamental theme of aerobic life. *Plant Physiology*, **141**, 312–322.

Halliwell, B. and Gutteridge, J. M. C. (1999). *Free Radicals in Biology and Medicine*. Oxford: Oxford University Press.

Hamada, N., Miyawaki, H. and Yamada, A. (1995). Distribution pattern of air pollution and epiphytic lichens in the Osaka Plain (Japan). *Journal of Plant Research*, **108**, 483–491.

Hamada, N., Tanahashi, T., Miyagawa, H. and Miyawaki, H. (2001). Characteristics of secondary metabolites from isolated lichen mycobionts. *Symbiosis*, **31**, 23–33.

Hammer, S. (2003). *Notocladonia*, a new genus in the Cladoniaceae. *Bryologist*, **106**, 162–167.

Hardy, R. W. F., Burns, R. C. and Holsten, R. D. (1973). Applications of the C_2H_2 reduction assay for measurement of N_2 fixation. *Soil Biology and Biochemistry*, **5**, 47–81.

Harper, K. T. and Marble, J. R. (1988). A role of nonvascular plants in management of semiarid rangelands. In *Vegetation Science Applications for Rangeland Analysis and Management*, ed. P. T. Tueller, pp. 135–169. London: Kluwer Academic.

Harris, G. B. (1971). The ecology of corticolous lichens. I. The zonation on oak and birch in South Devon. *Journal of Ecology*, **59**, 431–439.

Harrison, S. and Winchester, V. (2000). Nineteenth- and twentieth-century glacier fluctuations and climatic implications in the Arco and Colonia valleys, Hielo Patagónico Norte, Chile. *Arctic, Antarctic, and Alpine Research*, **32**, 55–63.

Hasegawa, M., Kishino, H. and Yano, K. (1985). Dating of the human-ape splitting by a molecular clock of mitochondrial DNA. *Journal of Molecular Evolution*, **22**, 160–174.

Hasenhüttl, G. and Poelt, J. (1978). Über die Brutkörner bei der Flechtengattung *Umbilicaria*. *Berichte der deutschen botanischen Gesellschaft*, **91**, 275–296.

Hasse, H. E. (1913). The lichen flora of southern California. *Contributions from the United States National Herbarium*, **17**, 1–132.

Hauck, M. and Paul, A. (2005). Manganese as a site factor for epiphytic lichens. *Lichenologist*, **37**, 409–423.

Hauck, M. and Spribille, T. (2005). The significance of precipitation and substrate chemistry for epiphytic lichen diversity in spruce-fir forests of the Salish Mountains, northwestern Montana. *Flora*, **200**, 547–562.

Hauck, M. and Zoller, T. (2003). Copper sensitivity of soredia of the epiphytic lichen *Hypogymnia physodes. Lichenologist*, **35**, 271–274.

Hauck, M., Helms, G. and Friedl, T. (2007). Photobiont selectivity in the epiphytic lichens *Hypogymnia physodes* and *Lecanora conizaeoides. The Lichenologist*, **39**, 195–204.

Hauck, M., Hesse, V., Jung, R., Zöller, T. and Runge, M. (2001). Long-distance transported sulphur as a limiting factor for the abundance of *Lecanora conizaeoides* in montane spruce forests. *Lichenologist*, **33**, 267–269.

Hauck, M., Hesse, V. and Runge, M. (2002a). Correlations between the Mn/Ca ratio in stemflow and epiphytic lichen abundance in a dieback-affected spruce forest of the Harz Mountains, Germany. *Flora*, **197**, 361–369.

Hauck, M., Mulack, C. and Paul, A. (2002b). Manganese uptake in the epiphytic lichens *Hypogymnia physodes* and *Lecanora conizaeoides. Environmental and Experimental Botany*, **48**, 107–117.

Hauck, M., Paul, A., Mulack, C., Fritz, E. and Runge, M. (2002c). Effects of manganese on the viability of vegetative diaspores of the epiphytic lichen *Hypogymnia physodes. Environmental and Experimental Botany*, **47**, 127–142.

Hauck, M., Paul, A. and Spribille, T. (2006). Uptake and toxicity of manganese in epiphytic cyanolichens. *Environmental and Experimental Botany*, **56**, 216–224.

Hawksworth, D. L. (1971). Lichens as a litmus for air pollution: a historical review. *International Journal of Environmental Studies*, **1**, 281–296.

Hawksworth, D. L. (1982). Secondary fungi in lichen symbioses: parasites, saprophytes and parasymbionts. *Journal of the Hattori Botanical Laboratory*, **52**, 357–366.

Hawksworth, D. L. (1983). A key to the lichen-forming, parasitic, parasymbiotic and saprophytic fungi occurring on lichens in the British Isles. *Lichenologist*, **15**, 1–44.

Hawksworth, D. L. (1985). Problems and prospects in the systematics of the Ascomycotina. *Proceedings of the Indian Academy of Sciences (Plant Sciences)*, **94**, 319–39.

Hawksworth, D. L. (1988a). The fungal partner. In *CRC Handbook of Lichenology*, Vol. 1, ed. M. Galun, pp. 35–38. Boca Raton: CRC Press.

Hawksworth, D. L. (1988b). The variety of fungal-algal symbioses, their evolutionary significance, and the nature of lichens. *Botanical Journal of the Linnean Society*, **96**, 3–20.

Hawksworth, D. L. (1988c). Effects of algae and lichen-forming fungi on tropical crops. In *Perspectives of Mycopathology*, ed. V. P. Agnihotri, K. A. Sarbhoy and D. Kumar, pp. 76–83. New Delhi: Malhorta Publishing House.

Hawksworth, D. L. (1988d). Conidiomata, conidiogenesis, and conidia. In *CRC Handbook of Lichenology*, Vol. 1, ed. M. Galun, pp. 181–193. Boca Raton: CRC Press.

Hawksworth, D. L. (1991). The fungal dimension of biodiversity: magnitude, significance, and conservation. *Mycological Research*, **95**, 641–655.

Hawksworth, D. L. (2002). Bioindication: calibrated scales and their utility. In *Monitoring with Lichens – Monitoring Lichens*, Nato Science Series IV: Earth and Environmental Sciences, ed. P. L. Nimis, C. Scheidegger and P. A. Wolseley, pp. 11–20. Dordrecht: Kluwer Academic.

Hawksworth, D. L. (2003). The lichenicolous fungi of Great Britain and Ireland: an overview and annotated checklist. *Lichenologist*, **35**, 191–232.

Hawksworth, D. L. and Hill, D. J. (1984). *The Lichen-forming Fungi.* Glasgow: Blackie.

Hawksworth, D. L. and Honegger, R. (1994). The lichen thallus: a symbiotic phenotype of nutritionally specialized fungi and its response to gall producers. In *Plant Galls: Organisms, Interactions, Populations,* ed. M. A. J. Williams, pp. 77–98. Oxford: Clarendon Press.

Hawksworth, D. L. and McManus, P. M. (1989). Lichen recolonization in London under conditions of rapidly falling sulphur dioxide levels, and the concept of zone skipping. *Botanical Journal of the Linnean Society,* **100,** 99–109.

Hawksworth, D. L. and McManus, P. M. (1992). Changes in the lichen flora on trees in Epping Forest through periods of increasing and then ameliorating sulphur dioxide air pollution. *Essex Naturalist,* **11,** 92–101.

Hawksworth, D. L. and Rose, F. (1970). Qualitative scale for estimating sulphur dioxide air pollution in England and Wales using epiphytic lichens. *Nature,* **227,** 145–148.

Hawksworth, D. L., Kirk, P. M., Sutton, B. C. and Pegler, D. N. (1995). *Ainsworth and Bisby's Dictionary of the Fungi.* 8th edn. Wallingford: CAB International.

Hawksworth, D. L., Lawton, R. M., Martin, P. G. and Stanley-Price, K. (1984). Nutritive value of *Ramalina duriaei* grazed by gazelles in Oman. *Lichenologist,* **16,** 93–94.

Hawksworth, D. L., Sutton. B. C. and Ainsworth, D. C. (1983). *Ainsworth and Bisby's Dictionary of the Fungi.* 7th edn. Kew: Commonwealth Mycological Institute.

Hayward, G. D. and Rosentreter, R. (1994). Lichens as nesting material for northern flying squirrels in the northern Rocky Mountains. *Journal of Mammalogy,* **75,** 663–673.

Heads, M. (1997). Regional patterns of biodiversity in New Zealand: one degree grid analysis of plant and animal distributions. *Journal of the Royal Society of New Zealand,* **27,** 337–354.

Heads, M. (1998). Biogeographic disjunction along the Alpine Fault, New Zealand. *Biological Journal of the Linnean Society,* **63,** 161–176.

Heads, M. (2001). Birds of paradise, biogeography and ecology in New Guinea: a review. *Journal of Biogeography,* **28,** 893–927.

Heads, M. (2004). What is a node? *Journal of Biogeography,* **31,** 1883–1891.

Heads, M. (2005). Towards a panbiogeography of the seas. *Biological Journal of the Linnean Society,* **84,** 675–723.

Heads, M. and Craw, R. (2004). The alpine fault biogeographic hypothesis revisited. *Cladistics,* **20,** 184–190.

Heber, U., Bilger, W. and Shuvalov, V. A. (2006a). Thermal energy dissipation in reaction centers and in the antenna of photosystem II protects desiccated poikilohydric mosses against photo-oxidation. *Journal of Experimental Botany,* **57,** 2993–3006.

Heber, U., Lange, O. L. and Shuvalov, V. A. (2006b). Conservation and dissipation of light energy as complementary processes: homoiohydric and poikilohydric autotrophs. *Journal of Experimental Botany,* **57,** 1211–1223.

Heibel, E., Lumbsch, H. T. and Schmitt, I. (1999). Genetic variation of *Usnea filipendula* (Parmeliaceae) populations in western Germany investigated by RAPDs suggest reinvasion from various sources. *American Journal of Botany,* **86,** 753–757.

Heidmarsson, S., Mattson, J. E., Moberg, R., *et al.* (1997). Classification of lichen photomorphs. *Taxon*, **46**, 519–520.

Helms, G. (2003). Taxonomy and symbiosis in associations of Physciaceae and *Trebouxia*. Ph.D. thesis. Göttingen: Albrecht-von-Haller Institute, University of Göttingen.

Helms, G., Friedl, T., Rambold, G. and Mayrhofer, H. (2001). Identification of photobionts from the lichen family Physciaceae using algal-specific ITS rDNA sequencing. *Lichenologist*, **33**, 73–86.

Henderson, A. (1999). Lichen dyes. An historical perspective. *Lees Museums and Galleries Review*, **2**, 30–34.

Henderson-Sellers, A. and Seaward, M. R. D. (1979). Monitoring lichen reinvasion of ameliorating environments. *Environmental Pollution*, **19**, 207–213.

Henriksson, E. and Pearson, L. C. (1981). Nitrogen fixation rate and chlorophyll content of the lichen *Peltigera canina* exposed to sulfur dioxide. *American Journal of Botany*, **68**, 680–684.

Henson, B. J., Hesselbrock, S. M., Watson, L. E., and Barnum, S. R. (2004). Molecular phylogeny of the heterocystous cyanobacteria (subsections IV and V) based on *nifD*. *International Journal of Systematic and Evolutionary Microbiology*, **54**, 493–497.

Henssen, A. (1963). Eine Revision der Flechtenfamilien Lichinaceae und Ephebaceae. *Symbolae Bototanicae Upsala*, **18**, 1–123.

Henssen, A., in cooperation with Keuck, G., Renner, B. and Vobis, G. (1981). The lecanoralean centrum. In *Ascomycete Systematics: The Lutrellian Concept*, ed. D. R. Reynolds, pp. 138–234. New York: Springer.

Henssen, A. (1986). The genus *Paulia* (Lichinaceae). *Lichenologist*, **18**, 201–229.

Henssen, A. (1995). The new lichen family *Gloeoheppiaceae* and its genera *Gloeoheppia*, *Pseudopeltula* and *Gudella* (Lichinales). *Lichenologist*, **27**, 261–290.

Henssen, A. and Jahns, H. M. (1973[1974]). *Lichenes. Eine Einführung in die Flechtenkunde*. Stuttgart: Thieme.

Henssen, A. and Tretiach, M. (1995). *Paulia glomerata*, a new epilithic species from Europe, and additional notes on some other *Paulia* species. *Nova Hedwigia*, **60**, 297–309.

Henssen, A., Büdel, B. and Titze, A. (1987). *Euopsis* and *Harpidium*, genera of the Lichinaceae (Lichenes) with rostrate asci. *Botanica Acta*, **101**, 49–55.

Henzler, T. and Steudle, E. (2000). Transport and metabolic degradation of hydrogen peroxide in *Chara coralline*: model calculations and measurements with the pressure probe suggest transport of H_2O_2 across water channels. *Journal of Experimental Botany*, **51**, 2053–2066.

Herrera-Campos, M. A., Lücking, R., Perez, R. E., *et al.* (2004). The foliicolous lichen flora of Mexico. V. Biogeographical affinities, altitudinal preferences, and an updated checklist of 293 species. *Lichenologist*, **36**, 309–327.

Hersoug, L. G. (1983). Lichen protein affinity towards walls of cultured and freshly isolated phycobionts and its relationship to cell wall cytochemistry. *FEMS Microbiology Letters*, **20**, 417–420.

Herzig, R. and Urech, M. (1991). Flechten als Bioindikatoren: Intergriertes biologisches Messsystem der Luftverschmutzung für das Schweizer Mittelland. *Bibliotheca Lichenologica*, **43**, 1–283.

Hesbacher, S., Fröbery, L., Baur, A., Baur, B. and Proksch, P. (1996). Chemical variation within and between individuals of the lichenized ascomycete *Tephromela atra*. *Biochemical Systematics and Ecology*, **24**, 603–609.

Hestmark, G. (1990). Thalloconidia in the genus *Umbilicaria*. *Nordic Journal of Botany*, **9**, 547–574.

Hestmark, G. (1992). Sex, size competition and escape – strategies of reproduction and dispersal in *Lasallia pustulata* (Umbilicariaceae, Ascomycetes). *Oecologia*, **92**, 305–312.

Hestmark, G. (1997). Species diversity and reproductive strategies in the family Umbilicariaceae on high equatorial mountains – with remarks on global patterns. *Bibliotheca Lichenologica*, **68**, 195–202.

Hestmark, G. (2000). The ecophysiology of lichen population biology. *Bibliotheca Lichenologica*, **75**, 397–403.

Hestmark, G., Schroeter, B. and Kappen, L. (1997). Intrathalline and size-dependent patterns of activity in *Lasallia pustulata* and their possible consequences for competitive interactions. *Functional Ecology*, **11**, 318–322.

Heywood, V. H. (ed.) (1995). *Global Biodiversity Assessment*. Cambridge: Cambridge University Press.

Hildreth, K. C. and Ahmadjian, V. (1981). A study of *Trebouxia* and *Pseudotrebouxia* isolated from different lichens. *Lichenologist*, **13**, 65–86.

Hill, D. J. (1971). Experimental study of the effect of sulfite on lichens with reference to atmospheric pollution. *New Phytologist*, **70**, 831–836.

Hill, D. J. (1976). The physiology of lichen symbiosis. In *Lichenology: Progress and Problems*, ed. D. H. Brown, D. L. Hawksworth and R. H. Bailey, pp. 457–496. London: Academic Press.

Hill, D. J. (1985). Changes in photobiont dimensions and numbers during co-development of lichen symbionts. In *Lichen Physiology and Cell Biology*, ed. D. H. Brown, pp. 303–317. New York: Plenum Press.

Hill, D. J. (1989). The control of the cell cycle in microbial symbionts. *New Phytologist*, **112**, 175–184.

Hill, D. J. and Smith, D. C. (1972). Lichen physiology XII. The "inhibition technique". *New Phytologist*, **71**, 15–30.

Hilmo, O. (1994). Distribution and succession of epiphytic lichens on *Picea abies* in a boreal forest, central Norway. *Lichenologist*, **26**, 149–169.

Hilmo, O. and Holien, H. (2002). Epiphytic lichen response to the edge environment in a boreal *Picea abies* forest in central Norway. *Bryologist*, **105**, 48–56.

Hilmo, O. and Ott, S. (2002). Juvenile development of the cyanolichen *Lobaria scrobiculata* and the green algal lichen *Platismatia glauca* and *Platismatia norvegica* in a boreal *Picea abies* forest. *Plant Biology*, **4**, 273–280.

Hilmo, O. and Sastad, S. M. (2001). Colonization of old-forest lichens in a young and an old boreal *Picea abies* forest: an experimental approach. *Biological Conservation*, **102**, 251–259.

Hilmo, O., Holien, H. and Hytteborn, H. (2005). Logging strategy influences colonization of common chlorolichens on branches of *Picea abies*. *Ecological Applications*, **15**, 983–996.

Hirt, R. P., Logsdon, J., Doolittle, W. F. and Embley, T. M. (1999). Microsporidia are related to Fungi: evidence from the largest subunit of RNA polymerase II and other proteins. *Proceedings of the National Academy of Sciences, USA*, **96**, 580–585.

Hiserodt, R. D., Swijter, D. F. H. and Mussinan, C. J. (2000). Identification of atranorin and related potential allergens in oakmoss absolute by high-performance liquid chromatography-tandem mass spectrometry using negative ion atmospheric pressure chemical ionization. *Journal of Chromatography A*, **888**, 103–111.

Hitch, C. J. B. and Millbank, J. W. (1975). Nitrogen metabolism in lichens. VI. The blue-green phycobiont content, heterocyst frequency and nitrogenase activity in *Peltigera* species. *New Phytologist*, **74**, 473–476.

Hitch, C. J. B. and Millbank, J. W. (1976). Nitrogen metabolism in lichens. VII. Nitrogenase activity and heterocyst frequency in lichens with blue-green phycobionts. *New Phytologist*, **75**, 239–244.

Hocking, D., Kuchar, P., Plambeck, J. A. and Smith, R. A. (1978). The impact of gold smelter emissions on vegetation and soils of a sub-arctic forest–tundra transition ecosystem. *Air Pollution Control Association Journal*, **28**, 133–137.

Högnabba, F. (2006). Molecular phylogeny of the genus *Stereocaulon* (Stereocaulaceae, lichenized Ascomycetes). *Mycological Research*, **110**, 1080–1092.

Holopainen, T. (1983). Ultrastructural changes in epiphytic lichen *Bryoria capillaris* and *Hypogymnia physodes* in central Finland. *Annales Botanici Fennici*, **19**, 39–52.

Holopainen, T. (1984). Types and distribution of ultrastructural symptoms in epiphytic lichens in several urban and industrial environments in Finland. *Annales Botanici Fennici*, **21**, 219–229.

Holopainen, T. and Kärenlampi, L. (1984). Injuries to lichen ultrastructure caused by sulphur dioxide fumigations. *New Phytologist*, **98**, 285–294.

Holopainen, T. and Kärenlampi, L. (1985). Characteristic ultrastructural symptoms caused in lichens by experimental exposure to nitrogen compounds and fluorides. *Annales Botanici Fennici*, **22**, 333–342.

Holopainen, T. and Kauppi, M. (1989). A comparison of light, fluorescence and electron microscopic observations in assessing the SO_2 injury of lichens under different moisture conditions. *Lichenologist*, **21**, 119–134.

Holt, N. E., Tigmantas, D., Valkunas, L., *et al.* (2005). Carotenoid cation formation and the regulation of photosynthetic light harvesting. *Science*, **307**, 433–436.

Holub, S. M. and Lajtha, K. (2003). Mass loss and nitrogen dynamics during the decomposition of a ^{15}N-labeled N_2-fixing epiphytic lichen, *Lobaria oregana*. *Canadian Journal of Botany*, **81**, 698–705.

Holub, S. M. and Lajtha, K. (2004). The fate and retention of organic and inorganic ^{15}N-nitrogen in an old-growth forest soil in Western Oregon. *Ecosystems*, **7**, 368–380.

Honegger, R. (1978a). Ascocarpontogenie, Ascusstruktur und -funktion bei Vertretern der Gattung Rhizocarpon. *Berichte der deutschen botanischen Gesellschaft*, **91**, 579–594.

Honegger, R. (1978b). The ascus apex in lichenized fungi. I. The *Lecanora-*, *Peltigera-* and *Teloschistes*-types. *Lichenologist*, **10**, 47–67.

Honegger, R. (1980). The ascus apex in lichenized fungi. II. The *Rhizocarpon*-type. *Lichenologist*, **12**, 157–172.

Honegger, R. (1982a). The ascus apex in lichenized fungi. III. The *Pertusaria*-type. *Lichenologist*, **14**, 205–217.

Honegger, R. (1982b). Ascus structure and function, ascospore delimitation, and phycobiont cell wall types associated with the Lecanorales (lichenized ascomycetes). *Journal of the Hattori Botanical Laboratory*, **52**, 417–429.

Honegger, R. (1984a). Scanning electron microscopy of the contact site of conidia and trichogynes in *Cladonia furcata*. *Lichenologist*, **16**, 11–19.

Honegger, R. (1984b). Ultrastructural studies on conidiomata, conidiophores, and conidiogenous cells in six lichen-forming Ascomycetes. *Canadian Journal of Botany*, **62**, 2081–2093.

Honegger, R. (1985). Ascus structure and ascospore formation in the lichen-forming *Chaenotheca chrysocephala* (Caliciales). *Sydowia, Annales Mycologici Series II*, **38**, 146–157.

Honegger, R. (1986a). Ultrastructural studies in lichens. I. Haustorial types and their frequencies in a range of lichens with trebouxioid phycobionts. *New Phytologist*, **103**, 785–795.

Honegger, R. (1986b). Ultrastructural studies in lichens. II. Mycobiont and photobiont cell wall surface layers and adhering crystalline lichen products in four Parmeliaceae. *New Phytologist*, **103**, 797–808.

Honegger, R. (1991a). Functional aspects of the lichen symbiosis. *Annual Review of Plant Physiology and Plant Molecular Biology*, **42**, 553–578.

Honegger, R. (1991b). Fungal evolution: Symbioses and morphogenesis. In *Symbiosis, a Source of Evolutionary Innovation*, ed. L. Margulis and R. Fester, pp. 319–340. Cambridge: Massachusetts Institute of Technology Press.

Honegger, R. (1991c). Haustoria-like structures and hydrophobic cell wall surface layers in lichens. In *Electron Microscopy of Plant Pathogens*, ed. K. Mendgen and D. E. Lesemann, pp. 277–290. Berlin: Springer.

Honegger, R. (1992). Lichens: mycobiont-photobiont relationships. In *Algae and Symbioses. Plants, Animals, Fungi, Viruses, Interactions Explored*, ed. W. Reisser, pp. 255–275. Bristol: Biopress.

Honegger, R. (1993). Developmental biology of lichens. *New Phytologist*, **125**, 659–677.

Honegger, R. (1995). Experimental studies with foliose macrolichens: fungal responses to spatial disturbance at the organismic level and to spatial problems at the cellular level. *Canadian Journal of Botany*, **73**, 569–578.

Honegger, R. (1997). Metabolic interactions at the mycobiont-photobiont interface in lichens. In *Plant Relationships, The Mycota*, Vol. V, Part A, ed. G. C. Carroll and P. Tudzynski, pp. 209–221. Berlin: Springer.

Honegger, R. (1998). The lichen symbiosis – what is so spectacular about it? *Lichenologist*, **30**, 193–212.

Honegger, R. (2000). Great discoveries in bryology and lichenology – Simon Schwendener (1829–1919) and the Dual Hypothesis of Lichens. *Bryologist*, **103**, 307–313.

Honegger, R. (2001). The symbiotic phenotype of lichen-forming ascomycetes. In *Fungal Associations*, Vol. IX: *The Mycota*, ed. B. Hock, pp. 165–188. Berlin: Springer.

Honegger, R. (2006). Water relations in lichens. In *Fungi in the Environment*, ed. G. M. Gadd, S. C. Watkinson and P. Dyer, pp. 185–200. Cambridge: Cambridge University Press.

Honegger, R. and Bartnicki-Garcia, S. (1991). Cell wall structure and composition of cultured mycobionts from the lichen *Cladonia macrophylla*, *Cladonia caespiticia*, and *Physcia stellaris* (Lecanorales, Ascomycetes). *Mycological Research*, **95**, 905–914.

Honegger, R. and Haisch, A. (2001). Immunocytochemical location of the (1 → 3) (1 → 4)-beta-glucan lichenin in the lichen-forming ascomycete *Cetraria islandica* (Icelandic moss). *New Phytologist*, **150**, 739–746.

Honegger, R. and Zippler, U. (2007). Mating systems in representatives of the Parmeliaceae, Ramalinaceae and Physciaceae (Lecanoromycetes, lichen-forming ascomycetes). *Mycological Research*, **11**, 424–432.

Honegger, R., Peter, M. and Scherrer, S. (1996). Drought-stress induced structural alterations at the mycobiont-photobiont interface in a range of foliose macrolichens. *Protoplasma*, **190**, 221–232.

Honegger, R., Zippler, U., Gansner, H. and Scherrer, S. (2004). Mating systems in the genus *Xanthoria* (lichen-forming ascomycetes). *Mycological Research*, **108**, 480–488.

Hopwood, D. A. (1997). Genetic contribution to understanding polyketide synthases. *Chemical Reviews*, **97**, 2465–2498.

Horstmann, J. L., Denison, W. C. and Silvester, W. B. (1982). $^{15}N_2$ fixation and molybdenum enhancement of acetylene reduction by *Lobaria* spp. *New Phytologist*, **92**, 235–241.

Houdijk, A. L. F. M. and Roelofs, J. G. M. (1991). Deposition of acidifying and eutrophicating substances in Dutch forest. *Acta Botanica Neerlandica*, **40**, 245–255.

Huebert, D. B., L'Hirondelle, S. J. and Addison, P. A. (1985). The effects of sulphur dioxide on net CO_2 assimilation in the lichen *Evernia mesomorpha* Nyl. *New Phytologist*, **100**, 643–651.

Huelsenbeck, J. P. and Ronquist, F. (2001). Bayesian inference of phylogenetic trees. *Bioinformatics*, **17**, 754–755.

Humphries, C. J. and Ebach, M. C. (2004). Biogeography on a dynamic earth. In *Frontiers of Biogeography: New Directions in the Geography of Nature*, ed. M. V. Lomolino and L. R. Heaney, pp. 67–86. Sunderland: Sinauer Associates.

Humphries, C. J. and Parenti, L. R. (1999). *Cladistic Biogeography: Interpreting Patterns of Plant and Animal Distributions.* 2nd edn, Oxford Biogeography Series 12. Oxford: Oxford University Press.

Humphries, C. J., Williams, P. H. and Vane-Wright, R. I. (1995). Measuring biodiversity value for conservation. *Annual Reviews of Ecology and Systematics*, **26**, 93–111.

Huneck, S. (1999). The significance of lichens and their metabolites. *Die Naturwissenschaften*, **86**, 559–570.

Huneck, S. (2001). New results on the chemistry of lichen substances. In *Progress in the Chemistry of Organic Products*, ed. W. Herz, H. Falk, G. W. Kirby and R. E. Moore, pp. 1–276. New York: Springer.

Huneck, S. and Schreiber, K. (1972). Wachstumsregulatorische Eigenschaften von Flechten- und Moos-Inhaltstoffen. *Phytochemistry*, **11**, 2429–2434.

Huneck, S. and Yoshimura, I. (1996). *Identification of Lichen Substances*. Springer: Berlin.

Huneck, S., Bothe, H.-K. and Richter, W. (1990). Über den Metallgehalt von Flechten von Kupferschieferhalden der Umgebung von Mansfeld. *Herzogia*, **8**, 295–304.

Huovinen, K., Hiltunen R. and von Schantz, M. (1985). A high performance liquid chromatographic method for the analyses of lichen compounds from the genera *Cladina* and *Cladonia*. *Acta Pharmaceutica Fennica*, **94**, 99–112.

Huss, V. A. R., Frank, C., Hartmann, E. C. *et al.* (1999). Biochemical taxonomy and molecular phylogeny of the genus *Chlorella* sensu lato (*Chlorophyta*). *Journal of Phycology*, **35**, 587–598.

Huss, V. A. R. and Sogin, M. L. (1990). Phylogenetic position of some *Chlorella* species within the Chlorococcales based upon complete small-subunit ribosomal RNA sequences. *Journal of Molecular Evolution*, **31**, 432–442.

Huss-Danell, K. (1977). Nitrogen fixation by *Stereocaulon paschale* under field conditions. *Canadian Journal of Botany*, **55**, 585–592.

Huss-Danell, K. (1978). Seasonal variation in the capacity for nitrogenase activity in the lichen *Stereocaulon paschale*. *New Phytologist*, **81**, 89–98.

Huss-Danell, K. (1979). The cephalodia and their nitrogenase activity in the lichen *Stereocaulon paschale*. *Zeitschrift für Pflanzenphysiologie*, **95**, 431–440.

Hyvärinen, M. and Crittenden, P. D. (1998a). Relationships between atmospheric nitrogen inputs and the vertical nitrogen and phosphorus concentration gradients in the lichen *Cladonia portentosa*. *New Phytologist*, **140**, 519–530.

Hyvärinen, M. and Crittenden, P. D. (1998b). Growth of the cushion-forming lichen, *Cladonia portentosa*, at nitrogen-polluted and unpolluted heathland sites. *Environmental and Experimental Botany*, **40**, 67–76.

Hyvärinen, M. and Crittenden, P. D. (2000). ^{33}P translocation in the thallus of the mat-forming lichen *Cladonia portentosa*. *New Phytologist*, **145**, 281–288.

Hyvärinen, M., Härdling, R. and Tuomi, J. (2002). Cyanobacterial lichen symbiosis: the fungal partner as an optimal harvester. *Oikos*, **98**, 498–504.

Hyvärinen, M., Roitto, M., Ohtonen, R. and Markkola, A. (2000). Impact of wet deposited nickel on the cation content of a mat-forming lichen *Cladina stellaris*. *Environmental and Experimental Botany*, **43**, 211–218.

Hyvärinen, M., Walter, B. and Koopmann, R. (2003). Impact of fertilization on phenol content and growth rate of *Cladina stellaris*: a test of the carbon-nutrient balance hypothesis. *Oecologia*, **134**, 176–181.

Ihda, T. A., Nakano, T., Yoshimura, I. and Iwatsuki, Z. (1993). Phycobionts isolated from Japanese species of *Anzia* (lichenes). *Archiv für Protistenkunde*, **143**, 163–172.

Ihlen, P. G. and Ekman, S. (2002). Outline of phylogeny and character evolution in *Rhizocarpon* (Rhizocarpaceae, lichenized Ascomycota) based on nuclear ITS and mitochondrial SSU ribosomal DNA sequences. *Biological Journal of the Linnean Society, London*, **77**, 535–546.

Ingólfsdóttir, K. (2002). Molecules of interest: usnic acid. *Phytochemistry*, **61**, 729–736.

Innes, J. L. (1983). Lichenometric dating of debris-flow deposits in the Scottish Highlands. *Earth Surface Processes and Landforms*, **8**, 579–588.

Innes, J. L. (1985). Lichenometry. *Progress in Physical Geography*, **9**, 187–254.

Innes, J. L. (1988). The use of lichens in dating. In *Handbook of Lichenology*, Vol. 3, ed. M. Galun, pp. 75–91. Boca Raton: CRC Press.

Insarov, G. and Schroeter, B. (2002). Lichen monitoring and climate change. In *Monitoring with Lichens – Monitoring Lichens*, ed. P. L. Nimis, C. Scheidegger and P. A. Wolseley, pp. 183–201. Dordrecht: Kluwer Academic.

Jaag, O. and Thomas, E. (1934). Neue Untersuchungen über die Flechte *Epigloea bactrospora* Zukal. *Berichte der schweierischen botanischen Gesellschaft*, **34**, 77–89.

Jahns, H. M. (1970). Untersuchungen zur Entwicklungsgeschichte der Cladoniaceen, unter besonderer Berücksichtigung des Podetien-Problems. *Nova Hedwigia*, **20**, 1–177

Jahns, H. M. (1984). Morphology, reproduction and water relations – a system of morphogenetic interactions in *Parmelia saxatilis*. *Nova Hedwigia*, **79**, 715–37.

Jahns, H. M. (1987). New trends in developmental morphology of the thallus. *Bibliotheca Lichenologica*, **25**, 17–33.

Jahns, H. M. (1988). The lichen thallus. In *CRC Handbook of Lichenology*, Vol. 1, ed. M. Galun, pp. 95–143. Boca Raton: CRC Press.

Jahns, H. M., Tuiz-Dubiel, A. and Blank, L. (1976). Hygroskopische Bewegungen der Sorale von *Hypogymnia physodes*. *Herzogia*, **4**, 15–23.

James, P. W. and Henssen, A. (1976). The morphological and taxonomic significance of cephalodia. In *Lichenology: Progress and Problems*, ed. D. H. Brown, D. L. Hawksworth and R. H. Bailey, pp. 27–77. London: Academic Press.

James, P. W., Hawksworth, D. L. and Rose, F. (1977). Lichen communities in the British Isles: a preliminary conspectus. In *Lichen Ecology*, ed. M. R. D. Seaward, pp. 295–413. London: Academic Press.

Jancey, R. (1966). The application of numerical methods of data analysis to the genus *Phyllota* (Benth.) in New South Wales. *Australian Journal of Botany*, **14**, 131–149.

Janex-Favre, M. C. and Ghaleb, M. I. (1986). L'ontogenie et la structure des apothecies du *Xanthoria parietina* (L.) Beltr. (discolichen). *Cryptogamie, Bryologie et Lichenologie*, **7**, 457–478.

Jansova, I. and Soldan, Z. (2006). The habitat factors that affect the composition of bryophyte and lichen communities on fallen logs. *Preslia*, **78**, 67–86.

Jarmuszkiewicz, W. (2001). Uncoupling proteins in mitochondria of plants and some microorganisms. *Acta Biochimica Polonica*, **48**, 145–155.

Jennings, D. (1995). *The Physiology of Fungal Nutrition*. Cambridge: Cambridge University Press.

Jensen, M. (2002). Measurement of chlorophyll fluorescence in lichens. In *Protocols in Lichenology*, ed. I. Kranner, R. P. Beckett and A. K. Varma, pp. 135–151. Berlin: Springer.

Jensen, M. and Kricke, R. (2002). Chlorophyll fluorescence measurements in the field: assessment of the vitality of large numbers of lichen thalli. In *Monitoring with Lichens – Monitoring Lichens*, ed. P. L. Nimis, C. Scheidegger and P. A. Wolseley, pp. 327–332. Nato Science Series IV: Earth and Environmental Sciences. Dordrecht: Kluwer Academic.

Jensen, M., Linke, K., Dickhäuser, A. and Feige, G. B. (1999). The effect of agronomic photosystem-II herbicides on lichens. *Lichenologist*, **31**, 95–103.

John, D. M., Whitton, B. A. and Brook, A. J. (2002). *The Freshwater Algal Flora of the British Isles*. Cambridge: Cambridge University Press.

John, E. and Dale, M. R. T. (1991). Determinants of spatial pattern in saxicolous lichen communities. *Lichenologist*, **23**, 227–236.

John, V. (1996). Preliminary catalogue of lichenized and lichenicolous fungi of Mediterranean Turkey. *Bocconea*, **6**, 173–216.

Johnson, L. R. and John, D. M. (1990). Observations on *Dilabifilium* (class Chlorophyta, order Chaetophorales *sensu strictu*) and allied genera. *British Phycological Journal*, **25**, 53–61.

Johnson, P. N. and Galloway, D. J. (2002). *Lichens and Their Conservation Needs in New Zealand*. Landcare Research Contract Report: LC0102/132. Dunedin: Landcare Research.

Johnston, J. (2001). *Siphulella*. *Flora of Australia*, **58A**, 22–23.

Jones, D. (1988). Lichens and pedogenesis. In *CRC Handbook of Lichenology*, Vol. 3, ed. M. Galun, pp. 109–124. Boca Raton: CRC Press.

Jones, D., Wilson, M. J. and Laundon, J. R. (1982). Observations on the location and form of lead in *Stereocaulon vesuvianum*. *Lichenologist*, **14**, 281–286.

Jørgensen, P. M. (1996). The oceanic element in the Scandinavian lichen flora revisited. *Symbolae Botanicae Upsalienses*, **31**, 297–317.

Jørgensen, P. M. (1997). Further notes on hairy *Leptogium* species. *Symbolae Botanicae Upsalienses*, **32**, 113–130.

Jørgensen, P. M. (1998). What shall we do with the blue-green counterparts? *Lichenologist*, **30**, 351–356.

Jørgensen, P. M. (2000a). On the sorediate counterparts of the lichen *Fuscopannaria leucosticta*. *Bryologist*, **103**, 104–107.

Jørgensen, P. M. (2000b). New or interesting *Parmeliella* species from the Andes and Central America. *Lichenologist*, **32**, 139–147.

Jørgensen, P. M. (2001). The lichen genus *Erioderma* (Pannariaceae) in China and Japan. *Annales Botanici Fennici*, **38**, 259–264.

Jørgensen, P. M. (2005). A new Atlantic species in *Fuscopannaria*, with a key to its European species. *Lichenologist*, **37**, 221–225.

Jørgensen, P. M. and Arvidsson, L. (2002). The lichen genus *Erioderma* (Pannariaceae) in Ecuador and neighbouring countries. *Nordic Journal of Botany*, **22**, 87–114.

Jørgensen, P. M. and Jahns, H. M. (1987). *Muhria*, a remarkable new lichen genus from Scandinavia. *Notes of the Royal Botanical Garden Edinburgh*, **44**, 581–599.

Jovan, S. and Carlberg, T. (2007). Nitrogen content of *Letharia vulpina* tissue from forests of the Sierra Nevada, California: geographic patterns and relationships to ammonia estimates and climate. *Environmental Monitoring and Assessment*, **129**, 243–251.

Jovan, S. and McCune, B. (2004). Regional variation in epiphytic macrolichen communities in northern and central California forests. *Bryologist*, **107**, 328–339.

Jovan, S. and McCune, B. (2005). Air-quality bioindication in the greater Central Valley of California, with epiphytic macrolichen communities. *Ecological Applications*, **15**, 1712–1726.

Jovan, S. and McCune, B. (2006). Using epiphytic macrolichen communities for biomonitoring ammonia in forests of the Greater Sierra Nevada, California. *Water, Air, and Soil Pollution*, **170**, 69–93.

Kaasalainen, S. and Rautiainen, M. (2005). Hot spot reflectance signatures of common boreal lichens. *Journal of Geophysical Research*, **110**, 15–24.

Kaiser, M. A. and Debbrecht, F. J. (1977). Qualitative and quantitative analysis of gas chromatography. In *Modern Practice of Gas Chromatography*, ed. R. Crob, pp. 151–211. New York: John Wiley.

Kalb, K. (1987). Brasilianische Flechten. 1. Die Gattung *Pyxine*. *Bibliotheca Lichenologica*, **24**, 1–89.

Kallio, P. (1974). Nitrogen fixation in subarctic lichens. *Oikos*, **25**, 194–198.

Kallio, P., Suhonen, S. and Kallio, S. (1972). The ecology of nitrogen fixation in *Nephroma arcticum* and *Solorina crocea*. *Reports from the Kevo Subarctic Research Station*, **9**, 7–14.

Kallio, S. (1978). On the effect of forest fertilizers on nitrogenase activity in two subarctic lichens. In *Environmental Role of Nitrogen-fixing Blue-green Algae and Asymbiotic Bacteria*, ed. U. Granhall, pp. 217–224. Stockholm: Swedish Natural Science Research Council.

Kandler, O. (1987). Lichen and conifer recolonization in Munich's cleaner air. In *Symposium of the Commission of the European Communities on "Effects of Air Pollution on Terrestrial and Aquatic Ecosystems"* 18–22 May 1987, ed. P. Mathy, pp. 1–7. Brussels: European Commission.

Kantvilas, G. (2000). Conservation of Tasmanian lichens. *Mitteilungen der Eidgenössischen Forschungsanstalt für Wald, Schnee und Landschaft*, **75**, 357–367.

Kantvilas, G. (2002). Studies on the lichen genus *Siphula* Fr. *Bibliotheca Lichenologica*, **82**, 37–53.

Kantvilas, G. (2004). *Steinia. Flora of Australia* **56A**, 1–3.

Kantvilas, G. and Jarman, S. J. (1999). Lichens of rainforest in Tasmania and south-eastern Australia. *Flora of Australia Supplementary Series*, **9**, 1–212.

Kantvilas, G. and Jarman, S. J. (2006). Recovery of lichens after logging: preliminary results from Tasmania's wet forests. *Lichenologist*, **38**, 383–394.

Kantvilas, G. and McCarthy, P. M. (2004). *Hueidea. Flora of Australia* **56A**, 182–183.

Kantvilas, G., Elix, J. A. and Jarman, S. J. (2002). Tasmanian lichens, identification, distribution and conservation status. I. Parmeliaceae. *Flora of Australia Supplementary Series* **15**, 1–274.

Kappen, L. (1974). Response to extreme environments. In *The Lichens*, ed. V. Ahmadjian and M. E. Hale, pp. 311–380. New York: Academic Press.

Kappen, L. (1985). Vegetation and ecology of ice-free areas of northern Victoria Land, Antarctica. *Polar Biology*, **4**, 213–225.

Kappen, L. (1988). Ecophysiological relationships in different climatic regions. In *CRC Handbook of Lichenology*, Vol. 2, ed. M. Galun, pp. 37–99. Boca Raton: CRC Press.

Kappen, L. (1993). Lichens in the antarctic region. In *Antarctic Microbiology*, ed. E. I. Friedmann, pp. 433–490. New York: Wiley-Liss.

Kappen, L. (2000). Some aspects of the great success of lichens in Antarctica. *Antarctic Science*, **12**, 314–324.

Kappen, L. (2004). The diversity of lichens in Antarctica, a review and comments. *Bibliotheca Lichenologica*, **88**, 331–343.

Kappen, L. and Valladares, F. (1999). Opportunistic growth and desiccation tolerance: the ecological success of poikilohydrous autotrophs. In *Handbook of Functional Plant Ecology*, ed. F. I. Pugnaire and F. Valldares, pp. 121–194. New York: Marcel Dekker.

Kappen, L., Breuer, M. and Bölter, M. (1991). Ecological and physiological investigations in continental Antarctic cryptogams. 3. Photosynthetic production of *Usnea sphacelata*: diurnal courses, models, and the effect of photoinhibition. *Polar Biology*, **11**, 393–401.

Kappen, L., Schroeter, B., Green, T. G. A. and Seppelt, R. D. (1998a). Chlorophyll *a* fluorescence and CO_2 exchange on *Umbilicaria aprina* under extreme light stress in the cold. *Oecologia*, **113**, 325–331.

Kappen, L., Schroeter, B., Green, T. G. A. and Seppelt, R. D. (1998b). Microclimatic conditions, meltwater moistening, and the distributional pattern of *Buellia frigida* on rock in a southern continental Antarctic habitat. *Polar Biology*, **19**, 101–106.

Kappen, L., Sommerkorn, M. and Schroeter, B. (1995). Carbon acquisition and water relations of lichens in polar regions – potentials and limitations. *Lichenologist*, **27**, 531–545.

Kardish, N., Silberstein, L., Fleminger, G. and Galun, M. (1991). Lectin from the lichen *Nephroma laevigatum* Ach. Localization and function. *Symbiosis*, **11**, 47–62.

Kärnefelt, I. (1989). Morphology and phylogeny in the Teloschistales. *Cryptogamic Botany*, **1**, 147–203.

Kärnefelt, I. (1990). Evidence of a slow evolutionary change in the speciation of lichens. *Bibliotheca Lichenologica*, **38**, 291–306.

Kärnefelt, E. I. and Thell, A. (1995). Genotypical variation and reproduction in natural populations of *Thamnolia*. *Bibliotheca Lichenologica*, **58**, 213–234.

Karnieli, A., Kokaly, R., West, N. E. and Clark, R. N. (2001). Remote sensing of biological soil crusts. In *Biological Soil Crusts: Structure, Function and Management*, ed. J. Belnap and O. L. Lange, pp. 431–455. Berlin: Springer.

Kauff, F. and Büdel, B. (2005). Ascoma ontogeny and apothecial anatomy in the Gyalectaceae (Ostropales, Ascomycota) support the re-establishment of the Coenogoniaceae. *Bryologist*, **108**, 272–281.

Kauff, F. and Lutzoni, F. (2002). Phylogeny of the Gyalectales and Ostropales (Ascomycota, Fungi): among and within order relationships based on nuclear ribosomal RNA small and large subunits. *Molecular Phylogenetics and Evolution*, **25**, 138–156.

Keller, N. P. and Hohn, T. M. (1997). Metabolic pathway gene clusters in filamentous fungi. *Fungal Genetics and Biology*, **21**, 17–21.

Kelly, B. B. and Becker, V. E. (1975). Effects of light intensity and temperature on nitrogen fixation by *Lobaria pulmonaria*, *Sticta weigelii*, *Leptogium cyanescens* and *Collema subfurvum*. *Bryologist*, **78**, 350–355.

Kelly, B. C. and Gobas, F. A. P. C. (2001). Bioaccumulation of persistent organic pollutants in lichen-caribou-wolf food chains of Canada's central and western Arctic. *Environmental Science and Technology*, **35**, 325–334.

Kerfin, W. and Boger, P. (1982). Light-induced hydrogen evolution by blue-green algae (Cyanobacteria). *Physiologia Plantarum*, **54**, 93–98.

Kershaw, K. A. (1974). Dependence of the level of nitrogenase activity on the water content of the thallus in *Peltigera canina*, *P. evansiana*, *P. polydactyla*, and *P. praetextata*. *Canadian Journal of Botany*, **52**, 1423–1427.

Kershaw, K. A. (1977). Physiological-environmental interactions in lichens. II. The pattern of net photosynthetic acclimation of *Peltigera canina* (L.) Willd. var. *praetextata* (Floreke in Somm.) Hue, and *P. polydactyla* (Neck.) Hoff. *New Phytologist*, **79**, 377–390.

Kershaw, K. A. (1985). *Physiological Ecology of Lichens*. Cambridge: Cambridge University Press.

Kershaw, K. A. and Larson, D. W. (1974). Studies on lichen-dominated systems. IX. Topographic influences on microclimate and species distribution. *Canadian Journal of Botany*, **52**, 1935–1945.

Kershaw, K. A. and Looney, J. H. H. (1985). *Quantitative and Dynamic Plant Ecology*, 3rd edn. London: Edward Arnold.

Kershaw, K. A. and Rouse, W. R. (1971). Studies on lichen dominated ecosystems. I. The water relation of *Cladonia alpestris* in spruce-lichen woodland in northern Ontario. *Canadian Journal of Botany*, **49**, 1389–1399.

Kershaw, M. J. and Talbot, N. J. (1998). Hydrophobins and repellents: proteins with fundamental roles in fungal morphogenesis. *Fungal Genetics and Biology*, **23**, 18–33.

Kershaw, M. J., Thornton, C., Wakley, G. and Talbot, N. J. (2005). Four conserved intramolecular disulphide linkages are required for secretion and cell wall localization of a hydrophobin during fungal morphogenesis. *Molecular Microbiology*, **56**, 117–125.

Kets, E., Galinski, E., de Wit, M., de Bont, J. and Heipieper, H. (1996). Mannitol, a novel bacterial compatible solute in *Pseudomonas putida* S12. *Journal of Bacteriology*, **178**, 6665–6670.

Kidd, D. M. and Ritchie, M. G. (2006). Phylogeographic information systems: putting the geography into phylogeography. *Journal of Biogeography*, **33**, 1851–1865.

Kieft, T. L. and Ruscetti, T. (1990). Characterization of biological ice nuclei from a lichen. *Journal of Bacteriology*, **172**, 3519–3523.

Kinoshita, Y. (1993). *The Production of Lichen Substances for Pharmaceutical Use by Lichen Tissue Culture*. Osaka: Nippon Paint Publication.

Kirk, P. M., Cannon, P. F., David, J. C. and Stalpers, J. A. (2001). *Ainsworth and Bisby's Dictionary of the Fungi*. 9th edn. Wallingford, UK: CAB International.

Kirkpatrick, R. C., Long, Y. C., Zhong, T. and Xia, L. (1998). Social organization and range use in the Yunnan snub-nosed monkey *Rhinopithecus bieti*. *International Journal of Primatology*, **19**, 13–51.

Knops, J. M.H, Nash, T. H., III, Boucher, V. L. and Schlesinger, W. L. (1991). Mineral cycling and epiphytic lichens: implications at the ecosystem level. *Lichenologist*, **23**, 309–321.

Knops, J. M. H., Nash, T. H., III and Schlesinger, W. H. (1996). The influence of epiphytic lichens on the nutrient cycling of an oak woodland. *Ecological Monographs*, **66**, 159–179.

Knowles, R. D., Pastor, J. and Biesboer, D. D. (2006). Increased soil nitrogen associated with dinitrogen-fixing, terricolous lichens of the genus *Peltigera* in northern Minnesota. *Oikos*, **114**, 37–48.

Koch, J. and Kilian, R. (2005). 'Little Ice Age' glacier fluctuations, Gran Campo Nevado, southernmost Chile. *Holocene*, **15**, 20–28.

Köck, M., Schlee, D. and Metzger, U. (1985). Sulfite-induced changes of oxygen metabolism in the action of superoxide dismutase in *Euglena gracilis* and *Trebouxia* sp. *Biochemie und Physiologie der Pflanzen*, **180**, 213–224.

Koga, S., Echigo, A. and Nunomura, K. (1966). Physical properties of cell water in partially dried *Saccharomyces cerevisiae*. *Biophysics Journal*, **6**, 665–674.

Komárek, J. and Anagnostidis, K. (1998). *Cyanoprokaryota. 1. Teil: Chroococcales*. Jena: Gustav Fischer.

Komárek, J. and Anagnostidis, K. (2005). *Cyanoprokaryota. 2. Teil: Oscillatoriales*. München: Elsevier.

Kondratyuk, S. and Kärnefelt, I. (1997). *Josefpoeltia* and *Xanthomendoxa*, two new genera in the Teloschistaceae (lichenized Ascomycotina). *Bibliotheca Lichenologica*, **68**, 19–44.

Kong, F. X., Hu, W., Chao, S. Y., Sang, W. L. and Wang, L. S. (1999). Physiological responses of the lichen *Xanthoparmelia mexicana* to oxidative stress of SO_2. *Environmental and Experimental Botany*, **42**, 201–209.

Kopecky, J., Azarkovich, M., Pfündel, E. E., Shuvalov, V. A. and Heber, U. (2005). Thermal dissipation of light energy is regulated differently and by different mechanisms in lichens and higher plants. *Plant Biology*, **7**, 156–167.

Koptsik, S. V., Koptsik, G. N. and Meryashkina, L. V. (2004). Ordination of plant communities in forest biogeocenoses under conditions of air pollution in the northern Kola Peninsula. *Russian Journal of Ecology*, **35**, 190–199.

Korf, R. P. (1973). Discomycetes and Tuberales. In *The Fungi*. Vol. IVA: *A Taxonomic Review with Keys: Ascomycetes and Fungi Imperfecti*, ed. G. C. Ainsworth, F. K. Sparrow and A. S. Sussman, pp. 249–319. New York: Academic Press.

Kostner, B. and Lange, O. L. (1986). Epiphytische Flechten in bayerischen Waldschadensgebieten des nordlichen Alpenraumes: Floristisch-soziologische Untersuchungen und Vitalitätstests durch Photosynthesemessungen. *Berichte der Akademie für Naturschutz und Landschaftspflege*, **10**, 185–210.

Kovacik, L. and Pereira, A. B. (2001). Green alga *Prasiola* and its lichenized form *Mastodia tesselata* in Antarctic environment: general aspects. *Nova Hedwigia, Beiheft*, **123**, 465–478.

Kranner, I. and Birtić, S. (2005). A modulating role for antioxidants in desiccation tolerance. *Integrative and Comparative Biology*, **45**, 734–740.

Kranner, I. and Grill, D. (1994). Rapid changes of the glutathione status and the enzymes involved in the reduction of glutathione-disulfide during the initial stage of wetting of lichens. *Cryptogamic Botany*, **4**, 203–206.

Kranner, I. and Lutzoni, F. (1999). Evolutionary consequences of transition to a lichen symbiotic state and physiological adaptation to oxidative damage associated with poikilohydry. In *Plant Response to Environmental Stress: From Phytohormones to Genome Reorganisation*, ed. H. R. Lerner, pp. 591–628. New York: Marcel Dekker.

Kranner, I., Beckett, R., Hochman, A. & Nash, III, T. H. (2008). Desiccation tolerance in lichens: a review. *Bryologist*, **111** (in press).

Kranner, I., Cram, W. J., Zorn, M., *et al.* (2005). Antioxidants and photoprotection in a lichen as compared with its isolated symbiotic partners. *Proceedings of the National Academy of Sciences, USA*, **102**, 3141–3146.

Kranner, I., Zorn, M., Turk, B., *et al.* (2003). Biochemical traits of lichens differing in relative desiccation tolerance. *New Phytologist*, **160**, 167–176.

Kroken, S. and Taylor, J. W. (2000). Phylogenetic species, reproductive mode, and specificity of the green alga *Trebouxia* forming lichens with the fungal genus *Letharia*. *Bryologist*, **103**, 645–660.

Kroken, S. and Taylor, J. W. (2001). A gene geneology approach to recognize phylogenetic species boundaries in the lichenized fungus *Letharia*. *Mycologia*, **93**, 38–53.

Kroken, S., Glass, N. L., Taylor, J. W., Yoder, O. C. and Turgeon, B. G. (2003). Phylogenomic analysis of type I polyketide synthase genes in pathogenic and saprobic ascomycetes. *Proceedings of the National Academy of Sciences, USA*, **100**, 15 670–15 675.

Kurina, L. M. and Vitousek, P. M. (1999). Controls over the accumulation and decline of a nitrogen-fixing lichen, *Stereocaulon vulcani* on young Hawaiian lava flows. *Journal of Ecology*, **87**, 784–799.

Kurina, L. M. and Vitousek, P. M. (2001). Nitrogen fixation rates of *Stereocaulon vulcani* on young Hawaiian lava flows. *Biogeochemistry*, **55**, 179–194.

Kytöviita, M. M. and Crittenden, P. D. (1994). Effects of simulated acid rain on nitrogenase activity (acetylene reduction) in the lichen *Stereocaulon paschale* (L.) Hoffm., with special reference to nutritional aspects. *New Phytologist*, **128**, 263–271.

Kytöviita, M. M. and Crittenden, P. D. (2002). Seasonal variation in growth rate in *Stereocaulon paschale*. *Lichenologist*, **34**, 533–537.

Laaksovirta, K. and Olkkonen, H. (1979). Effect of air pollution on epiphytic lichen vegetation and element contents of a lichen and pine needles at Valkeakoski, S. Finland. *Annales Botanici Fennici*, **16**, 285–296.

Laaksovirta, K., Olkkonen, H. and Alakijala, P. (1976). Observations on the lead content of lichen and bark adjacent to a highway in southern Finland. *Environmental Pollution*, **11**, 247–255.

Lakatos, M. (2002). Ökologische Untersuchungen wuchsformbedingter Verbreitungsmuster von Flechten im tropischen Regenwald. Ph.D. thesis. Kaiserslautern: University of Kaiserslautern.

Lakatos, M., Rascher, U. and Büdel, B. (2006). Functional characteristics of corticolous lichens in the understory of a tropical lowland rain forest. *New Phytologist*, **172**, 679–695.

Lambers, H. (1985). Respiration in intact plants and tissues: its regulation and dependence on environmental factors, metabolism and invaded organisms. In *Higher Plant Respiration*, ed. R. Douce and D. A. Day, pp. 418–465. Berlin: Springer.

Lambers, H., Chapin, F. S., III and Pons, T. L. (1998). Photosynthesis, respiration, and long distance transport. In *Plant Physiological Ecology*, ed. H. Lambers, F. S. Chapin III and T. L. Pons, pp. 10–95. Berlin: Springer.

Lang, G. E., Reiners, W. A. and Heier, R. K. (1976). Potential alteration of precipitation chemistry by epiphytic lichens. *Oecologia*, **25**, 229–241.

Lang, G. E., Reiners, W. A. and Pike, L. H. (1980). Structure and biomass of epiphytic lichen communities of balsam fir forests in New Hampshire. *Ecology*, **61**, 541–550.

Lange, O. L. (1953). Hitze- und Trockenresistenz der Flechten in Beziehung zu ihrer Verbreitung. *Flora*, **140**, 39–97.

Lange, O. L. (1965). Der CO_2-Gaswechsel von Flechten bei tiefen Temperaturen. *Planta*, **64**, 1–19.

Lange, O. L. (1969). Experimentell-ökologische Untersuchungen an Flechten der Negev-Wüste. I. CO_2-Gaswechsel von *Ramalina maciformis* (Del.) Bory unter kontrollierten Bedingungen im Laboratorium. *Flora*, **158**, 324–359.

Lange, O. L. (1980). Moisture content and CO_2 exchange of lichens. I. Influence of temperature on moisture-dependent net photosynthesis and dark respiration in *Ramalina maciformis*. *Oecologia*, **45**, 82–87.

Lange, O. L. (2002). Photosynthetic productivity of the epilithic lichen *Lecanora muralis*: long-term field monitoring of CO_2 exchange and its physiological interpretation. I. Dependence of photosynthesis on water content, light, temperature, and CO_2 concentration from laboratory measurements. *Flora*, **197**, 233–249.

Lange, O. L. (2003a). Photosynthetic productivity of the epilithic lichen *Lecanora muralis*: long-term field monitoring of CO_2 exchange and its physiological interpretation. II. Diel and seasonal patterns of net photosynthesis and respiration. *Flora*, **198**, 55–70.

Lange, O. L. (2003b). Photosynthetic productivity of the epilithic lichen *Lecanora muralis*: long-term field monitoring of CO_2 exchange and its physiological interpretation. III. Diel, seasonal, and annual carbon budgets. *Flora*, **198**, 277–292.

Lange, O. L. and Bertsch, A. (1965). Photosynthese der Wüstenflechte *Ramalina maciformis* nach Wasserdampfaufnahme aus dem Luftraum. *Naturwissenschaften*, **52**, 215–216.

Lange, O. L. and Green, T. G. A. (2003). Photosynthetic performance of a foliose lichen of biological soil crust communities: long-term monitoring of the CO_2 exchange of *Cladonia convoluta* under temperate habitat conditions. *Bibliotheca Lichenologica*, **86**, 257–280.

Lange, O. L. and Green, T. G. A. (2005). Lichens show that fungi can acclimate their respiration to seasonal changes in temperature. *Oecologia*, **142**, 11–19.

Lange, O. L. and Green, T. G. A. (2006). Nocturnal respiration in lichens in their natural habitat is not affected by preceding diurnal net photosynthesis. *Oecologia*, **148**, 396–404.

Lange, O. L. and Metzner, H. (1965). Lichtabhängiger Kohlenstoff-Einbau in Flechten bei tiefen Temperaturen. *Naturwissenschaften*, **52**, 191.

Lange, O. L. and Tenhunen, J. D. (1981). Moisture content and CO_2 exchange of lichens. II. Depression of net photosynthesis in *Ramalina maciformis* at high water content is caused by increased thallus carbon dioxide diffusion resistance. *Oecologia*, **51**, 426–429.

Lange, O. L. and Wagenitz, G. (2004). Vernon Ahmadjian introduced the term "chlorolichen". *Lichenologist*, **36**, 171.

Lange, O. L. and Ziegler, H. (1963). Der Schwermetallgehalt von Flechten aus dem Acarosporetum sinopicae auf Erzschlackenhalden des Harzes. q. Eisen und Kupfer. *Mitteilungen der floristischsoziologischen Arbeitsgemeinschaft, neue Folge*, **10**, 156–183.

Lange, O. L., Bilger, W., Rimke, S. and Schreiber, U. (1989). Chlorophyll fluorescence of lichens containing green and blue-green algae during hydration by water vapor uptake and by addition of liquid water. *Botanica Acta*, **102**, 306–313.

Lange, O. L., Büdel, B., Heber, U., et al. (1993a). Temperate rainforest lichens in New Zealand: high thallus water content can severely limit photosynthetic CO_2 exchange. *Oecologia*, **95**, 303–313.

Lange, O. L., Büdel, B., Meyer, A. and Kilian, E. (1993b). Further evidence that activation of net photosynthesis by dry cyanobacterial lichens requires liquid water. *Lichenologist*, **25**, 175–189.

Lange, O. L., Büdel, B., Meyer, A., Zellner, H. and Zotz, G. (2000). Lichen carbon gain under tropical conditions: water relations and CO_2 exchange of three *Leptogium* species of a lower montane rainforest in Panama. *Flora*, **195**, 172–190.

Lange, O. L., Büdel, B., Zellner, H., Zotz, G. and Meyer, A. (1994). Field measurements of water relations and CO_2 exchange of the tropical, cyanobacterial basidiolichen *Dictyonema glabratum* in a Panamanian rainforest. *Botanica Acta*, **107**, 279–290.

Lange, O. L., Green, T. G. A. and Heber, U. (2001). Hydration-dependent photosynthetic production of lichens: what do laboratory studies tell us about field performance? *Journal of Experimental Botany*, **52**, 2033–2042.

Lange, O. L., Green, T. G. A., Melzer, B., Meyer, A. and Zellner, H. (2006). Water relations and CO_2 exchange of the terrestrial lichen *Teloschistes capensis* in the Namib fog desert: measurements during two seasons in the field and under controlled conditions. *Flora*, **201**, 268–280.

Lange, O. L., Green, T. G. A. and Reichenberger H. (1999b). The response of lichen photosynthesis to external CO_2 concentration and its interaction with thallus water-status. *Journal of Plant Physiology*, **154**, 157–166.

Lange, O. L., Green, T. G. A. and Ziegler, H. (1988). Water status related photosynthesis and carbon isotope discrimination in species of the lichen genus *Pseudocyphellaria* with green or blue-green photobionts and in photosymbiodemes. *Oecologia*, **75**, 494–501.

Lange, O. L., Hahn, S. C., Meyer, A. and Tenhunen, J. D. (1998). Upland tundra in the foothills of the Brooks Range, Alaska, U.S.A.: lichen long-term photosynthetic CO_2 uptake and net carbon gain. *Arctic and Alpine Research*, **30**, 252–261.

Lange O. L., Kilian E. and Ziegler, H. (1986). Water vapour uptake and photosynthesis of lichens: performance differences in species with green and blue-green algae as phycobionts. *Oecologia*, **71**, 104–110.

Lange, O. L., Kilian, E. and Ziegler, H. (1990b). Photosynthese von Blattflechten mit hygroskopischen Thallusbewegungen bei Befeuchtung durch Wasserdampf oder mit flüssigem Wasser. *Bibliotheca Lichenologica*, **38**, 311–323.

Lange, O. L., Leisner, J. M. R. and Bilger, W. (1999a). Chlorophyll fluorescence characteristics of the cyanobacterial lichen *Peltigera rufescens* under field conditions. II. Diel and annual distribution of metabolic activity and possible mechanisms to avoid photoinhibition. *Flora*, **194**, 413–430.

Lange, O. L., Meyer, A., Zellner, H., Ullmann, I. and Wessels, D. C. J. (1990a). Eight days in the life of a desert lichen: water relations and photosynthesis of *Teloschistes capensis* in the coastal fog zone of the Namib Desert. *Madoqua*, **17**, 17–30.

Lange, O. L., Reichenberger, H. and Meyer, A. (1995). High thallus water content and photosynthetic CO_2 exchange of lichens. Laboratory experiments with soil crust species from local xerothermic steppe formations in Franconia, Germany. In *Flechten Follmann. Contributions to Lichenology in Honour of Gerhard Follmann*, ed. F. J. A. Daniëls, M. Schulz, and J. Peine, pp. 139–153. Cologne: Geobotanical and Phytotaxonomical Study Group, University of Cologne.

Lange, O. L., Reichenberger, H. and Walz, H. (1997). Continuous monitoring of CO_2 exchange of lichens in the field: short-term enclosure with an automatically operating cuvette. *Lichenologist*, **29**, 259–274.

Lange, O. L., Tenhunen, J. D., Harley, P. C. and Walz, H. (1985). Method for field measurements of CO_2-exchange. The diurnal changes in net photosynthesis and photosynthetic capacity of lichens under mediterranean climatic conditions. In *Lichen Physiology and Cell Biology*, ed. D. H. Brown, pp. 23–39. New York: Plenum Press.

Larcher, W. (2003). *Physiological Plant Ecology*. Berlin: Springer.

Larocque, S. J. and Smith, D. J. (2004). Calibrated *Rhizocarpon* spp. growth curve for the Mount Waddington area, British Columbia coast mountains, Canada. *Arctic, Antarctic, and Alpine Research*, **36**, 407–418.

Larson, D. W. (1983). The pattern of production within individual *Umbilicaria* lichen thalli. *New Phytologist*, **94**, 409–419.

Larson, D. W. (1984). Thallus size as a complicating factor in the physiological ecology of lichens. *New Phytologist*, **97**, 87–97.

Larson, D. W. (1987). The absorption and release of water by lichens. *Bibliotheca Lichenologica*, **25**, 351–360.

Larson, D. W. and Carey, C. K. (1986). Phenotypic variation with "individual" lichen thalli. *American Journal of Botany*, **73**, 214–223.

Larson, D. W. and Kershaw, K. A. (1974). Acclimation in arctic lichens. *Nature*, **254**, 421–423.

Larson, D. W. and Kershaw, K. A. (1975). Studies on lichen-dominated systems. XIII. Seasonal and geographical variation of net CO_2 exchange of *Alectoria ochroleuca*. *Canadian Journal of Botany*, **53**, 2598–2607.

Larson, D. W., Matthes-Sears, U. and Nash, T. H., III (1985). The ecology of *Ramalina menziesii*. I. Geographical variation in form. *Canadian Journal of Botany*, **63**, 2062–2068.

Laufer, Z., Beckett, R. P. and Minibayeva, F. V. (2006a). Co-occurrence of the multicopper oxidases tyrosinase and laccase in lichens in sub-order Peltigerineae. *Annals of Botany*, **98**, 1035–1042.

Laufer, Z., Beckett, R. P., Minibayeva, F. V., Lüthje, S. and Böttger, M. (2006b). Occurrence of laccases in lichenized ascomycetes of the Peltigerineae. *Mycological Research*, **110**, 846–853.

Laundon, J. R. (1978). *Haematomma* chemotypes form fused thalli. *Lichenologist*, **10**, 221–225.

Lawrey, J. D. (1984). *Biology of Lichenized Fungi*. New York: Praeger.

Lawrey, J. D. (1986). Biological role of lichen substances. *Bryologist*, **89**, 111–122.

Lawrey, J. and Diederich, P. (2003). Lichenicolous fungi: interactions, evolution, and biodiversity. *Bryologist*, **106**, 80–120.

Laxen, D. P. H. and Thompson, M. N. A. (1987). Sulphur dioxide in Greater London, 1931–1985. *Environmental Pollution*, **43**, 103–114.

LeBlanc, F. and De Sloover, J. (1970). Relation between industrialization and the distribution and growth of epiphytic lichens and mosses in Montreal. *Canadian Journal of Botany*, **48**, 1485–1496.

LeBlanc, F., Rao, D. N. and Comeau, G. (1972). Indices of atmospheric purity and fluoride pollution pattern in Arvida, Quebec. *Canadian Journal of Botany*, **50**, 991–998.

LeBlanc, F., Robitaille, G. and Rao, D. N. (1974). Biological response of lichens and bryophytes to environmental pollution in the Murdochville Copper Mine area, Quebec. *Journal of the Hattori Botanical Laboratory*, **38**, 405–433.

Lechowicz, M. J. (1982). Ecological trends in lichen photosynthesis. *Oecologia*, **53**, 330–336.

Lee, D. E., Lee, W. G. and Mortimer, N. (2001). Where and why have all the flowers gone? Depletion and turnover in the New Zealand Cenozoic angiosperm flora in relation to palaeography and climate. *Australian Journal of Botany*, **49**, 341–356.

Legaz, M. E., Fontaniella, B., Millanes, A. M. and Vicente, C. (2004). Secreted arginases from phylogenetically far-related lichen species act as cross-recognition factors for two different algal cells. *European Journal of Cell Biology*, **83**, 435–446.

Leisner, J. M. R., Green, T. G. A. and Lange, O. L. (1997). Photobiont activity of a temperate crustose lichen: long-term chlorophyll fluorescence and CO_2 exchange measurements in the field. *Symbiosis*, **23**, 165–182.

Leprince, O., McKersie, B. D. and Hendry, G. A. (1993). The mechanisms of desiccation tolerance in developing seeds. *Seed Science Research*, **3**, 231–246.

Leuckert, C., Ahmadjian, V., Culberson, C. F. and Johnson, A. (1990). Xanthones and depsidones of the lichen *Lecanaora dispersa* in nature and of its mycobiont in culture. *Mycologia*, **82**, 370–378.

Leverenz, J. and Jarvis, P. G. (1979). Photosynthesis in Sitka spruce. VIII. The effects of light flux density and direction on the rate of net photosynthesis and the stomatal conductance of needles. *Journal of Applied Ecology*, **16**, 919–932.

Leverenz, J. W., Falk, S., Pilström, C.-M. and Samuelsson, G. (1990). The effects of photoinhibition on the photosynthetic light-response curve of green plant cells (*Chlamydomonas reinhardtii*). *Planta*, **182**, 161–168.

Lewis, D. H. (1973). Concepts in fungal nutrition and the origin of parasitism and mutualism. *Biological Reviews*, **48**, 261–278.

Lewis, D. H. and Smith, D. C. (1967). Sugar alcohols (polyols) in fungi and green plants. I. Distribution, physiology and metabolism. *New Phytologist*, **66**, 143–184.

Lewis, L. A. and McCourt, R. M. (2004). Green algae and the origin of land plants. *American Journal of Botany*, **91**, 1535–1556.

Lewis Smith, R. I. (1995). Colonization by lichens and the development of lichen-dominated communities in the maritime Antarctic. *Lichenologist*, **27**, 473–483.

Lex, M., Silvester, W. B. and Stewart, W. D. P. (1972). Photorespiration and nitrogenase activity in the blue-green alga, *Anabaena cylindrica*. *Proceedings of the Royal Society of London B*, **180**, 87–102.

Li, Y. [M.] (2006). Seasonal variation of diet and food availability in a group of Sichuan snub-nosed monkeys in Shennongjia Nature Reserve, China. *American Journal of Primatology*, **68**, 217–233.

Liberatore, S., Garibotti, G. and Calvelo, S. (2002). Phytogeography of Argentinean lichens. *Bibliotheca Lichenologica*, **82**, 221–234.

Lidén, K. and Gustafsson, M. (1967). Relationships and seasonal variation of ^{137}Cs in lichen, reindeer and man in northern Sweden 1961–1965. In *Radioecological Concentration Processes*, ed. B. Aberg and F. P. Hungate, pp. 193–208. Oxford: Pergamon Press.

Lilly, V. G. and Barnett, H. L. (1951). *Physiology of the Fungi*. New York: McGraw-Hill Book Co.

Lindblom, L. and Ekman, S. (2006). Genetic variation and population differentiation in the lichen-forming ascomycete *Xanthoria parietina* on the island Storfosna, central Norway. *Molecular Ecology*, **15**, 1545–1559.

Linder, M. B., Szilvay, G. R., Nakari-Setälä, T. and Penttilä, M. E. (2005). Hydrophobins: the protein-amphiphiles of filamentous fungi. *FEMS Microbiology Reviews*, **29**, 877–896.

Lines, C. E. M., Ratcliffe, R. G., Rees, T. A. V. and Southon, T., E. (1989). A ^{13}C NMR study of photosynthate transport and metabolism in the lichen *Xanthoria calcicola* Oxner. *New Phytologist*, **111**, 447–456.

Link, S. O. and Nash, T. H., III (1984a). Ecophysiological studies of the lichen, *Parmelia praesignis* Nyl. Population variation and the effect of storage conditions. *New Phytologist*, **96**, 249–256.

Link, S. O. and Nash, T. H., III (1984b). A mathematical description of the effect of resaturation on net photosynthesis in the lichen, *Parmelia praesignis* Nyl. *New Phytologist*, **96**, 257–262.

Link, S. O., Nash, T. H., III and Driscoll, M. (1985). CO_2 exchange in lichens: towards a mechanistic model. In *Lichen Physiology and Cell Biology*, ed. D. H. Brown, pp. 77–91. New York: Plenum Press.

Linnaeus, C. (1753). *Species plantarum*, Vol. 2. Stockholm: L. Salvi.

Lipscomb, D. L., Farris, J. S., Källersjö, M. and Tehler, A. (1998). Support, ribosomal sequences, and the phylogeny of the eukaryotes. *Cladistics*, **14**, 303–38.

Litterski, B. and Ahti, T. (2004). World distribution of selected European *Cladonia* species. *Symbolae Botanicae Upsalienses*, **34**, 205–236.

Litterski, B. and Otte, V. (2002). Biogeographical research on European species of selected lichen genera. *Bibliotheca Lichenologica*, **82**, 83–90.

Liu, Y. J. and Hall, B. D. (2004). Body plan evolution of ascomycetes, as inferred from an RNA polymerase II phylogeny. *Proceedings of the National Academy of Sciences, USA*, **101**, 4507–4512.

Llimona, X. and Hladun, N. L. (2001). Checklist of the Lichens and lichenicolous Fungi of the Iberian Peninsula and the Balearic Islands. *Bocconea*, **14**, 1–581.

Lodenius, M. and Laaksovirta, K. (1979). Mercury content of *Hypogymnia physodes* and pine needles affected by a chlor-alkali works in Kuusankoski, SE Finland. *Annales Botanici Fennici*, **16**, 7–10.

Loewus, F. A. (1999). Biosynthesis and metabolism of ascorbic acid in plants and of analogs of ascorbic acid in fungi. *Phytochemistry*, **52**, 193–210.

Lohtander, K., Oksanen, I. and Rikkinen, J. (2003). Genetic diversity of green algal and cyanobacterial photobionts in *Nephroma* (Peltigerales). *Lichenologist*, **4**, 325–339.

Lomolino, M. V., Riddle, B. R. and Brown, J. H. (2006). *Biogeography*. 3rd edn. Sunderland: Sinauer Associates.

Longton, R. E. (1988). *Biology of Polar Bryophytes and Lichens*. Cambridge: Cambridge University Press.

López-Bautista, J. M. and Chapman, R. L. (2003). Phylogenetic affinities of the Trentepohliales inferred from small-subunit rDNA. *International Journal of Systematic and Evolutionary Microbiology*, **53**, 2099–2106.

Lorenz, M., Friedl, T. and Day, J. G. (2005). Perpetual maintenance of actively metabolizing microalgal cultures. In *Algal Culturing Techniques: A Book for All Phycologists*, ed. R. A. Anderson, pp. 145–156. Elsevier.

Lorenzini, G., Landi, U., Loppi, S. and Nali, C. (2003). Lichen distribution and bioindicator tobacco plants give discordant response: a case study from Italy. *Environmental Monitoring and Assessment*, **82**, 243–264.

Loso, M. G. and Doak, D. F. (2006). The biology behind lichenometric dating curves. *Oecologia*, **147**, 223–229.

Louwhoff, S. H. J. J. (2001). Biogeography of *Hypotrachyna, Parmotrema* and allied genera (Parmeliaceae) in the Pacific islands. *Bibliotheca Lichenologica*, **78**, 223–246.

Lücking, R. (1997). The use of foliicolous lichens as bioindicators in the tropics, with special reference to the microclimate. *Abstracta Botanica*, **21**, 99–116.

Lücking, R. (2003). Takhtajan's floristic regions and foliicolous lichen biogeography: a compatibility analysis. *Lichenologist* **35**, 33–54.

Lücking, R. and Bernecker-Lücking, A. (2002). Distance, dynamics, and diversity in tropical rainforests: an experimental approach using foliicolous lichens on artificial leaves. I. Growth performance and succession. *Ecotropica*, **8**, 1–13.

Lücking, R. and Kalb, K. (2001). New Caledonia, foliicolous lichens and island biogeography. *Bibliotheca Lichenologica*, **78**, 247–273.

Lücking, R., Sérusiaux, E. and Vezda, A. (2005). Phylogeny and systematics of the lichen family Gomphillaceae (Ostropales) inferred from cladistic analysis of phenotype data. *Lichenologist*, **37**, 123–170.

Lücking, R., Wirth, V., Ferraro, L. and Caceres, M. E. S. (2003). Foliicolous lichens from Valdivian temperate rain forest of Chile and Argentina: evidence of an austral element, with the description of seven new taxa. *Global Ecology and Biogeography*, **12**, 21–36.

Lumbsch, H. T. (1998). The use of metabolic data in lichenology at the species and subspecific levels. *Lichenologist*, **30**, 357–367.

Lumbsch, H. T. and Elix, J. A. (1985). A new species of the lichen genus *Diploschistes* from Australia. *Plant Systematics and Evolution*, **150**, 275–279.

Lumbsch, H. T. and Kothe, H. W. (1988). Anatomical features of *Chondropsis semiviridis* (Nyl.) Nyl. in relation to its vagrant habit. *Lichenologist*, **20**, 25–29.

Lumbsch, H. T., del Prado, R. and Kantvilas, G. (2005). *Gregorella*, a new genus to accommodate *Moelleropsis humida* and a molecular phylogeny of Arctomiaceae. *Lichenologist*, **37**, 291–302.

Lumbsch, H. T., Schmitt, I., Döring, H. and Wedin, M. (2001a). ITS sequence data suggest variability of ascus types and support ontogenetic characters as phylogenetic discriminators in the Agyriales (Ascomycota). *Mycological Research*, **105**, 265–274.

Lumbsch, H. T., Schmitt, I., Döring, H. and Wedin, M. (2001b). Molecular systematics supports the recognition of an additional order of Ascomycota: the Agyriales. *Mycological Research*, **105**, 16–23.

Lumbsch, H. T., Schmitt, I., Lücking, R., Wiklund, E. and Wedin, M. (2007). The phylogenetic placement of Ostropales within Lecanoromycetes (Ascomycota) revisited. *Mycological Research*, **111**, 257–267.

Lumbsch, H. T., Schmitt, I., Palice, Z., Wiklund, E. and Wedin, M. (2004). Supraordinal phylogenetic relationships of Lecanoromycetes based on a Bayesian analysis of combined nuclear and mitochondrial sequences. *Molecular Phylogenetics and Evolution*, **31**, 822–832.

Lumbsch, H. T., Wirtz, N., Lindemuth, R. and Schmitt, I. (2002). Higher level phylogenetic relationships of Euascomycetes (Pezizomycotina) inferred from a combined analysis of nuclear and mitochondrial sequence data. *Mycological Progress*, **1**, 57–70.

Lutzoni, F., Kauff, F., Cox, C., *et al.* (2004). Assembling the fungal tree of life: progress, classification, and evolution of subcellular traits. *American Journal of Botany*, **91**, 1446–1480.

Lutzoni, F., Pagel, M. and Reeb, V. (2001). Major fungal lineages are derived from lichen symbiotic ancestors. *Nature*, **411**, 937–940.

MacCracken, J. G., Alexander, L. E. and Uresk, D. W. (1983). An important lichen of southeastern Montana rangelands. *Journal of Range Management*, **36**, 35–37.

MacDonald, G. M. (2003). *Biogeography: Space, Time, and Life*. New York: John Wiley.

MacFarlane, J. D. and Kershaw, K. A. (1977). Physiological-environmental interactions in lichens. IV. Seasonal changes in the nitrogenase activity in *Peltigera canina* (L.) Willd. var. *praetextata* (Floerke in Somm.) Hue, and *P. canina* (L.) Willd. var. *rufescens* (Weiss) Mudd. *New Phytologist*, **69**, 403–408.

MacFarlane, J. D. and Kershaw, K. A. (1980). Physiological-environmental interactions in lichens. IX. Thermal stress and lichen ecology. *New Phytologist*, **84**, 669–685.

MacFarlane J. D. and Kershaw, K. A. (1982). Physiological-environmental interactions in lichens. XIV. The environmental control of glucose movement from alga to fungus in *Peltigera polydactyla*, *P. rufescens*, and *Collema furfuraceum*. *New Phytologist*, **91**, 93–101.

MacGinitie, H. (1937). The flora of the Weaverville beds of Trinity County, California, with descriptions of the plant-bearing beds. In *Eocene Flora of Western America*, pp. 83–151. Publication 465. Washington: Carnegie Institution of Washington.

MacKenzie, T. D. B., Król, M., Huner, N. P. A. and Campbell, D. A. (2002). Seasonal changes in chlorophyll fluorescence quenching and the induction and capacity of the photoprotective xanthophyll cycle in *Lobaria pulmonaria*. *Canadian Journal of Botany*, **80**, 255–261.

MacKenzie, T. D. B., MacDonald, T. M., Dubois, L. A. and Campbell, D. A. (2001). Seasonal changes in temperature and light driven acclimation of photosynthetic physiology and macromolecular content in *Lobaria pulmonaria*. *Planta*, **214**, 57–66.

Madelin, M. F. (1968). Parasitism on other fungi and lichens. In *The Fungi*, Vol. III: *The Fungal Population*, ed. G. C. Ainsworth and A. S. Sussman, pp. 253–269. New York: Academic Press.

Magan, N. (1997). Fungi in extreme environments. In *Environmental and Microbial Relationships*, Vol. IV: *The Mycota*, ed. D. Wicklow and B. Soderstrom, pp. 99–114. Berlin: Springer.

Mägdefrau, K. (1957). Flechten und Moose im baltischen Bernstein. *Berichte der deutschen botanischen Gesellschaft*, **9**, 433–435.

Máguas, C. and Griffiths, H. (2003). Applications of stable isotopes in plant ecology. *Progress in Botany*, **64**, 472–505.

Máguas, C., Valladares, F. and Brugnoli, E. (1997). Effects of thallus size on morphology and physiology of foliose lichens: new findings with a new approach. *Symbiosis*, **23**, 149–164.

Majerus, M. E. N. (1998). *Melanism: Evolution in Action*. Oxford: Oxford University Press.

Makkonen, S., Hurri, R. and Hyvärinen, M. (2007). Differential responses of lichen symbionts to enhanced nitrogen and phosphorous availability: an experiment with *Cladina stellaris*. *Annals of Botany*, **99**, 877–884.

Malcolm, W. M. and Galloway, D. J. (1997). *New Zealand Lichens. Checklist, Key, and Glossary*. Wellington: Museum of New Zealand Te Papa Tongarewa.

Malhotra, S. S. and Khan, A. A. (1983). Sensitivity to SO$_2$ of various metabolic processes in an epiphytic lichen, *Evernia mesomorpha*. *Biochemie und Physiologie der Pflanzen*, **178**, 121–130.

Manodori, A. M. and Melis, A. (1984). Photochemical apparatus organization in *Anacystis nidulans* (Cyanophyceae). Effect of CO$_2$ concentration during cell growth. *Plant Physiology*, **74**, 67–71.

Margot, J. (1973). Experimental study of the effects of sulphur dioxide on the soredia of *Hypogymnia physodes*. In *Air Pollution and Lichens*, ed. B. W. Ferry, M. S. Baddeley and D. L. Hawksworth, pp. 314–329. Toronto: University of Toronto Press.

Margules, C. R. and Pressey, R. L. (2000). Systematics conservation planning. *Nature*, **405**, 243–253.

Margulis, L. and Fester, R. (eds.) (1991). *Symbiosis as a Source of Evolutionary Innovation: Speciation and Morphogenesis*. Cambridge: Massachusetts Institute of Technology Press.

Marin, B. and Melkonian, M. (1999). Mesostigmatophyceae, a new class of streptophyte green algae revealed by SSU rRNA sequence comparison. *Protist*, **150**, 399–417.

Marsh, J. E. and Nash, T. H., III (1979). Lichens in relation to the Four Corners Power Plant in New Mexico. *Bryologist*, **82**, 20–28.

Marshall, W. A. (1996). Aerial dispersal of lichen soredia in the maritime antarctic. *New Phytologist*, **134**, 523–530.

Marti, J. (1983). Sensitivity of lichen phycobionts to dissolved air pollutants. *Canadian Journal of Botany*, **61**, 1647–1653.

Martin, D., Ciulla, R. and Roberts, M. (1999). Osmoadaptation in Archaea. *Applied and Environmental Microbiology*, **65**, 1815–1825.

Masterson, C. L. and Murphy, P. M. (1984). The acetylene reduction technique. In *Current Developments in Biological Nitrogen-fixation*, ed. N. S. Subba Rao, pp. 8–33. Baltimore: Edward Arnold.

Matthes-Sears, U. and Nash, T. H., III (1986). The ecology of *Ramalina menziesii*. V. Estimation of gross carbon and thallus hydration source from diurnal measurements and climatic data. *Canadian Journal of Botany*, **64**, 1698–1702.

Matthews, J. A. (2005). 'Little Ice Age' glacier variations in Jotunheimen, southern Norway: a study in regionally controlled lichenometric dating of recessional moraines with implications for climate and lichen growth rates. *Holocene*, **15**, 1–19.

Mattox, K. R. and Stewart, K. D. (1984). Classification of the green algae: a concept based on comparative cytology. In *Systematics of the Green Algae*, ed. D. E. G. Irvine and D. M. John, pp. 29–72. London: Academic Press.

Mattson, J. E. (1991). Protein banding patterns in some American and European species of *Cetraria*. *Bryologist*, **94**, 261–269.

Mayaba, N. and Beckett, R. P. (2001). The effect of desiccation on the activities of antioxidant enzymes in lichens from habitats of contrasting water status. *Symbiosis*, **31**, 113–121.

McCall, K. K. and Martin, C. E. (1991). Chlorophyll concentrations and photosynthesis in three forest understorey mosses in northeastern Kansas. *Bryologist*, **94**, 25–29.

McCarroll, D. (1993). Modelling late-Holocene snow-avalanche activity, incorporating a new approach to lichenometry. *Earth Surface Processes and Landforms*, **18**, 527–539.

McCarroll, D., Shakesby, R. and Matthews, J. A. (1998). Spatial and temporal patterns of late Holocene rockfall activity on a Norwegian talus slope: a lichenometric and simulation-modeling approach. *Arctic and Alpine Research*, **30**, 51–60.

McCarthy, D. P. (1999). A biological basis for lichenometry? *Journal of Biogeography*, **26**, 379–386.

McCarthy, P. M. (2001). *Polyblastia. Flora of Australia*, **58A**, 171–172.

McCarthy, P. M. (2004). *Maronina. Flora of Australia*, **56A**, 62–63.

McCarthy, P. M. (2006). *Checklist of the Lichens of Australia and its Island Territories*. Australian Biological Resources Study, Canberra. Version 6 April 2006. Online: www.anbg.gov.au/abrs/lichenlist/introduction.html.

McCarthy, P. M. and Healey, J. A. (1978). Dispersal of lichen propagules by slugs. *Lichenologist*, **10**, 131–132.

McCune, B. (1988). Lichen communities along O_3 and SO_2 gradients in Indianapolis. *Bryologist*, **91**, 223–228.

McCune, B. (1994). Using epiphytic litter to estimate epiphyte biomass. *Bryologist*, **97**, 396–401.

McCune, B. and Daly, W. J. (1994). Consumption and decomposition of lichen litter in a temperate coniferous rainforest. *Lichenologist*, **26**, 67–71.

McCune, B. and Grace, J. B. (2002). *Analysis of Ecological Communities*. Gleneden Beach, OR: MjM Software.

McCune, B., Berryman, S. D., Cissel, J. H. and Gitelman, A. I. (2003). Use of a smoother to forecast occurrence of epiphytic lichens under alternative forest management plans. *Ecological Applications*, **13**, 1110–1123.

McDowall, R. M. (2004). What biogeography is: a place for process. *Journal of Biogeography*, **31**, 345–351.

McEvoy, M. (2006). Acclimation of the photobiont and mycobiont partners in lichens to high solar radiation. Ph.D. thesis. Ås, Norway: Department of Ecology and Natural Resource Management. Norwegian University of Life Sciences.

McGlone, M. S. (2005). Goodbye Gondwana. *Journal of Biogeography*, **32**, 739–740.

McGlone, M. S., Duncan, R. P. and Heenan P. B. (2001). Endemism, species selection and the origin and distribution of the vascular plant flora of New Zealand. *Journal of Biogeography*, **28**, 199–216.

McKersie, B. D. and Lesham, Y. Y. (1994). *Stress and Stress Coping in Cultivated Plants.* Dordrecht: Kluwer.

McNabb, D. H. and Geist, J. M. (1979). Acetylene reduction assay of symbiotic N_2 fixation under field conditions. *Ecology*, **60**, 1070–1072.

Meier, F. A., Scherrer, S. and Honegger, R. (2002). Faecal pellets of lichenivorous mites contain viable cells of the lichen-forming ascomycete *Xanthoria parietina* and its green algal photobiont, *Trebouxia arboricola*. *Biological Journal of the Linnean Society*, **76**, 259–268.

Melkonian, M. (1990). Chlorophyte orders of uncertain affinities: Order Microthamniales. In *Handbook of Protoctista*, ed. L. Margulis, J. O. Corliss, M. Melkonian and D. J. Chapman, pp. 652–654. Boston: Jones and Bartlett.

Melkonian, M. and Peveling, E. (1988). Zoospore ultrastructure in species of *Trebouxia* and *Pseudotrebouxia* (Chlorophyta). *Plant Systematics and Evolution*, **158**, 183–210.

Miadlikowska, J. and Lutzoni, F. (2004). Phylogenetic classification of peltigeralean fungi (Peltigerales, Ascomycota) based on ribosomal RNA small and large subunits. *American Journal of Botany*, **91**, 449–464.

Miadlikowska, J., Arnold, A. E., Hofstetter, V. and Lutzoni, F. (2004a). High diversity of cryptic fungi inhabiting healthy lichen thalli in a temperate and tropical forest. In *Lichens in Focus*, ed. T. Randlane and A. Saag, p. 43. Tartu: Tartu University Press.

Miadlikowska, J., Arnold, A. and Lutzoni, F. (2004b). Diversity of cryptic fungi inhabiting healthy lichen thalli in a temperate and tropical forest. *Ecological Society of America Annual Meeting*, **89**, 349–350.

Miao, V. P. W., Rabenau, A. and Lee, A. (1997). Cultural and molecular characterization of photobionts of *Peltigera membranacea*. *Lichenologist*, **29**, 571–587.

Micallef, A. and Colls, J. J. (1999). Analysis of long-term measurements of airborne concentrations of sulphur dioxide and $SO_4{}^{2-}$ in the rural United Kingdom. *Environmental Monitoring and Assessment*, **57**, 277–290.

Mies, B. and Printzen, C. (1997). Notes on the lichens of Socotra (Yemen, Indian Ocean). *Bibliotheca Lichenologica*, **68**, 223–239.

Mietzsch, E., Lumbsch, H. T. and Elix, J. A. (1993). Notice: a new computer program for the identification of lichen substances. *Mycotaxon*, **47**, 475–479.

Mikhailova, I. N. and Scheidegger, C. (2001). Early development of *Hypogymnia physodes* (L.) Nyl. in response to emissions from a copper smelter. *Lichenologist*, **33**, 527–538.

Millbank, J. W. (1972). Nitrogen metabolism in lichens. IV. The nitrogenase activity of the *Nostoc* phycobiont in *Peltigera canina*. *New Phytologist*, **71**, 1–10.

Millbank, J. W. (1974). Nitrogen metabolism in lichens. V. The forms of nitrogen released by the blue-green phycobiont in *Peltigera* spp. *New Phytologist*, **73**, 1171–1181.

Millbank, J. W. (1976). Aspects of nitrogen metabolism in lichens. In *Lichenology: Progress and Problems*, ed. D. H. Brown, D. L. Hawksworth and R. H. Bailey, pp. 441–455. London: Academic Press.

Millbank, J. W. (1981). The assessment of nitrogen fixation and throughput by lichens. I. The use of a controlled environment chamber to relate acetylene reduction estimates to fixation. *New Phytologist*, **89**, 647–655.

Millbank, J. W. (1982a). Nitrogenase and hydrogenase in cyanophilic lichens. *New Phytologist*, **92**, 221–228.

Millbank, J. W. (1982b). The assessment of nitrogen fixation and throughput by lichens. III. Losses of nitrogenous compounds by *Peltigera membranacea, P. polydactyla* and *Lobaria pulmonaria* in simulated rainfall episodes. *New Phytologist*, **92**, 229–234.

Millbank, J. W. and Olsen, J. D. (1986). The assessment of nitrogen fixation and throughput by lichens. IV. Nitrogen losses from *Peltigera membranacea* (Ach.) Nyl. in autumn, winter and spring. *New Phytologist*, **104**, 643–651.

Miller, G. H. (1973). Variations in lichen growth from direct measurements: preliminary curves for *Alectoria minuscula* from eastern Baffin Island, N. W. T., Canada. *Arctic and Alpine Research*, **5**, 33–42.

Miller, J. E. and Brown, D. H. (1999). Studies of ammonia uptake and loss by lichens. *Lichenologist*, **31**, 85–93.

Miller, P. R. and McBride, J. R. (eds.) (1998). *Oxidant Air Pollution Impacts in the Montane Forests of Southern California*. Ecological Studies 134. New York: Springer.

Miller, P. R., Longbotham, G. J. and Longbotham, C. R. (1983). Sensitivity of selected western conifers to ozone. *Plant Disease*, **67**, 1113–1115.

Miszalski, Z. and Niewiadomska, E. (1993). Comparison of sulphite oxidation mechanisms in three lichen species. *New Phytologist*, **123**, 345–349.

Mitchell, R. J., Truscot, A. M., Leith, I. D., *et al.* (2005). A study of the epiphytic communities of Atlantic oak woods along an atmospheric nitrogen deposition gradient. *Journal of Ecology*, **93**, 482–492.

Moberg, R. (1990). *Waynea*, a new lichen genus in the Bacidiaceae from California. *Lichenologist*, **22**, 249–252.

Modenesi, P. (1993). An SEM study of injury symptoms in *Parmotrema reticulatum* treated with paraquat or growing in sulphur dioxide-polluted air. *Lichenologist*, **25**, 423–433.

Mohr, F., Ekman, S. and Heegaard, E. (2004). Evolution and taxonomy of the marine *Collemopsidium* species (lichenized Ascomycota) in north-west Europe. *Mycological Research*, **108**, 515–532.

Molina, M. C., Crespo, A., Vicente, C. and Elix, J. A. (2003). Differences in the composition of phenolics and fatty acids of cultured mycobiont and thallus of *Physconia distorta*. *Plant Physiology and Biochemistry*, **41**, 175–180.

Mollenhauer, D. (1992). *Geosiphon pyriforme*. In *Algae and Symbioses: Plants, Animals, Fungi, Viruses, Interactions Explored*, ed. W. Reisser, pp. 339–351. Bristol: Biopress.

Mollenhauer, D., Mollenhauer, R. and Kluge, M. (1996). Studies on initiation and development of the partner association in *Geosiphon pyriforme* (Kütz.) v. Wettstein, a unique endocytobiotic system of a fungus (Glomales) and the cyanobacterium *Nostoc punctiforme* (Kütz.) Hariot. *Protoplasma*, **193**, 3–9.

Möller, C. and Dreyfuss, M. M. (1996). Microfungi from Antarctic lichens, mosses and vascular plants. *Mycologia*, **88**, 922–933.

Møller, I. M. (2001). Plant mitochondria and oxidative stress. Electron transport, NADPH turnover and metabolism of reactive oxygen species. *Annual Review of Plant Physiology and Plant Molecular Biology*, **52**, 561–591.

Monnet, F., Bordas, F., Deluchat, V., *et al.* (2005). Use of the aquatic lichen *Dermatocarpon luridum* as bioindicator of copper pollution: accumulation and cellular distribution tests. *Environmental Pollution*, **138**, 455–461.

Montalvo, A. M. and Ellstrand, N. C. (2001). Non-local transplantation and outbreeding depression in the subshrub *Lotus scoparius* (Fabaceae). *American Journal of Botany*, **88**, 258–269.

Montieth, J. L. (1977). Climate and the efficiency of crop production in Britain. *Philosophical Transactions of the Royal Society of London B*, **281**, 277–294.

Morrone, J. J. (2005). Cladistic biogeography: identity and place. *Journal of Biogeography*, **32**, 1281–1284.

Mosbach, K. (1969). Biosynthesis of lichen substances, products of a symbiotic association. *Angewandte Chemie, International Edition*, **8**, 240–250.

Moser, T. J., Nash, T. H., III and Clark, W. D. (1980). Effects of a long-term sulfur dioxide fumigation on arctic caribou forage lichens. *Canadian Journal of Botany*, **58**, 2235–2240.

Moser, T. J., Nash, T. H., III and Link, S. O. (1983a). Diurnal gross photosynthetic patterns and potential seasonal CO_2 assimilation in *Cladonia stellaris* and *Cladonia rangiferina*. *Canadian Journal of Botany*, **61**, 642–655.

Moser, T. J., Nash, T. H., III and Olafsen, A. G. (1983b). Photosynthetic recovery in arctic caribou forage lichens following a long-term field sulfur dioxide fumigation. *Canadian Journal of Botany*, **61**, 367–370.

Moxham, T. H. (1980). Lichens and perfume manufacture. *Bulletin of the British Lichen Society*, **47**, 1–2.

Muir, D. C. G., Segstro, M. D., Welbourn, P. M., *et al.* (1993). Patterns of accumulation of airborne organochlorine contaminants in lichens from the Upper Great Lakes Region of Ontario. *Environmental Science and Technology*, **27**, 1201–1210.

Muir, P. S., Shirazi, A. M. and Patrie, J. (1997). Seasonal growth dynamics in the lichen *Lobaria pulmonaria*. *Bryologist*, **100**, 458–464.

Mukhtar, A., Garty, J. and Galun, M. (1994). Does the lichen alga *Trebouxia* occur free-living in nature? – Further immunological evidence. *Symbiosis*, **17**, 247–253.

Munne-Bosch, S. and Alegre, L. (2002). The function of tocopherols and tocotrienols in plants. *Critical Reviews in Plant Science*, **21**, 31–57.

Muñoz, J., Felicísmo, A., Cabezas, F., Burgaz, A. and Martínez, I. (2004). Wind as long-distance dispersal vehicle in the Southern Hemisphere. *Science*, **304**, 1144–1147.

Murashige, T. and Skoog, F. (1962). A revised medium for rapid growth and bioassays with tobacco tissue cultures. *Physiologia Plantarum*, **15**, 473–494.

Murtagh, G. J., Dyer, P. S. and Crittenden, P. D. (2000). Reproductive systems – sex and the single lichen. *Nature*, **404**, 564.

Murtagh, G. J., Dyer, P. S., McClure, P. C. and Crittenden, P. D. (1999). Use of randomly amplified polymorphic DNA markers as a tool to study variation in lichen-forming fungi. *Lichenologist*, **31**, 257–267.

Myachi, S., Nakayama, O., Yokohama, *et al.* (1989). *World Catalogue of Algae*, 2nd. edn. Tokyo: Japan Scientific Societies Press.

Myllys, L., Högnabba, F., Lohtander, K., *et al.* (2005). Phylogenetic relationships of Stereocaulaceae based on simultaneous analysis of beta-tubulin, GAPDH and SSU rDNA sequences. *Taxon*, **54**, 605–618.

Myllys, L, Stenroos, S., Thell, A. and Ahti, T. (2003). Phylogeny of bipolar *Cladonia arbuscula* and *Cladonia mitis* (Lecanorales, Euascomycetes). *Molecular Phylogenetics and Evolution*, **27**, 58–69.

Naef, A., Roy, B. A., Kaiser, R. and Honegger, R. (2002). Insect-mediated reproduction of systemic infections by *Puccinia arrhenatheri* on *Berberis vulgaris*. *New Phytologist*, **154**, 717–730.

Nakano, T., Handa, S. and Takeshita, S. (1991). Some corticolous algae from the Taishaku-kyô Gorge, western Japan. *Nova Hedwigia*, **52**, 427–451.

Nakayama, T., Marin, B., Kranz, H. D. *et al.* (1998). The basal position of scaly green flagellates among the green algae (Chlorophyta) is revealed by analyses of nuclear-encoded SSU rRNA sequences. *Protist*, **149**, 367–380.

Nash, T. H., III (1971). Lichen sensitivity to hydrogen fluoride. *Bulletin of the Torrey Botanical Club*, **98**, 103–106.

Nash, T. H., III (1972). Simplification of the Blue Mountain lichen communities near a zinc factory. *Bryologist*, **75**, 315–324.

Nash, T. H., III (1973). Sensitivity of lichens to sulfur dioxide. *Bryologist*, **76**, 333–339.

Nash, T. H., III (1975). Influence of effluents from a zinc factory on lichens. *Ecological Monographs*, **45**, 183–196.

Nash, T. H., III (1988). Correlating fumigation studies with field effects. In *Lichens, Bryophytes and Air Quality* ed. T. H. Nash III and V. Wirth, pp. 201–216. Bibliotheca Lichenologica 30. Berlin: J. Cramer.

Nash, T. H., III (1989). Metal tolerance in lichens. In *Heavy Metal Tolerance in Plants: Evolutionary Aspects*, ed. A. J. Shaw, pp. 119–131. Boca Raton: CRC Press.

Nash, T. H., III (1996). Photosynthesis, respiration, productivity and growth. In *Lichen Biology*, ed. T. H. Nash III, pp. 88–120. Cambridge: Cambridge University Press.

Nash, T. H., III and Gries, C. (2002). Lichens as bioindicators of sulfur dioxide. *Symbiosis*, **33**, 1–21.

Nash, T. H., III and Lange, O. L. (1988). Responses of lichens to salinity: concentration and time-course relationships and variability among Californian species. *New Phytologist*, **109**, 361–367.

Nash, T. H, III, and Moser, T. J. (1982). Vegetational and physiological patterns of lichens in North American deserts. *Journal of the Hattori Botanical Laboratory*, **53**, 331–336.

Nash, T. H. III, and Olafsen, A. G. (1995). Climate change and the ecophysiological response of Arctic lichens. *Lichenologist*, **27**, 559–565.

Nash, T. H., III and Riddell, J. (2006). Historical perspectives and new opportunities in the use of lichens as air pollutant monitors. Botany 2006 Symposium Abstracts, p. 51. St. Louis: Botanical Society of America.

Nash, T. H., III and Sigal, L. L. (1979). Gross photosynthetic response of lichens to short-term ozone fumigations. *Bryologist*, **82**, 280–285.

Nash, T. H., III and Sigal, L. L. (1980). Sensitivity of lichens to air pollution with an emphasis on oxidant air pollutants. In *Proceedings of the Symposium on Effects of Air Pollution on Mediterranean and Temperate Forest Ecosystems, June 22–27, 1980, Riverside, California, U.S.A.* Gen. Tech. Rep. PSW-43, ed. P. R. Miller (prin. coord.), pp. 117–124. Berkley: Pacific Southwest Forest and Range Experiment Station, Forest Service, U.S. Department of Agriculture.

Nash, T. H., III, Moser, T. J. and Link, S. O. (1980). Nonrandom variation of gas exchange within arctic lichens. *Canadian Journal of Botany*, **58**, 1181–1186.

Nash, T. H., III, Moser, T. J., Link, S. O., *et al.* (1983). Lichen photosynthesis in relation to CO_2 concentration. *Oecologia*, **58**, 52–56.

Nash, T. H., III, Reiner, A., Demmig-Adams, B., *et al.* (1990). The effect of atmospheric desiccation and osmotic water stress on photosynthesis and dark respiration of lichens. *New Phytologist*, **116**, 269–276.

Nash, T. H., III, Ryan, B. D., Diederich, P., Gries, C. and Bungartz, F., eds., (2004). *Lichen Flora of the Greater Sonoran Desert Region.* Vol. 2. Tempe: Lichens Unlimited.

Nash, T. H., III, Ryan, B. D., Gries, C. and Bungartz, F. (2002). *Lichen Flora of the Greater Sonoran Desert Region.* Vol. 1. Tempe: Lichens Unlimited.

Nash, T. H., III, Thomas, M. A., Hoober, J. K., Gries, C. and Zheng, S. X. (2001). Free amino acids in lichens and their symbionts. In *Lichenological Contributions in Honour of Jack Elix*, ed. P. M. McCarthy, G. Kantvilas and S. H. J. J. Louwhoff, pp. 313–319. Bibliotheca Lichenologica 78. Berlin: J. Cramer.

Nash, T. H., III, White, S. L. and Marsh, J. E. (1977). Lichen and moss distribution and biomass in hot desert ecosystems. *Bryologist*, **80**, 470–479.

Nathan, R. (2005). Long-distance dispersal research: building a network of yellow brick roads. *Diversity and Distributions* **11**, 125–130.

Nelsen, M. P. & Gargas, A. (2006). Actin type I introns offer potential for increasing phylogenetic resolution in *Asterochloris* (Chlorophyta: Trebouxiophyceae). *The Lichenologist*, **38**, 435–440.

Nevo, E., Apelbaum-Elkahar, I., Garty, J. and Beiles, A. (1997). Natural selection caused microscale allozyme diversity in wild barley and a lichen at "Evolution Canyon", Mt. Carmel, Israel. *Heredity*, **78**, 373–382.

Newmaster, S. T., Bell, F. W. and Vitt, D. H. (1999). The effects of glyphosate and triclopyr on common bryophytes and lichens in northwestern Ontario. *Canadian Journal of Forest Research*, **29**, 1101–1111.

Newsham, K. K., Low, M. N. R., McLeod, A. R., Greenslade, P. D. and Emmett, B. C. (1997). Ultraviolet-B radiation influences the abundance and distribution of phylloplane fungi on pedunculate oak (*Quercus robur*). *New Phytologist*, **136**, 287–297.

Nicholson, T. P., Rudd, B. A. M., Dawson, M., *et al.* (2001). Design and utility of oligonucleotide gene probes for fungal polyketide synthases. *Chemistry and Biology*, **8**, 57–178.

Nieboer, E. and Richardson, D. H. S. (1980). The replacement of the nondescript term 'heavy metals' by a biologically and chemically significant classification of metal ions. *Environmental Pollution*, **1**, 3–26.

Nieboer, E. and Richardson, D. H. S. (1981). Lichens as monitors of atmospheric deposition. In *Atmospheric Pollutants in Natural Waters*, ed. S. J. Eisenreich, pp. 339–388. Ann Arbor: Ann Arbor Science.

Nieboer, E., MacFarlane, J. D. and Richardson, D. H. S. (1984). Modification of plant cell buffering capacities by gaseous air pollutants. In *Gaseous Air Pollutants and Plant Metabolism*, ed. M. J. Kozsol and F. R. Whatley, pp. 313–330. London: Butterworths.

Nieboer, E., Richardson, D. H. S., Lavoie, P. and Padovan, D. (1979). The role of metal-ion binding in modifying the toxic effects of sulphur dioxide on the lichen *Umbilicaria muhlenbergii*. I. Potassium efflux studies. *New Phytologist*, **82**, 621–632.

Nieboer, E., Richardson, D. H. S., Puckett, K. J. and Tomassini, F. D. (1976). The phytotoxicity of sulphur dioxide in relation to measurable responses in lichens. In *Effects of Air Pollutants on Plants*, ed. T. A. Mansfield, pp. 61–85. Cambridge: Cambridge University Press.

Nieboer, E., Richardson, D. H. S. and Tomassini, F. D. (1978). Mineral uptake and release by lichens: an overview. *Bryologist*, **81**, 226–246.

Nieboer, E., Tomassini, F. D., Puckett, K. J. and Richardson, D. H. S. (1977). A model for the relationship between gaseous and aqueous concentrations of sulphur dioxide in lichen exposure studies. *New Phytologist*, **79**, 157–162.

Nikonov, A. A. and Shebalina, T. Y. (1979). Lichenometry and earthquake age determination in central Asia. *Nature*, **280**, 675–677.

Nimis, P. L. (1996). Towards a checklist of Mediterranean lichens. *Bocconea*, **6**, 5–17.

Nimis, P. L. and Martellos, S. (2003). A second checklist of the lichens of Italy, with a thesaurus of synonyms. *Museo Regionale di Scienze Naturali Monografie*, **4**, 1–192.

Nimis, P. L. and Poelt, J. (1987). The lichens and lichenicolous fungi of Sardinia (Italy). An annotated list. *Studia Geobotanica Trieste*, **7** (suppl.1), 1–269.

Nimis, P. L., Castello, M. and Perotti, M. (1990). Lichens as biomonitors of sulphur dioxide pollution in La Spezia (Northern Italy). *Lichenologist*, **22**, 333–344.

Nimis, P. L., Pinna, D. and Salvadori, O. (1992). *Licheni e Conservazione dei Monumenti*. Bologna: Cooperativa Libraria Universitaria Editrice Bologna.

Nimis, P. L., Scheidegger, C. and Wolseley, P. A., eds. (2002). *Monitoring with Lichens – Monitoring Lichens*. Nato Science Series IV: Earth and Environmental Sciences 7. Dordrecht: Kluwer Academic.

Noctor, G. and Foyer, C. (1998). Ascorbate and glutathione: keeping active oxygen under control. *Annual Review of Plant Physiology and Plant Molecular Biology*, **49**, 249–279.

Noeske, O., Läuchli, A., Lange, O. L., Vieweg, G. H. and Ziegler, H. (1970). Konzentration und Kokalisierung von Schwermetallen in Flechten der

Erzschlackenhalden des Harzes. *Berichte der deutschen botanischen Gesellschaft*, **4**, 67–79.

Nordberg, M. L. and Allard, A. (2002). A remote sensing methodology for monitoring cover. *Canadian Journal of Remote Sensing*, **28**, 262–274.

Notcutt, G. and Davies, F. (1993). Dispersion of gaseous volcanogenic fluoride, island of Hawaii. *Journal of Volcanology and Geothermal Research*, **56**, 125–131.

Notcutt, G. and Davies, F. (1999). Biomonitoring of volcanogenic fluoride, Furnas Caldera, Sao Miguel, Azores. *Journal of Volcanology and Geothermal Research*, **92**, 209–214.

Nriagu, J. O. and Pacyna, J. (1988). Quantitative assessment of worldwide contamination of air, water and soils by trace metals. *Nature*, **333**, 134–139.

Nyati, S. (2006). Photobiont diversity in *Teloschistaceae*. Ph.D. thesis, Mathematisch-Naturwissenschaftliche Fakultät, Universität Zürich.

Nybakken, L., Solhaug, K. A., Bilger, W. and Gauslaa, Y. (2004). The lichens *Xanthoria elegans* and *Cetraria islandica* maintain a high protection against UV-B radiation in Arctic habitats. *Oecologia*, **140**, 211–216.

Nylander, W. (1866). Circa novum in studio Lichenum criterium chemicum. *Flora*, **49**, 198–201.

Obermayer, W. and Poelt, J. (1992). Contributions to the knowledge of the lichen flora of the Himalayas. III. On *Lecanora somervellii* Paulson (lichenised Ascomycotina, Lecanoraceae). *Lichenologist*, **24**, 111–117.

Oberwinkler, F. (1984). Fungus-alga interactions in basidiolichens. *Nova Hedwigia*, **79**, 739–774.

Oberwinkler, F. (2001). Basidiolichens. In *Fungal Associations, The Mycota*, Vol. IX, ed. B. Hock, pp. 211–225. Berlin: Springer.

O'Brien, H., Miadlikowska, J. and Lutzoni, F. (2005). Assessing host specialization in symbiotic cyanobacteria associated with four closely related species of the lichen fungus *Peltigera*. *European Journal of Phycology*, **40**, 363–378.

Ochiai, E. (1977). *Bioinorganic Chemistry: An Introduction*. Boston: Allyn and Bacon.

Ockinger, E., Niklasson, M. and Nilsson, S. G. (2005). Is local distribution of the epiphytic lichen *Lobaria pulmonaria* limited by dispersal capacity or habitat quality? *Biodiversity and Conservation*, **14**, 759–773.

Ogren, E. (1993). Convexity of the photosynthetic light-response curve in relation to intensity and direction of light during growth. *Plant Physiology*, **101**, 1013–1019.

O'Hare, G. P. and Williams, P. (1975). Some effects of sulphur dioxide flow on lichens. *Lichenologist*, **7**, 116–120.

Ohmura, Y., Kawachi, M., Kasai, F. and Watanabe, M. (2006). Genetic combinations of symbionts in a vegetatively reproducing lichen, *Parmotrema tinctorum*, based on ITS rDNA sequences. *Bryologist*, **109**, 43–59.

Olafsen, A. G. (1989). Nitrogen and carbon fixation in two Arctic lichens, *Stereocaulon tomentosum* and *Peltigera canina*. M.S. thesis. Tempe: Arizona State University.

Oliver, A. E., Hinchab, D. K. and Crowe, J. H. (2002). Looking beyond sugars: the role of amphiphilic solutes in preventing adventitious reactions in anhydrobiotes at low water contents. *Comparative Biochemistry and Physiology Part A*, **131**, 515–525.

Oliver, A. E., Leprince, O., Wolkers, W. W., *et al.* (2001). Non-disaccharide-based mechanisms of protection during drying. *Cryobiology*, **43**, 151–167.

Olmez, I., Gulovali, M. C. and Gordon, G. E. (1985). Trace element concentrations in lichens near a coal-fired power plant. *Atmospheric Environment*, **19**, 1663–1669.

O'Neill, A. L. (1994). Reflectance spectra of microphytic soil crusts in semi-arid Australia. *International Journal of Remote Sensing*, **15**, 675–681.

Opanowicz, M. and Grube, M. (2004). Photobiont genetic variation in *Flavocetraria nivalis* from Poland (Parmeliaceae, lichenized Ascomycetes). *The Lichenologist*, **36**,

Öquist, G., Brunes, L. and Hällgren, J.-E. (1982). Photosynthetic efficiency of *Betula pendula* acclimated to different quantum flux densities. *Plant Cell and Environment*, **5**, 9–15.

Orange, A., James, P. W. and White, F. J. (2001). *Microchemical Methods for the Identification of Lichens*. London: British Lichen Society.

Ott, S. (1987a). The juvenile development of lichen thalli from vegetative diaspores. *Symbiosis*, **3**, 57–74.

Ott, S. (1987b). Reproductive strategies in lichens. *Bibliotheca Lichenologica*, **25**, 81–93.

Ott, S. and Zvoch, I. (1992). Ethylene production by lichens. *Lichenologist*, **24**, 73–80.

Ott, S., Krieg, T., Spanier, U. and Schieleit, P. (2000a). Phytohormones in lichens with emphasis on ethylene biosynthesis and functional aspects on lichen symbiosis. *Phyton*, **40**, 83–94.

Ott, S., Schröder, T. and Jahns, H. M. (2000b). Colonization strategies and interactions of lichens on twigs. *Bibliotheca Lichenologica*, **75**, 445–455.

Ott, S., Treiber, K. and Jahns, H. M. (1993). The development of regenerative thallus structures. *Botanical Journal of the Linnean Society*, **113**, 61–76.

Otte, V., Esslinger, T. L. and Litterski, B. (2002). Biogeographical research on European species of the lichen genus *Physconia*. *Journal of Biogeography*, **29**, 1125–1141.

Otte, V., Esslinger, T. L. and Litterski, B. (2005). Global distribution of the European species of the lichen genus *Melanelia* Essl. *Journal of Biogeography*, **32**, 1221–1241.

Øvstedal, D. O. and Lewis Smith, R. I. (2001). *The Lichens of Antarctica and South Georgia: A Guide to their Identification and Ecology*. Cambridge: Cambridge University Press.

Pakarinen, P. and Häsänen, E. (1983). Mercury concentrations of bog mosses and lichens. *Suo*, **34**, 17–20.

Palacios, D., Parrilla, G. and Zamorano, J. J. (1999). Paraglacial and postglacial debris flows on a Little Ice Age terminal moraine: Jamapa Glacier, Pico de Orizaba (Mexico). *Geomorphology*, **28**, 95–118.

Palice, Z. and Printzen, C. (2004). Genetic variability in tropical and temperate populations of *Trapeliopsis glaucolepidea*: evidence against long-range dispersal in a lichen with a disjunct distribution. *Mycotaxon*, **90**, 43–54.

Palmer, H. E., Hanson, W. C., Griffin, B. I. and Roesch, W. C. (1963). Cesium-137 in Alaskan eskimos. *Science*, **142**, 64–66.

Palmer, R. J., Jr. and Friedmann, E. I. (1990). Water relations, thallus structure and photosynthesis in Negev Desert lichens. *New Phytologist*, **116**, 597–603.

Palmqvist, K. (1993). Photosynthetic CO_2 use efficiency in lichens and their isolated photobionts: the possible role of a CO_2 concentrating mechanism in cyanobacterial lichens. *Planta*, **191**, 48–56.

Palmqvist, K. (2000). Carbon economy in lichens. *New Phytologist*, **148**, 11–36.

Palmqvist, K. (2002). Carbon metabolism in cyanobacterial lichens. In *Cyanobacteria in Symbiosis*, ed. A. N. Rai, B. Bergman and U. Rasmussen, pp. 73–96. Amsterdam: Kluwer Academic.

Palmqvist, K. and Dahlman, L. (2006). Responses of the green algal foliose lichen *Platismatia glauca* to increased nitrogen supply. *New Phytologist*, **171**, 343–356.

Palmqvist, K. and Sundberg, B. (2000). Light use efficiency of dry matter gain in five macro-lichens: relative impact of microclimate and species-specific traits. *Plant Cell and Environment*, **23**, 1–14.

Palmqvist, K., de los Ríos, A., Ascaso, C. and Samuelsson, G. (1997). Photosynthetic carbon acquisition in the lichen photobionts *Coccomyxa* and *Trebouxia* (Chlorophyta). *Physiologia Plantarum*, **101**, 67–76.

Palmqvist, K., Campbell, D., Ekblad, A. and Johansson, H. (1998). Photosynthetic capacity in relation to nitrogen content and its partitioning in lichens with different photobionts. *Plant, Cell and Environment*, **21**, 361–372.

Palmqvist, K., Dahlman, L., Valladares, F., *et al.* (2002). CO_2 exchange and thallus nitrogen across 75 contrasting lichen associations from different climate zones. *Oecologia*, **133**, 295–306.

Palmqvist, K., Samuelsson, G. and Badger, M. R. (1994). Photobiont-related differences in carbon acquisition among green-algal lichens. *Planta*, **195**, 70–79.

Pannewitz, S., Green, T. G. A., Schlensog, M., *et al.* (2006). Photosynthetic performance of *Xanthoria mawsonii* C. W. Dodge in coastal habitats, Ross Sea region, continental Antarctica. *Lichenologist*, **38**, 67–81.

Pannewitz, S., Schlensog, M., Green, T. G. A., Sancho, L. G. and Schroeter, B. (2003). Are lichens active under snow in continental Antarctica? *Oecologia*, **135**, 30–38.

Parguey-Leduc, A. and Janex-Favre, M. (1981). The ascocarps of ascohymenial pyrenomycetes. In *Ascomycete Systematics: The Luttrellian Concept*, ed. D. R. Reynolds, pp. 102–123. New York: Springer.

Parmasto, E. (2004). Integrating molecular and morphological data in the systematics of fungi. In *Lichens in Focus*, ed. T. Randlane and A. Saag, p. 5. Tartu: Tartu University Press.

Peake, J. F. and James, P. W. (1967). Lichens and mollusca. *Lichenologist*, **3**, 425–428.

Pearce, C. (1997). Biologically active fungal metabolites. *Advances in Applied Microbiology*, **44**, 1–80.

Pearson, L. C. and Henriksson, E. (1981). Air pollution damage to cell membranes in lichens. II. Laboratory experiments. *Bryologist*, **84**, 515–520.

Pearson, L. C. and Skye, E. (1965). Air pollution affects patterns of photosynthesis in *Parmelia sulcata*, a corticolous lichen. *Science*, **148**, 1600–1602.

Peiser, G. and Yang, S. F. (1985). Biochemical and physiological effects of SO_2 on nonphotosynthetic processes in plants. In *Sulfur Dioxide and Vegetation: Physiology, Ecology, and Policy Issues*, ed. W. E. Winner, H. A. Mooney and R. A. Goldstein, pp. 148–161. Stanford: Stanford University Press.

Pennycook, S. R. and Galloway, D. J. (2004). Checklist of New Zealand "Fungi". In *Fungi of New Zealand*, Vol. 1: *Introduction to Fungi of New Zealand*, ed. E. H. C. McKenzie, pp. 401–488. Hong Kong: Fungal Diversity Press.

Pérez, F. L. (1994). Vagrant cryptogams in a paramo of the high Venezuelan Andes. *Flora*, **189**, 263–276.

Perkins, D. F. (1992). Relationship between fluoride contents and loss of lichens near an aluminium works. *Water, Air, and Soil Pollution*, **64**, 503–510.

Perkins, D. F. and Millar, R. O. (1987). Effects of airborne fluoride emissions near an aluminium works in Wales: part 2 – saxicolous lichens growing on rocks and walls. *Environmental Pollution*, **48**, 185–196.

Peršoh, D. (2004). Diversity of lichen inhabiting fungi in the *Letharietum vulpinae*. In *Lichens in Focus*, ed. T. Randlane and A. Saag, p. 34. Tartu: Tartu University Press.

Peršoh, D., Beck, A. and Rambold, G. (2004). The distribution of ascus types and photobiontal selection in Lecanoromycetes (Ascomycota) against the background of a revised SSU nrDNA phylogeny. *Mycological Progress*, **3**, 103–121.

Peterson, E. B. (2000). An overlooked fossil lichen (Lobariaceae). *Lichenologist*, **32**, 298–300.

Peterson, E. B. and McCune, B. (2003). The importance of hotspots for lichen diversity in forests of western Oregon. *Bryologist*, **106**, 246–256.

Peterson, R. B. and Burris, R. H. (1976). Conversion of acetylene reduction rates in natural populations of blue-green algae. *Analytical Biochemistry*, **73**, 404–410.

Petrini, O., Hake, U. and Dreyfuss. M. M. (1990). An analysis of fungal communities isolated from fruticose lichens. *Mycologia*, **82**, 444–451.

Petzold, D. E. and Goward, S. N. (1988). Reflectance spectra of subarctic lichens. *Remote Sensing of Environment*, **24**, 481–492.

Pfanz, H., Martinoia, E, Lange, O. L. and Heber, U. (1987). Flux of SO_2 into leaf cells and cellular acidification by SO_2. *Plant Physiology*, **85**, 928–933.

Piercey-Normore, M. D. (2004). Selection of algal genotypes by three species of lichen fungi in the genus *Cladonia*. *Canadian Journal of Botany*, **82**, 947–961.

Piercey-Normore, M. D. (2006). The lichen-forming ascomycete *Evernia mesomorpha* associates with multiple genotypes of *Trebouxia jamesii*. *New Phytologist*, **169**, 331–344.

Piercey-Normore, M. D. and DePriest, P. T. (2001). Algal switching among lichen symbioses. *American Journal of Botany*, **88**, 1490–1498.

Pike, L. H. (1978). The importance of epiphytic lichens in mineral cycling. *Bryologist*, **81**, 247–257.

Pike, L. H. (1981). Estimation of lichen biomass and production with special reference to the use of ratios. In *The Fungal Community: Its Organization and Role in the Ecosystem*, ed. D. Wicklow and G. Carroll, pp. 533–552. New York: Marcel Dekker.

Pinna, D., Salvadori, O. and Tretiach, M. (1998). An anatomical investigation of calcicolous endolithic lichens from the Trieste karst (NE Italy). *Plant Biosystems*, **132**, 183–195.

Pintado, A. and Sancho, L. G. (2002). Ecological significance of net photosynthesis activation by water vapour uptake in *Ramalina capitata* from rain-protected habitats in central Spain. *Lichenologist*, **34**, 403–413.

Pišút, I. and Lisicka, E. (1985). A study of cryptogamic epiphytes on an oak trunk in the vicinity of Bratislava in the years 1973-1983. *Ekologia (CSSR)*, **4**, 225-234.

Plakunova, O. V. and Plakunova, V. G. (1987). Ultrastructure of *Cladina stellaris* lichen components in normal conditions and at SO_2 environment pollution. *Izvestiya Akademii Nauk SSR, Seriya Biologicheskaya*, **1987**, 361-369.

Platt, J. L. and Spatafora, J. W. (2000). Evolutionary relationships of nonsexual lichenized fungi: molecular phylogenetic hypotheses for the genera *Siphula* and *Thamnolia* from SSU and LSU rDNA. *Mycologia*, **92**, 475-487.

Plessl, A. (1963). Über die Beziehungen von Pilz und Alge im Flechtenthallus. *Österreichische Botanische Zeitschrift*, **110**, 194-269.

Poelt, J. (1970). Das Konzept der Artenpaare bei den Flechten. *Vorträge aus dem Gesamtgebeit der Botanik herausgegeben von der deutschen botanischen Gesellschaft, Berlin*, **4**, 187-198.

Poelt, J. (1972). Die Taxonomische Behandlung von Artenpaaren bei den Flechten. *Botaniska Notiser*, **125**, 77-81.

Poelt, J. (1980). *Physcia opuntiella* und die Lebensform der sprossenden Flechten. *Flora*, **169**, 22-31.

Poelt, J. (1985). Über auf Moosen parasitierende Flechten. *Sydowia, Annales Mycologici Series II*, **38**, 241-254.

Poelt, J. (1986). Morphologie der Flechten - Fortschritte und Probleme. *Berichte der deutschen botanischen Gesellschaft*, **99**, 3-29.

Poelt, J. (1989). Die Entstehung einer Strauchflechte aus einem Formenkreis krustiger Verwandter. *Flora*, **183**, 65-72.

Poelt, J. (1993). La riproduzione asessuale nei licheni. *Notiziario della Società Lichenologica Italiana*, **6**, 9-28.

Poelt, J. (1994). On lichenized asexual diaspores in foliose lichens - a contribution towards a more differentiated nomenclature (Lichens, Lecanorales). *Cryptogamic Botany*, **5**, 150-162.

Poelt, J. and Doppelbauer, H. (1956). Über parasitische Flechten. *Planta*, **46**, 467-480.

Poelt, J. and Mayrhofer, H. (1988). Über Cyanotrophie bei Flechten. *Plant Systematics and Evolution*, **158**, 265-281.

Poelt, J. and Obermayer, W. (1990a). Über Thallosporen bei einigen Krustenflechten. *Herzogia*, **8**, 273-288.

Poelt, J. and Obermayer, W. (1990b). Lichenisierte Bulbillen als Diasporen bei der Basidiolichene *Multiclavula vernalis* spec. coll. *Herzogia*, **6**, 289-294.

Poelt, J. and Vězda, A. (1990). Über kurzlebige Flechten. *Bibliotheca Lichenologica*, **38**, 377-394.

Pöggeler, S. (1999). Phylogenetic relationships between mating-type sequences from homothallic and heterothallic ascomycetes. *Current Genetics*, **36**, 222-231.

Poinar, G., Peterson, E. and Platt, J. (2000). Fossil *Parmelia* in New World amber. *Lichenologist*, **32**, 263-269.

Pole, M. (1993). Keeping in touch: vegetation prehistory on both sides of the Tasman. *Australian Systematic Botany*, **6**, 387-397.

Pole, M. (1994). The New Zealand flora – entirely long-distance dispersal? *Journal of Biogeography*, **21**, 625–635.

Ponzetti, J. M. and McCune, B. P. (2001). Biotic soil crusts of Oregon's shrub steppe: community composition in relation to soil chemistry, climate, and livestock activity. *Bryologist*, **104**, 212–225.

Popp, M. and Smirnoff, N. (1995). Polyol accumulation and metabolism during water deficit. In *Environment and Plant Metabolism*, ed. N. Smirnoff, pp. 199–215. Oxford: BIOS Scientific Publishers.

Potts, M. (1994). Desiccation tolerance in prokaryotes. *Microbiology Reviews*, **58**, 755–805.

Potts, M. and Bowman, M. (1985). Sensitivity of *Nostoc commune* UTEX 584 (Cyanobacteria) to water stress. *Archiv für Microbiologie*, **141**, 51–56.

Pressel, S., Ligrone, R. and Duckett, J. G. (2006). The effects of de- and rehydration on food-conducting cells in the moss *Polytrichum formosum* Hedw.: a cytological study. *Annals of Botany*, **98**, 67–76.

Price, S. and Long, S. P. (1989). An *in vivo* analysis of the effect of SO_2 fumigation on photosynthesis in *Zea mays*. *Physiologica Plantarum*, **76**, 193–200.

Prillinger, H. (1982). Zur genetischen Kontrolle und Evolution der sexuellen Fortplanzung und Heterothallie der Chitinpilzen. *Zeitschrift für Mykologie*, **48**, 297–324.

Printzen, C. and Ekman, S. (2002). Genetic variability and its geographical distribution in the widely disjunct *Cavernularia hultenii*. *Lichenologist*, **34**, 101–111.

Printzen, C. and Ekman, S. (2003). Local population subdivision in the lichen *Cladonia subcervicornis* as revealed by mitochondrial cytochrome oxidase subunit 1 intron sequences. *Mycologia*, **95**, 399–406.

Printzen, C. and Kantvilas, G. (2004). *Hertelidea*, genus novum Stereocaulaearum (Ascomycetes lichenisati). *Bibliotheca Lichenologica*, **88**, 539–553.

Printzen, C. and Lumbsch, H. T. (2000). Molecular evidence for the diversification of extant lichens in the late Cretaceous and Tertiary. *Molecular Phylogenetics and Evolution*, **17**, 379–387.

Printzen, C., Ekman, S. and Tønsberg, T. (2003). Phylogeography of *Cavernularia hultenii*: evidence of slow genetic drift in a widely disjunct lichen. *Molecular Ecology*, **12**, 1473–1486.

Proctor, M. C. F., Ligrone, R. and Duckett, J. G. (2006). Desiccation tolerance in the moss *Polytrichum formosum* Hedw.: physiological and fine-structural changes during desiccation and recovery. *Annals of Botany*, **99**, 75–93.

Puckett, K. J. (1976). The effect of heavy metals in some aspects of lichen physiology. *Canadian Journal of Botany*, **54**, 2695–2703.

Puckett, K. J. (1978). *Element Levels in Lichens from the Northwest Territories*. Report ARQA-56-76. Downsview: Atmospheric Environment Service, Environment Canada.

Puckett, K. J. (1988). Bryophytes and lichens as monitors of metal deposition. *Bibliotheca Lichenologica*, **30**, 231–267.

Puckett, K. J. and Burton, M. A. S. (1981). The effect of trace elements on lower plants. In *Effect of Heavy Metal Pollution on Plants*, Vol. 2: *Metals in the Environment*, ed. N. W. Lepp, pp. 213–238. London: Applied Science Publishers.

Puckett, K. J. and Finegan, E. J. (1980). An analysis of the element content of lichens from the Northwest Territories, Canada. *Canadian Journal of Botany*, **58**, 2073–2089.

Puckett, K. J., Nieboer, E., Flora, W. P. and Richardson, D. H. S. (1973). Sulphur dioxide: its effect on photosynthetic ^{14}C fixation in lichens and suggested mechanisms of phytotoxicity. *New Phytologist*, **72**, 141–154.

Puckett, K. J., Richardson, D. H. S., Flora, W. P. and Nieboer, E. (1974). Photosynthetic ^{14}C fixation by the lichen *Umbilicaria muhlenbergii* (Ach.) Tuck. following short exposures to aqueous sulphur dioxide. *New Phytologist*, **73**, 1183–1192.

Puckett, K. J., Tomassini, F. D., Nieboer, E. and Richardson, D. H. S. (1977). Potassium efflux by lichen thalli following exposure to aqueous sulphur dioxide. *New Phytologist*, **79**, 135–145.

Punz, W. (1979). Der Einfluss isolierter und kombinierter Schadstoffe auf die Flechtenphotosynthese. *Photosynthetica*, **13**, 428–433.

Purvis, A., Gittleman, J. L. and Brooks, T., eds. (2005). *Phylogeny and Conservation*. Cambridge: Cambridge University Press.

Purvis, O. W. (1984). The occurrence of copper oxalate in lichens growing on copper sulphide-bearing rocks in Scandinavia. *Lichenologist*, **16**, 197–204.

Purvis, O. W. (2000). *Lichens*. London: Natural History Museum and Washington: Smithsonian Institution.

Purvis, O. W., Coppins, B. J., Hawksworth, D. J., James, P. W. and Moore, M. D. (1992). *The Lichen Flora of Great Britain and Ireland*. London: Natural History Publications.

Purvis, O. W., Elix, J. A. Broomhead, J. A. and Jones, G. C. (1987). The occurrence of copper-norstictic acid in lichens from cupriferous substrata. *Lichenologist*, **19**, 193–203.

Qui, B. S. and Gao, K. S. (2001). Photosynthetic characteristics of the terrestrial blue-green alga *Nostoc flagelliforme*. *European Journal of Phycology*, **36**, 147–156.

Quilhot, W., Fernández, E., Rubio, C., Goddard, M. and Hidalgo, M. E. (1998). Lichen secondary products and their importance in environmental studies. In *Lichenology in Latin America: History, Current Knowledge and Applications*, ed. M. P. Marcelli and M. R. D. Seaward, pp. 171–179. São Paulo: CETESB.

Quispel, A. (1960). Respiration of lichens. In *Pflanzenatmung einschliesslich Gärung und Säurestoffwechsel (Handbuch der Pflanzenphysiologie vol XII/2)*, ed. J. Wolf, pp. 455–460. Berlin: Springer.

Rai, A. N. (1988). Nitrogen metabolism. In *Handbook of Lichenology*, Vol. I, ed. M. Galun, pp. 201–237. Boca Raton: CRC Press.

Rai, A. N. (2002). Cyanolichens: nitrogen metabolism. In *Cyanobacteria in Symbiosis*, ed. A. N. Rai, B. Bergman and U. Rasmussen, pp. 97–115. Dordrecht: Kluwer Academic.

Rai, A. N., Rowell, P. and Stewart, W. D. P. (1980). NH_4^+ assimilation and nitrogenase regulation in the lichen *Peltigera aphthosa* Willd. *New Phytologist*, **85**, 545–555.

Rai, A. N., Rowell, P. and Stewart, W. D. P. (1981). Nitrogenase activity and dark CO_2 fixation in the lichen *Peltigera aphthosa* Willd. *Planta*, **151**, 256–264.

Rai, A. N., Rowell, P. and Stewart, D. P. (1983). Mycobiont-cyanobiont interactions during dark nitrogen fixation by the lichen *Peltigera aphthosa*. *Physiologia Plantarum*, **57**, 285–290.

Rambold, G. and Triebel, D. (1992). The inter-lecanoralean associations. *Bibliotheca Lichenologica*, **48**, 3–201.

Rambold, G., Friedl, T. and Beck, A. (1998). Photobionts in lichens: possible indicators of phylogenetic relationships? *Bryologist*, **101**, 392–397.

Ramstad, S. and Hestmark, G. (2001). Population structure and size-dependent reproductive effort in *Umbilicaria sporochroa*. *Mycologia*, **93**, 453–458.

Rancan, F., Rosan, S., Boehm, K., *et al.* (2002). Protection against UVB irradiation by natural filters extracted from lichens. *Journal of Photochemistry and Photobiology B: Biology*, **68**, 133–139.

Randlane, T. and Saag, A. (1998). Synopsis of the genus *Nephromopsis* (Fam. *Parmeliaceae*, lichenized *Ascomycota*). *Cryptogamie, Bryologie et Lichénologie*, **19**, 175–191.

Randlane, T., Saag, A. and Thell, A. (1997). A second updated world list of cetrarioid lichens. *Bryologist*, **100**, 109–122.

Rao, D. N. and LeBlanc, F. (1966). Effects of sulfur dioxide on the lichen alga with special reference to chlorophyll. *Bryologist*, **69**, 69–75.

Raven, J. A. (1992). Energy and nutrient acquisition by autotrophic symbioses and their asymbiotic ancestors. (Review). *Symbiosis*, **14**, 33–60.

Raven, J. A., Johnston, A. M., Handley, L. L. and McInroy, S. G. (1990). Transport and assimilation of inorganic carbon by *Lichina pygmea* under emersed and submersed conditions. *New Phytologist*, **114**, 407–417.

Reeb, V., Lutzoni, F. and Roux, C. (2004). Contribution of RPB2 to multilocus phylogenetic studies of the euascomycetes (Pezizomycotina, Fungi) with special emphasis on the lichen-forming Acarosporaceae and evolution of polyspory. *Molecular Phylogenetics and Evolution*, **32**, 1036–1060.

Reich, P. B., Ellsworth, D. S., Walters, M. B., *et al.* (1999). Generality of leaf trait relationships: a test across six biomes. *Ecology*, **80**, 1955–1969.

Reich, P. B., Walters, M. B., Ellsworth, D. S., *et al.* (1998). Relationships of leaf dark respiration to leaf nitrogen, specific leaf area and leaf life-span: a test across biomes and functional groups. *Oecologia*, **114**, 471–482.

Reiners, W. A. and Olson, R. K. (1984). Effects of canopy components on throughfall chemistry: an experimental analysis. *Oecologia*, **63**, 320–330.

Reinke, J. (1894–1896). Abhandlungen über Flechten. *Jahrbücher wissenschaftliche Botanik*, **26, 28, 29**.

Reisser, W. (1992). Endosymbiotic associations of algae with freshwater protozoa and invertebrates. In *Algae and Symbioses: Plants, Animals, Fungi, Viruses, Interactions Explored*, ed. W. Reisser, pp. 1–19. Bristol: Biopress.

Reiter, R. and Türk, R. (2000a). Investigations on the CO_2 exchange of lichens in the alpine belt. I. Comparative patterns of net CO_2 exchange in *Cladonia mitis*, *Thamnolia vermicularis* and *Umbilicaria cylindrica*. *Bibliotheca Lichenologica*, **75**, 333–351.

Reiter, R. and Türk, R. (2000b). Investigations on the CO_2 exchange of lichens in the alpine belt. II. Comparative patterns of net CO_2 exchange in *Cetraria islandica* and *Flavocetraria nivalis*. *Phyton (Austria)*, **40**, 161–177.

Renhorn, K. E., Esseen, P.-A., Palmqvist, K. and Sundberg, B. (1997). Growth and vitality of epiphytic lichens. I. Responses to microclimate along a forest edge-interior gradient. *Oecologia*, **109**, 1–9.

Rennenberg, H. (1984). The fate of excess sulfur in higher plants. *Annual Review of Plant Physiology*, **35**, 121–153.

Rennenberg, H. and Polle, A. (1994). Metabolic consequences of atmospheric sulphur influx into plants. In *Plant Responses to the Gaseous Environment*, ed. R. G. Alscher and A. R. Wellburn, pp. 165–180. London: Chapman and Hall.

Reutimann, P. and Scheidegger, C. (1987). Importance of lichen secondary products in food choice of two oribatic mites (*Acari*) in an alpine meadow ecosystem. *Journal of Chemical Ecology*, **13**, 363–369.

Rhoades, F. M. (1977). Growth rates of *Lobaria oregana* as determined from sequential photographs. *Canadian Journal of Botany*, **55**, 2226–2233.

Rhoades, F. M. (1981). Biomass of epiphytic lichens and bryophytes on *Abies lasiocarpa* on a Mt. Baker lava flow. *Bryologist*, **84**, 39–47.

Rhoades, F. M. (1983). Distribution of thalli in a population of the epiphytic lichen *Lobaria oregana* and a model of population dynamics and production. *Bryologist*, **86**, 309–331.

Richardson, D. H. S. (1974). *The Vanishing Lichens*. New York: Hafner Press.

Richardson, D. H. S. (1988). Medicinal and other economic aspects of lichens. In *Handbook of Lichenology*, Vol. 3, ed. M. Galun, pp. 93–108. Boca Raton: CRC Press.

Richardson, D. H. S. (1991). Lichens and man. In *Frontiers in Mycology*, ed. D. L. Hawksworth, pp. 187–210. Kew: CAB International.

Richardson, D. H. S. (1999). War in the world of lichens: parasitism and symbiosis as exemplified by lichens and lichenicolous fungi. *Mycological Research*, **6**, 641–650.

Richardson, D. H. S. and Cameron, R. P. (2004). Cyanolichens: their response to pollution and possible management strategies for their conservation in northeastern North America. *Northeastern Naturalist*, **11**, 1–22.

Richardson, D. H. S. and Nieboer, E. (1983). Ecophysiological responses of lichens to sulphur dioxide. *Journal of the Hattori Botanical Laboratory*, **54**, 331–351.

Richardson, D. H. S. and Smith, D. C. (1966). The physiology of the symbiosis in *Xanthoria aureola* (Ach.) Erichs. *Lichenologist*, **3**, 202–206.

Richardson, D. H. S. and Smith, D. C. (1968). Lichen physiology. IX. Carbohydrate movement from the *Trebouxia* symbiont of *Xanthoria aureola*. *New Phytologist*, **67**, 61–68.

Richardson, D. H. S. and Young, C. M. (1977). Lichens and vertebrates. In *Lichen Ecology*, ed. M. R. D. Seaward, pp. 121–144. London: Academic Press.

Richardson, D. H. S., Beckett, P. J. and Nieboer, E. (1980). Nickel in lichens, bryophytes, fungi and algae. In *Nickel in the Environment*, ed. J. O. Nriagu, pp. 367–406. New York: John Wiley.

Richardson, D. H. S., Hill, D. J. and Smith, D. C. (1968). Lichen physiology. XI. The role of the alga in determining the pattern of carbohydrate movement between lichen symbionts. *New Phytologist*, **67**, 469–486.

Richardson, D. H. S., Nieboer, E., Lavoie, P. and Padovan, D. (1979). The role of metal-ion binding in modifying the toxic affects of sulphur dioxide on the lichen *Umbilicaria muhlenbergii*. II. ^{14}C-fixation studies. *New Phytologist*, **82**, 633–643.

Richardson, D. H. S., Nieboer, E., Lavoie, P. and Padovan, D. (1984). Anion accumulation by lichens. I. The characteristics and kinetics of arsenate uptake by *Umbilicaria muhlenbergii*. *New Phytologist*, **96**, 71–82.

Riddle, B. R. (2005). Is biogeography emerging from its identity crisis? *Journal of Biogeography*, **32**, 185–186.

Riddle, B. R. and Hafner, D. J. (2004). The past and future roles of phylogeography. In *Frontiers of Biogeography: New Directions in the Geography of Nature*, ed. M. V. Lomolino and L. R. Heaney, pp. 93–110. Sunderland: Sinauer Associates.

Rikkinen, J. (2002). Cyanolichens: an evolutionary overview. In *Cyanobacteria in Symbioses*, ed. A. N. Rai, B. Bergamn, and U. Rasmussen, pp. 31–72. Dordrecht: Kluwer Academic Publishers.

Rikkinen, J. (2003). Calicioid lichens from European tertiary amber. *Mycologia*, **95**, 1032–1036.

Rikkinen, J., Oksanen, I. and Lohtander, K. (2002). Lichen guilds share related cyanobacterial symbionts. *Science*, **297**, 357.

Rindi, F. and Guiry, M. D. (2003). Composition and distribution of subaerial algal assemblages in Galway City, western Ireland. *Cryptogamie Algologie*, **24**, 245–267.

Roberts, B. A. and Thompson, L. K. (1980). Lichens as indicators of fluoride emission from a phosphorous plant, Long Harbour, Newfoundland, Canada. *Canadian Journal of Botany*, **58**, 2218–2228.

Robinson, C. H. (2001). Cold adaptation in Arctic and Antarctic fungi. *New Phytologist*, **151**, 341–353.

Rogers, R. W. (1990). Ecological strategies of lichens. *Lichenologist*, **22**, 149–162.

Rogers, R. W. (1992). Lichen ecology and biogeography. *Flora of Australia* **54**, 30–42.

Rollin, E. M., Milton, E. J. and Roche, P. (1994). The influence of weathering and lichen cover on the reflectance spectra of granitic rocks. *Remote Sensing of Environment*, **50**, 194–199.

Rolstad, J., Gjerde, I., Storaunet, K. O. and Rolstad, E. (2001). Epiphytic lichens in Norwegian coastal spruce forest: historical logging and present forest structure. *Ecological Applications*, **11**, 421–436.

Romagni, J. G., Thomas, M. A., Gries, C. and Nash, T. H., III (1997). Sulfite reductase activity in six lichen species as a response to fumigations with sulfur dioxide. *Supplement to Plant Physiology*, **114**, 58.

Romagni, J. G., Thomas, M. A. and Nash, T. H., III (1998). Detoxification of SO_2 in lichens: total glutathione. *Supplement to Plant Physiology*, **115**, 89.

Romeike, J., Friedl, T., Helms, G. and Ott, S. (2002). Genetic diversity of algal and fungal partners in four species of *Umbilicaria* (lichenized ascomycetes) along a transect of the Antarctic peninsula. *Molecular Biology and Evolution*, **19**, 1209–1217.

Rorat, T. (2006). Plant dehydrins: tissue location, structure and function. *Cell Molecular Biology Letters*, **11**, 536–556.

Rose, C. I. and Hawksworth, D. L. (1981). Lichen recolonization in London's cleaner air. *Nature*, **289**, 289–292.

Rose, F. (1976). Lichenological indicators of age and environmental quality in woodlands. In *Lichenology: Progress and Problems*, ed. D. H. Brown, D. L. Hawksworth and R. H. Bailey, pp. 279–307. London: Academic Press.

Rosentreter, R. (1993). Vagrant lichens in North America. *Bryologist*, **96**, 333–338.

Rosentreter, R. and Eldridge, D. J. (2002). Monitoring biodiversity and ecosystem function: grasslands, deserts, and steppe. In *Monitoring with Lichens – Monitoring Lichens*. Nato Science Series IV: Earth and Environmental Sciences, ed. P. L. Nimis, C. Scheidegger and P. A. Wolseley, pp. 223–237. Dordrecht: Kluwer Academic.

Roser, D. J., Mellick, D. R., Ling, H. U. and Seppelt, R. D. (1992). Polyol and sugar content of terrestrial plants from continental Antarctica. *Antarctic Science*, **4**, 413–420.

Ross, L. J. (1982). Lichens on coastal live oak in relation to ozone. M. S. Thesis. Arizona State University, Tempe, Arizona.

Ross, L. J. and Nash, T. H., III (1983). Effect of ozone on gross photosynthesis of lichens. *Environmental and Experimental Botany*, **23**, 71–77.

Rosso, A. L., McCune, B. and Rambo, T. R. (2000). Ecology and conservation of a rare, old-growth-associated canopy lichen in a silvicultural landscape. *Bryologist*, **103**, 117–127.

Roux, C. (1981). Étude écologique et phytosociologique des peuplements lichéniques saxicoles–calcicoles du sud-est de la France. *Bibliotheca Lichenologica*, **15**, 1–557.

Rubio, C., Fernández, E., Hidalgo, M. E. and Quilhot, W. (2002). Effects of solar UV-B radiation in the accumulation of rhizocarpic acid in a lichen species from alpine zones of Chile. *Boletín, Sociedad Chilena de Química*, **47**, 67–72.

Ruchty, A., Rosso, A. L. and McCune, B. (2001). Changes in epiphyte communities as the shrub, *Acer circinatum*, develops and ages. *Bryologist*, **104**, 272–281.

Rundel, P. W. (1969). Clinal variation in the production of usnic acid in *Cladonia subtenuis* along light gradients. *Bryologist*, **72**, 40–44.

Rundel, P. W. (1978). The ecological role of secondary lichen substances. *Biochemical Systematics and Ecology*, **6**, 157–170.

Rundel, P. W. (1982). Water uptake by organs other than roots. In *Physiological Plant Ecology*. Vol. II: *Water Relations and Carbon Assimilation*, ed. O. L. Lange, P. S. Nobel, C. B. Osmond and H. Ziegler, pp. 111–134. Encyclopedia of Plant Physiology 12B. Berlin: Springer.

Rundel, P. W. (1988). Water relations. In *CRC Handbook of Lichenology*, Vol. 2, ed. M. Galun, pp. 17–36. Boca Raton: CRC Press.

Ruoss, E. (1987). Species differentiation in a group of reindeer lichens (*Cladonia* subg. *Cladina*). *Bibliotheca Lichenologica*, **25**, 197–206.

Rychert, R. C. and Skujins, J. (1974). Nitrogen fixation by blue-green algae–lichen crusts in the Great Basin Desert. *Proceedings of the Soil Science Society of America*, **38**, 768–771.

Salisbury, F. B. and Ross, C. W. (1992). *Plant Physiology*. 4th edn. Belmont: Wadsworth Publishing.

Sancho, L. G. and Kappen, L. (1989). Photosynthesis and water relations and the role of anatomy in Umbilicaricaeae (Lichenes) from central Spain. *Oecologia*, **81**, 473–480.

Sancho, L. G. and Pintado, A. (2004). Evidence of high annual growth rate for lichens in the maritime Antarctic. *Polar Biology*, **27**, 312–319.

Sancho, L. G., de la Torre, R., Horneck, G., *et al.* (2007). Lichens survive in space: Results from the 2005 LICHENS experiment. *Astrobiology*, **7**, 443–454.

Sancho, L. G., Pintado, A., Blanquer, J. M., Raggio, J. and Vilches, R. (2004). Lichen morphology, thallus water content and photosynthetic performance. Looking for a single trait. In *Book of Abstracts of the 5th IAL Symposium: Lichens in Focus*, ed. T. Randlane and A. Saag, p. 47. Tartu: Tartu University Press.

Sancho, L. G., Pintado, A., Green, T. G. A., Pannewitz, S. and Schroeter, B. (2003). Photosynthetic and morphological variation within and among populations of the Antarctic lichen *Umbilicaria aprina*: implications of the thallus size. *Bibliotheca Lichenologica*, **86**, 299–311.

Sancho, L. G., Pintado, A., Valladares, F., Schroeter, B. and Schlensog, M. (1997). Photosynthetic performance of cosmopolitan lichens in the maritime Antarctic. *Bibliotheca Lichenologica*, **67**, 197–210.

Sancho, L. G., Schroeter, B. and Del-Prado, R. (2000a). Ecophysiology and morphology of the globular erratic lichen *Aspicilia fruticulosa* (Eversm.) Flag. from central Spain. *Bibliotheca Lichenologica*, **75**, 137–147.

Sancho, L. G., Valladares, F., Schroeter, B. and Kappen, L. (2000b). Ecophysiology of Antarctic versus temperate populations of a bipolar lichen: the key role of the photosynthetic partner. In *Antarctic Ecosystems: Models for Wider Ecological Understanding*, ed. W. Davison, C. H. Williams and P. Broady, pp. 190–194. Christchurch: New Zealand Natural Sciences Publications.

Sanders, W. B. (1989). Growth and development of the reticulate thallus in the lichen *Ramalina menziesii*. *American Journal of Botany*, **76**, 666–678.

Sanders, W. B. (2001a). Preliminary light microscope observations of fungal and algal colonization and lichen thallus initiation on glass slides placed near foliicolous lichen communities within a lowland tropical forest. *Symbiosis*, **31**, 85–94.

Sanders, W. B. (2001b). Lichens: the interface between mycology and plant morphology. *BioScience*, **51**, 1025–1035.

Sanders, W. B. (2005). Observing microscopic phases of lichen life cycles on transparent substrata placed *in situ*. *Lichenologist*, **37**, 373–382.

Sanders, W. B. and Lücking, R. (2002). Reproductive strategies, relichenization and thallus development observed *in situ* in leaf-dwelling lichen communities. *New Phytologist*, **155**, 425–435.

Sanders, W. B., Moe, R. L. and Ascaso, C. (2004). The intertidal marine lichen formed by the pyrenomycete fungus *Verrucaria tavaresiae* (Ascomycotina) and the brown alga *Petroderma maculiforme* (Phaeophyceae): thallus organization and symbiont interaction. *American Journal of Botany*, **91**, 511–522.

Sanmartín, I. and Ronquist, F. (2004). Southern Hemisphere biogeography inferred
    by event-based models: plant versus animal patterns. *Systematic Biology*, **53**, 216–243.

Santesson, J. (1969). Chemical studies on lichens. 10. Mass spectrometry on lichens.
    Arkiv för Chemie, **30**, 363–377.

Santesson, R. (1939). Amphibious pyrenolichens I. *Arkiv för Botanik*, **29A**, 1–67.

Santesson, R. (1952). Foliicolous lichens. I. A revision of the taxonomy
    of the obligately foliicolous, lichenized fungi. *Symbolae Botanicae Upsalienses*,
    12, 1–590.

Santesson, R., Moberg, R., Nordin, A., Tønsberg, T. and Vitikainen, O. (2004). *Lichen-
    forming and Lichenicolous Fungi of Fennoscandia*. Uppsala: Museum of Evolution,
    Uppsala University.

Sanz, M.-J., Gries, C. and Nash, T. H., III (1992). Dose-response relationships for SO_2
    fumigations in the lichens *Evernia prunastri* (L.) Ach. and *Ramalina fraxinea* (L.) Ach.
    New Phytologist, **122**, 313–319.

Schaper, T. and Ott, S. (2003). Photobiont selectivity and interspecific interactions in
    lichen communities. I. Culture experiments with the mycobiont *Fulgensia
    bracteata*. *Plant Biology*, **5**, 441–450.

Scheidegger, C. (1985). Systematische Studien zur Krustenflechte *Anzina carneonivea*
    (Trapeliacae, Lecanorales). *Nova Hedwigia*, **41**, 191–218.

Scheidegger, C. (1993). A revision of saxicolous species of the genus *Buellia* De Not. and
    formerly included genera in Europe. *Lichenologist*, **25**, 315–364.

Scheidegger, C. (1994a). Low-temperature scanning electron microscopy: the
    localization of free and perturbed water and its role in the morphology of the
    lichen symbionts. *Cryptogamic Botany*, **4**, 290–299.

Scheidegger, C. (1994b). Reproductive strategies in *Vezdaea* (Lecanorales, lichenized
    Ascomycetes): a low-temperature scanning electron microscopy study of a
    ruderal species. *Cryptogamic Botany*, **5**, 163–171.

Scheidegger, C. (1998). *Erioderma pedicellatum*: a critically endangered lichen species.
    Species, **30**, 68–69.

Scheidegger, C. and Schroeter, B. (1995). Effects of ozone fumigation on epiphytic
    macrolichens: ultrastructure, CO_2 gas exchange and chlorophyll fluorescence.
    Environmental Pollution, **88**, 345–354.

Scheidegger, C., Schroeter, B. and Frey, B. (1995a). Structural and functional processes
    during water vapour uptake and desiccation in selected lichens with green algal
    photobionts. *Planta*, **197**, 399–409.

Scheidegger, C., Wolseley, P. A. and Thor, G., eds., (1995b). Conservation biology of
    lichenised fungi. *Mitteilungen der Eidgenössischen Forschungsanstalt für Wald, Schnee
    und Landschaft*, **70**, 1–173.

Scherrer, S. and Honegger, R. (2003). Inter- and intraspecific variation of homologous
    hydrophobin (H1) gene sequences among *Xanthoria* spp. (lichen-forming
    ascomycetes). *New Phytologist*, **158**, 375–389.

Scherrer, S., de Vries, O. M. H., Dudler, R., Wessels, J. G. H. and Honegger, R. (2000).
    Interfacial self-assembly of fungal hydrophobins of the lichen-forming ascomycetes
    Xanthoria parietina and *X. ectaneoides*. *Fungal Genetics and Biology*, **30**, 81–93.

Scherrer, S., Haisch, A. and Honegger, R. (2002). Characterization and expression of XPH1, the hydrophobin gene of the lichen-forming ascomycete *Xanthoria parietina*. *New Phytologist*, **154**, 175–184.

Scherrer, S., Zippler, U. and Honegger, R. (2005). Characterisation of the mating-type locus in the genus *Xanthoria* (lichen-forming ascomycetes, Lecanoromycetes). *Fungal Genetics and Biology*, **42**, 976–988.

Schlee, D., Kandzia, R., Tintemann, H. and Türk, R. (1995). Activity of superoxide dismutase and malondialdehyde content in lichens along an altitude profile. *Phyton*, **35**, 233–242.

Schlensog, M., Pannewitz, S., Green, T. G. A. and Schroeter, B. (2004). Metabolic recovery of continental antarctic cryptogams after winter. *Polar Biology*, **27**, 399–408.

Schlensog, M., Schroeter, B. and Green, T. G. A. (2000). Water dependent photosynthetic activity of lichens from New Zealand: differences in the green algal and the cyanobacterial thallus parts of photosymbiodemes. *Bibliotheca Lichenologica*, **75**, 149–160.

Schlensog, M., Schroeter, B., Sancho, L. G., Pintado, A. and Kappen, L. (1997). Effect of strong irradiance on photosynthetic performance of the melt-water dependent cyanobacterial lichen *Leptogium puberulum* (Collemataceae) Hue from the maritime Antarctic. *Bibliotheca Lichenologica*, **67**, 235–246.

Schmitt, I. and Lumbsch, H. T. (2004). Molecular phylogeny of the Pertusariaceae supports secondary chemistry as an important systematic character set in lichen-forming ascomycetes. *Molecular Phylogenetics and Evolution*, **33**, 43–55.

Schmitt, I., Lumbsch, H. T. and Søchting, U. (2003). Phylogeny of the lichen genus *Placopsis* and its allies based on Bayesian analyses of nuclear and mitochondrial sequences. *Mycologia*, **95**, 827–835.

Schmitt, I., Martin, M. P., Kautz, S. and Lumbsch, T. H. (2005a). Diversity of non-reducing polyketide synthase genes in the Pertusariales (lichenized Ascomycota). A phylogenetic perspective. *Phytochemistry*, **66**, 1241–1253.

Schmitt, I., Messuti, M. I., Feige, G. B. and Lumbsch, H. T. (2001). Molecular data support rejection of the generic concept in the Coccotremataceae (Ascomycota). *Lichenologist*, **33**, 315–321.

Schmitt, I., Mueller, G. M. and Lumbsch, H. T. (2005b). Ascoma morphology is homoplaseous and phylogenetically misleading in some pyrenocarpous lichens. *Mycologia*, **97**, 362–374.

Schmitt, I., Yamamoto, Y. and Lumbsch, H T. (2006). Phylogeny of Pertusariales (Ascomycotina): resurrection of Ochrolechiaceae and a new circumscription of Megasporaceae. *Journal of the Hattori Botanical Laboratory*, **100**, 753–764.

Schmull, M. and Hauck, M. (2003). Element microdistribution in the bark of *Abies balsamea* and *Picea rubens* and its impact on epiphytic lichen abundance on Whiteface Mountain, New York. *Flora*, **198**, 293–303.

Schreiber, U., Bilger, W. and Neubauer C. (1994). Chlorophyll fluorescence as a nonintrusive indicator for rapid assessment of in vivo photosynthesis. In

Ecophysiology of Photosynthesis ed. E.-D. Schulze and M. M. Caldwell, pp. 49–70. Ecological Studies 100. Berlin: Springer.

Schroeter, B. and Scheidegger, C. (1995). Water relations in lichens at subzero temperatures: structural changes and carbon dioxide exchange in the lichen *Umbilicaria aprina* from continental Antarctica. *New Phytologist*, **131**, 273–285.

Schroeter, B., Green, T. G. A., Kappen, L. and Seppelt, D. (1994). Carbon dioxide exchange at subzero temperatures. Field measurements on *Umbilicaria aprina* in Antarctica. *Cryptogamic Botany*, **4**, 233–241.

Schroeter, B., Kappen, L., Schulz, F. and Sancho, L. G. (2000). Seasonal variation in the carbon balance of lichens in the maritime Antarctic: long-term measurements of photosynthetic activity in *Usnea aurantiaco-atra*. In *Antarctic Ecosystems: Model for Wider Ecological Understanding*, ed. W. Davison, C. Howard-Williams and P. Broady, pp. 220–224. Christchurch: Caxton Press.

Schroeter, B., Schulz, F. and Kappen, L. (1997). Hydration-related spatial and temporal variation of photosynthetic activity in Antarctic lichens. In *Antarctic Communities: Species, Structure and Survival*, ed. B. Battaglia, J. Valencia and D. W. H. Walton, pp. 221–225. Cambridge: Cambridge University Press.

Schulz, M. (1995). Protein and ubiquitin conjugate patterns of *Peltigera horizontalis* (Huds.) Baumg. during desiccation and rehydration. In *Flechten Follmann. Contributions to Lichenology in Honour of Gerhard Follmann*, ed. F. J. A. Daniels, M. Schultz and J. Peine, pp. 87–96. Cologne: Botanical Institute, University of Cologne.

Schulze, E.-D., Beck, E. and Müller-Hohenstein, K. (2002). *Pflanzenökologie*. Heidelberg: Spektrum Akademischer.

Schulze, E.-D. and Chapin, F. S., III (1987). Plant specialization to environments of different resource availability. In *Potentials and Limitations of Ecosystem Analysis*, ed. E.-D. Schulze and H. Zwolfer, pp. 120–148. Berlin: Springer.

Schüssler, A. (2002). Molecular phylogeny, taxonomy, and evolution of *Geosiphon pyriformis* and arbuscular mycorrhizal fungi. *Plant and Soil*, **244**, 75–83.

Schüssler, A., Schnepf, E., Mollenhauer, D. and Kluge, M. (1995). The fungal bladders of the endocyanosis *Geosiphon pyriforme*, a *Glomus*-related fungus: cell wall permeability indicates a limiting pore radius of only 0.5 nm. *Protoplasma*, **185**, 131–139.

Schüssler, A., Schwarzott, D. and Walker, C. (2001). A new fungal phylum, the Glomeromycota: phylogeny and evolution. *Mycological Research*, **105**, 1413–1421.

Schuster, G., Ott, S. and Jahns, H. M. (1985). Artificial cultures of lichens in the natural environment. *Lichenologist*, **17**, 247–253.

Schwartzman, D. W. and Volk, T. (1989). Biotic enhancement of weathering and the habitability of Earth. *Nature*, **340**, 457–460.

Schwendener, S. (1867). Über die wahre Natur der Flechtengonidien. *Verhandlungen der schweizerischen naturforschenden Gesellschaft*, **57**, 9–11.

Schwendener, S. (1869). *Die Algentypen der Flechtengonidien*. Basel: Schultze.

Scotter, G. W. (1965). Chemical composition of forage lichens from northern Saskatchewan as related to use by barren-ground caribou in the taiga of northern Canada. *Canadian Journal of Plant Sciences*, **45**, 246–250.

Seaward, M. R. D. (1973). Lichen ecology of the Scunthorpe heathlands. I. Mineral accumulation. *Lichenologist*, **5**, 423–433.

Seaward, M. R. D. (1976). Performance of *Lecanora muralis* in an urban environment. In *Lichenology: Progress and Problems*, ed. D. H. Brown, D. L. Hawksworth and R. H. Bailey, pp. 323–357. London: Academic Press.

Seaward, M. R. D. (1982a). Lichen ecology of changing urban environments. In *Urban Ecology*, ed. R. Bornkamm, J. A. Lee and M. R. D. Seaward, pp. 181–189. Oxford: Blackwell Scientific.

Seaward, M. R. D. (1982b). Principles and priorities of lichen conservation. *Journal of the Hattori Botanical Laboratory*, **52**, 401–406.

Seaward, M. R. D. (1988). Contribution of lichens to ecosystems. In *CRC Handbook of Lichenology*, Vol. 2, ed. M. Galun, pp. 107–129. Boca Raton: CRC Press.

Seaward, M. R. D. (1993). Lichens and sulphur dioxide air pollution: field studies. *Environmental Reviews*, **1**, 73–91.

Seaward, M. R. D. (1996a). Checklist of Tunisian lichens. *Bocconea*, **6**, 115–148.

Seaward, M. R. D. (1996b). Lichens and the environment. In *A Century of Mycology*, ed. B. C. Sutton, pp. 293–320. Cambridge: Cambridge University Press.

Seaward, M. R. D. (1997). Urban deserts bloom: a lichen renaissance. *Bibliotheca Lichenologica*, **67**, 297–309.

Seaward, M. R. D. (ed.) (1998). *Lichen Atlas of the British Isles*. London: British Lichen Society.

Seaward, M. R. D. (2004). The use of lichens for environmental impact assessment. *Symbiosis*, **37**, 293–305.

Seaward, M. R. D. and Aptroot, A. (2003). Lichens of Silhouette Island (Seychelles). *Bibliotheca Lichenologica*, **86**, 423–439.

Seaward, M. R. D. and Coppins, B. J. (2004). Lichens and hypertrophication. *Bibliotheca Lichenologica*, **88**, 561–572.

Seaward, M. R. D. and Edwards, H. G. M. (1997). Biological origin of major chemical disturbances on ecclesiastical architecture studied by Fourier Transform Raman spectroscopy. *Journal of Raman Spectroscopy*, **28**, 691–696.

Seaward, M. R. D. and Letrouit-Galinou, M. (1991). Lichens return to the Jardin du Luxembourg after an absence of almost a century. *Lichenologist*, **23**, 181–186.

Sedelnikova, N. V. and Cheremisin, D. V. (2001). The use of lichens for dating of petroglyphs. *Siberian Journal of Ecology*, **8**, 479–481.

Sensen, M. and Richardson, D. H. S. (2002). Mercury levels in lichens from different host trees around a chlor-alkali plant in New Brunswick, Canada. *Science of the Total Environment*, **293**, 31–45.

Sérusiaux, E. (1985). Goniocysts, goniocystangia and *Opegrapha lambinonii* and related species. *Lichenologist*, **17**, 1–25.

Sérusiaux, E. (1986). The nature and origin of campylidia in lichenized fungi. *Lichenologist*, **18**, 1–35.

Sérusiaux, E. (1989). *Liste Rouge des Macrolichens dans la Communauté Européenne*. Liège: Centre des Recherches sur les Lichens.

Seymour, F. A., Crittenden, P. D., Dickinson, M. J., *et al.* (2005*a*). Breeding systems in the lichen-forming fungal genus *Cladonia*. *Fungal Genetics and Biology*, **42**, 554–563.

Seymour, F. A., Crittenden, P. D. and Dyer, P. S. (2005*b*). Sex in the extremes: lichen-forming fungi. *Mycologist*, **19**, 51–58.

Sharma, P., Bergman, B., Hallbom, L. and Hofsten, A. (1982). Ultrastructural changes of *Nostoc* of *Peltigera canina* in presence of SO_2. *New Phytologist*, **92**, 573–579.

Sheridan, R. P. (1979). Impact of emissions from coal-fired electricity generating facilities on N_2-fixing lichens. *Bryologist*, **82**, 54–58.

Shibata, S. (1973). Some aspects of lichen chemotaxonomy. In *Chemistry in Botanical Classification*, Vol. 25, ed. G. Bendz and J. Santesson, pp. 241–249. Stockholm: Nobel Symposia: Medicine and Natural Sciences.

Shibata, S. (1992). Studies on some lichen metabolites and their development. *Journal of Japanese Botany*, **67**, 63–71.

Shibuya, M., Ebizuka, Y., Noguchi, H., Iitaka, Y. and Sankawa, U. (1983). Inhibition of prostaglandin biosynthesis by 4-O-methylcryptochlorophaeic acid: synthesis of monomeric arylcarboxylic acids for inhibitory activity testing and X-ray analysis of 4-O-methylcryptochlorophaeic acid. *Chemical Pharmaceutical Bulletin of Tokyo*, **31**, 407–413.

Shuvalov, V. A. and Heber, U. (2003). Photochemical reactions in dehydrated photosynthetic organisms, leaves, chloroplasts, and photosystem II particles: reversible reduction of pheophytin and chlorophyll and oxidation of β-carotene. *Chemical Physics*, **294**, 227–237.

Sigal, L. L. and Johnston, J. W., Jr. (1986). Effects of acidic rain and ozone on nitrogen fixation and photosynthesis in the lichen *Lobaria pulmonaria* (L.) Hoffm. *Environmental and Experimental Botany*, **26**, 59–64.

Sigal, L. L. and Nash, T. H., III (1983). Lichen communities on conifers in southern California: an ecological survey relative to oxidant air pollution. *Ecology*, **64**, 1343–1354.

Sigal, L. L. and Taylor, O. C. (1979). Preliminary studies of the gross photosynthetic response of lichens to peroxyacetylnitrite fumigations. *Bryologist*, **82**, 564–575.

Silberstein, L., Siegel, B. Z., Siegel, S. M., Mukhtar, A. and Galun, M. (1996*a*). Comparative studies on *Xanthoria parietina*, a pollution-resistant lichen, and *Ramalina duriaei*, a sensitive species. I. Effects of air pollution on physiological processes. *Lichenologist*, **28**, 355–365.

Silberstein, L., Siegel, B. Z., Siegel, S. M., Mukhtar, A. and Galun, M. (1996*b*). Comparative studies on *Xanthoria parietina*, a pollution-resistant lichen, and *Ramalina duriaei*, a sensitive species. II. Evaluation of possible air pollution-protection mechanisms. *Lichenologist*, **28**, 367–383.

Sillett, S. C. and Goslin, M. N. (1999). Distribution of epiphytic macrolichens in relation to remnant trees in a multiple-age Douglas-fir forest. *Canadian Journal of Forest Research*, **29**, 1204–1215.

Sillett, S. C., McCune, B., Peck, J. E., Rambo, T. R. and Ruchty, A. (2000). Dispersal limitations of epiphytic lichens result in species dependent on old-growth forests. *Ecological Applications*, **10**, 789–799.

Simpson, T. J. (1995). Polyketide biosynthesis. *Chemistry and Industries*, **1995**, 407–411.

Sinnemann, S. J., Andrésson, Ó. S., Brown, D. W. and Miao, V. P. (2000). Cloning and heterologous expression of *Solorina crocea* pyrG. *Current Genetics*, **37**, 333–338.

Sipman, H. J. M. (1994). Foliicolous lichens on plastic tape. *Lichenologist*, **26**, 311–312.

Sipman, H. J. M. (1997). Observations on the foliicolous lichen and bryophyte flora in the canopy of a semi-deciduous tropical forest. *Abstracta Botanica*, **21**, 153–161.

Sipman, H. J. M. (2002). The significance of the northern Andes for lichens. *Botanical Review*, **68**, 88–99.

Sipman, H. J. M. (2006a). Diversity and biogeography of lichens in neotropical montane oak forests. In *Ecology and Conservation of Neotropical Oak Forests*, ed. M. Kappelle, pp. 69–81. Ecological Studies 185. Berlin: Springer.

Sipman, H. J. M. (2006b). Identification key and literature guide to the genera of lichenized fungi (Lichens) in the Neotropics (provisional version). Online: www.bgbm.org/sipman/keys/neokeyA.htm.

Sipman, H. J. M. and Harris, R. C. (1989). Lichens. In *Tropical Rain Forest Ecosystems*, ed. H. Lieth and M. J. A. Werger, pp. 303–309. Amsterdam: Elsevier.

Skujins, J. and Klubek, B. (1978). Nitrogen fixation and cycling by blue-green algae-lichen-crusts in arid rangeland soils. *Ecological Bulletin (Stockholm)*, **26**, 164–171.

Skulachev, V. P. (1998). Uncoupling: new approaches to an old problem of bioenergetics. *Biochimica Biophysica Acta*, **1363**, 100–124.

Skult, H. (1984). The *Parmelia omphaloides* (Ascomycetes) complex in Eastern Fennoscandia. *Annales Botanici Fennici*, **21**, 117–142.

Skye, E. (1968). Lichens and air pollution: a study of cryptogamic epiphytes and environment in the Stockholm region. *Acta Phytogeographica Suecica*, **52**, 8–123.

Slocum, R. D., Ahmadijan, V. and Hildreth, K. C. (1980). Zoosporogenesis in *Trebouxia gelatinosa*: ultrastrucutral potential for zoospore release and implications for the lichen association. *Lichenologist*, **12**, 173–187.

Sluiman, H. J. (1989). The green algal class Ulvophyceae – an ultrastructural survey and classification. *Cryptogamic Botany*, **1**, 83–94.

Sluiman, H. J., Kouwets, F. A. C. and Blommers, P. C. J. (1989). Classification and definition of cytokinetic patterns in Green Algae: sporulation versus (vegetative) cell division. *Archiv für Protistenkunde*, **137**, 277–90.

Smith, D. C. and Douglas, A. (1987). *The Biology of Symbiosis*. London: Edward Arnold.

Smith, D. C. and Molesworth, S. (1973). Lichen physiology. XIII. Effects of rewetting of dry lichens. *New Phytologist*, **72**, 525–533.

Smith, E. C. and Griffiths, H. (1996). The occurrence of the chloroplast pyrenoid is correlated with the activity of a CO_2 concentrating mechanism and carbon isotope discrimination in lichens and bryophytes. *Planta*, **198**, 6–16.

Snelgar, W. P. (1981). The ecophysiology of New Zealand forest lichens with special reference to carbon dioxide exchange. Ph.D. thesis, Waikato University, Hamilton, New Zealand.

Snelgar, W. P. and Green, T. G. A. (1980). Carbon dioxide exchange in lichens. II. Low carbon dioxide compensation levels and lack of apparent photorespiratory activity in some lichens. *Bryologist*, **83**, 505–507.

Snelgar, W. P. and Green, T. G. A. (1981). Carbon dioxide exchange in lichens: apparent photorespiration and the possible role of CO_2 refixation in some members of the Stictaceae (Lichenes). *Journal of Experimental Botany*, **32**, 661–668.

Snelgar, W. P. and Green, T. G. A. (1982). Growth rates of Stictaceae lichens in New Zealand beech forests. *Bryologist*, **85**, 301–306.

Snelgar, W. P., Green, T. G. A. and Beltz, C. K. (1981a). Carbon dioxide exchange in lichens: estimation of internal thallus CO_2 transport resistances. *Physiologia Plantarum*, **52**, 417–422.

Snelgar, W. P., Green, T. G. A. and Wilkins, A. L. (1981b). Carbon dioxide exchange in lichens. I. Resistances to CO_2 uptake at different thallus water contents. *New Phytologist*, **88**, 353–361.

Søchting, U. (1995). Lichens as monitors of nitrogen deposition. *Cryptogamic Botany*, **5**, 264–269.

Søchting, U. and Johnsen, I. (1978). Lichen transplants as biological indicators of SO_2 air pollution in Copenhagen. *Bulletin of Environmental Contamination and Toxicology*, **19**, 1–7.

Solhaug, K. A. and Gauslaa, Y. (1996). Parietin, a photoprotective secondary product of the lichen *Xanthoria parietina*. *Oecologia*, **108**, 412–418.

Solhaug, K. A. and Gauslaa, Y. (2004a). Testing of ecological roles of secondary lichen compounds. In *Lichens in Focus*, ed. T. Randlane and A. Saag, p. 40. Tartu: Tartu University Press.

Solhaug, K. A. and Gauslaa, Y. (2004b). Photosynthates stimulate the UV-B induced fungal anthraquinone synthesis in the foliose lichen *Xanthoria parietina*. *Plant Cell and Environment*, **27**, 167–176.

Solhaug, K. A., Gauslaa, Y., Nybakken, L. and Bilger, W. (2003). UV-induction of sun-screening pigments in lichens. *New Phytologist*, **158**, 91–100.

Solomina, O. and Calkin, P. E. (2003). Lichenometry as applied to moraines in Alaska, U.S.A., and Kamchatka, Russia. *Arctic, Antarctic, and Alpine Research*, **35**, 129–143.

Sommerkorn, M. (2000). The ability of lichens to benefit from natural CO_2 enrichment under a spring snow-cover: a study with two arctic-alpine species from contrasting habitats. *Bibliotheca Lichenologica*, **75**, 365–380.

Sonesson, M., Osborne, C. and Sandberg, G. (1994). Epiphytic lichens as indicators of snow depth. *Arctic and Alpine Research*, **26**, 159–165.

Sonesson, M., Schipperges, B. and Carlsson, B. Å. (1992). Seasonal patterns of photosynthesis in alpine and subalpine populations of the lichen *Nephroma arcticum*. *Oikos*, **65**, 3–12.

Spiro, B., Morrisson, J. and Purvis, O. W. (2002). Sulphur isotopes in lichens as indicators of sources. In *Monitoring with Lichens – Monitoring Lichens*. Nato Science Series IV: Earth and Environmental Sciences, ed. P. L. Nimis, C. Scheidegger and P. A. Wolseley, pp. 311–315. Dordrecht: Kluwer Academic.

St. Clair, L. L. and Seaward, M. R. D., eds. (2004). *Biodeterioration of Stone Surfaces*. Dordrecht: Kluwer.

St. Clair, L. L., Johansen, J. R. and Rushforth, S. R. (1993). Lichens of soil crust communities in the intermountain area of the western United States. *Great Basin Naturalist*, **53**, 5–12.

Staiger, B. (2002). Die Flechtenfamilie Graphidaceae. Studien in Richtung einer natürlicheren Gliederung. *Bibliotheca Lichenologica*, **85**, 1–526.

Staiger, B., Kalb, K. and Grube, M. (2006). Phylogeny and phenotypic variation in the lichen family Graphidaceae (Ostropomycetidae, Ascomycota). *Mycological Research*, **110**, 765–772.

Stålfelt, M. G. (1939). Der Gasaustausch der Flechten. *Planta*, **29**, 11–31.

Stamatakis, A. (2006). RAxML-VI-HPC: maximum likelihood-based phylogenetic analyses with thousands of taxa and mixed models. *Bioinformatics*, **22**, 2688–2690.

Stamp, N. (2004). Can the growth-differentiation balance hypothesis be tested rigorously? *Oikos*, **107**, 439–448.

Steinkötter, J., Bhattacharya, D., Semmelroth, I., Bibeau, C. and Melkonian, M. (1994). Prasinophytes form independent lineages within the Chlorophyta: evidence from ribosomal RNA sequence comparisons. *Journal of Phycology*, **30**, 340–345.

Steinnes, E. and Krog, H. (1977). Mercury, arsenic and selenium fall-out from an industrial complex studied by means of lichen transplants. *Oikos*, **28**, 160–164.

Stenroos, S. (1993). Taxonomy and distribution of the lichen family Cladoniaceae in the Antarctic and peri-Antarctic regions. *Cryptogamic Botany*, **3**, 310–344.

Stenroos, S. and DePriest, P. T. (1998). SSU rDNA phylogeny of cladoniiform lichens. *American Journal of Botany*, **85**, 1548–1559.

Stenroos, S., Feuerer, T. and Ahti, T. (2002a). Chilean Cladoniaceae online. *Mitteilungen aus dem Institut für Allgemeine Botanik Hamburg*, **30-32**, 241–251.

Stenroos, S., Hyvönen, J., Myllys, L, Thell, A. and Ahti, T. (2002b). Phylogeny of the genus *Cladonia* s. lat. (Cladoniaceae, Ascomycetes) inferred from molecular, morphological, and chemical data. *Cladistics*, **18**, 237–278.

Stenroos, S., Stocker-Wörgötter, E., Yoshimura, I., *et al.* (2003). Culture experiments and DNA sequence data confirm the identity of *Lobaria* photomorphs. *Canadian Journal of Botany*, **81**, 232–247.

Stewart, W. D. P. (1980). Some aspects of structure and function in N_2-fixing cyanobacteria. *Annual Reviews in Microbiology*, **34**, 497–536.

Stewart, W. D. P. and Rodgers, G. A. (1978). Studies on the symbiotic blue-green algae of *Anthoceros*, *Blasia*, and *Peltigera*. In *Environmental Role of Nitrogen-fixing Blue-green Algae and Asymbiotic Bacteria*, ed. U. Granhall, pp. 247–259. Stockholm: Swedish Natural Science Research Council.

Stewart, W. D. P. and Rowell, P. (1975). Effects of L-methionine DL-sulphoximine on the assimilation of newly fixed NH_3, C_2H_2 reduction and heterocyst production in *Anabaena cylindrica*. *Biochemical Biophysical Research Communications*, **65**, 846–856.

Stewart, W. D. P. and Rowell, P. (1977). Modifications of nitrogen-fixing algae in lichen symbioses. *Nature (London)*, **265**, 371–372.

Stewart, W. D. P., Fitzgerald, G. P. and Burris, R. H. (1967). *In situ* studies on N_2 fixation using the acetylene reduction technique. *Proceedings of the National Academy of Sciences, USA*, **58**, 2071–2088.

Stocker, O. (1927). Physiologische und ökologische Untersuchungen an Laub- und Strauchflechten. *Flora*, **21**, 334–415.

Stocker-Wörgötter, E. (1995). Experimental cultivation of lichens and lichen symbionts. *Canadian Journal of Botany*, **73**, S579–S589.

Stocker-Wörgötter, E. (2001). Experimental lichenology and microbiology of lichens: culture experiments, secondary chemistry of cultured mycobionts, resynthesis and thallus morphogenesis. *Bryologist*, **104**, 576–581.

Stocker-Wörgötter, E. (2002*a*). Resynthesis of Photosymbiodemes. In *Protocols in Lichenology: Culturing, Biochemistry, Ecophysiology and Use in Biomonitoring* (Springer Lab Manual), ed. I. Kranner, R. P. Beckett and A. K. Varma, pp. 47–60. Berlin: Springer.

Stocker-Wörgötter, E. (2002*b*). Analysis of secondary compounds in cultured mycobionts. In *Protocols in Lichenology: Culturing, Biochemistry, Ecophysiology and Use in Biomonitoring* (Springer Lab Manual), ed. I. Kranner, R. P. Beckett and A. K. Varma, pp. 296–306. Berlin: Springer.

Stocker-Wörgötter, E. (2005). Approaches to a biotechnology of lichen-forming fungi: induction of polyketide pathways and chemosyndromes in axenically cultured mycobionts. *Recent Research Developments in Phytochemistry*, **9**, 115–131.

Stocker-Wörgötter, E. (2008). Metabolic diversity of lichen-forming ascomycetous fungi: culturing, polyketide and shikimate metabolite production, and PKS genes. *Natural Product Reports*, **25**, 188–200.

Stocker-Wörgötter, E. and Elix, J. A. (2004). Experimental studies of lichenized fungi: formation of rare depsides and dibenzofurans by the cultured mycobiont of *Bunodophoron patagonicum* (Sphaerophoraceae, lichenized Ascomycota). In *Contributions to Lichenology, Festschrift in Honour of Hannes Hertel*, ed. P. Döbbeler and G. Rambold, pp. 659–669. Bibliotheca Lichenologica 88. Berlin: J. Cramer.

Stocker-Wörgötter, E. and Elix, J. A. (2006). Morphogenetic strategies and induction of secondary metabolite biosynthesis in cultured lichen-forming Ascomycota, as exemplified by *Cladia retipora* (Labill.) Nyl. and *Dactylina arctica* (Richards.) Nyl. *Symbiosis*, **41**, 9–20.

Stocker-Wörgötter, E., Elix, J. A. and Grube, M. (2004). Secondary chemistry of lichen-forming fungi: chemosyndromic variation and DNA-analyses of cultures and chemotypes in the *Ramalina farinacea* complex. *Bryologist*, **107**, 152–162.

Stone, D. F. (1989). Epiphytic succession on *Quercus garryana* branches in the Willamette Valley of western Oregon. *Bryologist*, **92**, 81–94.

Stulen, I. and De Kok, L. J. (1993). Whole plant regulation of sulfur-metabolism – a theoretical approach and comparison with current ideas on regulation of nitrogen metabolism. In *Sulfur Nutrition and Assimilation in Higher Plants*, ed. L. J. De Kok, I. Stulen, H. Rennenberg, C. Brunold and W. E. Rauser, pp. 77–91. The Hague: SPB Academic Publishing.

Sugiyama, K., Kurokawa, S. and Okada, G. (1976). Studies on lichens as a bioindicator of air pollution. I. Correlation of distribution of *Parmelia tinctorum* with SO_2 air pollution. *Japanese Journal of Ecology*, **26**, 209–212.

Sun, H. J. and Friedmann, E. I. (2005). Communities adjust their temperature optima by shifting producer-to-consumer ratio, shown in lichens as models. I. Experimental verification. *Microbial Ecology*, **49**, 528–535.

Sun, H. J., DePriest, P. T., Gargas, A., Rossman, A. Y. and Friedmann, E. I. (2002). *Pestalotiopsis maculans*: a dominant parasymbiont in North American lichens. *Symbiosis*, **33**, 215–226.

Sundberg, B., Ekblad, A, Näsholm, T. and Palmqvist, K. (1999a). Lichen respiration in relation to active time, nitrogen and ergosterol concentrations. *Functional Ecology*, **13**, 119–125.

Sundberg, B., Lundberg, P., Ekblad, A. and Palmqvist, K. (1999b). *In vivo* ^{13}C NMR spectroscopy of carbon fluxes in four diverse lichens. In *Physiological Ecology of Lichen Growth*, by B. Sundberg. Ph.D. thesis paper VI. Umeå: Umeå University.

Sundberg, B., Näsholm, T. and Palmqvist, K. (2001). The effect of nitrogen on growth and key thallus components in the two tripartite lichens, *Nephroma arcticum* and *Peltigera aphthosa*. *Plant, Cell and Environment*, **24**, 517–527.

Sundberg, B., Palmqvist, K., Esseen, P.-A. and Renhorn, K.-E. (1997). Growth and vitality of epiphytic lichens. II. Modelling of carbon gain using field and laboratory data. *Oecologia*, **109**, 10–18.

Suryanarayanan, T. S., Thirunavukkarasu, N., Hariharan, G. N. and Balaji, P. (2005). Occurrence of non-obligate microfungi inside lichen thalli. *Sydowia*, **57**, 120–130.

Svenning, M. M., Eriksson, T. and Rasmussen, U. (2005). Phylogeny of symbiotic cyanobacteria within the genus *Nostoc* based on 16S rDNA sequence analyses. *Archiv für Mikrobiologie*, **183**, 19–26.

Swanson, A. and Fahselt, D. (1997). Effects of ultraviolet light on the polyphenolics of *Umbilicaria americana* Poelt & Nash. *Canadian Journal of Botany*, **75**, 284–289.

Syers, J. K. and Iskander, I. K. (1973). Pedogenetic significance of lichens. In *The Lichens*, ed. V. Ahmadjian and M. E. Hale, pp. 225–248. New York: Academic Press.

Szabo, I., Bergantino, E. and Giacometti, G. M. (2005). Light and oxygenic photosynthesis: energy dissipation as a protection mechanism against photo-oxidation. *EMBO Reports*, **6**, 629–634.

Takala, K., Kaurenen, P. and Olkkonen, H. (1978). Fluorine content of two lichen species in the vicinity of a fertilizer factory. *Annales Botanici Fennici*, **15**, 158–167.

Takala, K., Olkkonen, H., Ikonen, J., Jääskeläinen, J. and Puumalainen, P. (1985). Total sulphur contents of epiphytic and terricolous lichens in Finland. *Annales Botanici Fennici*, **22**, 91–100.

Tanabe, Y., Watanabe, M. M. and Sugiyama, J. (2002). Are Microsporidia really related to Fungi?: a reappraisal based on additional gene sequences from basal fungi. *Mycological Research*, **106**, 1380–1391.

Tapper, R. (1976). Dispersal and changes in the local distribution of *Evernia prunastri* and *Ramalina farinacea*. *New Phytologist*, **77**, 725–734.

Tarhanen, S., Holopainen, T. and Oksanen, J. (1997). Ultrastructural changes and electrolyte leakage from ozone fumigated epiphytic lichens. *Annals of Botany*, **80**, 611–621.

Taylor, O. C. (1969). Importance of peroxyacetyl nitrate (PAN) as a phytotoxic air pollutant. *Journal of the Air Pollution Control Association*, **19**, 347–351.

Taylor, R. J. and Bell, M. (1983). Effects of SO_2 on the lichen flora in an industrial area, northwest Whatcom County, Washington. *Northwest Science*, **57**, 157–166.

Taylor, T. N., Hass, H. and Kerp, H. (1997). A cyanolichen from the Lower Devonian Rhynie Chert. *American Journal of Botany*, **84**, 992–1004.

Taylor, T. N., Hass, H., Remy, W. and Kerp, H. (1995b). The oldest fossil lichen. *Nature*, **378**, 244.

Taylor, T. N., Remy, W., Hass, H. and Kerp, H. (1995a). Fossil arbuscular mycorrhizae from the early Devonian. *Mycologia*, **87**, 560–573.

Taylor, W. A., Free, C., Boyce, C., Helgemo, R. and Ochoada, J. (2004). SEM analysis of *Spongiophyton* interpreted as a fossil lichen. *International Journal of Plant Sciences*, **165**, 875–881.

Tehler, A. (1982). The species pair concept in lichenology. *Taxon*, **31**, 708–714.

Tehler, A. (1988). A cladistic outline of the Eumycota. *Cladistics*, **4**, 227–277.

Tehler, A. (1990). A new approach to the phylogeny of Euascomycetes with a cladistic outline of Arthoniales focussing on Roccellaceae. *Canadian Journal of Botany*, **68**, 2458–2592.

Tehler, A. (1995). Morphological data, molecular data and total evidence in phylogenetic analysis. *Canadian Journal of Botany*, **73**, S667–S676.

Tehler, A. (1996). Systematics, phylogeny and classification. In *Lichen Biology*, ed. T. H. Nash III, pp. 217–239. Cambridge: Cambridge University Press.

Tehler, A. and Irestedt, M. (2007). Parallel evolution of lichen growth forms in the family Roccellaceae (Arthoniales, Ascomycota). *Cladistics*, **23**, 432–454.

Tehler, A., Dahlkild, Å., Eldenäs, P. and Feige, G. B. (2004). The phylogeny and taxonomy of Macaronesian, European and Mediterranean *Roccella* (Roccellaceae, Arthoniales). *Symbolae Botanicae Upsalienses*, **34**, 405–428.

Tehler, A., Farris, J. S., Lipscomb, D. L. and Källersjö, M. (2000). Phylogenetic analyses of the fungi based on large rDNA data sets. *Mycologia*, **92**, 459–474.

Tehler, A., Little, D. P. and Farris, J. S. (2003). The full-length phylogenetic tree from 1551 ribosomal sequences of chitinous fungi, Fungi. *Mycological Research*, **107**, 901–916.

Tel-Or, E. and Stewart, W. D. P. (1976). Photosynthetic electron transport, ATP synthesis and nitrogenase activity in isolated heterocysts of *Anabaena cylindrica*. *Biochimica et Biophysica Acta*, **423**, 189–195.

Tel-Or, E. and Stewart, W. D. P. (1977). Photosynthetic components of activities of nitrogen-fixing isolated heterocysts of *Anabaena cylindrica*. *Proceedings of the Royal Society of London B*, **198**, 61–96.

Théau, J. and Duguay, C. R. (2003). Mapping lichen changes in the summer range of the George River caribou herd (Québec-Labrador, Canada) using Landsat imagery (1976–1998). *Rangifer*, **24**, 31–49.

Théau, J. and Duguay, C. R. (2004). Lichen mapping in the summer range of the George River caribou herd using Landsat TM imagery. *Canadian Journal of Remote Sensing*, **30**, 867–881.

Théau, J., Peddle, D. R. and Duguay, C. R. (2005). Mapping lichen in a caribou habitat of northern Quebec, Canada, using an enhancement-classification method and spectral mixture analysis. *Remote Sensing of Environment*, **94**, 232–243.

Thell, A. and Goward, T. (1996). The new cetrarioid genus *Kaernefeltia* and related groups in the Parmeliaceae (lichenized Ascomycotina). *Bryologist*, **99**, 125–136.

Thell, A., Feuerer, T. Kärnefelt, I., Myllys, L. and Stenroos, S. (2004a). Monophyletic groups within the Parmeliaceae identified by IST rDNA, β-tubulin and GAPDH sequences. *Mycological Progress*, **3**, 297–314.

Thell, A., Goward, T., Randlane, T., Kärnefewlt, E. I. and Saag, A. (1995). A revision of the North American lichen genus *Ahtiana* (Parmeliaceae). *Bryologist*, **98**, 596–605.

Thell, A., Randlane, T. and Saag, A. (2005). A new circumscription of the lichen genus *Nephromopsis* (Parmeliaceae, lichenized Ascomycetes). *Mycological Progress*, **4**, 303–316.

Thell, A., Westberg, M. and Kärnefelt, I. (2004b). Biogeography of the lichen family Parmeliaceae in the Nordic countries with taxonomic remarks. *Symbolae Botanicae Upsalienses*, **34**, 429–452.

Thomas, J., Wolk, C. P., Shaffer, P. W., Austin, S. M. and Galonsky, A. (1975). The initial organic product of fixation of ^{15}N-labeled nitrogen gas by the blue-green alga *Anabaena cylindrica*. *Biochemical Biophysical Research Communication*, **67**, 501–507.

Thomas, M. A. (1999). Effect of sulfur dioxide and ozone on glutathione reductase and superoxide dismutase in lichens. Ph.D. dissertation. Tempe: Arizona State University.

Thomas, M. A., Romagni, J. G., Gries, C. and Nash, T. H., III (1997). The effects of sulfur dioxide exposure on glutathione reductase activity in the cyanolichen *Peltigera canina*. *Supplement to Plant Physiology*, **114**, 57.

Thomson, J. W. (1972). Distributional patterns in American Arctic lichens. *Canadian Journal of Botany*, **50**, 1135–1156.

Tibell, L. (1984). A reappraisal of the taxonomy of Caliciales. *Nova Hedwigia, Beiheft*, **79**, 597–713.

Tibell, L. (1998). Practice and prejudice in lichen classification. *Lichenologist*, **30**, 439–453.

Tibell, L. (1999). Calicioid lichens and fungi. *Nordic Lichen Flora*, **1**, 20–71.

Tibell, L. (2001). Photobiont association and molecular phylogeny of the lichen genus *Chaenotheca*. *Bryologist*, **104**, 191–198.

Tibell, L. (2003). *Tholurna dissimilis* and generic delimitations in *Caliciaceae* inferred from nuclear IST and LSU rDNA phylogenies (Lecanorales, lichenized Ascomycetes). *Mycological Research*, **107**, 1403–1418.

Tibell, L. and Beck, A. (2001). Morphological variation, photobiont association and ITS phylogeny of *Chaenotheca phaeocephala* and *C. subroscida* (Coniocybaceae, lichenized Ascomycetes). *Nordic Journal of Botany*, **22**, 651–660.

Timdal, E. (1991). A monograph of the genus *Toninia* (Lecideaceae, Ascomycetes). *Opera Botanica*, **110**, 1–137.

Timdal, E. and Tønsberg, T. (2006). *Psoroma paleaceum* comb. nov. the only hairy *Psoroma* in northern Europe. *Graphis Scripta*, **18**, 54–57.

Timoney, K. P. and Marsh, J. E. (2004). Lichen trimlines in northern Alberta: establishment, growth rates, and historic water levels. *Bryologist*, **107**, 429–440.

Tomassini, F. D., Lavoie, P., Puckett, K. J., Nieboer, E. and Richardson, D. H. S. (1977). The effect of time of exposure to sulphur dioxide on potassium loss from and photosynthesis in the lichen, *Cladina rangiferina* (L.) Harm. *New Phytologist*, **79**, 147–155.

Tomassini, F. D., Puckett, K. J., Nieboer, E., Richardson, D. H. S. and Grace, B. (1976). Determination of copper, iron, nickel, and sulphur by x-ray fluorescence in lichens from the Mackenzie Valley, Northwest Territories, and the Sudbury District, Ontario. *Canadian Journal of Botany*, **54**, 1591–1603.

Tomitani, A., Knoll, A. H., Cavanaugh, C. M. and Ohna, T. (2006). The evolutionary diversification of cyanobacteria: molecular-phylogenetic and paleontological perspectives. *Proceedings of the National Academy of Sciences, USA*, **103**, 5442–5447.

Tønsberg, T. (1992). The sorediate and isidiate, corticolous, crustose lichens in Norway. *Sommerfeltia*, **14**, 1–331.

Tønsberg, T. and Holtan-Hartwig. J. (1983). Phycotype pairs in *Nephroma*, *Peltigera* and *Lobaria* in Norway. *Nordic Journal of Botany*, **3**, 681–688.

Topham, P. B. (1977). Colonization, growth, succession and competition. In *Lichen Ecology*, ed. M. R. D. Seaward, pp. 31–68. London: Academic Press.

Tormo, R., Recio, D., Silva, I. and Muñoz, A. (2001). A quantitative investigation of airborne algae and lichen soredia obtained from pollen traps in south-west Spain. *European Journal of Phycology*, **36**, 385–390.

Trembley, M. L., Ringli, C. and Honegger, R. (2002*a*). Differential expression of hydrophobins DGH1, DGH2 and DGH3 and immunolocalization of DGH1 in strata of the lichenized basidiocarp of *Dictyonema glabratum*. *New Phytologist*, **154**, 185–195.

Trembley, M. L., Ringli, C. and Honegger, R. (2002*b*). Hydrophobins DGH1, DGH2, and DGH3 in the lichen-forming basidiomycete *Dictyonema glabratum*. *Fungal Genetics and Biology*, **35**, 247–259.

Trembley, M. L., Ringli, C. and Honegger, R. (2002*c*). Morphological and molecular analysis of early stages in the resynthesis of the lichen *Baeomyces rufus*. *Mycological Research*, **106**, 768–776.

Tretiach, M. and Carpanelli, A. (1992). Chlorophyll content and morphology as factors influencing the photosynthetic rate of *Parmelia caperata*. *Lichenologist*, **24**, 81–90.

Triebel, D. and Rambold, G. (1988). *Cecidonia* und *Phacopsis* (Lecanorales): zwei lichenicole Pilzgattungen mit cecidogenen Arten. *Nova Hedwigia*, **47**, 279–309.

Tschermak, E. (1941). Untersuchungen über die Beziehungen von Pilz und Alge im Flechtenthallus. *Österreichische Botanische Zeitschrift*, **90**, 233–307.

Tschermak-Woess, E. (1976). Algal taxonomy and the taxonomy of lichens: the phycobiont of *Verrucaria adriatica*. In *Lichenology: Progress and Problems*, ed. D. H. Brown, D. L. Hawksworth and R. H. Bailey, pp. 79–87. Orlando: Academic Press.

Tschermak-Woess, E. (1978). *Myrmecia reticulata* as a phycobiont and free-living *Trebouxia* – the problem of *Stenocybe septata*. *Plant Systematics and Evolution*, **129**, 185–208.

Tschermak-Woess, E. (1980). *Chaenothecopsis consociata* – kein parasitischer oder parasymbiotischer Pilz, sondern lichenisiert mit *Dictyochloropsis symbiontica*, spec. nova. *Plant Systematics and Evolution*, **136**, 287–306.

Tschermak-Woess, E. (1988). The algal partner. In *CRC Handbook of Lichenology*, Vol. 1, ed. M. Galun, pp. 39–92. Boca Raton: CRC Press.

Tschermak-Woess, E. and Poelt, J. (1976). *Vezdea*, a peculiar lichen genus, and its phycobiont. In *Lichenology: Progress and Problems*, ed. D. H. Brown, D. L. Hawksworth and R. H. Bailey, pp. 89–105. Orlando: Academic Press.

Tuba, Z., Csintalan, Z., Szente, K., Nagy, Z. and Grace, J. (1998). Carbon gains by desiccation-tolerant plants at elevated CO_2. *Functional Ecology*, **12**, 39–44.

Tuba, Z., Proctor, M. C. F. and Takács, Z. (1999). Desiccation-tolerant plants under elevated air CO_2: a review. *Zeitschrift für Naturforschung, Section C*, **54**, 788–796.

Tupa, D. D. (1974). An investigation of certain chaetophoralean algae. *Nova Hedwigia*, **46**, 1–155.

Turgeon, B. and Yoder, O. (2000). Proposed nomenclature for mating type genes in filamentous ascomycetes. *Fungal Genetics and Biology*, **31**, 1–5.

Türk, R. and Christ, R. (1980). Untersuchungen des CO_2-Gaswechsels von Flechtenexplantaten zur Indikation von SO_2-Belastung im Stadtgebiet von Salzburg. In *Bioindikation auf subzellularen und zellular Ebene (Bioindikation 2)*, ed. R. Schubert and J. Schuh, pp. 39–45. Halle-Wittenberg: Martin-Luther-Universitat.

Türk, R. and Wirth, V. (1975). The pH dependence of SO_2 damage to lichens. *Oecologia*, **19**, 285–291.

Türk, R., Wirth, V. and Lange, O. L. (1974). CO_2-Gaswechsel-Untersuchungen zur SO_2-Resistenz von Flechten. *Oecologia*, **15**, 33–64.

Turner, S., Huang, T. C., and Chaw, S.-M. (2001). Molecular phylogeny of nitrogen-fixing unicellular cyanobacteria. *Botanical Bulletin of Academia Sinica*, **42**, 181–186.

Turner, W. B. and Aldridge, D. C. (1983). *Fungal Metabolites II*. London: Academic Press.

US EPA (1996). *Air Quality Criteria for Ozone and Related Photochemical Oxidants*. Vol. II. EPA/600/p-93/004bF. Research Triangle Park, N.C. National Center for Environmental Assessment, Office of Research and Development.

Vainio, E. A. (1890). Étude sur classification naturelle et morphologie des lichens du Brésil. *Acta Societatis pro Fauna et Flora Fennica, Helsinki*, **7**, 1–256.

Valladares, F., Sancho, L. G. and Ascaso, C. (1997). Water storage in the lichen family Umbilicariaceae. *Botanica Acta*, **111**, 1–9.

van den Hoek, C., Jahns, H. M. and Mann, D. G. (1993). *Algen*, 3. Auflage. Stuttgart: Thieme.

van der Eerden, L., de Vries, W. and van Dobben, H. (1998). Effects of ammonia deposition on forests in The Netherlands. *Atmospheric Environment*, **32**, 525–532.

van Dobben, H. (1993) Vegetation as a monitor for deposition of nitrogen and acidity. Ph.D. dissertation, University of Utrecht.

van Dobben, H. F. (1996). Decline and recovery of epiphytic lichens in an agricultural area in The Netherlands (1900–1988). *Nova Hedwigia*, **62**, 477–485.

van Dobben, H. F. and de Bakker, A. J. (1996). Re-mapping epiphytic lichen biodiversity in The Netherlands: effects of decreasing SO_2 and increasing NH_3. *Acta Botanica Neerlandica*, **45**, 55–71.

van Dobben, H. F. and ter Braak, C. J. F. (1998). Effects of atmospheric NH_3 on epiphytic lichens in The Netherlands: the pitfalls of biological monitoring. *Atmospheric Environment*, **32**, 551–557.

van Dobben, H. F. and ter Braak, C. J. F. (1999). Ranking of epiphytic lichen sensitivity to air pollution using survey data: a comparison of indicator scales. *Lichenologist*, **31**, 27–39.

van Herk, C. M. (2002). Epiphytes on wayside trees as an indicator of eutrophication in the Netherlands. In *Monitoring with Lichens – Monitoring Lichens*, ed. P. L. Nimis, C. Scheidegger and P. A. Wolseley, pp. 285–289. Nato Science Series IV: Earth and Environmental Sciences. Dordrecht: Kluwer Academic.

van Herk, C. M., Aptroot, A. and van Dobben, H. F. (2002). Long-term monitoring in the Netherlands suggests that lichens respond to global warming. *Lichenologist*, **34**, 141–154.

Vězda, A. (1979). Flechtensystematische Studien. XI. Beiträge zur Kenntnis der Familie Asterothyriaceae (Discolichenes). *Folia Geobotanica Phytotaxonomica Bohemoslovaca, Praha*, **14**, 43–94.

Vězda, A. (1980). Foliikole Flechten aus Zaire. Die Arten der Sammelgattungen *Catillaria* und *Bacidia*. *Folia Geobotanica Phytotaxonomica Bohemoslovaca, Praha*, **15**, 75–94.

Villeneuve, J. P. and Holm, E. (1984). Atmospheric background of chlorinated hydrocarbons studied in Swedish lichens. *Chemosphere*, **13**, 1133–1138.

Virtala, M. (1992). Optimal harvesting of a plant–herbivore system: lichen and reindeer in northern Finland. *Ecological Modelling*, **60**, 233–255.

Vitikainen, O. (1998). Taxonomic notes on neotropical species of *Peltigera*. In *Lichenology in Latin America: History, Current Knowledge and Applications*, ed. M. P. Marcelli and M. R. D. Seaward, pp. 135–139. São Paulo: CETESB.

Vitikainen, O. (2001). William Nylander (1822–1899) and lichen chemotaxonomy. *Bryologist*, **104**, 263–267.

Vitousek, P. M., Walker, L. R., Whittaker, L. D., Mueller-Dombois, D. and Matson, P. A. (1987). Biological invasion by *Myrica faga* alters ecosystem development in Hawaii. *Science*, **238**, 802–804.

Vobis, G. and Hawksworth, D. L. (1981). Conidial lichen-forming fungi. In *The Biology of Conidial Fungi*, ed. G. T. Cole and B. Kendrick, pp. 245–273. New York: Academic Press.

Vogel, H. J. (1964). Distribution of lysine pathways among fungi: evolutionary implications. *American Naturalist*, **98**, 435–446.

Vogel, S. (1955). "Niedere Fensterpflanzen" in der südafrikanischen Wüste. Eine ökologische Schilderung. *Beiträge zur Biologie der Pflanzen*, **31**, 45–135.

von Arb, C., Mueller, C., Ammann, K. and Brunold, C. (1990). Lichen physiology and air pollution. II. Statistical analysis of the correlation between SO_2, NO_2, NO and O_3, and chlorophyll content, net photosynthesis, sulphate uptake and protein synthesis of *Parmelia sulcata* Taylor. *New Phytologist*, **115**, 431–437.

Vráblíková, H., McEvoy, M., Solhaug, K. A., Barták, M. and Gauslaa, Y. (2006). Annual variation in photoacclimation and photoprotection of the photobiont in the foliose lichen *Xanthoria parietina*. *Journal of Photochemistry and Photobiology B: Biology*, **83**, 151–162.

Wachtmeister, C. A. (1956). Identification of lichen acids by paper chromatography, *Botaniser Notiser*, **109**, 313–324.

Wainright, P. O., Hinkle, G., Sogin, M. L. and Stickel, S. K. (1993). Monophyletic origins of the Metazoa: an evolutionary link with fungi. *Science*, **260**, 340–342.

Walker, D. A., Webber, P. J., Everett, K. R. and Brown, J. (1978). Effects of crude and diesel oil spills on plant communities at Prudhoe Bay, Alaska, and the derivation of oil spill sensitivity maps. *Arctic*, **31**, 242–259.

Walker, F. J. (1985). The lichen genus *Usnea* subgenus *Neuropogon*. *Bulletin of the British Museum (Natural History)*, **13**, 1–130.

Walker, T. R., Crittenden, P. D. and Young, S. D. (2003). Regional variation in the chemical composition of winter snow pack and terricolous lichens in relation to sources of acid emissions in the Usa River basin, northeast European Russia. *Environmental Pollution*, **125**, 401–412.

Walser, J.-C., Holderegger, R., Gugerli, F., Hoebee, S. E. and Scheidegger, C. (2005). Microsatellites reveal regional population differentiation and isolation in *Lobaria pulmonaria*, an epiphytic lichen. *Molecular Ecology*, **14**, 457–468.

Walser, J.-C., Sperisen, C., Soliva, M. and Scheidegger, C. (2003). Fungus-specific microsatellite primers of lichens: application for the assessment of genetic variation on different spatial scales in *Lobaria pulmonaria*. *Fungal Genetics and Biology*, **40**, 72–82.

Warén, H. (1918–19) [1920]. Reinkulturen von Flechtengonidien. *Öfversigt af Finska Vetenskaps-Societetens Förhandlingar (Helsingfors)*, **61**, 1–79.

Waring, R. H. and Schlesinger, W. H. (1985). *Forest Ecosystems: Concepts and Management*. Orlando: Academic Press.

Warren, S. D. and Eldridge, D. J. (2001). Biological soil crusts and livestock in arid ecosystems: are they compatible?. In *Biological Soil Crusts: Structure, Function and Management*, ed. J. Belnap and O. L. Lange, pp. 401–415. Berlin: Springer.

Washburn, S. (2005 [2006]). The Epiphytic Macrolichens of the Greater Cincinnati, Ohio, Metropolitan Area. M.S. thesis. Cincinnati: University of Cincinnati.

Washburn, S. (2006). Ozone exposure indices correlated with lichen abundance data in greater Cincinnati metropolitan area, Ohio. Botany 2006 Abstracts, p. 51. St. Louis: Botanical Society of America.

Watanabe, A. (1960). List of algal strains in collection at the Institute of Applied Microbiology, University of Tokyo. *Journal of General and Applied Microbiology*, **6**, 283–292.

Waterbury, J. B. and Stanier, R. Y. (1978). Patterns of growth and development in pleurocapsalean cyanobacteria. *Microbiological Reviews*, **42**, 2–44.

Waters, J. M. and Craw, D. (2006). Goodby Gondwana? New Zealand biogeography, geology, and the problem of circularity. *Systematic Biology*, **55**, 351–356.

Weber, W. A. (1977). Environmental modification and lichen taxonomy. In *Lichen Ecology*, ed. M. R. D. Seaward, pp. 9–29. London: Academic Press.

Weber, W. A. (2003). The middle Asian element in the southern Rocky Mountain flora of the western United States: a critical biogeographical review. *Journal of Biogeography*, **30**, 649–685.

Wedin, M. (1995). The lichen family Sphaerophoraceae (Caliciales, Ascomycotina) in temperate areas of the Southern Hemisphere. *Symbolae Botanicae Upsalienses*, **31**, 1–102.

Wedin, M. and Döring, H. (1999). The phylogenetic relationship of the Spaerophoraceae, *Austropeltum* and *Neophyllis* (lichenized Ascomycota) inferred by SSU rDNA sequences. *Mycological Research*, **109**, 1131–1137.

Wedin, M. and Wiklund, E. (2004). The phylogenetic relationships of Lecanorales suborder Peltigerineae revisited. *Symbolae Botanicae Upsalienses*, **34**, 469–475.

Wedin, M., Döring, H. and Ekman, S. (2000a). Molecular phylogeny of the lichen families Cladoniaceae, Sphaerophoraceae, and Stereocaulaceae (Lecanorales, Ascomycotina). *Lichenologist*, **32**, 171–187.

Wedin, M., Döring, H. and Gilenstam, G. (2004). Saprotrophy and lichenization as options for the same fungal species on different substrata: environmental plasticity and fungal lifestyles in the *Stictis-Conotrema* complex. *New Phytologist*, **164**, 459–465.

Wedin, M., Döring, H., Könberg, K. and Gilenstam, G. (2005a). Generic delimitations in the family Stictidaceae (Ostropales, Ascomycota): the *Stictis-Conotrema* problem. *Lichenologist*, **37**, 67–75.

Wedin, M., Döring, H., Nordin, A. and Tibell, L. (2000b). Small subunit rDNA phylogeny shows the lichen families Caliciaceae and Physciaceae (Lecanorales, Ascomycotina) to form a monophyletic group. *Canadian Journal of Botany*, **78**, 246–254.

Wedin, M., Wiklund, E., Crewe, A., et al. (2005b). Phylogenetic relationships of Lecanoromycetes (Ascomycota) as revealed by analyses of mtSSU and nLSU rDNA sequence data. *Mycological Research*, **109**, 159–172.

Wedin, M., Wiklund, E. and Jørgensen, P. M. (2007). Massalongiaceae, fam. nov., an overlooked monophyletic group among the cyanobacterial lichens (Peltigerales, Lecanoromycetes, Ascomycota). *Lichenologist*, **39**, 61–67.

Wein, R. W. and Speer, J. E. (1975). Lichen biomass in Acadian and boreal forests of Cape Breton Island, Nova Scotia. *Bryologist*, **78**, 328–333.

Weissman, J. C. and Benemann, J. R. (1977). Hydrogen products by nitrogen-starved cultures of *Anabaena cylindrica*. *Applied Environmental Microbiology*, **33**, 123–131.

Weissman, L., Garty, J. and Hochman, A. (2005a). Rehydration of the lichen *Ramalina lacera* results in production of reactive oxygen species and nitric oxide and a decrease in antioxidants. *Applied and Environmental Microbiology*, **71**, 2121–2129.

Weissman, L., Garty, J. and Hochman, A. (2005b). Characterization of enzymatic antioxidants in the lichen *Ramalina lacera* and their response to rehydration. *Applied and Environmental Microbiology*, **71**, 6508–6514.

Werth, S., Wagner, H. H., Holderegger, J. M., Kalwij, J. and Scheidegger, C. (2005). Genetic diversity of an old-forest associated lichen is affected by stand-replacing disturbances. In *Dispersal and Persistence of an Epiphytic Lichen in a Dynamic Pasture-woodland Landscape*. Ph.D. dissertation, by S. Werth, pp. 23–48. Bern: Universität Bern.

Wessels, D. C. J. and Schoeman, P. (1988). Mechanism and rate of weathering of Clarens sandstone by an endolithic lichen. *South African Journal of Science*, **84**, 274–277.

Wessels, D. C. J. and Wessels, L. A. (1991). Erosion of biogenically weathered Clarens sandstone by lichenophagous bagworm larvae (Lepidoptera: Psychidae). *Lichenologist*, **23**, 283–291.

Wessels, J. G. H. (1999). Fungi in their own right. *Fungal Genetics and Biology*, **27**, 134–145.

Wetmore, C. M. (1973). Multiperforate septa in lichens. *New Phytologist*, **72**, 535–538.

White, F. J. and James, P. W. (1985). A new guide to microchemical techniques for the identification of lichen substances. *British Lichen Society Bulletin*, **57** (supp.), 1–41.

Whiteford, J. and Spanu, P. (2002). Hydrophobins and the interactions between fungi and plants. *Molecular Plant Pathology*, **3**, 391–400.

Whiton, J. C. and Lawrey, J. D. (1984). Inhibition of crustose lichen spore germination by lichen acids. *Bryologist*, **87**, 42–43.

Wiklund, E. and Wedin, M. (2003). The phylogenetic relationships of the cyanobacterial lichens in the Lecanorales suborder Peltigerineae. *Cladistics*, **19**, 419–431.

Will-Wolf, S. (1980). Structure of corticolous lichen communities before and after exposure to emissions from a "clean" coal-fired generating station. *Bryologist*, **83**, 281–295.

Will-Wolf, S. (2002). Monitoring regional status and trends in forest health with lichen communities: the United States Forest Service approach. In *Monitoring with Lichens – Monitoring Lichens*, ed. P. L. Nimis, C. Scheidegger and P. A. Wolseley, pp. 353–357. Nato Science Series IV: Earth and Environmental Sciences. Dordrecht: Kluwer Academic Publishers.

Wilmotte, A. and Golubić, S. (1991). Morphological and genetic criteria in the taxonomy of Cyanophyta/Cyanobacteria. *Algological Studies*, **64**, 1–24.

Winchester, V. and Harrison, S. (1994). A development of the lichenometric method applied to the dating of glacially influenced debris flows in southern Chile. *Earth Surface Processes and Landforms*, **19**, 137.

Winkworth, R. C., Wagstaff, S. J., Glenny, D. and Lockhart, P. J. (2002). Plant dispersal N.E.W.S. from New Zealand. *Trends in Ecology and Evolution*, **17**, 514–520.

Winkworth, R. C., Wagstaff, S. J., Glenny, D. and Lockhart, P. J. (2005). Evolution of the New Zealand mountain flora: origins, diversification and dispersal. *Organisms, Diversity and Evolution*, **5**, 237–247.

Winner, W. E., Atkinson, C. J. and Nash, T. H., III. (1988). Comparisons of SO_2 absorption capacities of mosses, lichens, and vascular plants in diverse habitats. *Bibliotheca Lichenologica*, **30**, 217–230.

Wirth, V. (1972). Die Silikatflechten – Gemeinschaften im ausseralpinen Zentraleuropa. *Dissertationes Botanicae*, **17**, 1–306.

Wirth, V. (1987). The influence of water relations on lichen SO_2-resistance. In *Progress and Problems in Lichenology in the Eighties*, ed. E. Peveling, pp. 347–350. Bibliotheca Lichenologica 25. Berlin-Stuttgart: J. Cramer.

Wirth, V. (1993). Trenwende bei der Ausbreitung der anthropogen geförderten Flechte *Lecanora conizaeoides*? *Phytocoenologia*, **23**, 625–636.

Wirth, V. (1995) *Die Flechten Baden-Würtembergs*. Verbreitungsatlas. Vols. I and II. Stuttgart: Eugen Ulmer.

Wirth, V. (2001). Zeiberwerte von Flechten. *Scripta Geobotanica*, **18**, 221–243.

Wirtz, N., Lumbsch, H. T., Green, T. G. A., *et al.* (2003). Lichen fungi have low cyanobiont selectivity in maritime Antarctica. *New Phytologist*, **160**, 177–183.

Wirtz, N., Printzen, C., Sancho, L. G. and Lumbsch, H. T. (2006). The phylogeny and classification of *Neuropogon* and *Usnea* (Parmeliaceae, Ascomycota) revisited. *Taxon*, **55**, 367–376.

Wise, M. J. and Tunnacliffe, A. (2004). POPP the question: what do LEA proteins do? *Trends in Plant Science*, **9**, 13–17.

Wolf, J. H. D. (1993). Diversity patterns and biomass of epiphytic bryophytes and lichens along an altitudinal gradient in the northern Andes. *Annals of the Missouri Botanical Garden*, **80**, 928–960.

Wolk, C. P., Ernst, A. and Elhai, J. (1994). Heterocyst metabolism and development. In *Molecular Genetics of Cyanobacteria*, ed. D. Bryant, pp. 769–823. Dordrecht: Kluwer Academic Publishing.

Wolken, G. J. (2006). High-resolution multispectral techniques for mapping former Little Ice Age terrestrial ice cover in the Canadian High Arctic. *Remote Sensing of Environment*, **101**, 104–114.

Wolken, G. J., England, J. H. and Dyke, A. S. (2005). Re-evaluating the relevance of vegetation trimlines in the Canadian Arctic as an indicator of Little Ice Age paleoenvironments. *Arctic*, **58**, 341–353.

Wolseley, P. A. (1995). A global perspective on the status of lichens and their conservation. *Mitteilungen der Eidgenössischen Forshungsanstalt für Wald, Schnee und Landschaft*, **70**, 11–22.

Wolseley, P. A. (2002). Using corticolous lichens of tropical forests to assess environmental changes. In *Monitoring with Lichens – Monitoring Lichens*, ed. P. L. Nimis, C. Scheidegger and P. A. Wolseley, pp. 373–378. Nato Science Series IV: Earth and Environmental Sciences. Dordrecht: Kluwer Academic Publishers.

Wolseley, P. A., James, P. W., Theobald, M. R. and Sutton, M. A. (2006). Detecting changes in epiphytic lichen communities at sites affected by atmospheric ammonia from agricultural sources. *Lichenologist*, **38**, 161–176.

Wösten, H. A. B. (2001). Hydrophobins: multipurpose proteins. *Annual Reviews in Microbiology*, **55**, 625–646.

Wösten, H. A. B., Devries, O. M. H. and Wessels, J. G. H. (1993). Interfacial self-assembly of a fungal hydrophobin into a hydrophobic rodlet layer. *Plant Cell*, **5**, 1567–1574.

Yahr, R., Vilgalys, R. and DePriest, P. T. (2004). Strong fungal specificity and selectivity for algal symbionts in Florida scrub *Cladonia* lichens. *Molecular Ecology*, **13**, 3367–3378.

Yahr, R., Vilgalys, R. and DePriest, P. T. (2006). Geographic variation in algal partners of *Cladonia subtenuis* (Cladoniaceae) highlights the dynamic nature of a lichen symbiosis. *New Phytologist*, **171**, 847–860.

Yamamoto, Y. (1990). *Studies of Cell Aggregates and the Production of Natural Pigments in Plant Cell Culture*. Osaka: Nippon Paint Publication.

Yamamoto, Y., Kinoshita, Y. and Yoshimura, I. (2002). Culture of thallus fragments and redifferentiation of lichens. In *Protocols in Lichenology: Culturing, Biochemistry, Ecophysiology and Use in Biomonitoring* (Springer Lab Manual), ed. I. Kranner, R. P. Beckett and A. K. Varma, pp. 34–46. Berlin: Springer.

Yoshimura, I. (1998). *Lobaria* in Latin America: taxonomic, geographic and evolutionary aspects. In *Lichenology in Latin America: History, Current Knowledge and Applications*, ed. M. P. Marcelli and M. R. D. Seaward, pp. 129–134. São Paulo: CETESB.

Yoshimura, I. and Arvidsson, L. (1994). Taxonomy and chemistry of the *Lobaria crenulata* group in Ecuador. *Acta Botanica Fennica*, **150**, 223–233.

Yoshimura, I., Yamamoto, Y., Nakano, T. and Finnie, J. (2002). Isolation and culture of lichen photobionts and mycobionts. *Protocols in Lichenology: Culturing, Biochemistry, Ecophysiology and Use in Biomonitoring* (Springer Lab Manual), ed. I. Kranner, R. P. Beckett and A. K. Varma, pp. 3–33. Berlin: Springer.

Young, K. (2005). Hardy lichen shown to survive in space. *New Scientist*, November 10th, 2005.

Yuan, X., Xiao, S. and Taylor, T. N. (2005). Lichen-like symbiosis 600 million years ago. *Science*, **308**, 1017–1020.

Yun, S., Berbee, M., Yoder, O. and Turgeon, B. (1999). Evolution of the fungal self-fertile reproductive life style from self-sterile ancestors. *Proceedings of the National Academy of Sciences, USA*, **96**, 5592–5597.

Yurchenko, E. and Golubkov, V. (2003). The morphology, biology, and geography of a necrotrophic basidiomycete *Athelia arachnoidea* in Belarus. *Mycological Progress*, **2**, 275–284.

Zachariassen, K. E. and Kristiansen, E. (2000). Ice nucleation and antinucleation in nature. *Cryobiology*, **41**, 257–279.

Zahlbruckner, A. (1907). *Die naturlichen Pflanzenfamilien, I Teil, 1 Apt*, ed. A. Engler and K. Prantl. Leipzig: Borntraeger.

Zahlbruckner, A. (1926). *Die naturlichen Pflanzenfamilien*, Vol. 8, ed. A. Engler and K. Prantl. Leipzig: Borntraeger.

Zambrano, A., Nash, T. H., III and Gries, C. (2000). Responses of *Ramalina farinacea* (L.) Ach. to transplanting in southern California and to gaseous formaldehyde. *Bibliotheca Lichenologica*, **75**, 219–230.

Zavarzina, A. and Zavarzin, A. (2006). Laccase and tyrosinase activity in lichens. *Microbiology (Rus)*, **75**, 630–641.

Zeitler, I. (1954). Untersuchungen über die Morphologie, Entwicklungsgeschichte und Systematik von Flechtengonidien. *Österreichische Botanische Zeitschrift*, **101**, 453–487.

Zhu, Y. Y., Machleder, E. M., Chenchik, A., Li, R. and Siebert, P. M. (2001). Reverse transcriptase template switching: a SMART™ approach for full-length cDNA library construction. *BioTechniques*, **30**, 892–897.

Ziegler, I. (1977). Sulfite action on ribulosediphosphate carboxylase in the lichen *Pseudevernia furfuracea*. *Oecologia*, **29**, 63–66.

Zoller, S. and Lutzoni, F. (2003). Slow algae, fast fungi: exceptionally high nucleotide substitution rate differences between lichenized fungi *Omphalina* and their symbiotic green algae *Coccomyxa*. *Molecular Phylogenetics and Evolution*, **29**, 629–640.

Zoller, S., Lutzoni, F. and Scheidegger, C. (1999). Genetic variation within and among populations of the threatened lichen *Lobaria pulmonaria* in Switzerland and implications for its conservation. *Molecular Ecology*, **8**, 2049–2059.

Zopf, W. (1897). Über Nebensymbiose (Parasymbiose). *Berichte der deutschen Botanischen Gesellschaft*, **15**, 90–92.

Zopf, W. (1907). *Die Flechtenstoffe in chemischer, botanischer, pharmakologischer, und technischer Beziehung*. Jena: Gustav Fischer.

Zotz, G. (1999). Altitudinal changes in diversity and abundance of non-vascular epiphytes in the tropics – an ecophysiological explanation. *Selbyana*, **20**, 256–260.

Zotz, G. and Winter, K. (1994). Photosynthesis and carbon gain of the lichen, *Leptogium azureum*, in a lowland tropical forest. *Flora*, **189**, 179–186.

Zotz, G., Büdel, B., Meyer, A., Zellner, H. and Lange, O. L. (1998). *In situ* studies of water relations and CO_2 exchange of the tropical macrolichen, *Sticta tomentosa*. *New Phytologist*, **139**, 525–535.

Zotz, G., Schultz, S. and Rottenberger, S. (2003). Are tropical lowlands a marginal habitat for macrolichens? Evidence from a field study with *Parmotrema endosulphureum* in Panama. *Flora*, **198**, 71–77.

Zukal, H. (1895). Morphologische und biologische Untersuchungen über die Flechten (II. Abhandlung). *Sitzungsberichte der Kaiserlichen Akademie der Wissenschaften Wien, Math.-naturw. Cl.*, **104**/1, 1–68.

Taxon index
(excluding animals and vascular plants)

Subject index

Printed in the United States
By Bookmasters